Hematologic Cancers: From Molecular Pathobiology to Targeted Therapeutics

Cancer Growth and Progression

Volume 14

For further volumes:
http://www.springer.com/series/5721

Jianguo Tao • Eduardo Sotomayor
Editors

Hematologic Cancers: From Molecular Pathobiology to Targeted Therapeutics

Springer

Editors
Jianguo Tao
Department of Hematopathology
and Laboratory Medicine
H. Lee Moffitt Cancer Center
and Research Institute
University of South Florida College
of Medicine
Tampa, FL, USA

Eduardo Sotomayor
Department of Malignant Hematology
H. Lee Moffitt Cancer Center
and Research Institute
University of South Florida College
of Medicine
Tampa, FL, USA

ISBN 978-94-007-9581-5 ISBN 978-94-007-5028-9 (eBook)
DOI 10.1007/978-94-007-5028-9
Springer Dordrecht Heidelberg New York London

Softcover reprint of the hardcover 1st edition 2012

Printed on acid-free paper

Springer is part of Springer Science+Business Media (www.springer.com)

Contents

B Cell Growth, Differentiation and Malignancies

Jianguo Tao and Chih-Chi Andrew Hu

Abstract

The primary function of a B cell (or lymphocyte) is to produce large quantities of secreted immunoglobulin (also known as antibody) to fight against bacteria, viruses and other foreign insults to the human body. Each B cell makes only one distinct immunoglobulin which recognizes a cognate antigen. It is estimated that B cells in the human body can produce as many as 10^{11} different antibodies. Thus, each B cell must undergo a series of differentiation, selection and maturation processes before it is endowed with the ability to produce a functional immunoglobulin to represent in the large and diverse antibody repertoire. While insufficient B cells and insufficient antibody production can thus lead to infections, uncontrolled growth of B cells can lead to leukemia and lymphoma. In this article, our discussion will focus on transcription factors and signaling molecules that involve in normal B cell development and differentiation. These molecules, when mutated or not tightly regulated, will contribute to the formation of B cell malignancies.

J. Tao, M.D., Ph.D.
Hematopathology and Laboratory Medicine, Hematologic Malignancies,
H. Lee Moffitt Cancer Center & Research Institute, Tampa, FL, USA

C.-C.A. Hu, Ph.D. (✉)
Immunology Program, H. Lee Moffitt Cancer Center & Research Institute,
12902 Magnolia Drive, Tampa, FL 33612, USA
e-mail: Chih-Chi.Hu@moffitt.org

J. Tao and E. Sotomayor (eds.), *Hematologic Cancers: From Molecular Pathobiology to Targeted Therapeutics*, Cancer Growth and Progression 14,
DOI 10.1007/978-94-007-5028-9_1, © Springer Science+Business Media Dordrecht 2012

B Cell Development

B cell development begins in the bone marrow in an antigen-independent manner. In the pro-B cell stage, the rearrangement of the immunoglobulin heavy chain variable gene regions occurs with the initial joining of a D_H gene segment to a J_H gene segment, followed by the later rearrangement of a V_H gene segment to the joined DJ_H. Such a process is followed by the pre-B cell phase in which the subsequent rearrangement of the immunoglobulin light chain gene occurs by joining a V_L and a J_L segments. The successfully rearrangement of V, D and J gene segments is the major contribution of RAG-1 and RAG-2 enzymes. Once the nascent B cells express functional, not overtly self-reactive B cell receptors (BCR), they migrate to the spleen, lymph nodes and peritoneal cavity as mature naïve B cells. Further development of naïve mature B cells requires an encounter with foreign antigens in an either T cell-dependent or T cell-independent manner so that they can differentiate into antibody-secreting plasma cells. Plasma cells can be found in the spleen, lymph nodes, infection sites and blood. Some antigen-activated B cells become memory cells ready for the next run of antigen challenges. Some plasma cells can colonize the long-lived niches in the bone marrow for sustained immunoglobulin production and secretion. Of the various stages of B cell differentiation, the generation and maintenance of plasma cells represents a poorly charted territory. *In vivo*, terminally differentiated plasma cells move to specialized niches at anatomical sites where they are difficult to access. They also do not survive well *ex vivo*. Protocols for *in vitro* differentiation all involve the use of cytokines and mitogens, but these only allow the yield of antibody-secreting B cells in culture without full differentiation to the plasma cell stage. The lack of antigen-induced B cell differentiation protocol *in vitro* also hinders studies to link BCR-initiated signal transduction to the regulation of transcription factor expression in the B cell nucleus.

To differentiate into a plasma cell, a mature B cell must reprogram itself by tuning the expression levels of a set of transcription factors important for B cell differentiation. These include decreased expression of Pax5 (paired box gene 5) and BCL6 (B-cell lymphoma 6) and increased expression of IRF4 (interferon regulatory factor 4), PRDM1 (positive regulatory domain zinc finger protein 1; or Blimp-1 (B lymphocyte-induced maturation protein-1) in mouse), and XBP-1 (X-box binding protein 1). A plasma cell, eventually, appears to be very different from a mature B cell by acquiring a cart-wheel heterochromatin pattern in its eccentric nucleus and expanding massively its endoplasmic reticulum (ER) for antibody production. Although a plasma cell initially produces only IgM and IgD antibodies, it can switch to express other isotypes (IgG, IgE or IgA) by recombining the immunoglobulin heavy chain variable regions to different heavy chain constant (C_H) region genes to acquire different effector functions. The activation-induced cytidine deaminase (AID) is responsible for isotype switching since AID deficiency completely blocks such a process. In addition, AID is also important for somatic hypermutation, a process that allows antigen-exposed B cells or plasma cells to introduce high-rate point mutations to the variable regions of the rearranged heavy chain and light chain genes to achieve immunoglobulin affinity maturation, resulting in an antibody with high affinity and efficiency in binding to its antigen. The occurrence of somatic hypermutation also requires proper signals from activated T cells. Since constant gene recombination and mutations are required for an antigen-exposed B cell to eventually make a functional antibody, it is not hard to imagine that any of these processes, when not tightly regulated or not confined to restricted regions of chromosomes, can introduce detrimental mutations, deletions, or chromosome translocations, leading to the occurrence of B cell malignancies.

B Cell Activation via the BCR

Plasma cell differentiation begins when a B cell encounters an antigen on its cell-surface BCR. A functional BCR consists of a membrane-bound IgM molecule and a disulfide-linked Igα/Igβ heterodimer. Upon antigen binding, the BCR is recruited into lipid rafts, where the GPI-anchored Lyn kinase activates the BCR via phosphorylation of the immunoreceptor tyrosine-based activation motifs (ITAM) on the Igα/Igβ heterodimer (Dykstra et al. 2003; Pierce 2002). Phosphorylated Igα/Igβ then recruits Syk (spleen tyrosine kinase) and other kinases, which, after phosphorylation, transduce signals to multiple downstream molecules, eventually leading to differentiation of the antigen-exposed B cell into a plasma cell. The role of the Syk kinase is pivotal as it leads to activation of Bruton's tyrosine kinase (Btk), phosphatidylinositol 3 kinase (PI-3K), Akt and NFκB, all of which can promote B cell survival (Pogue et al. 2000).

The Syk kinase can also connect the BCR signal transduction to the regulation of B cell-specific transcription factors during plasma cell differentiation. Syk phosphorylation in response to activation of the BCR can activate the extracellular signal-regulated kinase (ERK) via Ras/Raf-1/MEK (Jiang et al. 1998). Activation of ERK induces phosphorylation and degradation of BCL6 (Moriyama et al. 1997; Niu et al. 1998). BCL6 expression directly inhibits Blimp-1 (Tunyaplin et al. 2004); thus, its degradation would allow the expression of Blimp-1, a transcription factor required for plasma cell differentiation. Protocols that can turn B cells into full-blown plasma cells *in vitro* have not been established. Terminal differentiation of plasma cell *in vivo* requires signals from activation of the Toll-like receptors on the B cell surface as well as helping signals from T cells.

BCR Signal Transduction in B Cell Malignancies

BCR activation plays an important role in B cell malignancies as constitutive signaling through the BCR provides signals of survival and proliferation for B cell leukemia and lymphoma. Sequence analysis of immunoglobulin variable heavy chain (IgV_H) genes reveals two distinct types of chronic lymphocytic leukemia (CLL), one with somatically unmutated and the other with mutated IgV_H genes. IgV_H-unmutated CLL responds to BCR activation and undergoes high rate of proliferation, while IgV_H-mutated CLL is less responsive or unresponsive to BCR activation and less proliferative. These features determine the distinct clinical outcomes for these two types of CLL: The IgV_H-mutated CLL patients usually survive significantly longer than those diagnosed with the IgV_H-unmutated CLL. In addition to the BCR signal transduction initiated by Igα, Igβ and Syk, some CLL cells can express high levels of the ZAP-70 kinase (zeta-chain-associated protein kinase 70, normally expressed by T cells and natural killer cells) to strengthen this survival signal (Chen et al. 2002, 2005). In the diffuse large B cell lymphoma (DLBCL), the constitutive activation of the BCR, even in the absence of antigen, was also found critical for the survival and proliferation (Chen et al. 2008; Davis et al. 2010; Gururajan et al. 2006). Mutations frequently occur to the B cell receptor signaling components, Igα and Igβ, in the cases of CLL and DLBCL. Recently, a mutation of a critical tyrosine residue in the ITAM of Igβ was found in 18% of the activated B-cell-like DLBCL cases, and such a mutation contributes to increased BCR expression on the B cell surface, accounting for the strong BCR signaling required for B cell cancer survival (Davis et al. 2010).

B Cell Activation via Toll-Like Receptors (TLRs)

Functions of TLRs have been widely explored in dendritic cells and macrophages. As a semi-professional antigen-presenting cell, the B cell proliferates and differentiates to secrete antibodies in response to *in vitro* stimulation by lipopolysaccharide (LPS) (Coutinho et al. 1974) and the unmethylated CpG oligodeoxynucleotides (Krieg et al. 1995). LPS and CpG are recognized by cell surface pattern recognition receptors TLR4 and TLR9, respectively. Other than LPS and CpG, TLR ligands that can trigger B cells to respond include peptidoglycan and Pam_3CSK_4 (TLR1/2), bacterial lipoproteins and MALP2 (TLR2/6), dsRNA (TLR3), ssRNA and imidazoquinolines (TLR7 and TLR8), and profilin-like molecule (TLR11). Mouse B cells do not respond to flagellin, due to their lack of TLR5 expression (Genestier et al. 2007); and normal human B cells do not express TLR4 (Bourke et al. 2003), thus unresponsive to LPS. Other than enhancing antibody-mediated defense against infections (Meyer-Bahlburg et al. 2007), activation of TLRs in B cells also has important physiological functions in the immunoglobulin isotype switching (He et al. 2004) and the maintenance of memory B cells.

TLRs in B Cell Malignancies

Because dendritic cells can be activated upon stimulation via TLRs, some TLR agonists have been used in clinical trials to improve tumor antigen presentation and promote T cell activation (Krieg 2008). Since chronic infections and TLR ligands may promote growth of tumor cells, it is important to carefully investigate the functions of TLRs in malignant as compared to normal cells. For example, TLR9 has been proposed to be an attractive target for the treatment of CLL (Jahrsdorfer et al. 2005). However, while TLR9 activation by CpG leads to apoptosis of the IgV_H-mutated CLL cells, it encourages proliferation of the IgV_H-unmutated CLL cells (Longo et al. 2007). Thus, TLR9 activation will most likely aggravate the disease progress in IgV_H-unmutated CLL patients who already have predictably poor clinical outcomes. In addition, cancerous B cells can have a different expression profile of TLRs. For example, malignant human B cells express TLR4, which can be activated by putative endogenous ligands, including fibronectin, heparan sulfate, and heat shock proteins, which are released during cellular stress and tissue damage. Notably, multiple myeloma cells can express all kinds of TLRs and high levels of TLRs are correlated with poor disease prognosis.

Transcription Factors Required for B Cell Differentiation and Their Roles in B Cell Cancer

Pax5 and BCL6 are two important transcription factors that guard the naïve B cell phenotypes. When a B cell differentiates into a plasma cell, as a result of stimulation via its BCR and/or other differentiation-initiating signals, Pax5 and BCL6 decrease their expression levels to allow the expression of plasma cell differentiation-requiring transcription factors that include IRF4, Blimp1 and XBP-1 (Fig. 1a). Many of these transcription factors also play important roles in the formation of B cell cancer (Fig. 1b). While many other transcription factors are also important for B cell development and differentiation, the roles of the above five transcription factors as well as their crosstalk in B cell differentiation and B cell malignancies will be primarily discussed in this chapter.

Pax5 in B Cell Development

Pax5, belonging to the paired box (PAX) family of transcription factors, is important for the commitment of early lymphoid progenitors to the B cell lineage and for B cell development (Nutt et al. 1999). The Pax5 gene encodes the B-cell lineage specific activator protein (BSAP) that is expressed in pro-, pre-, and mature B cells, but not in plasma cells. In the early B cell development stages, Pax5 interacts with the RAG-1 and RAG-2 enzyme complex to assist the joining of the V_H gene to the DJ_H gene (Zhang et al. 2006), and it also regulates the expression of B cell-specific genes, which include Igα, CD19 and CD20. Hence, Pax5 deficiency blocks B cell development in its early stage. In addition, Pax5 also guards the naïve B cell phenotypes by binding to the promoters of the Blimp-1 and XBP-1 genes to suppress their expression (Mora-Lopez et al. 2007; Reimold et al. 1996).

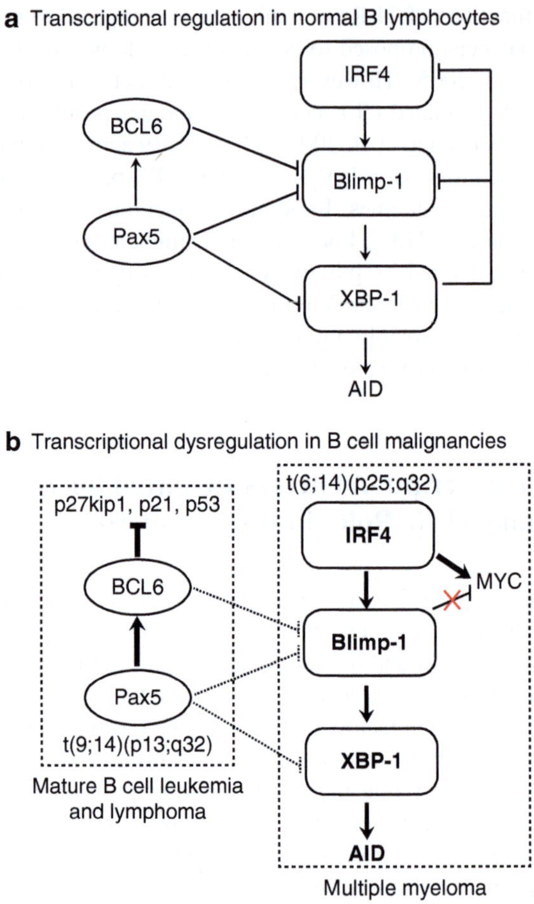

Fig. 1 (**a**) A model that illustrates the crosstalk between transcription factors in normal B cells. (**b**) A model that summarizes the contributions of transcription factors to the formation of mature B cell leukemia and lymphoma or multiple myeloma (plasma cell cancer). Increase expression is depicted by the boldface type. Please see text for details

The Roles of Pax5 in B Cell Malignancies

Alterations in Pax5 expression contribute to tumor formation. The Pax5 gene is located in human chromosome 9p13 region, which is involved in t(9;14)(p13;q32) translocations that occur to several types of lymphomas (Busslinger et al. 1996; Iida et al. 1996). Such translocation brings the potent intronic enhancer of the immunoglobulin μ heavy chain gene (E-mu) into close proximity of the Pax5 gene promoter, causing the overexpression of Pax5 and contributing to the pathogenesis of lymphomas. Pax5 is expressed in pre-B and mature B cell lymphoma and leukemia, but not in plasmacytoma or multiple myeloma, although alternatively spliced transcripts encoding different Pax5 isoforms are expressed in multiple myeloma (Borson et al. 2002). Pax5 is also a frequent target for abnormal somatic hypermutation in B cell lymphoma (Gaidano et al. 2003; Liso et al. 2006; Pasqualucci et al. 2001).

BCL6 in B Cell Development

The transcriptional repressor BCL-6 regulates B cell fate by preventing terminal differentiation of B lymphocytes into plasma cells. BCL6 is a Krueppel zinc finger transcription repressor that contains a POZ (poxvirus and zinc finger) domain. Together with transcriptional corepressors like NCoR (Nuclear receptor corepressor)-1 and 2, and BCoR (BCL-6-interacting corepressor), BCL6 represses genes necessary for the terminal differentiation of B cells. Importantly, BCL6 blocks the transactivation ability of IRF-4 (Gupta et al. 1999) and inhibits the expression of Blimp-1 (Tunyaplin et al. 2004). Both IRF-4 and Blimp-1 are required for plasma cell differentiation. In addition, BCL6 also functions with the transcription factor, Myc-interacting zinc finger protein 1, to suppress the cell cycle arrest gene and facilitate proliferative expansion of B cells in the germinal centers during the immune response (Phan et al. 2005). BCL6 also directly suppresses the expression of CD80, a costimulatory receptor involved in the B and T cell interaction in the germinal center, critical for the development of T cell-mediated antibody responses (Niu et al. 2003). Thus indeed BCL6-deficient mice do not develop germinal centers in their spleens upon immunization (Dent et al. 1997; Fukuda et al. 1997; Ye et al. 1997).

BCL6 in B Cell Malignancies

Just like immunoglobulin genes, BCL6 is a frequent target for somatic hypermutation when examined in normal germinal center B cells and memory B cells (Pasqualucci et al. 1998; Shen et al. 1998). Overexpression of BCL6 due to chromosomal translocations and hypermutations can lead to the development of DLBCL. In some DLBCL cases, heterologous promoters derived from other chromosomes are juxtaposed to the BCL6 gene (Ye et al. 1993, 1995), dysregulating the expression of BCL6. In addition, mutations mediated by the AID enzyme occur in the promoter region or in the negative autoregulatory domain of the BCL6 gene in some DLBCL cases, causing the dysregulation of BCL6 (Migliazza et al. 1995; Wang et al. 2002).

BCL6 is involved in the development of B cell malignancies via its role as a transcription repressor. BCL6 represses the expression of cell cycle inhibitors like p27kip1 and p21, allowing rapid proliferation of malignant B cells (Phan et al. 2005; Sanchez-Beato et al. 1997; Shaffer et al. 2000). BCL6 also suppresses the expression of p53 (Phan and Dalla-Favera 2004), a tumor suppressor that can initiate DNA repair and cell growth arrest upon DNA damage. By doing so, BCL6 allows proliferation of B cell tumors even when they carry serious gene mutations. In addition, upregulated BCL6 expression can repress the expression of chemokines like CCL3 and CXCL10 by the tumor cells, and so prevents the chemotaxis of macrophages, natural killer cells and T cells (Shaffer et al. 2000). Together with its role in suppressing malignant B cells to express CD80 (Niu et al. 2003), a co-stimulatory ligand that binds to two T cell surface proteins, CD28 and CTLA4, BCL6 helps tumor cells to successfully evade possible interactions with immune surveillance cells that may cause their eradication. In fact, enforced expression of BCL6 driven by the immunoglobulin heavy chain promoter in mice recapitulates the pathogenesis of human DLBCL (Cattoretti et al. 2005).

IRF4 in B Cell Differentiation

Expression of IRF4 is restricted to immune cells, including B and T lymphocytes, dendritic cells and macrophages. IRF4 belongs to the interferon-regulatory factor (IRF) family of transcription factors, all of which share a specific DNA-binding domain that can bind to the regulatory elements in the promoter regions of interferon-inducible genes, for example, the promoter of the class I MHC gene. Different from other IRF family members, the expression of IRF4 is not induced by interferon, but by activation via the antigen receptor, Toll-like receptor or CD40 receptor (Grumont and Gerondakis 2000; Gupta et al. 1999; Mittrucker et al. 1997). Activation through these different pathways eventually leads to the binding of NFκB heterodimers to the IRF4 promoter. In addition, the transcription of IRF4 can be induced IL-4. In B cells costimulated with CD40 and IL-4, IRF-4 is upregulated and interacts with signal transducer and activator of transcription (STAT)-6 to drive the expression of IL-4-inducible genes (Gupta et al. 1999), contributing to B cell growth and differentiation.

IRF4 has a biphasic function in both pre-B cells and plasma cells. B cell development in the IRF-4-deficient mouse is blocked from maturing into the naïve follicular B cell (Mittrucker et al. 1997), suggesting that IRF4 plays an important part in early B cell development in the bone marrow. It turns out that IRF4 is expressed in pre-B cells in the bone marrow because it is required for the rearrangements of immunoglobulin light chain genes (Lu et al. 2003). Since IRF4 is expressed in high levels in multiple myeloma (a cancer of plasma cell) and plasma cells (Falini et al. 2000; Iida et al. 1997), the role of IRF4 in plasma cell differentiation has been implicated. IRF4 is indeed indispensable for plasma cell differentiation. When IRF4 was conditionally deleted from germinal center B cells, such B cells fail to differentiate into plasma cells (Klein et al. 2006). To promote plasma cell differentiation, IRF-4 binds to the first intron of BCL6, leading to downregulation of BCL6 (Saito et al. 2007). BCL6's repression on Blimp-1 is thus relieved, and plasma cell differentiation can be achieved. IRF4 also binds to the promoter and the fourth intron of Blimp-1 gene and directly upregulates the expression of Blimp-1 to promote plasma cell differentiation (Shaffer et al. 2008). In addition, IRF4 regulates the expression of AID, thus contributing to immunoglobulin isotype switching in plasma cells (Klein et al. 2006).

IRF4 in B Cell Malignancies

IRF4 by itself is capable of transforming cells in vitro (Iida et al. 1997). In multiple myeloma (plasma cell cancer), IRF4 is expressed at high levels as a result of the aberrant chromosomal translocation t(6;14)(p25;q32), in which the immunoglobulin heavy chain locus is juxtaposed to the IRF4 gene. IRF4 is also found indispensable for the survival of multiple myeloma since knockdown of IRF4 by RNA interference causes drastic cell death in multiple myeloma cell lines (Shaffer et al. 2008). Other than multiple myeloma, IRF4 is often expressed at abnormally elevated levels in many B cell lymphomas, including lymphoplasmacytoid lymphoma, and DLBCL (Falini et al. 2000; Rosenwald et al. 2003; Savage et al. 2003), and thus IRF4 is used as a diagnostic indicator for these lymphomas. Just like the expression of IRF4 is controlled by the NFκB pathway, the elevated levels of IRF4 in the malignant type activated B cell-like DLBCL was reported to be a result of constitutive activation of NFκB (Lam et al. 2005; Lenz et al. 2008a; Ngo et al. 2006). In addition, the overexpression of IRF4 can be a response to infections and transformation of B cells by the Epstein-Barr virus (EBV) as the EBV-encoded protein LMP1 can activate the NFκB pathway to induce the expression of IRF4 (Xu et al. 2008).

How does IRF4 function as an oncoprotein? In multiple myeloma, the dysregulated IRF4 is shown to directly bind to the promoter region of a strong proto-oncoprotein MYC and trigger its expression (Shaffer et al. 2008). Because IRF4 can bind to SPIB and induce SPIB's transactivation activity (Su et al. 1996), the dysregulated IRF4 may also function through the oncogenic activity of SPIB in activated B cell-like DLBCL (Lenz et al. 2007, 2008b).

PRDM1 in Plasma Cell Differentiation

PRDM1 (or the murine Blimp-1) is a transcription repressor that is absolutely required for the formation and prolonged maintenance of plasma cells (Angelin-Duclos et al. 2000; Shaffer et al. 2002; Shapiro-Shelef et al. 2003, 2005; Turner et al. 1994). To differentiate into plasma cells, Blimp-1 suppresses MYC expression to reduce the proliferation capability of B cells (Lin et al. 1997). Blimp-1 can also reduce the BCR signaling capability of B cells by repressing transcription factor SPIB (Shaffer et al. 2002), whose expression is important for normal BCR signal transduction (Garrett-Sinha et al. 1999). In addition, Blimp-1 binds to the Pax-5 promoter in vitro and in vivo and represses the Pax-5 promoter leading to downregulation of Pax-5 so that plasma cell differentiation can be achieved (Lin et al. 2002). Blimp-1 can also regulate the expression of XBP-1 (Shaffer et al. 2004), which upregulates the expression of endoplasmic reticulum-associated chaperones and phospholipids to prepare for the massive production and secretion of immunoglobulin in the endoplasmic reticulum of plasma cells (see below). B cells deficient of Blimp-1 fail to form immunoglobulin-secreting plasma cells (Shapiro-Shelef et al. 2003).

PRDM1/Blimp-1 and B Cell Malignancies

PRDM1/Blimp-1 does not trigger cancer formation; instead, it represses the expression of proto-oncoproteins like MYC, causing cell cycle arrest. PRDM1/Blimp-1 is expressed in abundant levels in human multiple myeloma and mouse plasmacytoma; however, the cell cycle-arresting function of PRDM1/Blimp-1 on MYC is often overcome by gene rearrangements and chromosomal translocations that occur to c-myc, leading to the high-level expression of MYC in the face of Blimp-1 (Kuehl and Bergsagel 2002; Park et al. 2005; Potter 2003). With regard to the tumor suppressor function of PRDM1/Blimp-1, it is not surprising that the PRDM1/Blimp-1 gene can be a target of genetic inactivation as a consequence of mutations, deletions or translocations in B cell malignancies. Indeed, the PRDM1 gene has been reported to be deleted or mutated in human activated B cell-like DLBCL (Pasqualucci et al. 2006; Tam et al. 2006; Tate et al. 2007).

IRE-1-Mediated XBP-1 Activation in B Cell Differentiation

XBP-1 is a basic-region leucine zipper type transcription factor belonging to the family of CREB/ATF (cyclic AMP response element binding proteins/activating transcription factors). The regulation of XBP-1 expression is unique, as the unspliced XBP-1 mRNA transcript is constantly translated into an inactive XBP-1 protein. XBP-1 activation requires splicing of XBP-1 mRNA in the cytoplasm by inositol-requiring enzyme-1 (IRE-1), a kinase/RNase protein containing an endoplasmic reticulum (ER) stress sensor domain facing the lumen of the ER, a transmembrane segment sitting on the ER membrane, and a kinase and an endoribonuclease domain in the cytoplasm (Calfon et al. 2002; Shen et al. 2001; Yoshida et al. 2001). IRE-1 activation is achieved by initial oligomerization of the sensor domains in the ER lumen, resulting in the kinase domains in the cytoplasm to be brought into close proximity for autophosphorylation, eventually contributing to the increased RNase activity of IRE-1. In mammalian cells, the unconventional XBP-1 mRNA splicing by IRE-1 leads to a shift in the reading frame of the XBP-1 mRNA and the synthesis of a larger, active XBP-1 protein. To synthesize the functional XBP-1 protein is therefore controlled by IRE-1 whose activation should correspond to the homeostatic states of the ER. Thus, conditions that can lead to the accumulation of misfolded proteins in the ER are believed to trigger IRE-1-mediated XBP-1 activation. Activated XBP-1 then translocates into the nucleus and upregulates the synthesis of chaperones and other proteins to restore ER function (Lee et al. 2003, 2005). XBP-1 also induces increased levels of membrane phospholipids and expansion of the ER via upregulation of phosphatidylcholine synthesis (Sriburi et al. 2004, 2007) and presumably other lipids as well.

XBP-1 is found to be important for plasma cell differentiation because XBP-1 expression is upregulated corresponding to plasma cell differentiation and XBP-1-deficient B cells fail to be induced to become terminally differentiated plasma cells that can sustain antibody secretion (Iwakoshi et al. 2003a, b; Reimold et al. 2001; Hu et al. 2009). Based on the role of XBP-1 in maintaining ER homeostasis, and the observation that increased immunoglobulin synthesis is accompanied by activation of XBP-1, it has been suggested that the failure of XBP-1-deficient B cells to become plasma cells may be due to the incapability of such B cells to deal with large amounts of newly synthesized immunoglobulin during B cell differentiation

(Calfon et al. 2002; Iwakoshi et al. 2003a, b; van Anken et al. 2003). Without XBP-1, some of this immunoglobulin would fail to fold correctly, thus leading to the accumulation of misfolded immunoglobulin in the ER and cell death. However, this view is not supported by data showing that XBP-1 is still activated in B cells that do not produce massive amounts of immunoglobulin (Hu et al. 2009), and that XBP-1-deficient B cells do not accumulate any detectable misfolded immunoglobulin, or other misfolded integral membrane or secretory proteins that travel through the ER (McGehee et al. 2009). So far, it is still unknown what triggers IRE-1-mediated XBP-1 activation during plasma cell differentiation.

Much research attention is centered on the roles of XBP-1 in plasma cell differentiation. However, it is interesting to note that XBP-1 also expresses in pro-B cells in the bone marrow (Brunsing et al. 2008), and naïve mature XBP-1-deficient B cells already exhibit different expression levels of a chemokine receptor (Hu et al. 2009). XBP-1 is also essential for the proper signal transduction through the BCR and the chemokine receptor, CXCR4 (Hu et al. 2009), most likely due to its essential function in forming a proficient signaling platform (e.g., lipid rafts) on the plasma membrane (McGehee et al. 2009). These suggest that XBP-1 may play other important roles in B cell biology outside the arena of promoting plasma cell differentiation.

Crosstalk Between XBP-1 and Other Transcription Factors in B Cell Differentiation

The crosstalk between XBP-1 and other transcription factors implicated in B cell differentiation has been a matter of debate (Kallies et al. 2007; Martins and Calame 2008; Shaffer et al. 2002). Microarray analyses on Blimp-1-deficient B cells suggested that XBP-1 expression requires Blimp-1, and thus placed XBP-1 downstream of Blimp-1 (Shaffer et al. 2002). However, XBP-1 activation is apparently normal in B cells from Blimp-$1^{GFP/GFP}$ mice, in which the Blimp-1 functional domain was replaced by green fluorescence protein (Kallies et al. 2007). XBP-1 can thus be activated also in the absence of Blimp-1. Using XBP-1-deficient B cells that are induced to differentiate, it is demonstrated that the deficiency of XBP-1 leads to an increase in the expression levels of Blimp-1 together with its upstream transcription regulator, IRF4, establishing an XBP-1-initiated transcription factor feedback loop (Hu et al. 2009). Thus normally, XBP-1 helps to attenuate the expression levels of IRF4 and Blimp-1. Since Blimp-1 expression is regulated by IRF4 (Sciammas et al. 2006; Shaffer et al. 2008), XBP-1 may modulate the Blimp-1 expression by regulating IRF4. Because XBP-1-deficent B cells still express IRF4 and Blimp-1, this would place XBP-1 downstream of IRF4 and Blimp-1. Given that XBP-1 activation still occurs in Blimp-1-deficient B cells (Kallies et al. 2007), perhaps XBP-1 and Blimp-1 act in parallel to regulate plasma cell differentiation, while XBP-1 can additionally inhibit the expression of Blimp-1. Although IRF4 and Blimp-1 are upregulated in XBP-1-deficient B cells, these B cells still fail to differentiate into plasma cells (Hu et al. 2009). In addition, the ability of IRF4 in regulating the expression of AID is also blocked when XBP-1 is absent. These findings suggest that the regulation of plasma cell differentiation by IRF4 and Blimp-1 requires XBP-1. The lack of XBP-1 does not affect the expression of Pax5 and BCL6.

IRE-1 and XBP-1 in B Cell Malignancies

The IRE-1/XBP-1 pathway has been implicated in cancer formation because IRE-1 mutations have been found in large scale cancer genome screens (Greenman et al. 2007). It is not known whether these are gain-of-function or loss-of-function mutations. However, enforced expression of XBP-1 in mouse B cell is sufficient to cause the syndrome of monoclonal gammopathy of undetermined significance (Carrasco et al. 2007), a predictor condition for multiple myeloma. Recently, multiple myeloma patients with low XBP-1 expression were also found to respond better to thalidomide treatments and these patients have a much better survival rate (Bagratuni et al. 2010), suggesting a similar role of XBP-1 in malignant progression of multiple myeloma. More recently, a novel small-molecule chemical inhibitor, STF-083010, that targets IRE-1 and downregulates the expression of XBP-1 has been shown effective in eliminating human multiple myeloma xenografts in mice with no apparent cytotoxicity to other non-plasma cell types (Papandreou et al. 2011). Inhibition of the IRE-1/XBP-1 pathway in CLL cells also leads to apoptosis (Kriss et al. 2012).

Crucial Cell Survival and Growth Pathways That Contributes to B Cell Malignancies—The NFκB and mTOR Pathways

The NFκB pathway is activated upon antigen stimulation of the BCR and provides a survival signal to mature B cells (Siebenlist et al. 2005). Engagement of a cognate antigen by the BCR in normal B cells leads to the assembly of the CBD protein complex which consists of CARD11, BCL10 and MALT1, and such a complex is known to activate the IκB kinase (IKK) which is responsible for phosphorylation of the inhibitory protein, IκB. The inhibitory IκB normally sequesters NFκB in the cytoplasm. The phosphorylation of IκB causes its degradation via proteasomal proteolysis, thus allowing translocation of NFκB into the nucleus to activate its target genes (Vallabhapurapu and Karin 2009). In multiple myeloma, NFκB is often found constitutively active, in part due to mutations occurring to the NFκB gene (Annunziata et al. 2007; Keats et al. 2007). Similarly in activated B cell-like DLBCL, constitutive activation of NFκB is observed and linked to chronic BCR activation via a spontaneous formation of BCR clusters on the cancer cell surface, mimicking antigen-stimulated B cells (Davis et al. 2010). Many target genes of NFκB are cell survival and proliferation factors (Vallabhapurapu and Karin 2009). Constitutive activation of NFκB in B cell malignancies has indeed been shown to cause the upregulated expression levels of cell cycle regulators (e.g., cyclin D1, cyclin D2 and MYC), anti-apoptotic proteins (e.g., BCL2 and BCL-XL) and proliferation-promoting cytokines (IL-6 and IL-10) (Davis et al. 2001; Lam et al. 2008). The expression of all these NFκB-regulated factors contributes to continuous proliferation of cancerous B cells.

The mTOR (mammalian target of rapamycin) was identified as a rapamycin-binding protein, based on the initial observation that rapamycin can inhibit progression through the G1 phase of the cell cycle in yeast and mammalian cells (Brown et al. 1994; Sabatini et al. 1994). Because mTOR contains a carboxyl terminal domain that is homologous to the catalytic domain of phosphatidylinositol 3-kinase (PI3K), it belongs to the family of the PI3K-related protein kinases (Gingras et al. 2001; Schmelzle and Hall 2000). The activation of mTOR leads to the increased biogenesis of the ribosomes, initiation of protein translation, expression of metabolism-related genes, and uptake of nutrients and amino acids; and inhibition of apoptosis and autophagy (Asnaghi et al. 2004; Wullschleger et al. 2006; Zeng and Kinsella 2008). Thus, mTOR contributes to cell growth and survival, and maintains a nutritional and bioenergenic state for the cell to enter the cell cycle. In light of these many crucial functions regulated by the mTOR protein, it is not hard to understand that the constitutive expression of mTOR can contribute to the growth and survival of B cell malignancies. Although rapamycin and its structural analogs have been used to target CLL (Decker et al. 2009; Ringshausen et al. 2005; Zanesi et al. 2006), multiple myeloma (Pene et al. 2002; Shi et al. 2002) and mantle cell lymphoma (Hipp et al. 2005; Peponi et al. 2006; Younes 2008), clinical benefits derived from using these mTOR inhibitors are not immediately clear. Targeting the mTOR pathway in B cell malignancies will benefit from studies focusing on the identification of the mechanisms that contribute to constitutive activation of the mTOR pathway in each different type of B cell malignancies.

References

Angelin-Duclos C, Cattoretti G, Lin KI, Calame K (2000) Commitment of B lymphocytes to a plasma cell fate is associated with Blimp-1 expression in vivo. J Immunol 165:5462–5471

Annunziata CM, Davis RE, Demchenko Y, Bellamy W, Gabrea A, Zhan F, Lenz G, Hanamura I, Wright G, Xiao W, Dave S, Hurt EM, Tan B, Zhao H, Stephens O, Santra M, Williams DR, Dang L, Barlogie B, Shaughnessy JD Jr, Kuehl WM, Staudt LM (2007) Frequent engagement of the classical and alternative NF-kappaB pathways by diverse genetic abnormalities in multiple myeloma. Cancer Cell 12:115–130

Asnaghi L, Calastretti A, Bevilacqua A, D'Agnano I, Gatti G, Canti G, Delia D, Capaccioli S, Nicolin A (2004) Bcl-2 phosphorylation and apoptosis activated by damaged microtubules require mTOR and are regulated by Akt. Oncogene 23:5781–5791

Bagratuni T, Wu P, Gonzalez de Castro D, Davenport EL, Dickens NJ, Walker BA, Boyd K, Johnson DC, Gregory W, Morgan GJ, Davies FE (2010) XBP1s levels are implicated in the biology and outcome of myeloma mediating different clinical outcomes to thalidomide-based treatments. Blood 116:250–253

Borson ND, Lacy MQ, Wettstein PJ (2002) Altered mRNA expression of Pax5 and Blimp-1 in B cells in multiple myeloma. Blood 100:4629–4639

Bourke E, Bosisio D, Golay J, Polentarutti N, Mantovani A (2003) The toll-like receptor repertoire of human B lymphocytes: inducible and selective expression of TLR9 and TLR10 in normal and transformed cells. Blood 102:956–963

Brown EJ, Albers MW, Shin TB, Ichikawa K, Keith CT, Lane WS, Schreiber SL (1994) A mammalian protein targeted by G1-arresting rapamycin-receptor complex. Nature 369:756–758

Brunsing R, Omori-Fitzpatrick S, Weber F, Bicknell AA, Friend L, Rickert R, Niwa M (2008) B and T cell development both involve activity of the unfolded protein response pathway. J Biol Chem 283(26):17954–17961

Busslinger M, Klix N, Pfeffer P, Graninger PG, Kozmik Z (1996) Deregulation of PAX-5 by translocation of the Emu enhancer of the IgH locus adjacent to two alternative PAX-5 promoters in a diffuse large-cell lymphoma. Proc Natl Acad Sci U S A 93:6129–6134

Calfon M, Zeng H, Urano F, Till JH, Hubbard SR, Harding HP, Clark SG, Ron D (2002) IRE1 couples endoplasmic reticulum load to secretory capacity by processing the XBP-1 mRNA. Nature 415:92–96

Carrasco DR, Sukhdeo K, Protopopova M, Sinha R, Enos M, Carrasco DE, Zheng M, Mani M, Henderson J, Pinkus GS, Munshi N, Horner J, Ivanova EV, Protopopov A, Anderson KC, Tonon G, DePinho RA (2007) The differentiation and stress response factor XBP-1 drives multiple myeloma pathogenesis. Cancer Cell 11:349–360

Cattoretti G, Pasqualucci L, Ballon G, Tam W, Nandula SV, Shen Q, Mo T, Murty VV, Dalla-Favera R (2005) Deregulated BCL6 expression recapitulates the pathogenesis of human diffuse large B cell lymphomas in mice. Cancer Cell 7:445–455

Chen L, Widhopf G, Huynh L, Rassenti L, Rai KR, Weiss A, Kipps TJ (2002) Expression of ZAP-70 is associated with increased B-cell receptor signaling in chronic lymphocytic leukemia. Blood 100:4609–4614

Chen L, Apgar J, Huynh L, Dicker F, Giago-McGahan T, Rassenti L, Weiss A, Kipps TJ (2005) ZAP-70 directly enhances IgM signaling in chronic lymphocytic leukemia. Blood 105:2036–2041

Chen L, Monti S, Juszczynski P, Daley J, Chen W, Witzig TE, Habermann TM, Kutok JL, Shipp MA (2008) SYK-dependent tonic B-cell receptor signaling is a rational treatment target in diffuse large B-cell lymphoma. Blood 111:2230–2237

Coutinho A, Gronowicz E, Bullock WW, Moller G (1974) Mechanism of thymus-independent immunocyte triggering. Mitogenic activation of B cells results in specific immune responses. J Exp Med 139:74–92

Davis RE, Brown KD, Siebenlist U, Staudt LM (2001) Constitutive nuclear factor kappaB activity is required for survival of activated B cell-like diffuse large B cell lymphoma cells. J Exp Med 194:1861–1874

Davis RE, Ngo VN, Lenz G, Tolar P, Young RM, Romesser PB, Kohlhammer H, Lamy L, Zhao H, Yang Y, Xu W, Shaffer AL, Wright G, Xiao W, Powell J, Jiang JK, Thomas CJ, Rosenwald A, Ott G, Muller-Hermelink HK, Gascoyne RD, Connors JM, Johnson NA, Rimsza LM, Campo E, Jaffe ES, Wilson WH, Delabie J, Smeland EB, Fisher RI, Braziel RM, Tubbs RR, Cook JR, Weisenburger DD, Chan WC, Pierce SK, Staudt LM (2010) Chronic active B-cell-receptor signalling in diffuse large B-cell lymphoma. Nature 463:88–92

Decker T, Sandherr M, Goetze K, Oelsner M, Ringshausen I, Peschel C (2009) A pilot trial of the mTOR (mammalian target of rapamycin) inhibitor RAD001 in patients with advanced B-CLL. Ann Hematol 88:221–227

Dent AL, Shaffer AL, Yu X, Allman D, Staudt LM (1997) Control of inflammation, cytokine expression, and germinal center formation by BCL-6. Science 276:589–592

Dykstra M, Cherukuri A, Sohn HW, Tzeng SJ, Pierce SK (2003) Location is everything: lipid rafts and immune cell signaling. Annu Rev Immunol 21:457–481

Falini B, Fizzotti M, Pucciarini A, Bigerna B, Marafioti T, Gambacorta M, Pacini R, Alunni C, Natali-Tanci L, Ugolini B, Sebastiani C, Cattoretti G, Pileri S, Dalla-Favera R, Stein H (2000) A monoclonal antibody (MUM1p) detects expression of the MUM1/IRF4 protein in a subset of germinal center B cells, plasma cells, and activated T cells. Blood 95:2084–2092

Fukuda T, Yoshida T, Okada S, Hatano M, Miki T, Ishibashi K, Okabe S, Koseki H, Hirosawa S, Taniguchi M, Miyasaka N, Tokuhisa T (1997) Disruption of the Bcl6 gene results in an impaired germinal center formation. J Exp Med 186:439–448

Gaidano G, Pasqualucci L, Capello D, Berra E, Deambrogi C, Rossi D, Maria Larocca L, Gloghini A, Carbone A, Dalla-Favera R (2003) Aberrant somatic hypermutation in multiple subtypes of AIDS-associated non-Hodgkin lymphoma. Blood 102:1833–1841

Garrett-Sinha LA, Su GH, Rao S, Kabak S, Hao Z, Clark MR, Simon MC (1999) PU.1 and Spi-B are required for normal B cell receptor-mediated signal transduction. Immunity 10:399–408

Genestier L, Taillardet M, Mondiere P, Gheit H, Bella C, Defrance T (2007) TLR agonists selectively promote terminal plasma cell differentiation of B cell subsets specialized in thymus-independent responses. J Immunol 178:7779–7786

Gingras AC, Raught B, Sonenberg N (2001) Regulation of translation initiation by FRAP/mTOR. Genes Dev 15:807–826

Greenman C, Stephens P, Smith R, Dalgliesh GL, Hunter C, Bignell G, Davies H, Teague J, Butler A, Stevens C, Edkins S, O'Meara S, Vastrik I, Schmidt EE, Avis T, Barthorpe S, Bhamra G, Buck G, Choudhury B, Clements J, Cole J, Dicks E, Forbes S, Gray K, Halliday K, Harrison R, Hills K, Hinton J, Jenkinson A, Jones D, Menzies A, Mironenko T, Perry J, Raine K, Richardson D, Shepherd R, Small A, Tofts C, Varian J, Webb T, West S, Widaa S, Yates A, Cahill DP, Louis DN, Goldstraw P, Nicholson AG, Brasseur F, Looijenga L, Weber BL, Chiew YE, DeFazio A, Greaves MF, Green AR, Campbell P, Birney E, Easton DF, Chenevix-Trench G, Tan MH, Khoo SK, Teh BT, Yuen ST, Leung SY, Wooster R, Futreal PA, Stratton MR (2007) Patterns of somatic mutation in human cancer genomes. Nature 446:153–158

Grumont RJ, Gerondakis S (2000) Rel induces interferon regulatory factor 4 (IRF-4) expression in lymphocytes: modulation of interferon-regulated gene expression by rel/nuclear factor kappaB. J Exp Med 191:1281–1292

Gupta S, Jiang M, Anthony A, Pernis AB (1999) Lineage-specific modulation of interleukin 4 signaling by interferon regulatory factor 4. J Exp Med 190:1837–1848

Gururajan M, Jennings CD, Bondada S (2006) Cutting edge: constitutive B cell receptor signaling is critical for basal growth of B lymphoma. J Immunol 176:5715–5719

He B, Qiao X, Cerutti A (2004) CpG DNA induces IgG class switch DNA recombination by activating human B cells through an innate pathway that requires TLR9 and cooperates with IL-10. J Immunol 173:4479–4491

Hipp S, Ringshausen I, Oelsner M, Bogner C, Peschel C, Decker T (2005) Inhibition of the mammalian target of rapamycin and the induction of cell cycle arrest in mantle cell lymphoma cells. Haematologica 90:1433–1434

Hu CC, Dougan SK, McGehee AM, Love JC, Ploegh HL (2009) XBP-1 regulates signal transduction, transcription factors and bone marrow colonization in B cells. EMBO J 28:1624–1636

Iida S, Rao PH, Nallasivam P, Hibshoosh H, Butler M, Louie DC, Dyomin V, Ohno H, Chaganti RS, Dalla-Favera R (1996) The t(9;14)(p13;q32) chromosomal translocation associated with lymphoplasmacytoid lymphoma involves the PAX-5 gene. Blood 88:4110–4117

Iida S, Rao PH, Butler M, Corradini P, Boccadoro M, Klein B, Chaganti RS, Dalla-Favera R (1997) Deregulation of MUM1/IRF4 by chromosomal translocation in multiple myeloma. Nat Genet 17:226–230

Iwakoshi NN, Lee AH, Glimcher LH (2003a) The X-box binding protein-1 transcription factor is required for plasma cell differentiation and the unfolded protein response. Immunol Rev 194:29–38

Iwakoshi NN, Lee AH, Vallabhajosyula P, Otipoby KL, Rajewsky K, Glimcher LH (2003b) Plasma cell differentiation and the unfolded protein response intersect at the transcription factor XBP-1. Nat Immunol 4:321–329

Jahrsdorfer B, Wooldridge JE, Blackwell SE, Taylor CM, Griffith TS, Link BK, Weiner GJ (2005) Immunostimulatory oligodeoxynucleotides induce apoptosis of B cell chronic lymphocytic leukemia cells. J Leukoc Biol 77:378–387

Jiang A, Craxton A, Kurosaki T, Clark EA (1998) Different protein tyrosine kinases are required for B cell antigen receptor-mediated activation of extracellular signal-regulated kinase, c-Jun NH2-terminal kinase 1, and p38 mitogen-activated protein kinase. J Exp Med 188:1297–1306

Kallies A, Hasbold J, Fairfax K, Pridans C, Emslie D, McKenzie BS, Lew AM, Corcoran LM, Hodgkin PD, Tarlinton DM, Nutt SL (2007) Initiation of plasma-cell differentiation is independent of the transcription factor Blimp-1. Immunity 26:555–566

Keats JJ, Fonseca R, Chesi M, Schop R, Baker A, Chng WJ, Van Wier S, Tiedemann R, Shi CX, Sebag M, Braggio E, Henry T, Zhu YX, Fogle H, Price-Troska T, Ahmann G, Mancini C, Brents LA, Kumar S, Greipp P, Dispenzieri A, Bryant B, Mulligan G, Bruhn L, Barrett M, Valdez R, Trent J, Stewart AK, Carpten J, Bergsagel PL (2007) Promiscuous mutations activate the noncanonical NF-kappaB pathway in multiple myeloma. Cancer Cell 12:131–144

Klein U, Casola S, Cattoretti G, Shen Q, Lia M, Mo T, Ludwig T, Rajewsky K, Dalla-Favera R (2006) Transcription factor IRF4 controls plasma cell differentiation and class-switch recombination. Nat Immunol 7:773–782

Krieg AM (2008) Toll-like receptor 9 (TLR9) agonists in the treatment of cancer. Oncogene 27:161–167

Krieg AM, Yi AK, Matson S, Waldschmidt TJ, Bishop GA, Teasdale R, Koretzky GA, Klinman DM (1995) CpG motifs in bacterial DNA trigger direct B-cell activation. Nature 374:546–549

Kriss CL, Pinilla-Ibarz JA, Mailloux AW, Powers JJ, Tang CH, Kang CW, Zanesi N, Epling-Burnette PK, Sotomayor EM, Croce CM, Del Valle JR, Hu CC (2012) Overexpression of TCL1 activates the endoplasmic reticulum stress response: a novel mechanism of leukemic progression in mice. Blood 120:1027–1038

Kuehl WM, Bergsagel PL (2002) Multiple myeloma: evolving genetic events and host interactions. Nat Rev Cancer 2:175–187

Lam LT, Davis RE, Pierce J, Hepperle M, Xu Y, Hottelet M, Nong Y, Wen D, Adams J, Dang L, Staudt LM (2005) Small molecule inhibitors of IkappaB kinase are selectively toxic for subgroups of diffuse large B-cell lymphoma defined by gene expression profiling. Clin Cancer Res 11:28–40

Lam LT, Wright G, Davis RE, Lenz G, Farinha P, Dang L, Chan JW, Rosenwald A, Gascoyne RD, Staudt LM (2008) Cooperative signaling through the signal transducer and activator of transcription 3 and nuclear factor-{kappa}B pathways in subtypes of diffuse large B-cell lymphoma. Blood 111:3701–3713

Lee AH, Iwakoshi NN, Glimcher LH (2003) XBP-1 regulates a subset of endoplasmic reticulum resident chaperone genes in the unfolded protein response. Mol Cell Biol 23:7448–7459

Lee AH, Chu GC, Iwakoshi NN, Glimcher LH (2005) XBP-1 is required for biogenesis of cellular secretory machinery of exocrine glands. EMBO J 24:4368–4380

Lenz G, Nagel I, Siebert R, Roschke AV, Sanger W, Wright GW, Dave SS, Tan B, Zhao H, Rosenwald A, Muller-Hermelink HK, Gascoyne RD, Campo E, Jaffe ES, Smeland EB, Fisher RI, Kuehl WM, Chan WC, Staudt LM (2007) Aberrant immunoglobulin class switch recombination and switch translocations in activated B cell-like diffuse large B cell lymphoma. J Exp Med 204:633–643

Lenz G, Davis RE, Ngo VN, Lam L, George TC, Wright GW, Dave SS, Zhao H, Xu W, Rosenwald A, Ott G, Muller-Hermelink HK, Gascoyne RD, Connors JM, Rimsza LM, Campo E, Jaffe ES, Delabie J, Smeland EB, Fisher RI, Chan WC, Staudt LM (2008a) Oncogenic CARD11 mutations in human diffuse large B cell lymphoma. Science 319:1676–1679

Lenz G, Wright GW, Emre NC, Kohlhammer H, Dave SS, Davis RE, Carty S, Lam LT, Shaffer AL, Xiao W, Powell J, Rosenwald A, Ott G, Muller-Hermelink HK, Gascoyne RD, Connors JM, Campo E, Jaffe ES, Delabie J, Smeland EB, Rimsza LM, Fisher RI, Weisenburger DD, Chan WC, Staudt LM (2008b) Molecular subtypes of diffuse large B-cell lymphoma arise by distinct genetic pathways. Proc Natl Acad Sci U S A 105:13520–13525

Lin Y, Wong K, Calame K (1997) Repression of c-myc transcription by Blimp-1, an inducer of terminal B cell differentiation. Science 276:596–599

Lin KI, Angelin-Duclos C, Kuo TC, Calame K (2002) Blimp-1-dependent repression of Pax-5 is required for differentiation of B cells to immunoglobulin M-secreting plasma cells. Mol Cell Biol 22:4771–4780

Liso A, Capello D, Marafioti T, Tiacci E, Cerri M, Distler V, Paulli M, Carbone A, Delsol G, Campo E, Pileri S, Pasqualucci L, Gaidano G, Falini B (2006) Aberrant somatic hypermutation in tumor cells of nodular-lymphocyte-predominant and classic Hodgkin lymphoma. Blood 108:1013–1020

Longo PG, Laurenti L, Gobessi S, Petlickovski A, Pelosi M, Chiusolo P, Sica S, Leone G, Efremov DG (2007) The Akt signaling pathway determines the different proliferative capacity of chronic lymphocytic leukemia B-cells from patients with progressive and stable disease. Leukemia 21:110–120

Lu R, Medina KL, Lancki DW, Singh H (2003) IRF-4,8 orchestrate the pre-B-to-B transition in lymphocyte development. Genes Dev 17:1703–1708

Martins G, Calame K (2008) Regulation and functions of Blimp-1 in T and B lymphocytes. Annu Rev Immunol 26:133–169

McGehee AM, Dougan SK, Klemm EJ, Shui G, Park B, Kim YM, Watson N, Wenk MR, Ploegh HL, Hu CC (2009) XBP-1-deficient plasmablasts show normal protein folding but altered glycosylation and lipid synthesis. J Immunol 183:3690–3699

Meyer-Bahlburg A, Khim S, Rawlings DJ (2007) B cell intrinsic TLR signals amplify but are not required for humoral immunity. J Exp Med 204:3095–3101

Migliazza A, Martinotti S, Chen W, Fusco C, Ye BH, Knowles DM, Offit K, Chaganti RS, Dalla-Favera R (1995) Frequent somatic hypermutation of the 5′ noncoding region of the BCL6 gene in B-cell lymphoma. Proc Natl Acad Sci U S A 92:12520–12524

Mittrucker HW, Matsuyama T, Grossman A, Kundig TM, Potter J, Shahinian A, Wakeham A, Patterson B, Ohashi PS, Mak TW (1997) Requirement for the transcription factor LSIRF/IRF4 for mature B and T lymphocyte function. Science 275:540–543

Mora-Lopez F, Reales E, Brieva JA, Campos-Caro A (2007) Human BSAP and BLIMP1 conform an autoregulatory feedback loop. Blood 110:3150–3157

Moriyama M, Yamochi T, Semba K, Akiyama T, Mori S (1997) BCL-6 is phosphorylated at multiple sites in its serine- and proline-clustered region by mitogen-activated protein kinase (MAPK) in vivo. Oncogene 14:2465–2474

Ngo VN, Davis RE, Lamy L, Yu X, Zhao H, Lenz G, Lam LT, Dave S, Yang L, Powell J, Staudt LM (2006) A loss-of-function RNA interference screen for molecular targets in cancer. Nature 441:106–110

Niu H, Ye BH, Dalla-Favera R (1998) Antigen receptor signaling induces MAP kinase-mediated phosphorylation and degradation of the BCL-6 transcription factor. Genes Dev 12:1953–1961

Niu H, Cattoretti G, Dalla-Favera R (2003) BCL6 controls the expression of the B7-1/CD80 costimulatory receptor in germinal center B cells. J Exp Med 198:211–221

Nutt SL, Heavey B, Rolink AG, Busslinger M (1999) Commitment to the B-lymphoid lineage depends on the transcription factor Pax5. Nature 401:556–562

Papandreou I, Denko NC, Olson M, Van Melckebeke H, Lust S, Tam A, Solow-Cordero DE, Bouley DM, Offner F, Niwa M, Koong AC (2011) Identification of an Ire1alpha endonuclease specific inhibitor with cytotoxic activity against human multiple myeloma. Blood 117:1311–1314

Park SS, Kim JS, Tessarollo L, Owens JD, Peng L, Han SS, Tae Chung S, Torrey TA, Cheung WC, Polakiewicz RD, McNeil N, Ried T, Mushinski JF, Morse HC 3rd, Janz S (2005) Insertion of c-Myc into Igh induces B-cell and plasma-cell neoplasms in mice. Cancer Res 65:1306–1315

Pasqualucci L, Migliazza A, Fracchiolla N, William C, Neri A, Baldini L, Chaganti RS, Klein U, Kuppers R, Rajewsky K, Dalla-Favera R (1998) BCL-6 mutations in normal germinal center B cells: evidence of somatic hypermutation acting outside Ig loci. Proc Natl Acad Sci U S A 95:11816–11821

Pasqualucci L, Neumeister P, Goossens T, Nanjangud G, Chaganti RS, Kuppers R, Dalla-Favera R (2001) Hypermutation of multiple proto-oncogenes in B-cell diffuse large-cell lymphomas. Nature 412:341–346

Pasqualucci L, Compagno M, Houldsworth J, Monti S, Grunn A, Nandula SV, Aster JC, Murty VV, Shipp MA, Dalla-Favera R (2006) Inactivation of the PRDM1/BLIMP1 gene in diffuse large B cell lymphoma. J Exp Med 203:311–317

Pene F, Claessens YE, Muller O, Viguie F, Mayeux P, Dreyfus F, Lacombe C, Bouscary D (2002) Role of the phosphatidylinositol 3-kinase/Akt and mTOR/P70S6-kinase pathways in the proliferation and apoptosis in multiple myeloma. Oncogene 21:6587–6597

Peponi E, Drakos E, Reyes G, Leventaki V, Rassidakis GZ, Medeiros LJ (2006) Activation of mammalian target of rapamycin signaling promotes cell cycle progression and protects cells from apoptosis in mantle cell lymphoma. Am J Pathol 169:2171–2180

Phan RT, Dalla-Favera R (2004) The BCL6 proto-oncogene suppresses p53 expression in germinal-centre B cells. Nature 432:635–639

Phan RT, Saito M, Basso K, Niu H, Dalla-Favera R (2005) BCL6 interacts with the transcription factor Miz-1 to suppress the cyclin-dependent kinase inhibitor p21 and cell cycle arrest in germinal center B cells. Nat Immunol 6:1054–1060

Pierce SK (2002) Lipid rafts and B-cell activation. Nat Rev Immunol 2:96–105

Pogue SL, Kurosaki T, Bolen J, Herbst R (2000) B cell antigen receptor-induced activation of Akt promotes B cell survival and is dependent on Syk kinase. J Immunol 165:1300–1306

Potter M (2003) Neoplastic development in plasma cells. Immunol Rev 194:177–195

Reimold AM, Ponath PD, Li YS, Hardy RR, David CS, Strominger JL, Glimcher LH (1996) Transcription factor B cell lineage-specific activator protein regulates the gene for human X-box binding protein 1. J Exp Med 183:393–401

Reimold AM, Iwakoshi NN, Manis J, Vallabhajosyula P, Szomolanyi-Tsuda E, Gravallese EM, Friend D, Grusby MJ, Alt F, Glimcher LH (2001) Plasma cell differentiation requires the transcription factor XBP-1. Nature 412:300–307

Ringshausen I, Peschel C, Decker T (2005) Mammalian target of rapamycin (mTOR) inhibition in chronic lymphocytic B-cell leukemia: a new therapeutic option. Leuk Lymphoma 46:11–19

Rosenwald A, Wright G, Leroy K, Yu X, Gaulard P, Gascoyne RD, Chan WC, Zhao T, Haioun C, Greiner TC, Weisenburger DD, Lynch JC, Vose J, Armitage JO, Smeland EB, Kvaloy S, Holte H, Delabie J, Campo E, Montserrat E, Lopez-Guillermo A, Ott G, Muller-Hermelink HK, Connors JM, Braziel R, Grogan TM, Fisher RI, Miller TP, LeBlanc M, Chiorazzi M, Zhao H, Yang L, Powell J, Wilson WH, Jaffe ES, Simon R, Klausner RD, Staudt LM (2003) Molecular diagnosis of primary mediastinal B cell lymphoma identifies a clinically favorable subgroup of diffuse large B cell lymphoma related to Hodgkin lymphoma. J Exp Med 198:851–862

Sabatini DM, Erdjument-Bromage H, Lui M, Tempst P, Snyder SH (1994) RAFT1: a mammalian protein that binds to FKBP12 in a rapamycin-dependent fashion and is homologous to yeast TORs. Cell 78:35–43

Saito M, Gao J, Basso K, Kitagawa Y, Smith PM, Bhagat G, Pernis A, Pasqualucci L, Dalla-Favera R (2007) A signaling pathway mediating downregulation of BCL6 in germinal center B cells is blocked by BCL6 gene alterations in B cell lymphoma. Cancer Cell 12:280–292

Sanchez-Beato M, Saez AI, Martinez-Montero JC, Sol Mateo M, Sanchez-Verde L, Villuendas R, Troncone G, Piris MA (1997) Cyclin-dependent kinase inhibitor p27KIP1 in lymphoid tissue: p27KIP1 expression is inversely proportional to the proliferative index. Am J Pathol 151:151–160

Savage KJ, Monti S, Kutok JL, Cattoretti G, Neuberg D, De Leval L, Kurtin P, Dal Cin P, Ladd C, Feuerhake F, Aguiar RC, Li S, Salles G, Berger F, Jing W, Pinkus GS, Habermann T, Dalla-Favera R, Harris NL, Aster JC, Golub TR, Shipp MA (2003) The molecular signature of mediastinal large B-cell lymphoma differs from that of other diffuse large B-cell lymphomas and shares features with classical Hodgkin lymphoma. Blood 102:3871–3879

Schmelzle T, Hall MN (2000) TOR, a central controller of cell growth. Cell 103:253–262

Sciammas R, Shaffer AL, Schatz JH, Zhao H, Staudt LM, Singh H (2006) Graded expression of interferon regulatory factor-4 coordinates isotype switching with plasma cell differentiation. Immunity 25:225–236

Shaffer AL, Yu X, He Y, Boldrick J, Chan EP, Staudt LM (2000) BCL-6 represses genes that function in lymphocyte differentiation, inflammation, and cell cycle control. Immunity 13:199–212

Shaffer AL, Lin KI, Kuo TC, Yu X, Hurt EM, Rosenwald A, Giltnane JM, Yang L, Zhao H, Calame K, Staudt LM (2002) Blimp-1 orchestrates plasma cell differentiation by extinguishing the mature B cell gene expression program. Immunity 17:51–62

Shaffer AL, Shapiro-Shelef M, Iwakoshi NN, Lee AH, Qian SB, Zhao H, Yu X, Yang L, Tan BK, Rosenwald A, Hurt EM, Petroulakis E, Sonenberg N, Yewdell JW, Calame K, Glimcher LH, Staudt LM (2004) XBP1, downstream of Blimp-1, expands the secretory apparatus and other organelles, and increases protein synthesis in plasma cell differentiation. Immunity 21:81–93

Shaffer AL, Emre NC, Lamy L, Ngo VN, Wright G, Xiao W, Powell J, Dave S, Yu X, Zhao H, Zeng Y, Chen B, Epstein J, Staudt LM (2008) IRF4 addiction in multiple myeloma. Nature 454:226–231

Shapiro-Shelef M, Lin KI, McHeyzer-Williams LJ, Liao J, McHeyzer-Williams MG, Calame K (2003) Blimp-1 is required for the formation of immunoglobulin secreting plasma cells and pre-plasma memory B cells. Immunity 19:607–620

Shapiro-Shelef M, Lin KI, Savitsky D, Liao J, Calame K (2005) Blimp-1 is required for maintenance of long-lived plasma cells in the bone marrow. J Exp Med 202:1471–1476

Shen HM, Peters A, Baron B, Zhu X, Storb U (1998) Mutation of BCL-6 gene in normal B cells by the process of somatic hypermutation of Ig genes. Science 280:1750–1752

Shen X, Ellis RE, Lee K, Liu CY, Yang K, Solomon A, Yoshida H, Morimoto R, Kurnit DM, Mori K, Kaufman RJ (2001) Complementary signaling pathways regulate the unfolded protein response and are required for C. elegans development. Cell 107:893–903

Shi Y, Gera J, Hu L, Hsu JH, Bookstein R, Li W, Lichtenstein A (2002) Enhanced sensitivity of multiple myeloma cells containing PTEN mutations to CCI-779. Cancer Res 62:5027–5034

Siebenlist U, Brown K, Claudio E (2005) Control of lymphocyte development by nuclear factor-kappaB. Nat Rev Immunol 5:435–445

Sriburi R, Jackowski S, Mori K, Brewer JW (2004) XBP1: a link between the unfolded protein response, lipid biosynthesis, and biogenesis of the endoplasmic reticulum. J Cell Biol 167:35–41

Sriburi R, Bommiasamy H, Buldak GL, Robbins GR, Frank M, Jackowski S, Brewer JW (2007) Coordinate regulation of phospholipid biosynthesis and secretory pathway gene expression in XBP-1(S)-induced endoplasmic reticulum biogenesis. J Biol Chem 282:7024–7034

Su GH, Ip HS, Cobb BS, Lu MM, Chen HM, Simon MC (1996) The Ets protein Spi-B is expressed exclusively in B cells and T cells during development. J Exp Med 184:203–214

Tam W, Gomez M, Chadburn A, Lee JW, Chan WC, Knowles DM (2006) Mutational analysis of PRDM1 indicates a tumor-suppressor role in diffuse large B-cell lymphomas. Blood 107:4090–4100

Tate G, Hirayama-Ohashi Y, Kishimoto K, Mitsuya T (2007) Novel BLIMP1/PRDM1 gene mutations in B-cell lymphoma. Cancer Genet Cytogenet 172:151–153

Tunyaplin C, Shaffer AL, Angelin-Duclos CD, Yu X, Staudt LM, Calame KL (2004) Direct repression of prdm1 by Bcl-6 inhibits plasmacytic differentiation. J Immunol 173:1158–1165

Turner CA Jr, Mack DH, Davis MM (1994) Blimp-1, a novel zinc finger-containing protein that can drive the maturation of B lymphocytes into immunoglobulin-secreting cells. Cell 77:297–306

Vallabhapurapu S, Karin M (2009) Regulation and function of NF-kappaB transcription factors in the immune system. Annu Rev Immunol 27:693–733

van Anken E, Romijn EP, Maggioni C, Mezghrani A, Sitia R, Braakman I, Heck AJ (2003) Sequential waves of functionally related proteins are expressed when B cells prepare for antibody secretion. Immunity 18:243–253

Wang X, Li Z, Naganuma A, Ye BH (2002) Negative autoregulation of BCL-6 is bypassed by genetic alterations in diffuse large B cell lymphomas. Proc Natl Acad Sci U S A 99:15018–15023

Wullschleger S, Loewith R, Hall MN (2006) TOR signaling in growth and metabolism. Cell 124:471–484

Xu D, Zhao L, Del Valle L, Miklossy J, Zhang L (2008) Interferon regulatory factor 4 is involved in Epstein-Barr virus-mediated transformation of human B lymphocytes. J Virol 82:6251–6258

Ye BH, Lista F, Lo Coco F, Knowles DM, Offit K, Chaganti RS, Dalla-Favera R (1993) Alterations of a zinc finger-encoding gene, BCL-6, in diffuse large-cell lymphoma. Science 262:747–750

Ye BH, Chaganti S, Chang CC, Niu H, Corradini P, Chaganti RS, Dalla-Favera R (1995) Chromosomal translocations cause deregulated BCL6 expression by promoter substitution in B cell lymphoma. EMBO J 14:6209–6217

Ye BH, Cattoretti G, Shen Q, Zhang J, Hawe N, de Waard R, Leung C, Nouri-Shirazi M, Orazi A, Chaganti RS, Rothman P, Stall AM, Pandolfi PP, Dalla-Favera R (1997) The BCL-6 proto-oncogene controls germinal-centre formation and Th2-type inflammation. Nat Genet 16:161–170

Yoshida H, Matsui T, Yamamoto A, Okada T, Mori K (2001) XBP1 mRNA is induced by ATF6 and spliced by IRE1 in response to ER stress to produce a highly active transcription factor. Cell 107:881–891

Younes A (2008) Therapeutic activity of mTOR inhibitors in mantle cell lymphoma: clues but no clear answers. Autophagy 4:707–709

Zanesi N, Aqeilan R, Drusco A, Kaou M, Sevignani C, Costinean S, Bortesi L, La Rocca G, Koldovsky P, Volinia S, Mancini R, Calin G, Scott CP, Pekarsky Y, Croce CM (2006) Effect of rapamycin on mouse chronic lymphocytic leukemia and the development of nonhematopoietic malignancies in Emu-TCL1 transgenic mice. Cancer Res 66:915–920

Zeng X, Kinsella TJ (2008) Mammalian target of rapamycin and S6 kinase 1 positively regulate 6-thioguanine-induced autophagy. Cancer Res 68:2384–2390

Zhang Z, Espinoza CR, Yu Z, Stephan R, He T, Williams GS, Burrows PD, Hagman J, Feeney AJ, Cooper MD (2006) Transcription factor Pax5 (BSAP) transactivates the RAG-mediated V(H)-to-DJ(H) rearrangement of immunoglobulin genes. Nat Immunol 7:616–624

Follicular Lymphoma: Recent Advances

Alyssa Bouska, SharathKumar Bagvati, Javeed Iqbal, Basem M. William, and Wing C. Chan

Introduction

Follicular lymphoma (FL) is a B-cell lymphoma that recapitulates the germinal center (GC)B-cell stage of differentiation and in most cases form neoplastic follicles that resemble the normal GC. Normal GCB-cells do not express the anti-apoptotic protein B-cell leukemia/lymphoma 2 (BCL2) and they readily undergo apoptosis if they do not have a high affinity antigen receptor that will transmit survival signals on recognition of the antigen. The vast majority of FL has a translocation that leads to an inappropriate constitutive expression of the BCL2 protein that protects the cell from apoptosis. This cell can survive for long periods in the GC environment and undergoes further genetic alterations that eventually establish the neoplastic clonal population as a FL. These critical secondary changes are of great interest in understanding the evolution of a pre-neoplastic clone with *BCL2* translocation to a malignant lymphoma.

Alyssa Bouska and SharathKumar Bagvati contributed equally.

A. Bouska • S. Bagvati • J. Iqbal
Department of Pathology and Microbiology,
University of Nebraska Medical Center, Omaha, NE, USA

B.M. William
Department of Internal Medicine, University of Nebraska Medical Center,
Omaha, NE, USA

W.C. Chan, M.D. (✉)
Department of Pathology and Microbiology, University of Nebraska Medical Center,
Omaha, NE, USA

Department of Pathology and Microbiology, Center for Research in Lymphoma
and Leukemia, University of Nebraska Medical Center,
983135 Nebraska Medical Center, Omaha, NE 68198-3135, USA
e-mail: jchan@unmc.edu

J. Tao and E. Sotomayor (eds.), *Hematologic Cancers: From Molecular Pathobiology to Targeted Therapeutics*, Cancer Growth and Progression 14,
DOI 10.1007/978-94-007-5028-9_2,

FL is generally an indolent tumor and it undergoes further genetic changes during its clinical course. In around 30% of the patients, the tumor undergoes transformation to a diffuse large B-cell lymphoma (DLBCL) or uncommonly a Burkitt-like lymphoma with marked worsening of prognosis. This subset of cases with more aggressive clinical behavior should be managed differently and prognostic biomarkers would be valuable. The treatment of FL has undergone substantial changes in recent years with an improvement of overall survival (OS). This chapter will discuss the biology and pathology of FL and recent clinical advances will also be briefly reviewed.

Pathobiology of Follicular Lymphoma

The tumor cell in FL arises from a GCB-cell that, in 85–90% of cases, has acquired a t(14;18) translocation; however, in about 10–15% of the cases, the t(14;18) is absent and the initiating event(s) in these FLs is unclear (Hoglund et al. 2004). This translocation is believed to be acquired during B-cell development since, in the vast majority of cases, the break at the *IgH* gene occurs at the J_h region. The *BCL2* gene is brought to the vicinity of the *IgH* enhancers leading to inappropriate expression of the anti-apoptotic protein BCL2. If this cell differentiates into a GCB-cell, it is protected from apoptosis in the germinal center reaction resulting in its clonal expansion and the accumulation of additional genetic abnormalities with time. Although crucial in the initiation of the malignant process, the t(14;18) translocation alone is not sufficient for the establishment of FL. In fact, the classical t(14;18)(q32;q21) translocation has been demonstrated in apparently healthy individuals and follow-up of these individuals failed to detect clinical FL (Limpens et al. 1995). Clearly, additional genetic alterations are necessary to establish a FL and, once established, a FL will continue to acquire additional genetic abnormalities with consequent changes in biologic and clinical behavior of the tumor.

In a cytogenetic model proposed by Hoglund et al. (2004), four cytogenetic abnormalities (del(6q), +add(1q), +der(18)t(14;18), and +7) have been identified as early events in the clonal evolution of t(14;18)-positive FL. These then lead to further downstream events of clonal progression in a temporal sequence of chromosomal aberrations that frequently overlap. Genomic gains are more frequent early in the course of the disease, while later stages are characterized more by genomic losses. Our observations confirm that+der(18)t(14;18) is an early and relatively unique event (d'Amore et al. 2008). Aside from that, there is a group of abnormalities that seems to provide multiple entry points in the further evolution of the t(14;18)-positive clone, dominated by −6q/i(6p) and +7 but also include+X, +12, 10q-, 1p- and +18. Unlike+der(18)t(14;18), the −6q/i(6p), +7, +X, +12 and 1p- anomalies are commonly paired with each other and with +5, +8,+11, and dup6p, suggesting these events may cooperate to establish the FL. It should be noted that, though infrequent, events other than the common ones mentioned above could also occur as early events or even as the sole cytogenetic abnormalities with t(14;18)(q32;q21). These observations suggest that functional alterations important in the establishment of FL may be accomplished by multiple genetic alterations; some of these alterations are preferred and they are represented as the frequent early abnormalities.

Cytogenetic studies, though useful, are of rather low resolution and it can be quite difficult to determine the origin of the aberrant genetic material. Innovative techniques have been developed to address these problems. Spectral karyotyping (Nanjangud et al. 2007) hybridizes labeled probes to individual chromosomes so that each chromosome can be recognized by its unique fluorescence spectrum. This is very helpful in identifying translocation partners and the origin of abnormally acquired genetic material in chromosomes. Comparative genomic hybridization (CGH) (Kallioniemi et al. 1992) was developed to study copy number abnormalities (CNAs) using DNA from tumor and normal tissues labeled with different fluorophores and hybridized to normal metaphases. While CGH is useful, the resolution is still very low and recently array CGH (aCGH) techniques (Ylstra et al. 2006; Coe et al. 2007) have been developed to markedly improve the resolution.

Several studies using the high resolution technique of aCGH have been reported and have helped to confirm known alterations detected by conventional CGH and karyotype analysis, but also have identified novel alterations undetectable by less sensitive techniques. An aCGH study of 25 FL cases by Cheung et al. was able to identify aberrant regions affecting single genes, and four genes (*CDKN2A, FHIT, KIT,* and *PTPRD*) occurred in more than one sample (Cheung et al. 2010a). The increased resolution of aCGH makes it useful for narrowing down the minimal common region of recurrent abnormalities, thus providing a better guide for teasing out which genes are important for FL. Furthermore, a related technique using SNP arrays makes analysis of copy neutral loss of heterozygosity (LOH) possible. Copy neutral LOH or copy number loss was found at 1p36 in half of the FL cases studied, suggesting multiple mechanisms of inactivating genes within the locus (Ross et al. 2007; Cheung et al. 2010a) (Fig. 1). Deletion at 1p36.22-p36.33 correlated with increased risk for transformation and poor survival (Cheung et al. 2009). This region contains the *TP53* family member *TP73*, though no mutations were found (Ross et al. 2007). LOH that was copy neutral (uniparental disomy (UPD)) or had copy loss was also found at high frequency on 6p, 9p, 10q, 12q, and 16p, (Fitzgibbon et al. 2007; Ross et al. 2007; Cheung et al. 2010a). UPD at 16p has been suggested to predict transformation (O'Shea et al. 2009). Homozygous deletions were found occasionally at 9p21 and genes deleted included important cell cycle regulators *CDKN2A/2B* (Ross et al. 2007; Cheung et al. 2010a). 15% of FL cases had 10q LOH or copy loss that included the *PTEN* locus (Ross et al. 2007). Deletions in 6q have been associated with poor prognosis, and two separate regions of deletion were identified on chromosome 6 at q13-15 and q23.3-24.1 (Ross et al. 2007). The genes encoding the NF-κB regulator TNFAIP3/A20 and apoptotic protein PERP are possible important genes in the 6q23.3-24.1 region (Ross et al. 2007; Cheung et al. 2009). Gains and amplification were found at 1q, 2p15-16, 6p, 7, 12, and X (Ross et al. 2007; Cheung et al. 2010a). Amplification was also frequently found on 8q, but interestingly the minimal region of amplification was 8q24.22-24.23 and did not include the *MYC* gene, but included genes *NIBP, KCNK9, PTK2/FAK, and PTP4A3* (Ross et al. 2007; Cheung et al. 2009). Thus aCGH studies, by characterizing common CNAs, will be

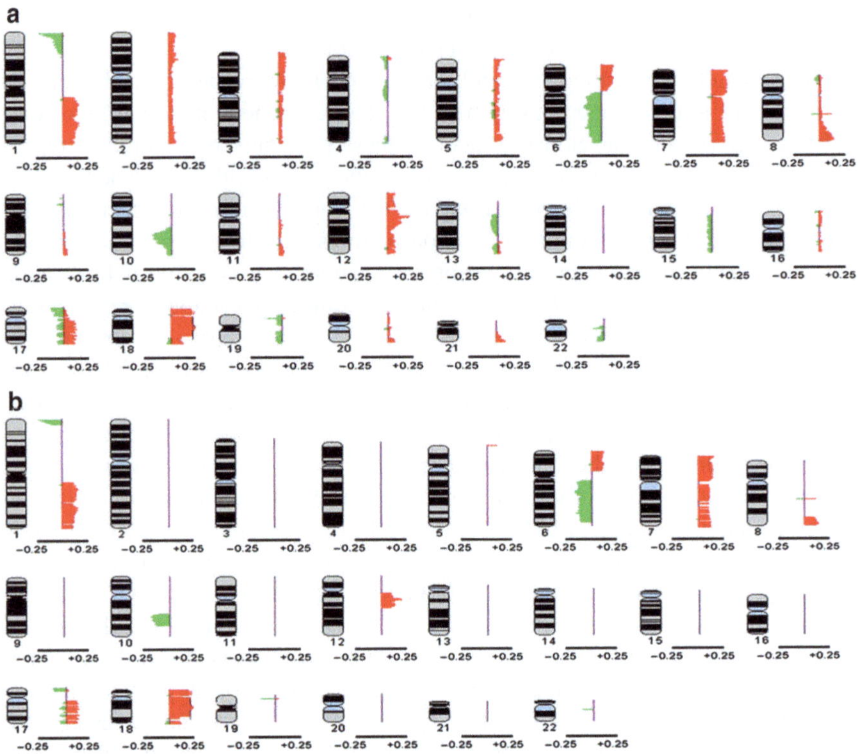

Fig. 1 Composite frequency ideogram plot of genome-wide copy number alterations in 106 diagnostic FL cases based on intersection analysis. (**a**) The frequencies of aberrations, represented by green signals for losses and red signals for gains, in the autosomes were derived from intersection analysis, where the union was taken between calls made visually by a cytogenetic pathologist and those determined by CNA-HMMer version 0.1. (**b**) Genetic losses or gains are represented by *green* and *red* signals, respectively. The *horizontal bar* below each ideogram represents gain and loss frequencies of +0.25 and −0.25, respectively (Reproduced with permission Cheung et al. (2009))

especially useful in identifying genes that contribute to FL initiation, progression and tumor transformation.

The B-cell receptor (BCR) continues to be expressed in the vast majority of FLs indicating that BCR signaling may still be important in tumor cell survival. Attempts to measure BCR signaling in FL cells by phospho protein specific-flow cytometry found that the BCR signaling cascade differs in FL cells compared to normal B cells (Irish et al. 2006). Another interesting finding is the detection of mutations in the surface immunoglobulin frequently resulting in sequence motifs for N-glycan addition. The added glycans terminate at high mannose that may interact with mannose receptors on stromal cellular components with resultant BCR signaling (Coelho et al. 2010).

Recently, a number of frequent mutations involving *CREBBP/EP300, MLL2, EZH2, and TNFRSF14* have been demonstrated (Cheung et al. 2010b; Morin et al.

2010a, b; Bodor et al. 2011; Pasqualucci et al. 2011). *CREBBP* and the closely related *EP300* encode acetyltransferases for both histone and non-histone proteins. Mutations affecting *CREBBP* are much more common, affecting over 30% of cases in one report (Pasqualucci et al. 2011). The mutations are inactivating and in most of the cases, it seems that haploinsufficiency is sufficient for the functional consequences. Monoallelic deletions are observed more commonly in DLBCL than in FL and appear to be mutually exclusive to mutations (Pasqualucci et al. 2011). The epigenetic modification of Histone 3 by trimethylation of lysine 27 (H3K27me3) represses gene expression, and EZH2, a subunit of the polycomb repressive complex 2 (PRC2) catalyzes the mono, di, and tri-methylation of H3K27. EZH2 and PRC2 can also recruit DNA methyltransferases to promote CpG methylation (Vire et al. 2006). Mutation affecting the Y641 codon in the SET domain of *EZH2* occurs in over 20% of GCB-DLBCL but at lower frequency in FL (about 7–10%) (Morin et al. 2010a; Bodor et al. 2011). This was initially considered an inactivating mutation but subsequent studies demonstrated that the mutated form is less efficient in catalyzing mono- and di-methylation, but much more efficient in catalyzing the addition of the third methyl group (Sneeringer et al. 2010). Thus, this mutation facilitates the synthesis of H3K9me3 and also explains why the wildtype *EZH2* is maintained. *EZH2* is expressed in normal GCB-cells and found by ChIP-on-CHIP experiments to associate with many genes in GCB-cells aside from its known embryonic stem cell targets (Velichutina et al. 2010). These genes include several key cell cycle-related tumor suppressor genes. Knocking down *EZH2* by siRNA induces acute cell cycle arrest at the G1/S transition. Changes in the activity of EZH2 due to the Y641 mutation probably alter the H3K9me3 profile of a number of key genes that promotes lymphomagenesis. No prognostic significance of this mutation was observed in FL patients carrying *EZH2* mutation (Bodor et al. 2011). MLL (mixed lineage leukemia)2 is a member of a family of DNA binding proteins that methylate H3K4 through the SET domain (Ansari and Mandal 2010). Loss of function mutation with resultant haploinsufficiency causes a rare syndrome (Kabuki Syndrome) characterized by multiple malformations and moderate intellectual impairment (Hannibal et al. 2011). Translocations involving *MLL* genes have been well recognized in the pathogenesis of acute leukemias (Krivtsov and Armstrong 2007), but recently loss of function mutations affecting *MLL2* have been described as a very frequent event in FL (Morin et al. 2010b). Presumably, reduced expression of a set of MLL2 targets in GCB-cells may promote their transformation into FL cells. The nonsynonymous *TNFRSF14* mutation (Cheung et al. 2010b) was observed in 18.3% FL patients and inferior OS and disease specific survival (DSS) were most pronounced in patients whose lymphomas contained *TNFRSF14* mutations and 1p36 deletions.

Promoter hypermethylation is a well known mechanism of gene inactivation in cancers, and studies of FLs have identified frequent promoter hypermethylation in genes such as the androgen receptor (Yang et al. 2003), *SHP1* (Koyama et al. 2003; Chim et al. 2004), *DBC1* (Gronbaek et al. 2008) and *ID4* (Hagiwara et al. 2007). Additionally, methylation of cell cycle regulators *CDKN2A*, *CDKN2B*, and *RB1* have been detected in some FLs (Chim et al. 2007) and may play a role in the transformation of FL into a more aggressive disease (Hayslip and Montero 2006).

Parallel clonal evolution is a common phenomenon in FL with multiple subclones emerging from a parental clone and even a single tumor site may be populated by one or more of these clones. As noted in other studies (Hoglund et al. 2004; Eide et al. 2010) and our study (d'Amore et al. 2008), clones from sequential biopsies were frequently different from that of the original biopsy. Molecular analyses examining the pattern of Ig gene mutation also confirm the cytogenetics observation (Carlotti et al. 2009; Eide et al. 2010). Evolution of one of the subclones may give rise to more refractory disease or a large cell lymphoma and this subclone may have a similar, but not matching, profile of genetic changes as the subclone detected from a previous biopsy.

Dave and co-workers have determined the gene expression profiles of 191 cases of FL (Dave et al. 2004). Interestingly, the major determinant of prognosis appears to be related to two gene expression signatures derived from the microenvironment (Fig. 2). These two signatures (immune response -1 (IR-1) and immune response -2 (IR-2)) can be combined to form a molecular predictor of survival for FL. The IR-1 signature contains a number of genes that are known T-cell transcripts, but the signature does not just reflect the presence of tumor infiltrating T-cells, and high expression of this signature is associated with better outcome. The IR-2 signature contains many transcripts expressed by myeloid and monocytic cells and high expression of this signature is associated with poor prognosis. When patients were grouped into quartiles on the basis of survival-predictor scores, patients in the top quartile had a median survival of 13.6 years, whereas those in the bottom quartile had a median survival of only 3.9 years. Thus, using this molecular prognosticator, a group of patients with a poor prognosis, and requiring special attention in management, can be identified. This study highlights the importance of the tumor microenvironment in influencing the prognosis of patients with FL and provides additional impetus in designing therapy that may alter the microenvironment in addition to targeting tumor cells. However, it should be pointed out that the failure to detect tumor associated predictors of survival in FL does not indicate that the tumor component has little or no influence on survival. This may be due to the reactive elements in many FLs that might have diluted the signal from the tumor. Furthermore, there are frequently multiple related subclones of follicular lymphoma in the same patient (d'Amore et al. 2008) and gene expression profiling (GEP) of a tumor from a certain site may not be predictive of survival if another clone at a site that is not biopsied is the determinant of survival in that patient. Tumor/host interaction, on the other hand, may be a more consistent characteristic and is therefore delineated as an important prognosticator in this study (Dave et al. 2004).

A comparison of the GEP of t(14;18) negative and positive cases shows that the former appears to be at a later stage of GCB-cell maturation with an enrichment of the activated B-cell-like and NF-kB signatures (Leich et al. 2009). This postulate is supported by the lower frequency of CD10 positivity and higher IRF4 positivity in the t(14;18) negative cases. The proliferation signature is also enriched in the t(14;18) negative group which agrees with the finding of higher Ki-67 positivity in this group. Differences in copy number abnormalities were also observed but no alterations are restricted to the t(14;18) negative group. On the other hand, several

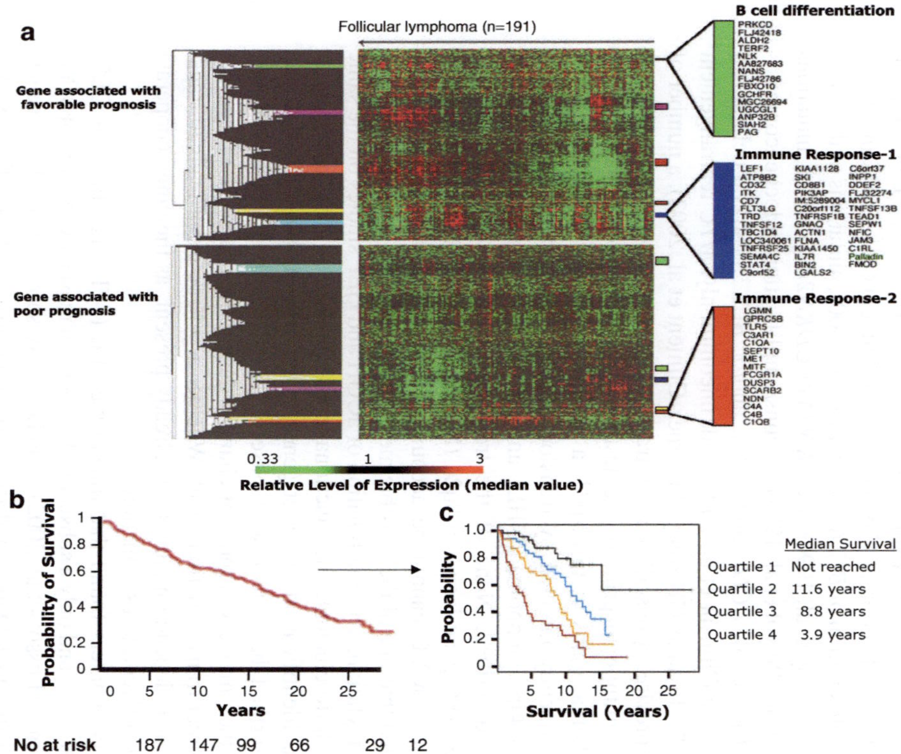

Fig. 2 Survival and genes associated with prognosis in follicular lymphoma. (**a**) The hierarchical clustering of survival-associated genes according to their expression in the training set of 95 follicular lymphoma biopsy specimens. The *dendrogram* shows the degree to which the expression pattern of each gene is correlated with that of the other genes; the *colored bars* represent sets of coordinately regulated genes, defined as gene-expression signatures. To the *right* of the dendrogram, the genes making up the immune response 1 and immune response 2 signatures that formed the survivor-predictor model are listed. (**b**) Kaplan-Meier survival curve for all the patients for whom these data were available. (**c**) Overall survival among the patients with biopsy specimens in the test set, according to the quartile of the survival-predictor score (SPS) (Reproduced with permission from Copyright 2003 Massachusetts Medical Society: Figure 1a and c)

gains and losses are only observed in the t(14;18) positive group, but only 10 t(14;18) negative cases were studied. However, gain or amplification of 18q21 was present in 32% of t(14;18) positive but not seen in any t(14;18) negative cases suggesting that this abnormality may indeed be restricted to the t(14;18) positive group (Leich et al. 2009).

Transformation occurs in about 30% of FLs, resulting in a DLBCL or uncommonly a Burkitt lymphoma-like tumor, with a rapidly progressive clinical course and short survival time. A variety of genetic changes have been implicated (Nagy et al. 2000) such as *MYC* overexpression (Lossos et al. 2002), *TP53* mutation (Sander et al. 1993) or deletion of 17p13 (Martinez-Climent et al. 2003), mutations in the 5′ untranslated region of *BCL6* (Szereday et al. 2000), mutations of the translocated *BCL2* (Matolcsy et al. 1996) or *CDKN2A* or *CDKN2B* deletions, mutations or hypermethylation (Pinyol et al. 1998) or gains/amplification of loci encompassing *REL/BCL11A* (2p16), *BCL6* (3q27-q29), and/or *BCL2* (18q21), deletion of 6q22-q24 (Martinez-Climent et al. 2003) and, uncommonly, *MYC* rearrangement (Yano et al. 1992). In an aCGH study (Martinez-Climent et al. 2003), a number of imbalances not previously described in FL transformation including gains of 9p23-p24, which is common in primary mediastinal B-cell lymphoma and classical Hodgkin lymphoma, gains of 6p12-p21, a site including oncogenes *PIM1* and *CCDN3* and frequently targeted by translocation and copy number increase in B-cell non-Hodgkin lymphoma (NHL) and multiple myeloma, and gains of 17q21.33 were described. None of the transformed cases had elevated copy numbers of the *MYC* gene locus on 8q24 (Martinez-Climent et al. 2003). Thus, over expression of MYC cannot be accounted for by gene amplification. The authors suggested that one of the possible mechanisms could be related to changes in promoter methylation. The combination of aCGH and GEP studies on the same patients may help to elucidate the functional consequences of the genetic alterations as well as identification of target genes in the abnormal loci. The genetic aberrations described have also been observed in FL without transformation, indicating that they may also represent changes during the course of FL and not necessarily a direct transformation event. Whether a genetic event leads to transformation may also depend on alterations already present, as they may cooperate with the event to effect a transformation.

Lossos and co-workers found that transformed FL (t-FL) has different GEPs from *de novo* DLBCL, and only a subset of t-FL show increased expression of MYC and its target genes (Lossos et al. 2002). Another study showed that t-FL is generally a GCB-cell-like malignancy, and transformation involves two pathways, one that is similar in proliferation rate to the precursor FL and the other that is characterized by higher proliferation and the presence of recognized oncogenic abnormalities including high expression of MYC and its target genes, *TP53* mutation, *CDNK2A* loss and *REL* amplification (Davies et al. 2007). In another study of FL transformation, 12 matched pairs of FL and DLBCL were examined and all the DLBCLs exhibited a GCB-cell-like profile (Elenitoba-Johnson et al. 2003). Several members of the *RAS* family, growth factors and cytokine receptors with known growth promoting activity (*C-MET, FGFR3, LTβR* and *PDGFRβ*) and *p38MAPK* were found to have high expression levels in the DLBCL. A three-gene predictor (*PLA2, PDGFRβ* and *Rab-6*) for transformation was proposed. The authors also showed that inhibition of p38MAPK blocked the growth of t(14;18)+cell lines, and inhibited growth of transplanted tumors in NOD-SCID mice, thereby suggesting that p38MAPK may be a promising target for transformed FL.

Pathology of Follicular Lymphoma

FL is a malignant neoplasm of the GCB-cells that is composed of cleaved (centrocytes) and non-cleaved (centroblasts) cells and shows, at least, a focal nodular pattern (Swerdlow et al. 2008). Diagnosis of FL requires appropriate morphological and immunohistochemical findings with or without molecular studies. FL commonly involves the lymph nodes, bone marrow, and spleen. Less common sites of involvement are the gastro intestinal tract, skin, and ocular adenexa, but in some cases could be the primary site of involvement. On microscopic examination, the key architectural and cytological findings in the lymph node includes (i) total or extensive loss of nodal architecture, (ii) even distribution of malignant lymphoid nodules throughout the node, (iii) back-to-back arrangement of nodules with little interposed tissue, (iv) overall uniformity in size and shape of nodules, (v) paucity or absence of reactive lymphoid cells (plasma cells and immunoblasts) in the inter-follicular areas, and (vi) absence of polarization and tingible body macrophages in the nodules (Fig. 3). These features are helpful in differentiating FL from reactive follicular hyperplasia of the lymph node as summarized in Table 1.

The t(14;18)(q32;q21) occurs in up to 90% of low grade FL and to a lesser percentage in high grade FL. Occasionally, *BCL2* translocation involves immunoglobulin light chains kappa or lambda (Tomita 2011). In the approximately 10% of FL cases which are t(14;18) negative, other mechanisms play a role in lymphomagenesis. One of the mechanisms could be rearrangement involving chromosome band 3q27 leading to deregulation of *BCL6*, a proto-oncogene (Gu et al. 2009). Grade 3b FL has the highest incidence of *BCL6* translocation (44%) compared to the lower frequency (13%) in low grade FL (Ott et al. 2002). Interestingly, t(14;18) negative FL also have a higher frequency (22%) of *BCL6* translocation. Two common breakpoint regions of *BCL6* have been described in lymphoma: (i) the major breakpoint region (MBR) involving the noncoding first exon and part of the first intron (Ye et al. 1993) and (ii) the alternative breakpoint region (ABR) located at 245–285 kb 5′ to BCL6. For FLs with t(14;18), *BCL6* translocation mainly occurs at the MBR (Butler et al. 2002). However, FL without t(14;18) show a higher frequency of ABR involvement (Gu et al. 2009). Both immunoglobulin heavy and light chains in FL show ongoing somatic mutations (Stamatopoulos et al. 1997; Matolcsy et al. 1999). These mutations may interfere with primer binding, and monoclonality by PCR assay can be negative in a higher proportion of cases of FL compared with tumors with few mutations such as CLL.

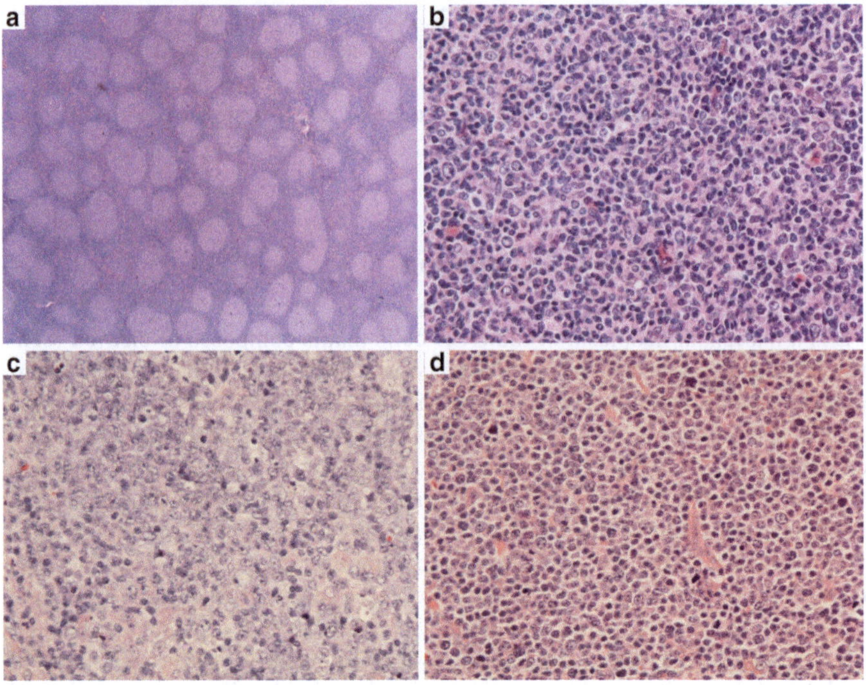

Fig. 3 (**a**) Follicular lymphoma (20×). The malignant follicles are closely packed, focally back to back with absence of mantle zone. (**b**) Follicular lymphoma, grade 1 showing mainly centrocytes with occasional centroblasts (400×). (**c**) Follicular lymphoma, grade 3, mainly consists of centroblasts with occasional centrocytes (400×). (**d**) Diffuse large cell lymphoma arising from a case of follicular lymphoma (400×)

Table 1 Differentiating features of follicular lymphoma from reactive follicular hyperplasia

Characteristics	Follicular lymphoma	Reactive follicular hyperplasia
Effacement of architecture	Present	Absent
Nodules outside of cortex	Present	Mainly in the cortex
Uniformity of nodules	Present	More variation in size and shape
Interposed tissue between nodules	Inconspicuous	More prominent
Tingible body macrophages	Absent/few	Present
Polarization of nodules	Absent	Present

Other frequent genetic abnormalities that can be seen in FL have been discussed in the section on pathobiology.

The neoplastic cells typically express B-cell markers (CD19, CD20, CD22, CD79a), BCL6, BCL2, CD10, and are negative for CD5, CD43 and CyclinD1 (Swerdlow et al. 2008). The percent positivity and intensity of staining of CD10, BCL6 and BCL2 varies depending on the grade and location of neoplastic cells such as follicular versus interfollicular. The intensity of staining is higher in the neoplastic

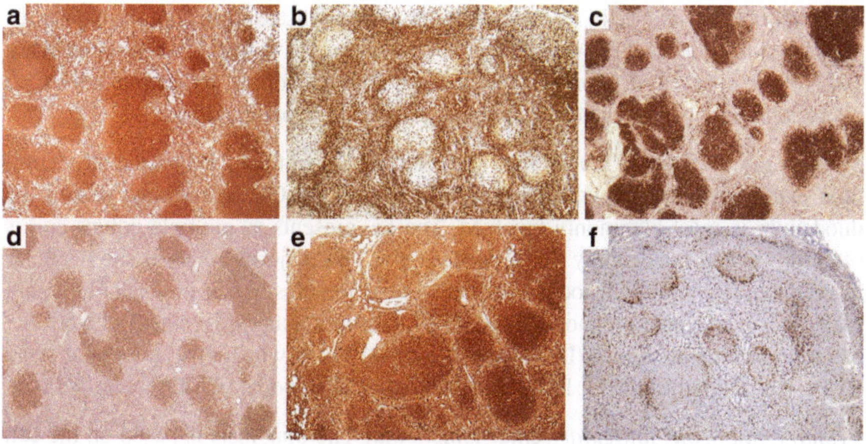

Fig. 4 Typical immunophenotype of follicular lymphoma. A case of follicular lymphoma showing CD20+ (**a**), CD3- (**b**), CD10+ (**c**), BCL6+ (**d**), and BCL2+ (**e**) staining and low number of Ki-67 positive cells (**f**) (100×)

nodules compared to inter-nodular areas. CD10 and BCL2 are positive in the vast majority of FL (Fig. 4). It should be noted that BCL2 may be falsely negative due to mutations that eliminate the epitope that the primary antibody binds to (Schraders et al. 2005). BCL2 staining is positive in a slightly higher percentage of low grade FL (90%) compared to high grade (80%) FL (Swerdlow et al. 2008). The follicular dendritic cell (FDC) meshwork is identified by CD21, CD23 or CD35 immunostaining and helps in determining the extent of follicular areas. Forty-two percent of grade 3B FL are MUM1 positive (Horn et al. 2011). FLs without t(14;18) tend to have higher grade histology, lower percentage of CD10 positivity and much lower incidence of BCL2 positivity (Leich et al. 2009). The proliferation index, determined by Ki-67 immunohistochemical staining is usually low (<20%) in low grade FL in comparison to higher grade FL (>20%). However, a minority of cases of low grade FL have a high Ki-67 index and a worse prognosis (Wang et al. 2005). Hence, Ki-67 may be used as an adjunct prognostic marker in low grade lymphoma.

Specific Variants of FL

Pediatric follicular lymphoma: FLs are extremely rare in the pediatric age group and comprise less that 6% of pediatric lymphomas (Frizzera and Murphy 1979; Lu et al. 2001). Unlike adult lymphomas which express BCL2 protein and possess t(14;18), pediatric FL generally do not have t(14;18) nor express BCL2 protein. The common sites of pediatric FL include tonsils, cervical and other lymph nodes, and the testis. Tonsils and lymph nodes are often associated with reactive hyperplasia, which is also negative for BCL2 protein expression and t(14;18). Therefore, careful examination of morphology is critical for differentiating pediatric FL from reactive hyperplasia.

The majority of pediatric FL presents with lower stage disease, but high grade features with large follicles and show excellent overall and disease free survival (Rosenberg 1977; Pinto et al. 1990; Montes-Moreno et al. 2010). There is suggestion that BCL2 expression may be associated with worse prognosis in pediatric FL.

Primary gastrointestinal lymphoma: The gastrointestinal tract (GIT) is the most common extranodal site of involvement by FL. Within the GIT, the second part of the duodenum is most commonly involved. The FL presents as multiple polyps and is often diagnosed incidentally during endoscopy. The majority of them show similar morphological, immunohistochemical and genetic features typical of nodal FL. It generally has an excellent prognosis. The GIT FL presents at an earlier stage, with 68% of cases presenting at stage I-II, compared to nodal FL in which 83% present at stage III-IV. Additionally, GIT FL has a better prognosis than nodal FL (Tari et al. 2011).

Primary cutaneous follicular center lymphoma: The diagnosis is based on the cell of origin rather than growth pattern and patients with primary cutaneous follicular center lymphoma have good prognosis irrespective of the cytology. Most of the cases are BCL2 negative and lack the t(14;18) and CD10 tends to be negative in cases with a diffuse growth pattern (Cerroni and Kerl 2001).

In situ follicular lymphoma: *In situ* FL is characterized by preservation of lymph node architecture and presence of occasional germinal centers which show strong BCL2 protein expression in part or all of the GCB-cells (CD10+, BCL6+). The rest of the lymph node shows reactive hyperplasia and absence of inter-follicular infiltration. The exact clinical significance of *in situ* FL is yet to be determined. In a series of 13 cases of *in situ* FL, only four developed subsequent FL after a median follow-up of 12 months (Montes-Moreno et al. 2010). Interestingly, five cases in this series showed association with other lymphoproliferative disorders like DLBCL, classic Hodgkin lymphoma, and splenic marginal zone lymphoma. In patients who develop subsequent FL elsewhere or found to have concurrent FL, the *in situ* FL may represent follicular colonization by overt FL. However, patients who do not progress soon into frank FLs require close clinical surveillance for the development of FL.

Prognostic Factors in FL

The majority of FLs are indolent but present with advanced stages, and are characterized by multiple relapses with shortening remissions. Transformation to aggressive lymphoma, mostly DLBCL, occurs in about 30% of cases and shows clinical heterogeneity with relation to OS. Based on the clinical, biochemical, and pathological findings, the Follicular Lymphoma International Prognostic Index (FLIPI) was developed for risk stratification of FL (Federico et al. 2009). With the advent of radioimmunotherapy and additional clinical and biochemical parameters, FLIPI has been revised to FLIPI2 (Federico et al. 2009). Based on the presence or absence of the number of risk factors in FLIPI, FL can be stratified into three risk groups, low (0–1 factors), intermediate (2 factors) and high (≥3 factors), with different OS (Table 2).

Table 2 Components of Follicular Lymphoma International Prognostic Index (FLIPI) 1 and 2

FLIPI	FLIPI2
Age (≤60 vs. >60 years)	Age (≤60 vs. >60 years)
Hemoglobin (≤120 vs. >120 g/L)	Hemoglobin (≤120 vs. >120 g/L)
Advanced stage disease (Grade I–II vs. III–IV)	Bone marrow involvement (uninvolved vs. involved)
Number of nodal areas (≤4 vs. >4)	Longest tumor diameter (≤6 vs. >6 cm)
Serum LDH levels (normal vs. elevated)	Serum beta 2-microglobulin (normal vs. elevated)

The role of cytogenetic and molecular findings in the prognostication of FL is not yet clear. Rambaldi et al. (2005) have reported the pretreatment quantitative evaluation of marrow involvement by t(14;18) positive cells by RT-PCR to predict outcome after chemotherapy. Their study showed that a low level positivity of BCL2/IgH cells in bone marrow at the time of diagnosis is associated with higher clinical and molecular response and event free survival. However, the above finding has not yet been validated in clinical trials.

Torlakovic et al. (2006) have reported the expression of PU.1 but not CD10 and BCL6, is associated with favorable outcome in FL. PU.1 is a transcription factor essential for myeloid and B lymphoid development. Mutations in *TP53* in various lymphomas are generally associated with inferior survival and resistance to chemotherapy. Heterozygous *TP53* mutations have been reported at diagnosis in FL (6%) and are associated with inferior survival (O'Shea et al. 2008). These mutations are mainly missense with an occasional splice site mutation (O'Shea et al. 2008). *TP53* mutations are associated with older age, higher IPI score, lower immune expression signature 1, and shorter progression free and overall survival (O'Shea et al. 2008). Sander et al. (1993) have studied serial biopsies of 34 patients with FL who eventually transformed into higher grade lymphoma for TP53 by immunohistochemistry, single strand conformation polymorphism (SSCP) analysis and sequencing. Their study showed that *TP53* mutations are associated in histological transformation in 25–30% of FL. They also showed that *TP53* mutation positive cells can be identified (by SSCP) in pre-transformed FL; however, these cells constituted a small percentage of tumor cells. It has been shown by aCGH that deletion of 6q25-27, 1p36.22-p36.33, 6q21-q24.3, 9p21 (*CDKN2A/B*), 6q25 and 6q26 are associated with inferior survival (Cheung et al. 2009; Schwaenen et al. 2009). It has also been reported that the acquired uniparental disomy (aUPD) of 1p36 is associated with inferior OS and aUPD of chromosome 16 is predictive of transformation (Schwaenen et al. 2009). Many of these findings have not been confirmed by large independent studies. Furthermore, the heterogeneity in treatment also complicates data interpretation.

The microenvironment plays an important role in FL. There is an active cross-talk between FL cells and its microenvironment that determines the clinical behavior, prognosis and response to specific therapy. Non-neoplastic cells such as T-cells, macrophages, and FDCs are associated with FL. It has been shown by GEP that the microenvironment plays a pivotal role in patient outcome in FL (Dave et al. 2004) as discussed earlier in the section on pathobiology.

Immunohistochemistry (IHC) and flow cytometry (FC) are routinely used techniques in the clinical laboratory for diagnostic and prognostic purpose. Several studies have been performed to discover prognostic biomarkers using these techniques. Studies evaluating tumor associated immune regulatory cells have shown variable results. Lee et al. (2006) have studied 59 patients with FL (34 lived less than 5 years vs. 25 lived more than 15 years) for various IHC markers- CD4, CD7, CD8, CD25, CD68, forkhead box protein P3 (FOXP3), T-cell intracellular antigen 1 (TIA-1) and Granzyme B. Their study showed that patients who lived more than 15 years had higher CD4+ and FOXP3+ cells in the perifollicular location. Wahlin et al. (2010) have studied 70 patients (poor vs. good outcome) with FL using IHC markers CD4, CD8, PD-1, and CD68. They reported that CD4+ cells were associated with poor prognosis in comparison with FL with higher CD8+ and CD68+ cells. In a different study, Wahlin et al. (2007) used FC to quantitate CD8+ cells in 139 patients with FL and showed that higher number of CD8+ lymphocytes predict better survival. Kelly et al. (2007) used CD68 and FOXP3 to assess lymphoma associated macrophages (CD68+) and regulatory T-cells (FOXP3+) in 69 patients with FL. They reported that increased extrafollicular macrophages (CD68+) and intrafollicular regulatory T cells (FOXP3+) were associated with shorter OS. Farina et al. (2005) assessed CD68+ lymphoma-associated macrophages (LAM) in 99 patients with FL and correlated their number with OS. In their study, 87 patients had less than 15 (median, 7; range, 1–14) LAM/high power field (hpf) and 12 patients had more than 15 (median, 20; range, 16–25) LAM/hpf. Patients with less than 15 LAM/hpf had a median OS of 16.3 years compared with an OS of 5 years in patients with more than 15 LAM/hpf. Butsch et al. (2011) evaluated plasmacytoid dendritic cells (pDC) in FL employing immunohistochemical stain for CD123. They have shown that increased number of pDC is associated with better OS and acts as an independent prognostic marker. Camacho et al. (2011) have studied expression of 28 immunohistochemical markers in 186 nodal FL cases to delineate specific markers associated with OS. In this study, higher expression of Cyclin E, MDM2, P21, IgD, BCL-xL, CD30, and E2F6 by lymphoma cells was associated with significantly shorter OS whereas expression of CD10 was associated with better OS. Furthermore, the number and impact of FDCs on OS in patients with FL was reported by Jin et al. (2011). They studied 102 FL cases and evaluated FDCs semi-quantitatively by immunohistochemistry (CD23, CD35 and CD54) and digital image analysis and found that the abundance of FDCs in FL do not act as a prognostic factor.

Clinical Presentation and Diagnosis

The standard Ann Arbor staging scheme with Cotswold modification (Table 3) is used to stage FL (Rosenberg 1977; Moormeier et al. 1990). The majority of the patients (70%) present with advanced (clinical stage III/IV) disease (Table 2). Most patients present with painless peripheral cervical, axillary, inguinal, or femoral lymphadenopathy,

Table 3 Ann Arbor System, with Cotswold modification, for staging of NHL (Rosenberg 1977; Moormeier et al. 1990)

Stage I	Involvement of a single lymph node region or of a single extralymphatic organ or site (IE)
Stage II	Involvement of two or more lymph node regions or lymphatic structures on the same side of the diaphragm alone or with involvement of limited, contiguous extralymphatic organ or tissue (IIE)
Stage III	Involvement of lymph node regions on both sides of the diaphragm, which may include the spleen (IIIS) or limited, contiguous extralymphatic organ or site (IIIE) or both (IIIES)
Stage IV	Diffuse or disseminated foci of involvement of one or more extralymphatic organs or tissues, with or without associated lymphatic involvement.
E-designation	Extranodal contiguous extension (ie, proximal or contiguous extranodal disease) that can be encompassed within an irradiation field appropriate for nodal disease of the same anatomic extent.
B-status	("B") symptoms or significant unexplained fever (>38°C), night sweats, or unexplained weight loss exceeding 10% of body weight during the 6 months prior to diagnosis.

typically long-standing, which can be waxing and waning. Hilar and mediastinal adenopathy may be present, yet bulky mediastinal disease is unusual.

Some patients present with asymptomatic large abdominal masses with or without evidence of gastrointestinal or urinary obstruction. Involvement of liver, spleen, and bone marrow is usual in patients with advanced disease, but involvement of other organs is rare. Despite widespread disease at diagnosis, most patients are asymptomatic except for nodal masses. Involvement of the central nervous system is rare yet compression of a peripheral nerve, or the spinal cord, by epidural masses is not unusual. Fever, drenching night sweats, and unintentional loss of weight are present in only 20% of patients. Children and adolescents with FL typically present with asymptomatic cervical lymphadenopathy and low stage disease (stage I/II). Diagnosis requires an excisional lymph node biopsy. Whole body CT scans and bone marrow core biopsies are required for accurate staging (Freedman and Aster 2010). The usefulness of positron emission tomography (PET) scans in imaging indolent lymphomas, like FL, remains unclear (Juweid et al. 2007).

Current Treatment Strategy

Treatment is usually tailored to the clinical circumstances; the major variables influencing treatment are patient age, performance status, associated co-morbidities, stage, histologic subtype, presence of symptoms, evidence of organ dysfunction related to the lymphoma, prognostic factors (using the FLIPI score), and patient preference. It is key to understand that most patients with FL will have an excellent prognosis with expected median survival of 10 years. Hence, it is of paramount importance not to cause unjustified harm to these patients with excessively toxic

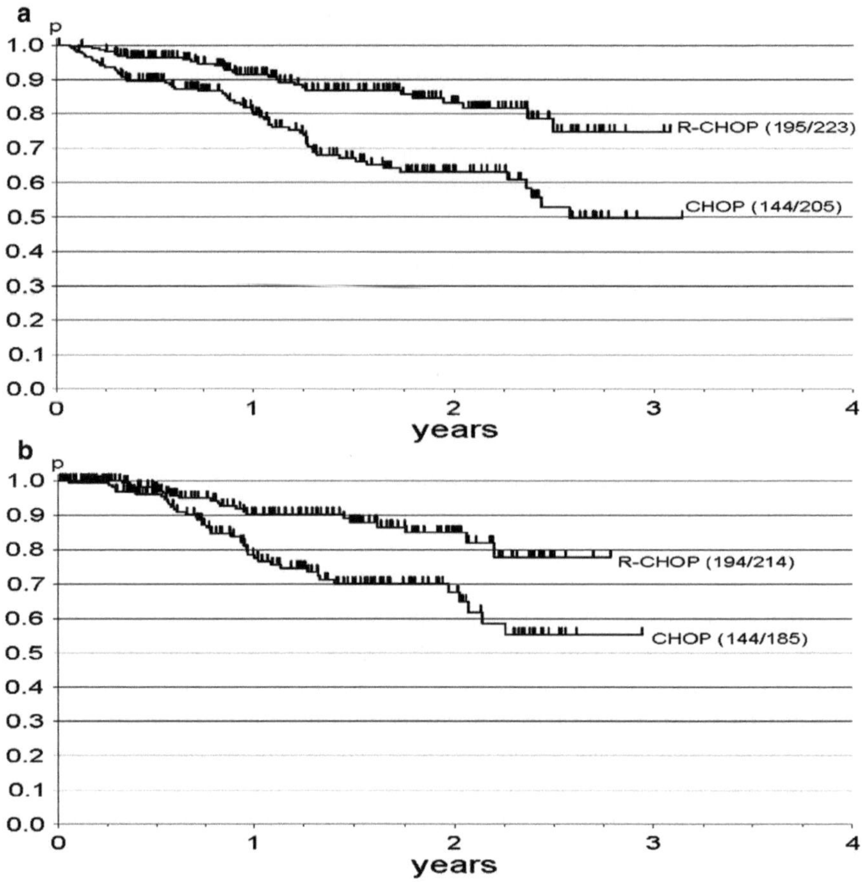

Fig. 5 Rituximab prolonged time to treatment failure (TTF) and duration of response when added to CHOP chemotherapy. (**a**) TTF after start of therapy with CHOP or R-CHOP. In the R-CHOP arm, only 28 of 223 patients experienced treatment failure versus 61 of 205 patients in the CHOP arm (P<.001). (**b**) Duration of response after CHOP and R-CHOP. In patients achieving CR or PR after initial therapy, a significantly lower relapse rate was observed after R-CHOP versus CHOP (P=.001) (Reproduced with permission Hiddemann et al. (2005))

treatment regimens, unless a clear survival, or quality of life advantage could be demonstrated in randomized controlled trials (Wilson and Armitage 2008).

Ritxuimab, an anti-CD20 monoclonal antibody, is the most important advance in the treatment of follicular lymphomas in the past 20 years and has remarkably changed treatment paradigms for this disease (Fig. 5). Radiotherapy is considered the standard of care for patients with early stage disease (stage I/II) and remains the most popular (Mac Manus and Hoppe 1996). There is data that early therapy is beneficial and may improve cure rate in these patients, but these studies were done in the pre-Rituximab era. Treatment is indicated in patients with advanced (stage III/IV) symptomatic disease or with evidence of organ/system dysfunction

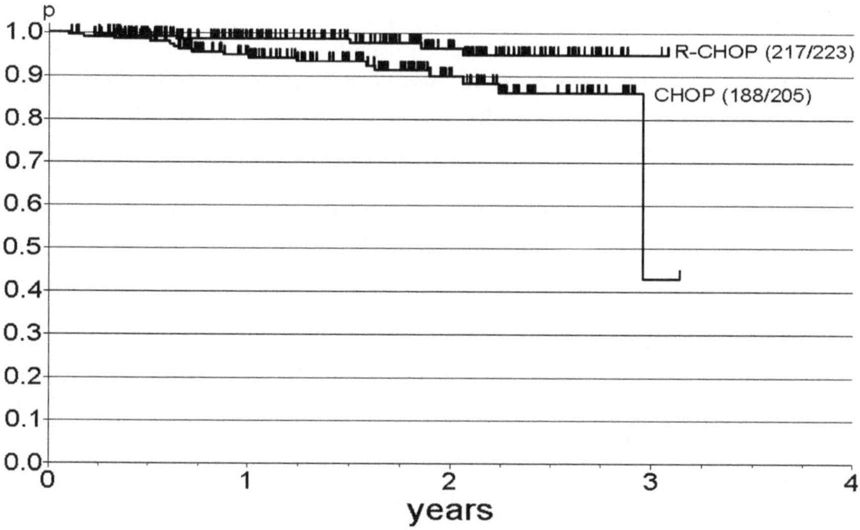

Fig. 6 OS after start of therapy for CHOP or R-CHOP. Median OS has not been reached in either group at the conclusion of the study. After 3 years, 6 patients in the R-CHOP arm have died versus 17 patients in the CHOP arm (P=.016) (Reproduced with permission Hiddemann et al. (2005))

secondary to the lymphoma. Numerous single agents and combinations are available for treatment of patients with advanced FL. Popular regimens include CVP and CHOP (cyclophosphamide, vincristine, and prednisone +/– doxorubicin), nucleoside analogue based regimens (fludarabine-based regimens being most popular). The choice of regimen is individualized to the aggressiveness of the disease and the severity of organ compromise (Wilson and Armitage 2008). Rituximab may be used as a single agent, with overall response rate of 73% (Hainsworth et al. 2000, 2002), or added to combination chemotherapy. Rituximab added to CHOP (R-CHOP) resulted in 60% reduction in the rate of treatment failure and significant improvement in OS (Hiddemann et al. 2005) (Fig. 6) The recently completed large PRIMA trial reported that maintenance Rituximab, given for 2 years to patients who initially responded to chemotherapy, resulted in doubling of the rate of progression-free survival (PFS) (Salles et al. 2010). Relapsed patients can be treated with the same agents used initially or with novel agents.

Autologous stem cell transplantation may be used as a salvage therapy in younger, and more fit, patients (Wilson and Armitage 2008). Radioimmunotherapy with anti-CD20 monoclonal antibodies coupled to radioisotopes like iodine-131 and yttrium-90 have superior activity and are approved for the treatment of relapsed patients with FL and there is emerging data that they can be used in the front line in selected patients or for "consolidation" after initial response to chemotherapy (William and Bierman 2010). Other available strategies include α-interferon, anti-idiotype vaccines, and allogeneic stem cell transplantation. The discussion of the details of all of these treatment options is beyond the scope of this chapter.

Patients with asymptomatic advanced disease, with no evidence of organ dysfunction associated with the FL, can be usually managed by watchful waiting based on no demonstrable advantage of early vs. delayed treatment in clinical trials (Young et al. 1988; Ardeshna et al. 2003). There is very recent data suggesting that early treatment with Rituximab in asymptomatic patients improves PFS and time to new treatment (Ardeshna et al. 2010). These data are encouraging, yet an OS benefit needs to be demonstrated before being considered practice-changing. Patients with grade 3 (especially grade 3b) FL are considered, for all practical purposes, to behave like aggressive NHL and should be treated with the same regimens employed for aggressive NHL (Ganti et al. 2006). Response to therapy is usually monitored by serial clinical assessments (decrease in size of palpable lymph nodes), CT scans, serum lactate dehydrogenase levels, and bone marrow biopsies (in certain occasions). Molecular remission can be assessed by PCR that detect the *BCL2/IGH* translocation or the specific *IGH* rearrangement present in the clone in blood or bone marrow, though clinical reports had been conflicting if achievement of molecular remission is related to improved clinical outcomes (van Oers et al. 2010).

References

Ansari KI, Mandal SS (2010) Mixed lineage leukemia: roles in gene expression, hormone signaling and mRNA processing. FEBS J 277(8):1790–1804

Ardeshna KM, Smith P et al (2003) Long-term effect of a watch and wait policy versus immediate systemic treatment for asymptomatic advanced-stage non-Hodgkin lymphoma: a randomised controlled trial. Lancet 362(9383):516–522

Ardeshna KM, Qian W et al (2010) An intergroup randomised trial of rituximab versus a watch and wait strategy in patients with Stage II, III, IV, asymptomatic, non-bulky follicular lymphoma (Grades 1, 2 and 3a). A preliminary analysis. ASH Annu Meet Abstr 116(21):6

Bodor C, O'Riain C et al (2011) EZH2 Y641 mutations in follicular lymphoma. Leukemia 25(4):726–729

Butler MP, Iida S et al (2002) Alternative translocation breakpoint cluster region 5′ to BCL-6 in B-cell non-Hodgkin's lymphoma. Cancer Res 62(14):4089–4094

Butsch R, Lukas Waelti S et al (2011) Intratumoral plasmacytoid dendritic cells associate with increased survival in patients with follicular lymphoma. Leuk Lymphoma 52(7):1230–1238

Camacho FI, Bellas C et al (2011) Improved demonstration of immunohistochemical prognostic markers for survival in follicular lymphoma cells. Mod Pathol 24(5):698–707

Carlotti E, Wrench D et al (2009) Transformation of follicular lymphoma to diffuse large B-cell lymphoma may occur by divergent evolution from a common progenitor cell or by direct evolution from the follicular lymphoma clone. Blood 113(15):3553–3557

Cerroni L, Kerl H (2001) Primary cutaneous follicle center cell lymphoma. Leuk Lymphoma 42(5):891–900

Cheung KJ, Shah SP et al (2009) Genome-wide profiling of follicular lymphoma by array comparative genomic hybridization reveals prognostically significant DNA copy number imbalances. Blood 113(1):137–148

Cheung KJ, Delaney A et al (2010a) High resolution analysis of follicular lymphoma genomes reveals somatic recurrent sites of copy-neutral loss of heterozygosity and copy number alterations that target single genes. Genes Chromosom Cancer 49(8):669–681

Cheung KJ, Johnson NA et al (2010b) Acquired TNFRSF14 mutations in follicular lymphoma are associated with worse prognosis. Cancer Res 70(22):9166–9174
Chim CS, Wong KY et al (2004) SOCS1 and SHP1 hypermethylation in mantle cell lymphoma and follicular lymphoma: implications for epigenetic activation of the Jak/STAT pathway. Leukemia 18(2):356–358
Chim CS, Wong KY et al (2007) Frequent epigenetic inactivation of Rb1 in addition to p15 and p16 in mantle cell and follicular lymphoma. Hum Pathol 38(12):1849–1857
Coe BP, Ylstra B et al (2007) Resolving the resolution of array CGH. Genomics 89(5):647–653
Coelho V, Krysov S et al (2010) Glycosylation of surface Ig creates a functional bridge between human follicular lymphoma and microenvironmental lectins. Proc Natl Acad Sci U S A 107(43):18587–18592
d'Amore F, Chan E et al (2008) Clonal evolution in t(14;18)-positive follicular lymphoma, evidence for multiple common pathways, and frequent parallel clonal evolution. Clin Cancer Res 14(22):7180–7187
Dave SS, Wright G et al (2004) Prediction of survival in follicular lymphoma based on molecular features of tumor-infiltrating immune cells. N Engl J Med 351(21):2159–2169
Davies AJ, Rosenwald A et al (2007) Transformation of follicular lymphoma to diffuse large B-cell lymphoma proceeds by distinct oncogenic mechanisms. Br J Haematol 136(2):286–293
Eide MB, Liestol K et al (2010) Genomic alterations reveal potential for higher grade transformation in follicular lymphoma and confirm parallel evolution of tumor cell clones. Blood 116(9): 1489–1497
Elenitoba-Johnson KS, Jenson SD et al (2003) Involvement of multiple signaling pathways in follicular lymphoma transformation: p38-mitogen-activated protein kinase as a target for therapy. Proc Natl Acad Sci U S A 100(12):7259–7264
Farinha P, Masoudi H et al (2005) Analysis of multiple biomarkers shows that lymphoma-associated macrophage (LAM) content is an independent predictor of survival in follicular lymphoma (FL). Blood 106(6):2169–2174
Federico M, Bellei M et al (2009) Follicular lymphoma international prognostic index 2: a new prognostic index for follicular lymphoma developed by the international follicular lymphoma prognostic factor project. J Clin Oncol 27(27):4555–4562
Fitzgibbon J, Iqbal S et al (2007) Genome-wide detection of recurring sites of uniparental disomy in follicular and transformed follicular lymphoma. Leukemia 21(7):1514–1520
Freedman A, Aster J (2010) Clinical manifestations, pathologic features, and diagnosis of follicular lymphoma. UptoDate. D. Basow. Waltham, Wolters Kluwer
Frizzera G, Murphy SB (1979) Follicular (nodular) lymphoma in childhood: a rare clinical-pathological entity. Report of eight cases from four cancer centers. Cancer 44(6):2218–2235
Ganti AK, Weisenburger DD et al (2006) Patients with grade 3 follicular lymphoma have prolonged relapse-free survival following anthracycline-based chemotherapy: the Nebraska lymphoma study group experience. Ann Oncol 17(6):920–927
Gronbaek K, Ralfkiaer U et al (2008) Frequent hypermethylation of DBC1 in malignant lymphoproliferative neoplasms. Mod Pathol 21(5):632–638
Gu K, Fu K et al (2009) t(14;18)-negative follicular lymphomas are associated with a high frequency of BCL6 rearrangement at the alternative breakpoint region. Mod Pathol 22(9):1251–1257
Hagiwara K, Nagai H et al (2007) Frequent DNA methylation but not mutation of the ID4 gene in malignant lymphoma. J Clin Exp Hematop 47(1):15–18
Hainsworth JD, Burris HA 3rd et al (2000) Rituximab monoclonal antibody as initial systemic therapy for patients with low-grade non-Hodgkin lymphoma. Blood 95(10):3052–3056
Hainsworth JD, Litchy S et al (2002) Rituximab as first-line and maintenance therapy for patients with indolent non-Hodgkin's lymphoma. J Clin Oncol 20(20):4261–4267
Hannibal MC, Buckingham KJ et al (2011) Spectrum of MLL2 (ALR) mutations in 110 cases of Kabuki syndrome. Am J Med Genet A 155(7):1511–1516
Hayslip J, Montero A (2006) Tumor suppressor gene methylation in follicular lymphoma: a comprehensive review. Mol Cancer 5:44

Hiddemann W, Kneba M et al (2005) Frontline therapy with rituximab added to the combination of cyclophosphamide, doxorubicin, vincristine, and prednisone (CHOP) significantly improves the outcome for patients with advanced-stage follicular lymphoma compared with therapy with CHOP alone: results of a prospective randomized study of the German low-grade lymphoma study group. Blood 106(12):3725–3732

Hoglund M, Sehn L et al (2004) Identification of cytogenetic subgroups and karyotypic pathways of clonal evolution in follicular lymphomas. Genes Chromosom Cancer 39(3):195–204

Horn H, Schmelter C et al (2011) Follicular lymphoma grade 3b is a distinct neoplasm according to cytogenetic and immunohistochemical profiles. Haematologica 96(9):1327–1334

Irish JM, Czerwinski DK et al (2006) Altered B-cell receptor signaling kinetics distinguish human follicular lymphoma B cells from tumor-infiltrating nonmalignant B cells. Blood 108(9): 3135–3142

Jin MK, Hoster E et al (2011) Follicular dendritic cells in follicular lymphoma and types of non-Hodgkin lymphoma show reduced expression of CD23, CD35 and CD54 but no association with clinical outcome. Histopathology 58(4):586–592

Juweid ME, Stroobants S et al (2007) Use of positron emission tomography for response assessment of lymphoma: consensus of the imaging subcommittee of international harmonization project in lymphoma. J Clin Oncol 25(5):571–578

Kallioniemi A, Kallioniemi OP et al (1992) Comparative genomic hybridization for molecular cytogenetic analysis of solid tumors. Science 258(5083):818–821

Kelley T, Beck R et al (2007) Biologic predictors in follicular lymphoma: importance of markers of immune response. Leuk Lymphoma 48(12):2403–2411

Koyama M, Oka T et al (2003) Activated proliferation of B-cell lymphomas/leukemias with the SHP1 gene silencing by aberrant CpG methylation. Lab Invest 83(12):1849–1858

Krivtsov AV, Armstrong SA (2007) MLL translocations, histone modifications and leukaemia stem-cell development. Nat Rev Cancer 7(11):823–833

Lee AM, Clear AJ et al (2006) Number of CD4+ cells and location of forkhead box protein P3-positive cells in diagnostic follicular lymphoma tissue microarrays correlates with outcome. J Clin Oncol 24(31):5052–5059

Leich E, Salaverria I et al (2009) Follicular lymphomas with and without translocation t(14;18) differ in gene expression profiles and genetic alterations. Blood 114(4):826–834

Limpens J, Stad R et al (1995) Lymphoma-associated translocation t(14;18) in blood B cells of normal individuals. Blood 85(9):2528–2536

Lossos IS, Alizadeh AA et al (2002) Transformation of follicular lymphoma to diffuse large-cell lymphoma: alternative patterns with increased or decreased expression of c-myc and its regulated genes. Proc Natl Acad Sci U S A 99(13):8886–8891

Lu D, Medeiros LJ et al (2001) Primary follicular large cell lymphoma of the testis in a child. Arch Pathol Lab Med 125(4):551–554

Mac Manus MP, Hoppe RT (1996) Is radiotherapy curative for stage I and II low-grade follicular lymphoma? Results of a long-term follow-up study of patients treated at Stanford University. J Clin Oncol 14(4):1282–1290

Martinez-Climent JA, Alizadeh AA et al (2003) Transformation of follicular lymphoma to diffuse large cell lymphoma is associated with a heterogeneous set of DNA copy number and gene expression alterations. Blood 101(8):3109–3117

Matolcsy A, Casali P et al (1996) Morphologic transformation of follicular lymphoma is associated with somatic mutation of the translocated Bcl-2 gene. Blood 88(10):3937–3944

Matolcsy A, Schattner EJ et al (1999) Clonal evolution of B cells in transformation from low- to high-grade lymphoma. Eur J Immunol 29(4):1253–1264

Montes-Moreno S, Castro Y et al (2010) Intrafollicular neoplasia/in situ follicular lymphoma: review of a series of 13 cases. Histopathology 56(5):658–662

Moormeier JA, Williams SF et al (1990) The staging of non-Hodgkin's lymphomas. Semin Oncol 17(1):43–50

Morin RD, Johnson NA et al (2010a) Somatic mutations altering EZH2 (Tyr641) in follicular and diffuse large B-cell lymphomas of germinal-center origin. Nat Genet 42(2):181–185

Morin RD, Mungall AJ et al (2010b) Mutations in MLL2 and MEF2B genes in follicular lymphoma and diffuse large B-cell lymphoma 2010. American Society of Hematology (ASH). https://ash.confex.com/ash/2010/webprogram/Paper30713.html

Nagy M, Balazs M et al (2000) Genetic instability is associated with histological transformation of follicle center lymphoma. Leukemia 14(12):2142–2148

Nanjangud G, Rao PH et al (2007) Molecular cytogenetic analysis of follicular lymphoma (FL) provides detailed characterization of chromosomal instability associated with the t(14;18)(q32;q21) positive and negative subsets and histologic progression. Cytogenet Genome Res 118(2–4):337–344

O'Shea D, O'Riain C et al (2008) The presence of TP53 mutation at diagnosis of follicular lymphoma identifies a high-risk group of patients with shortened time to disease progression and poorer overall survival. Blood 112(8):3126–3129

O'Shea D, O'Riain C et al (2009) Regions of acquired uniparental disomy at diagnosis of follicular lymphoma are associated with both overall survival and risk of transformation. Blood 113(10):2298–2301

Ott G, Katzenberger T et al (2002) Cytomorphologic, immunohistochemical, and cytogenetic profiles of follicular lymphoma: 2 types of follicular lymphoma grade 3. Blood 99(10):3806–3812

Pasqualucci L, Dominguez-Sola D et al (2011) Inactivating mutations of acetyltransferase genes in B-cell lymphoma. Nature 471(7337):189–195

Pinto A, Hutchison RE et al (1990) Follicular lymphomas in pediatric patients. Mod Pathol 3(3):308–313

Pinyol M, Cobo F et al (1998) p16(INK4a) gene inactivation by deletions, mutations, and hypermethylation is associated with transformed and aggressive variants of non-Hodgkin's lymphomas. Blood 91(8):2977–2984

Rambaldi A, Carlotti E et al (2005) Quantitative PCR of bone marrow BCL2/IgH+cells at diagnosis predicts treatment response and long-term outcome in follicular non-Hodgkin lymphoma. Blood 105(9):3428–3433

Rosenberg SA (1977) Validity of the Ann Arbor staging classification for the non-Hodgkin's lymphomas. Cancer Treat Rep 61(6):1023–1027

Ross CW, Ouillette PD et al (2007) Comprehensive analysis of copy number and allele status identifies multiple chromosome defects underlying follicular lymphoma pathogenesis. Clin Cancer Res 13(16):4777–4785

Salles GA, Seymour JF et al (2010) Rituximab maintenance for 2 years in patients with untreated high tumor burden follicular lymphoma after response to immunochemotherapy. J Clin Oncol Meet Abstr 28(15_Suppl):8004

Sander CA, Yano T et al (1993) p53 mutation is associated with progression in follicular lymphomas. Blood 82(7):1994–2004

Schraders M, de Jong D et al (2005) Lack of Bcl-2 expression in follicular lymphoma may be caused by mutations in the BCL2 gene or by absence of the t(14;18) translocation. J Pathol 205(3):329–335

Schwaenen C, Viardot A et al (2009) Microarray-based genomic profiling reveals novel genomic aberrations in follicular lymphoma which associate with patient survival and gene expression status. Genes Chromosom Cancer 48(1):39–54

Sneeringer CJ, Scott MP et al (2010) Coordinated activities of wild-type plus mutant EZH2 drive tumor-associated hypertrimethylation of lysine 27 on histone H3 (H3K27) in human B-cell lymphomas. Proc Natl Acad Sci U S A 107(49):20980–20985

Stamatopoulos K, Kosmas C et al (1997) Follicular lymphoma immunoglobulin kappa light chains are affected by the antigen selection process, but to a lesser degree than their partner heavy chains. Br J Haematol 96(1):132–146

Swerdlow SH, Campo E, Harris NL, Jaffe ES, Pileri SA, Stein H, Thiele J, Vardiman JW (2008) Follicular lymphoma. WHO classification of tumours of haematopoietic and lymphoid tissues. IARC, Lyon. WHO press

Szereday Z, Csernus B et al (2000) Somatic mutation of the 5′ noncoding region of the BCL-6 gene is associated with intraclonal diversity and clonal selection in histological transformation of follicular lymphoma. Am J Pathol 156(3):1017–1024

Tari A, Asaoku H et al (2011) Clinical features of gastrointestinal follicular lymphoma: comparison with nodal follicular lymphoma and gastrointestinal MALT lymphoma. Digestion 83(3):191–197

Tomita N (2011) BCL2 And MYC dual-hit lymphoma/leukemia. J Clin Exp Hematop 51(1):7–12

Torlakovic EE, Bilalovic N et al (2006) Prognostic significance of PU.1 in follicular lymphoma. J Pathol 209(3):352–359

van Oers MH, Tonnissen E et al (2010) BCL-2/IgH polymerase chain reaction status at the end of induction treatment is not predictive for progression-free survival in relapsed/resistant follicular lymphoma: results of a prospective randomized EORTC 20981 phase III intergroup study. J Clin Oncol 28(13):2246–2252

Velichutina I, Shaknovich R et al (2010) EZH2-mediated epigenetic silencing in germinal center B cells contributes to proliferation and lymphomagenesis. Blood 116(24):5247–5255

Vire E, Brenner C et al (2006) The Polycomb group protein EZH2 directly controls DNA methylation. Nature 439(7078):871–874

Wahlin BE, Sander B et al (2007) CD8+ T-cell content in diagnostic lymph nodes measured by flow cytometry is a predictor of survival in follicular lymphoma. Clin Cancer Res 13(2 Pt 1):388–397

Wahlin BE, Aggarwal M et al (2010) A unifying microenvironment model in follicular lymphoma: outcome is predicted by programmed death-1–positive, regulatory, cytotoxic, and helper T cells and macrophages. Clin Cancer Res 16(2):637–650

Wang SA, Wang L et al (2005) Low histologic grade follicular lymphoma with high proliferation index: morphologic and clinical features. Am J Surg Pathol 29(11):1490–1496

William BM, Bierman PJ (2010) I-131 tositumomab. Expert Opin Biol Ther 10(8):1271–1278

Wilson WH, Armitage JO (2008) Non-Hodgkin's lymphoma. In: Abeloff MDA, Armitage JO, Niederhuber JE, Kastan MB, Gillies McKenna WG (eds) Abeloff's clinical oncology, 4th edn. Churchill Livingstone, Philadelphia, pp 2371–2404

Yang H, Chen CM et al (2003) The androgen receptor gene is preferentially hypermethylated in follicular non-Hodgkin's lymphomas. Clin Cancer Res 9(11):4034–4042

Yano T, Jaffe ES et al (1992) MYC rearrangements in histologically progressed follicular lymphomas. Blood 80(3):758–767

Ye BH, Lista F et al (1993) Alterations of a zinc finger-encoding gene, BCL-6, in diffuse large-cell lymphoma. Science 262(5134):747–750

Ylstra B, van den Ijssel P et al (2006) BAC to the future! or oligonucleotides: a perspective for micro array comparative genomic hybridization (array CGH). Nucleic Acids Res 34(2):445–450

Young RC, Longo DL et al (1988) The treatment of indolent lymphomas: watchful waiting v aggressive combined modality treatment. Semin Hematol 25(2 Suppl 2):11–16

Chronic Lymphocytic Leukemia: From Pathobiology to Targeted Therapy

Javier Pinilla-Ibarz and Chih-Chi Andrew Hu

Abstract Chronic lymphocytic leukemia (CLL) is a heterogeneous type of malignant B cell cancer that occurs in older adults. CLL patients may stay free of cancer symptoms for a long period of time and require no treatment. However, malignant progression of CLL strikes indefinitely in some patients, and such CLL responds poorly to conventional chemotherapies and will eventually acquire chemoresistance. Currently, no efficient therapeutic options are available to prevent or stop CLL from progressing into an uncontrollable chemoresistant stage. CLL thus remains to be a challenging disease that requires research explorations for efficient therapeutic methods. CLL cells harbor gene deletions and mutations that continue to accumulate in the course of disease progression and chemotherapy. While it is difficult to directly correct genetic defects in CLL, we reviewed recent research advancements focusing on several crucial molecular pathways that support the proliferative potential of CLL. Since molecules in these pathways can contribute to proliferation, survival and chemoresistance of CLL, they are potentially useful therapeutic targets. We discussed promising methods that have been developed to target functions of these molecules. Encouraging results emerged from these continuing research efforts have provided a new hope for more efficient treatments and increased survival rates for CLL patients.

J. Pinilla-Ibarz, M.D., Ph.D.
Hematologic Malignancies, Immunology, H. Lee Moffitt Cancer Center & Research Institute, Tampa, FL, USA

C.-C.A. Hu, Ph.D. (✉)
Immunology Program, H. Lee Moffitt Cancer Center & Research Institute, 12902 Magnolia Drive, Tampa, FL 33612, USA
e-mail: Chih-chi.Hu@Moffitt.org

J. Tao and E. Sotomayor (eds.), *Hematologic Cancers: From Molecular Pathobiology to Targeted Therapeutics*, Cancer Growth and Progression 14,
DOI 10.1007/978-94-007-5028-9_3, © Springer Science+Business Media Dordrecht 2012

Introduction

Chronic lymphocytic leukemia (CLL) is the most common leukemia in the Western world accounting for 30% of adult leukemias. Typically present in older individuals, CLL is the result of accumulations of functionally incompetent mature B lymphocytes in the blood, bone marrow, and lymph nodes. Cell surface markers (e.g., CD5, CD19, CD23 and CD38) together with serum markers (e.g., thymidine kinase and β2-microglobulin) allow unambiguous clinical distinction of CLL from other B cell leukemia. Although many genetic aberrations of CLL have been described, it is unclear why these defects occur. In fact, there is no single chromosomal aberration that can account for all types of human CLL. Different CLL patient samples can harbor distinct genetic defects. While genetic defects might directly cause CLL, the progress of CLL also drives the occurrence of more genetic defects. Thus, the genetic defects in CLL are accumulating during disease progression and treatments.

While B cells in a healthy person can manufacture a polyclonal repertoire of antibodies that contribute to a functional immune system, a CLL patient accumulates a monoclonal population of B cells. CLL cells from different patients can be categorized into a limited number of clones by their expression of stereotyped B cell receptors (membrane-bound IgM/IgD). These B cell receptors are encoded from the same immunoglobulin heavy and light chain gene loci. For example, CLL cells have a higher frequency of using two heavy chain genes, IGHV3-21 and IGHV1-69, to encode their B cell receptors. The use of IGHV3-21 gene also suggests the poor clinical outcome of CLL. In addition, the more frequent use of some immunoglobulin genes in CLL cells than that in healthy B cells suggest that CLL cells develop from antigen-experienced B cells and the development of CLL might be driven by exposure to certain antigens. According to this hypothesis, repeated stimulation by some antigens may drive B cells to accumulate genetic alterations and to proliferate abnormally. These processes eventually contribute to the transformation of normal B cells into CLL. However, such a hypothesis requires support by more experimental evidence.

Genetic mutations in the B cell genome can lead to the development of CLL. The primary function of B cells is to make immunoglobulin against infection and against other pathological insults to our body. To produce a repertoire of over 10^{11} different B cells producing distinctive immunoglobulins, B cells employ gene rearrangement mechanisms such as chromosomal translocations to recombine different regions of the immunoglobulin genes. In the pro-B cell stage, rearrangement of the immunoglobulin heavy chain variable regions occurs with the initial joining of a D_H gene segment to a J_H gene segment, followed by the later rearrangement of a V_H gene segment to the joined DJ_H. Such a process is followed by the pre-B cell phase in which subsequent rearrangements of the immunoglobulin light chain occur by joining a V_L and a J_L. These gene rearrangement mechanisms, when not properly regulated, can cause cancerous transformation of B cells. In many B cell malignancies, genes that regulate the cell cycle are reported to be joined mistakenly to the immunoglobulin promoter region, leading to their overexpression and increased cell proliferation.

CLL cells rarely harbors such aberrant chromosomal rearrangements that are responsible for many other types of B cell leukemia and lymphoma; instead, CLL cells harbor more complicated genetic alterations that make CLL a very challenging type of B cell malignancy to manage clinically (Cotter and Auer 2007; Dohner et al. 2000; Karhu et al. 2003).

In addition to chromosomal rearrangements, B cells employ a gene-mutating mechanism assisted by an enzyme called activation-induced cytidine deaminase (AID). AID performs somatic hypermutation in the immunoglobulin genes allowing B cells to produce immunoglobulins of higher affinity to bind to antigens. Somatic hypermutation, if not confined to immunoglobulin genes, can activate oncogenes or inactivate tumor suppressors, and contribute to the transformation of B cells into B cell leukemia and lymphoma. In fact, elevated levels of AID are often observed in advanced CLL cases (Albesiano et al. 2003; Hancer et al. 2011).

CLL is still incurable and conventional therapies focus on the management of disease symptoms in CLL patients. Although CLL may respond to chemotherapy or chemo-immunotherapy regimens initially, CLL develops resistance to treatments within a short period of time. Relapsed CLL responds to treatments poorly, and allogenic stem cell transplantation is the only alternative for a small group of patients. Because CLL cells carry gene defects, it justifies research efforts in identifying new ways to correct their genetic alterations. To correct genetic alterations in CLL is a challenging task. Alternatively, the information of genetic alterations carried by CLL cells can be used to carefully categorize CLL into subgroups. A thorough research on CLL cells harboring similar genetic alterations will help to devise the most suitable treatment plan for each distinctive CLL subgroup. To expand the treatment options, it also becomes important to identify and investigate common molecular pathways shared by all types of CLL cells to support their survival and confer their chemoresistance. In this article, we will discuss about some of these pathways as well as the methods that can be used to target these pathways.

Chromosomal and Epigenetic Changes

Although chromosomal rearrangements are rare in CLL, gene aberrations are found in more than 80% CLL patients (Dohner et al. 2000), and these at least include trisomy 12 and chromosomal deletions in 13q, 11q and 17p, among other genetic defects. Trisomy 12 and 13q deletion are associated with early-stage CLL, and little is known about the contribution of trisomy 12 to the pathogenesis of CLL. Chromosomal analysis of the deletion at 13q leads to the discovery of two microRNA genes, miR-15a and miR-16-1. The loss of miR-15a and miR-16-1 expression leads to upregulation of an anti-apoptotic protein BCL2, accounting for the contribution of 13q deletion in CLL pathogenesis. Deletions in 11q and 17p (and TP53 mutations) are often found in more advanced CLL cases. Deletions in chromosome 11q result in the loss of a tumor suppressor gene, ataxia-telangiectasia-mutated (ATM), which encodes a protein kinase that can activate apoptosis in the cells that contain DNA double-strand breaks, and directly maintain DNA ends during the repair of double-strand breaks. On chromosome 17p encodes the p53 tumor suppressor, which has important roles in DNA repair, cell cycle arrest upon DNA damage, and induction of apoptosis. CLL cells that resist to chemotherapeutic treatments are often found to acquire mutations in their p53 gene (TP53).

TCL1 (T cell lymphoma 1) was first discovered as an oncoprotein contributing to the occurrence of T cell pro-lymphocytic leukemia. Overexpression of TCL-1 is a result of chromosomal translocations and inversions at 14q31.2 (Virgilio et al. 1994). Although such a chromosomal defect has not been found in B cell leukemia or lymphoma, TCL-1 expression is found in most human CLL patients (Herling et al. 2006; Kriss et al. 2012). TCL-1 overexpression in CLL cells has been shown to associate with unmutated IGHV (see below), higher expression levels of ZAP70 (zata-associated protein 70) and strong BCR activation, all of which are poor prognostic markers for CLL (Herling et al. 2006, 2009; Pekarsky et al. 2006). Abnormal epigenetic regulations (miRNA and DNA methylation) in CLL cells have been shown to account for the overexpression of TCL1 and ZAP70. Strikingly, forced expression of TCL-1 alone in mouse B cells can drive the development of CLL, providing an invaluable mouse model for investigating the disease (Bichi et al. 2002).

Microenvironment

CLL cells can create microenvironments in the lymphoid organs to support their growth and survival. By the recruitments of T cells, stromal cells and survival factors into the lymph nodes, CLL cells can create proliferation centers, called pseudofollicles. Surrounding these CLL pseudofollicles one often identifies the CD3+/CD40L+/CD4+ T cells, which can stimulate the proliferation of CLL cells as a result of CD40-CD40L interaction (Ghia et al. 2005; Granziero et al. 2001). CD40-CD40L interactions also cause CLL cells to express and secrete survivin at high levels in the pseudofollicles to inhibit apoptosis (Granziero et al. 2001). CLL cells after contacting with CD4+ T cells begin to produce CD38 (Patten et al. 2008). These CD38+ CLL cells become hyperproliferative and anti-apoptotic by interacting with blood-derived nurse-like cells which can produce the stromal cell-derived factor 1 (a.k.a. CXCL12) and TNF-like ligands (APRIL and BAFF) to stimulate the growth of CLL cells (Kern et al. 2004; Novak et al. 2002; Tsukada et al. 2002). Thus, the presence of CD38+ CLL cells often predicts poor clinical outcomes (Damle et al. 1999, 2007; Krober et al. 2002).

Targeting Microenvironments Created by CLL Cells

Thalidomide and lenalidomide are immunomodulatory agents that exert numerous cellular functions. In addition to countering the T cell function in forming growth-stimulating immunological synapses with CLL cells (Ramsay et al. 2008), they can stimulate the expansion of natural killer cells leading to tumor lysis. In addition, lenalidomide also upregulates surface expression of CD154 (CD40L) on CLL cells and normal B cells, but only sensitizes CLL cells to tumor necrosis factor-related apoptosis-inducing ligand (TRAIL)-mediated apoptosis (Lapalombella et al. 2010). Since the expression of CD154 in normal B cells enhances their functions in antibody production, thalidomide and lenalidomide can thus be applied to CLL patients to add on beneficial clinical outcomes.

Aggressive CLL Is a Consequence of Constant B Cell Receptor Activation

Sequence analysis of immunoglobulin variable heavy chain genes (IGHV) reveals two distinct types of CLL: IGHV-unmutated CLL versus IGHV-mutated CLL (Fais et al. 1998). IGHV-unmutated CLL is defined when the nucleotide variation at the immunoglobulin gene loci of the patient's CLL cells is not exceeding 2% when compared to the germline sequence in the same patient. The 2% variation rate is an empirical cutting point. A gray zone (fickle clinical outcomes) exists for patients whose CLL cells contain a sequence variation between 2 and 3% (Hamblin et al. 2008). Nevertheless, sequencing CLL patient's immunoglobulin genes has allowed medical doctors to predict distinct disease outcomes for CLL patients. The IGHV-mutated CLL patients indeed survive significantly longer than those diagnosed with the IGHV-unmutated CLL (Damle et al. 1999; Hamblin et al. 1999; Krober et al. 2002).

Aggressive IGHV-unmutated CLL cells are responsive to B cell receptor (BCR) activation, and such constitutive BCR signal transduction allows CLL cells to sustain their highly proliferative capability. IGHV-mutated CLL cells are less responsive to BCR activation and are thus less proliferative (Chen et al. 2005; Guarini et al. 2008; Lanham et al. 2003; Mockridge et al. 2007; Muzio et al. 2008). Since the aggressiveness of IGHV-unmutated CLL cells are associated with their capability to respond to BCR activation, it is reasonable to target molecular components in the BCR signal transduction pathway as efficient measures of therapeutic intervention. In normal B cells, BCR signal transduction is induced when the membrane-bound immunoglobulin IgM encounters its cognate antigen on the cell surface. Membrane-bound IgM has a short cytoplasmic domain, and so a functional BCR must contain an IgM and its associated disulfide-linked Igα/Igβ heterodimer. Both Igα and Igβ contain the immunoreceptor tyrosine-based activation motifs (ITAM) that can be phosphorylated by the GPI-anchored Lyn kinase. Phosphorylated Igα/Igβ then recruits the Syk kinase. The phosphorylation of the Syk kinase is followed by phosphorylation of a number of kinases that include Btk and PI3K. Activation of these kinases ensures the initiation of a series of distinct downstream signal transduction cascades, leading to B cell proliferation and differentiation. Similar BCR signal transduction pathways are believed to occur in CLL cells upon stimulation. Different from activation in normal B cells, ZAP70 (zeta-associated protein 70), an essential tyrosine kinase in T-cell signaling is frequently found overexpressed and activated in CLL cells. Although ZAP70 is not expressed by normal B cells, it has been shown to contribute to strong BCR response in malignant CLL cells.

Targeting B Cell Receptor Signaling for the Treatment of CLL

Selective small-molecule chemical inhibitors of kinases in the BCR signal transduction pathway have been developed and are currently being tested in early phase clinical trial in patients with relapsed or refractory CLL, opening a new pharmacopeia for CLL where for decades nucleoside analogs, alkylating agents and monoclonal antibodies have remained the mainstay of therapy. CAL-101 is an isoform-selective inhibitor of PI3Kδ that has been reported to exert therapeutic activity in refractory CLL with an overall response of 30% of the patients (Herman et al. 2010). Small-molecule inhibitors for the Syk kinase (Fostamatinib) and for the Btk kinase (PCI-32765) are also being tested in CLL patients with encouraging results and overall responses of 55–60% in the patients (Friedberg et al. 2010; Gandhi 2010).

Expression of Toll-Like Receptors in Normal B Cells and CLL Cells

In addition to their cognate antigens, B cells can also be activated by ligands for Toll-like receptors (TLRs). It has been established in cell culture system that lipopolysaccharide (LPS) or the unmethylated CpG oligodeoxynucleotides alone can stimulate mouse B cells to proliferate and differentiate into antibody-producing plasmablasts (Coutinho et al. 1974; Krieg et al. 1995). LPS and CpG are recognized by pattern recognition receptors TLR4 on cell surface and TLR9 in the endosomes, respectively. In addition to LPS and CpG, other TLR ligands that can trigger B cells to proliferate include peptidoglycan and Pam_3CSK_4 (TLR1/2), bacterial lipoproteins and MALP2 (TLR2/6), dsRNA (TLR3), ssRNA and imidazoquinolines (TLR7 and TLR8), and profilin-like molecules (TLR11). Mouse B cells do not respond to flagellin, due to the lack of TLR5 expression (Genestier et al. 2007). Normal human B cells do not express TLR4 (Bourke et al. 2003), and thus are unresponsive to LPS. Other than enhancing antibody-mediated defense against infections (Meyer-Bahlburg et al. 2007), activation of TLRs in B cells also has important physiological functions in the immunoglobulin isotype switching (He et al. 2004) and maintenance of memory B cells. Human CLL cells have been shown to express TLR7 and TLR9. Because human CLL cells are heterogeneous, TLR expression patterns in different types of human CLL cells have not been systematically examined.

Targeting TLRs for the Treatment of CLL

Because dendritic cells are activated upon stimulations that engage their TLRs, some TLR agonists have been used in clinical trials to improve tumor antigen presentation by dendritic cells to promote T cell-mediated eradication of cancer (Krieg 2008), which includes CLL. However, before TLR agonists can be used as therapeutics, expression patterns of TLRs and consequences of administration should be carefully evaluated since different types of CLL cells may respond differently to activation of TLRs. An example posits that even when the same TLR agonist is administered, strikingly opposite results can be observed in different types of human CLL cells. CpG-mediated activation of TLR9 causes the IGHV-mutated CLL cells to undergo apoptosis, but it promotes the IGHV-unmutated CLL cells to proliferate (Jahrsdorfer et al. 2005; Longo et al. 2007). Such results highlight the potential use of TLR agonists in some CLL patients, but a thorough investigation on the clinical consequences brought by TLR activation in each type of human CLL cells is required.

Roles of Chaperone Proteins in the Development and Treatment of CLL

Because CLL cells harbor genetic heterogeneities, it has been proposed that treatments should target respective genetic defects in patients. However, it might be possible to find a survival pathway that all CLL cells share, and target such a pathway for treatment. For example, tumor cells often upregulate chaperones to sustain their survival and to confer chemoresistance. GRP78/BiP and GRP94 have been targeted by a select group of inhibitors, leading to encouraging clinical outcomes in CLL treatments. For example, geldanamycin and herbimycin A were used as inhibitors for GRP94. Both inhibitors can induce apoptosis in CLL and can synergize with chlorambucil and fludarabine in killing a subset of CLL cells (Jones et al. 2004). GRP78/BiP downregulation by siRNA has also been shown to induce apoptosis in CLL (Rosati et al. 2010). Other than BiP siRNA, subtilase cytotoxin which specifically cleaves and inactivates GRP78/BiP has been used to disrupt the cytoprotective effect of GRP78/BiP in fenretinide- or bortezomib-treated melanoma (Martin et al. 2010). Because subtilase cytotoxin also blocks antibody secretion from normal B cells (Hu et al. 2009), a concern arises that such a treatment regimen may hinder B cell help in the elimination of tumors. CLL is a result of the abnormal expansion of a monoclonal B cell population, thus subtilase cytotoxin might serve as an attractive agent to be tested in the treatment of CLL.

References

Albesiano E, Messmer BT, Damle RN, Allen SL, Rai KR, Chiorazzi N (2003) Activation-induced cytidine deaminase in chronic lymphocytic leukemia B cells: expression as multiple forms in a dynamic, variably sized fraction of the clone. Blood 102:3333–3339

Bichi R, Shinton SA, Martin ES, Koval A, Calin GA, Cesari R, Russo G, Hardy RR, Croce CM (2002) Human chronic lymphocytic leukemia modeled in mouse by targeted TCL1 expression. Proc Natl Acad Sci U S A 99:6955–6960

Bourke E, Bosisio D, Golay J, Polentarutti N, Mantovani A (2003) The toll-like receptor repertoire of human B lymphocytes: inducible and selective expression of TLR9 and TLR10 in normal and transformed cells. Blood 102:956–963

Chen L, Apgar J, Huynh L, Dicker F, Giago-McGahan T, Rassenti L, Weiss A, Kipps TJ (2005) ZAP-70 directly enhances IgM signaling in chronic lymphocytic leukemia. Blood 105:2036–2041

Cotter FE, Auer RL (2007) Genetic alteration associated with chronic lymphocytic leukemia. Cytogenet Genome Res 118:310–319

Coutinho A, Gronowicz E, Bullock WW, Moller G (1974) Mechanism of thymus-independent immunocyte triggering. Mitogenic activation of B cells results in specific immune responses. J Exp Med 139:74–92

Damle RN, Wasil T, Fais F, Ghiotto F, Valetto A, Allen SL, Buchbinder A, Budman D, Dittmar K, Kolitz J, Lichtman SM, Schulman P, Vinciguerra VP, Rai KR, Ferrarini M, Chiorazzi N (1999) Ig V gene mutation status and CD38 expression as novel prognostic indicators in chronic lymphocytic leukemia. Blood 94:1840–1847

Damle RN, Temburni S, Calissano C, Yancopoulos S, Banapour T, Sison C, Allen SL, Rai KR, Chiorazzi N (2007) CD38 expression labels an activated subset within chronic lymphocytic leukemia clones enriched in proliferating B cells. Blood 110:3352–3359

Dohner H, Stilgenbauer S, Benner A, Leupolt E, Krober A, Bullinger L, Dohner K, Bentz M, Lichter P (2000) Genomic aberrations and survival in chronic lymphocytic leukemia. N Engl J Med 343:1910–1916

Fais F, Ghiotto F, Hashimoto S, Sellars B, Valetto A, Allen SL, Schulman P, Vinciguerra VP, Rai K, Rassenti LZ, Kipps TJ, Dighiero G, Schroeder HW Jr, Ferrarini M, Chiorazzi N (1998) Chronic lymphocytic leukemia B cells express restricted sets of mutated and unmutated antigen receptors. J Clin Invest 102:1515–1525

Friedberg JW, Sharman J, Sweetenham J, Johnston PB, Vose JM, Lacasce A, Schaefer-Cutillo J, De Vos S, Sinha R, Leonard JP, Cripe LD, Gregory SA, Sterba MP, Lowe AM, Levy R, Shipp MA (2010) Inhibition of Syk with fostamatinib disodium has significant clinical activity in non-Hodgkin lymphoma and chronic lymphocytic leukemia. Blood 115:2578–2585

Gandhi V (2010) Targeting kinases in CML CLL. Blood 116:1999–2000

Genestier L, Taillardet M, Mondiere P, Gheit H, Bella C, Defrance T (2007) TLR agonists selectively promote terminal plasma cell differentiation of B cell subsets specialized in thymus-independent responses. J Immunol 178:7779–7786

Ghia P, Circosta P, Scielzo C, Vallario A, Camporeale A, Granziero L, Caligaris-Cappio F (2005) Differential effects on CLL cell survival exerted by different microenvironmental elements. Curr Top Microbiol Immunol 294:135–145

Granziero L, Ghia P, Circosta P, Gottardi D, Strola G, Geuna M, Montagna L, Piccoli P, Chilosi M, Caligaris-Cappio F (2001) Survivin is expressed on CD40 stimulation and interfaces proliferation and apoptosis in B-cell chronic lymphocytic leukemia. Blood 97:2777–2783

Guarini A, Chiaretti S, Tavolaro S, Maggio R, Peragine N, Citarella F, Ricciardi MR, Santangelo S, Marinelli M, De Propris MS, Messina M, Mauro FR, Del Giudice I, Foa R (2008) BCR ligation induced by IgM stimulation results in gene expression and functional changes only in IgV H unmutated chronic lymphocytic leukemia (CLL) cells. Blood 112:782–792

Hamblin TJ, Davis Z, Gardiner A, Oscier DG, Stevenson FK (1999) Unmutated Ig V(H) genes are associated with a more aggressive form of chronic lymphocytic leukemia. Blood 94:1848–1854

Hamblin TJ, Davis ZA, Oscier DG (2008) Determination of how many immunoglobulin variable region heavy chain mutations are allowable in unmutated chronic lymphocytic leukaemia – long-term follow up of patients with different percentages of mutations. Br J Haematol 140:320–323

Hancer VS, Kose M, Diz-Kucukkaya R, Yavuz AS, Aktan M (2011) Activation-induced cytidine deaminase mRNA levels in chronic lymphocytic leukemia. Leuk Lymphoma 52:79–84

He B, Qiao X, Cerutti A (2004) CpG DNA induces IgG class switch DNA recombination by activating human B cells through an innate pathway that requires TLR9 and cooperates with IL-10. J Immunol 173:4479–4491

Herling M, Patel KA, Khalili J, Schlette E, Kobayashi R, Medeiros LJ, Jones D (2006) TCL1 shows a regulated expression pattern in chronic lymphocytic leukemia that correlates with molecular subtypes and proliferative state. Leukemia 20:280–285

Herling M, Patel KA, Weit N, Lilienthal N, Hallek M, Keating MJ, Jones D (2009) High TCL1 levels are a marker of B-cell receptor pathway responsiveness and adverse outcome in chronic lymphocytic leukemia. Blood 114:4675–4686

Herman SE, Gordon AL, Wagner AJ, Heerema NA, Zhao W, Flynn JM, Jones J, Andritsos L, Puri KD, Lannutti BJ, Giese NA, Zhang X, Wei L, Byrd JC, Johnson AJ (2010) Phosphatidylinositol 3-kinase-delta inhibitor CAL-101 shows promising preclinical activity in chronic lymphocytic leukemia by antagonizing intrinsic and extrinsic cellular survival signals. Blood 116:2078–2088

Hu CC, Dougan SK, Winter SV, Paton AW, Paton JC, Ploegh HL (2009) Subtilase cytotoxin cleaves newly synthesized BiP and blocks antibody secretion in B lymphocytes. J Exp Med 206:2429–2440

Jahrsdorfer B, Wooldridge JE, Blackwell SE, Taylor CM, Griffith TS, Link BK, Weiner GJ (2005) Immunostimulatory oligodeoxynucleotides induce apoptosis of B cell chronic lymphocytic leukemia cells. J Leukoc Biol 77:378–387

Jones DT, Addison E, North JM, Lowdell MW, Hoffbrand AV, Mehta AB, Ganeshaguru K, Folarin NI, Wickremasinghe RG (2004) Geldanamycin and herbimycin A induce apoptotic killing of B chronic lymphocytic leukemia cells and augment the cells' sensitivity to cytotoxic drugs. Blood 103:1855–1861

Karhu R, Tobin G, Thunberg U, Vilpo L, Sundstrom C, Knuutila S, Rosenquist R, Vilpo J (2003) More extensive genetic alterations in unmutated than in hypermutated cases of chronic lymphocytic leukemia. Genes Chromosom Cancer 37:417–420

Kern C, Cornuel JF, Billard C, Tang R, Rouillard D, Stenou V, Defrance T, Ajchenbaum-Cymbalista F, Simonin PY, Feldblum S, Kolb JP (2004) Involvement of BAFF and APRIL in the resistance to apoptosis of B-CLL through an autocrine pathway. Blood 103:679–688

Krieg AM (2008) Toll-like receptor 9 (TLR9) agonists in the treatment of cancer. Oncogene 27:161–167

Krieg AM, Yi AK, Matson S, Waldschmidt TJ, Bishop GA, Teasdale R, Koretzky GA, Klinman DM (1995) CpG motifs in bacterial DNA trigger direct B-cell activation. Nature 374:546–549

Kriss CL, Pinilla-Ibarz JA, Mailloux AW, Powers JJ, Tang CH, Kang CW, Zanesi N, Epling-Burnette PK, Sotomayor EM, Croce CM, Del Valle JR, Hu CC (2012) Overexpression of TCL1 activates the endoplasmic reticulum stress response: a novel mechanism of leukemic progression in mice. Blood 120:1027–1038

Krober A, Seiler T, Benner A, Bullinger L, Bruckle E, Lichter P, Dohner H, Stilgenbauer S (2002) V(H) mutation status, CD38 expression level, genomic aberrations, and survival in chronic lymphocytic leukemia. Blood 100:1410–1416

Lanham S, Hamblin T, Oscier D, Ibbotson R, Stevenson F, Packham G (2003) Differential signaling via surface IgM is associated with VH gene mutational status and CD38 expression in chronic lymphocytic leukemia. Blood 101:1087–1093

Lapalombella R, Andritsos L, Liu Q, May SE, Browning R, Pham LV, Blum KA, Blum W, Ramanunni A, Raymond CA, Smith LL, Lehman A, Mo X, Jarjoura D, Chen CS, Ford R Jr, Rader C, Muthusamy N, Johnson AJ, Byrd JC (2010) Lenalidomide treatment promotes CD154 expression on CLL cells and enhances production of antibodies by normal B cells through a PI3-kinase-dependent pathway. Blood 115:2619–2629

Longo PG, Laurenti L, Gobessi S, Petlickovski A, Pelosi M, Chiusolo P, Sica S, Leone G, Efremov DG (2007) The Akt signaling pathway determines the different proliferative capacity of chronic lymphocytic leukemia B-cells from patients with progressive and stable disease. Leukemia 21:110–120

Martin S, Hill DS, Paton JC, Paton AW, Birch-Machin MA, Lovat PE, Redfern CP (2010) Targeting GRP78 to enhance melanoma cell death. Pigment Cell Melanoma Res 23:675–682

Meyer-Bahlburg A, Khim S, Rawlings DJ (2007) B cell intrinsic TLR signals amplify but are not required for humoral immunity. J Exp Med 204:3095–3101

Mockridge CI, Potter KN, Wheatley I, Neville LA, Packham G, Stevenson FK (2007) Reversible anergy of sIgM-mediated signaling in the two subsets of CLL defined by VH-gene mutational status. Blood 109:4424–4431

Muzio M, Apollonio B, Scielzo C, Frenquelli M, Vandoni I, Boussiotis V, Caligaris-Cappio F, Ghia P (2008) Constitutive activation of distinct BCR-signaling pathways in a subset of CLL patients: a molecular signature of anergy. Blood 112:188–195

Novak AJ, Bram RJ, Kay NE, Jelinek DF (2002) Aberrant expression of B-lymphocyte stimulator by B chronic lymphocytic leukemia cells: a mechanism for survival. Blood 100:2973–2979

Patten PE, Buggins AG, Richards J, Wotherspoon A, Salisbury J, Mufti GJ, Hamblin TJ, Devereux S (2008) CD38 expression in chronic lymphocytic leukemia is regulated by the tumor microenvironment. Blood 111:5173–5181

Pekarsky Y, Santanam U, Cimmino A, Palamarchuk A, Efanov A, Maximov V, Volinia S, Alder H, Liu CG, Rassenti L, Calin GA, Hagan JP, Kipps T, Croce CM (2006) Tcl1 expression in chronic lymphocytic leukemia is regulated by miR-29 and miR-181. Cancer Res 66:11590–11593

Ramsay AG, Johnson AJ, Lee AM, Gorgun G, Le Dieu R, Blum W, Byrd JC, Gribben JG (2008) Chronic lymphocytic leukemia T cells show impaired immunological synapse formation that can be reversed with an immunomodulating drug. J Clin Invest 118:2427–2437

Rosati E, Sabatini R, Rampino G, De Falco F, Di Ianni M, Falzetti F, Fettucciari K, Bartoli A, Screpanti I, Marconi P (2010) Novel targets for endoplasmic reticulum stress-induced apoptosis in B-CLL. Blood 116:2713–2723

Tsukada N, Burger JA, Zvaifler NJ, Kipps TJ (2002) Distinctive features of "nurselike" cells that differentiate in the context of chronic lymphocytic leukemia. Blood 99:1030–1037

Virgilio L, Narducci MG, Isobe M, Billips LG, Cooper MD, Croce CM, Russo G (1994) Identification of the TCL1 gene involved in T-cell malignancies. Proc Natl Acad Sci U S A 91:12530–12534

Genetic and Environmental Determinants in Multiple Myeloma: Implications for Therapy

Kenneth H. Shain and William S. Dalton

Abstract Multiple myeloma is an incurable hematopoietic malignancy linked to more than 10,000 deaths annually. This dyscrasia is associated with significant co-morbidities as a result of the clonal expansion of bone marrow resident, immunoglobulin secreting, plasma cells. As will most cancers, myeloma tumorigenesis is a defined by both genetic transforming events and the growth and survival factors provided by the bone marrow microenvironment. The earliest genetic aspects of myeloma tumorigenesis have been shown to result from non-random translocations of genes that normally function as determinants of cell proliferation or cell survival to regions juxtaposed to active immunoglobulin heavy chain (IgH) enhancer elements, Secondary somatic mutations are then acquired that affect oncogenic signaling via Ras, Raf, NF-κB, FGFR3, or myc. Further, the loss of tumor suppressor function with deletion of the short of arm of chromosome 17 as later events. These events have been characterized over the last several decades and provide a model for the progression from the premalignant MGUS (monoclonal gammopathy of undetermined significance), to smoldering myeloma, to active myeloma, to extramedually plasma cell leukemia, and cells lines. Even in the setting of a genetic disease, it has long been appreciated that the bone marrow microenvironment also promotes multiple myeloma cell pathogenesis and has become an important target for therapeutic intervention. With increased understanding of the both the genetics of myeloma and the bone marrow niche numerous novel therapies are underdevelopment that will be discussed within this book chapter with the anticipation of improved clinical outcomes.

K.H. Shain, M.D., Ph.D.
Experimental Therapeutics Program, H. Lee Moffitt Cancer Center and Research Institute, Tampa, FL, USA

W.S. Dalton, Ph.D., M.D. (✉)
Experimental Therapeutics Program, Departments of Experimental Therapeutics and Oncologic Sciences, H. Lee Moffitt Cancer Center and Research Institute, University of South Florida, 12902 Magnolia Drive, Tampa 33612, FL, USA
e-mail: william.dalton@moffitt.org

J. Tao and E. Sotomayor (eds.), *Hematologic Cancers: From Molecular Pathobiology to Targeted Therapeutics*, Cancer Growth and Progression 14, DOI 10.1007/978-94-007-5028-9_4, © Springer Science+Business Media Dordrecht 2012

Introduction

Multiple Myeloma is a mortal hematologic malignancy of bone marrow resident plasma cells. Myeloma tumorigensis is a defined by both genetic transforming events and the growth and survival factors provided by the bone marrow microenvironment. Despite an incredible amount of research, myeloma remains incurable due to both genetic and environmental factors. This plasma cell disorder was diagnosed in 20,180 individuals in 2010 and linked to 10,650 deaths representing 2% of all cancers and 15% of hematopoietic malignancies (Rajkumar 2011). The a median age of diagnosis remains 69–71, with less than 2% of diagnoses in individuals younger than 40. The incidence is slighter higher in men than woman, in African American individuals than white, and in first-degree relatives of individuals with B cell malignancies. Risk for myeloma has been linked to a number of potential environmental sources including radiation exposure, agent orange exposure, pesticides, organic solvents, cosmetic hair dyes, viral hepatitides and most recently with exposure in younger "first responders" at Ground Zero following the World Trade Center tragedies (Moline et al. 2009). However, definitive links between the inciting etiology and myeloma remain enigmatic and controversial.

This plasma cell dyscrasia is associated with significant co-morbidities as a result of the clonal expansion of bone marrow resident, immunoglobulin secreting, plasma cells. Myeloma is associated with defined systemic manifestations leading to end organ damage. The presence of myeloma-induced hyper*c*alcemia, *r*enal insufficiency, *a*nemia, and/or *b*oney lytic lesions is a prerequisite for diagnosis and treatment of this plasma cell dyscrasia. An increased rate of serious *i*nfections is a frequently included as a fifth; thereby defining the CRAB-I criteria (Fig. 1). The presence of

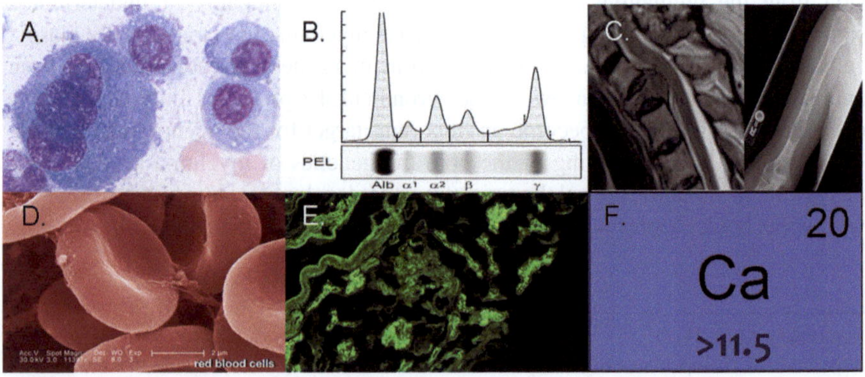

Fig. 1 Multiple myeloma. Multiple myeloma is an incurable malignancy of bone marrow resident plasma cells (**a**) that secrete a monoclonal immunoglobulin or "M-spike" on protein electrophoresis (**b**). Myeloma is characterized by classic stigmata of end organ damage including hyper***c***alcemia (**f**), ***r***enal insufficiency due to light chain deposition (**e**), ***a***nemia (**d**), and ***b***oney lytic lesions (**c**) or the CRAB criteria. A fifth criteria, increased ***i***nfections, is frequently considered (CRAB-I) (Kyle and Rajkumar 2009)

one or more of these five criteria necessitates a need for treatment. As one can imagine, from this list of CRAB-I criteria the presentation of myeloma varies widely; from asymptomatic, to fatigue (anemia), to altered mental status (hypercalcemia), to renal failure requiring hemodialysis, to skeletal fractures (lytic disease), and to life threatening infections (immune suppression). Together with mortal nature of multiple myeloma, the potential of these co-morbidities provides a significant impetus to devine new and better therapies to overcome this disease. To do so, decades of clinical and preclinical research has been carried to better understand the biology of myeloma as well as the bone marrow niche in which it resides.

Progression of Multiple Myeloma

Myeloma has been long understood to follow a model for the progression from the premalignant MGUS (monoclonal gammopathy of undetermined significance), smoldering multiple myeloma (SMM), active multiple myeloma (AMM), and to extramedullary plasma cell leukemia (PCL) (Fig. 2). Although individuals may present at any stage of disease, it is widely anticipated that myeloma progresses from the premalignant MGUS (Weiss et al. 2009). MGUS is a clonal plasma cell disorder categorized by a relatively low burden of clonal plasma cells (less then 3 g/dL and less than 10% bone marrow plasma cells (BMPC)) and the absence of end organ dysfunction (no CRAB-I). The incidence of MGUS increases with age from <5% in individuals less than 50 to nearly 8–9% in patients greater that 80 years old (Weiss et al. 2009; Bida et al. 2009). MGUS is a relatively benign *stage* with a rate of progression to AMM of approximately 1% per annum (Kyle et al. 2010). However, based on specific criteria individuals may be at greater risk than other for progression. The risk factors include (1) Monoclonal paraprotein (>1.5 g/dL), (2) abnormal serum free light chain ratio (SFLC: >1.62 or <0.26), and (3) non-IgG immunoglobulin. Utilization of these parameters patients can be subdivided into low, low-intermediate, high-intermediate, and high risk strata with 5, 21, 37, and 58% 20-year progression rates, respectively. More recently, analysis of (Dispenzieri et al. 2008) MGUS patient with multi-parameter flow cytometry revealed that a threshold of >95% aberrant plasma cells/BMPC, aneuploidy, and immunoparesis also provided prognostic progression rate data. MGUS patients should be monitored semi-annually, but treatment is not indicated. SMM is defined by an increase burden of disease with ≥3 g/dL of paraprotein or ≥10% BMPC, and the absence of end organ damage (no CRAB-I). Like MGUS, SMM can follow a relatively benign course. However, the rate of progression is approximately 10% per annum for the first 3–5 years (Blade et al. 2010). The risk then tapers to that of MGUS within 5–10 years. Risk factors have been identified that subdivide SMM patient in low, intermediate, and high risk groups based on the percent BMPC (>10%), M-spike (>3 g/dL), and abnormal SFLC ratio (>8 or <0.125) (Blade et al. 2010). The median time to progression to AMM is 10, 5.1, and 1.9 years with 1, 2, or 3 risk factors, respectively. Monitoring at 2–4 month intervals is recommended (Blade et al. 2010).

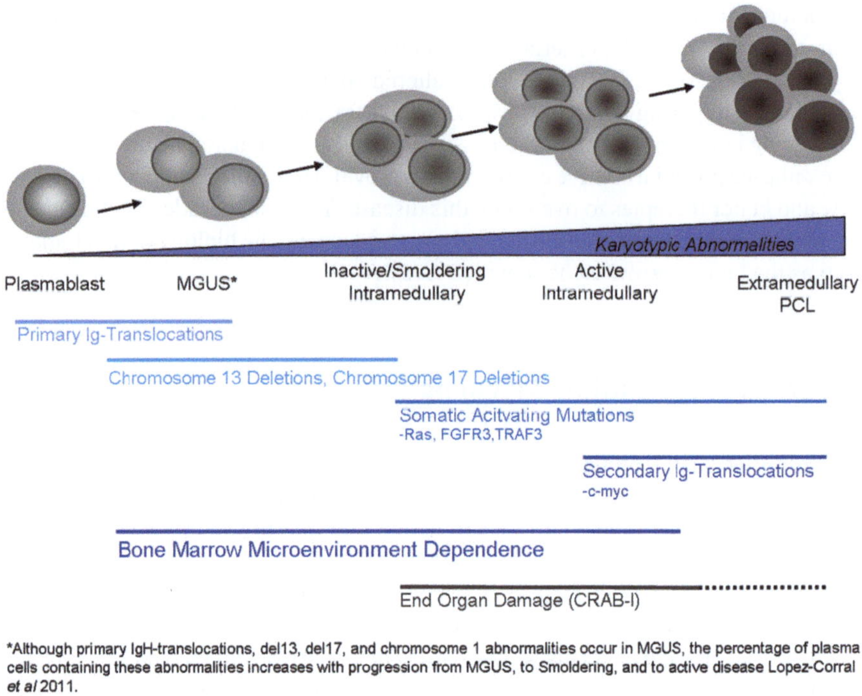

Fig. 2 A multi-step progression of MM. Myeloma follows a continuum from pre-malignant MGUS, to smoldering/indolent MM, to an active intramedullary stage, and finally to extramedullary disease (secondary Plasma Cell Leukemia (*PCL*)). Importantly, progression through this model parallels an accumulation of karyotypic abnormalities. Identification of a number of these karyotypic alterations have revealed a multi-step pathogenesis of MM. Examination of patient DNA has demonstrated a progression from primary Ig-translocations (cyclin D1 t(11q13), cyclin D3 (6p21), FGFR3 (4p16.3), and c-maf (16q23), chromosome 13 deletions, activating somatic mutations (Ras and FGFR mutations), and secondary Ig-translocations or insertions (c-myc). Importantly, these are the most prevalent mutations observed. This progression also parallels the changes in the requirement of prosurvival and proliferative effectors of the bone marrow niche as well as the onset of end organ damage CRAB-I)

For individuals with high or intermediate risk SMM (as defined above), prevention trials should be recommended if available. Lastly, arguments can be made to initiate treatment on SMM patients with greater than 60% BMPC even in the absence of CRAB-I because of the extremely short time to progression (Rajkumar 2011). Active multiple myeloma generally refers to individuals with evidence of end organ damage as defined by the CRAB-I criteria.

PCL (secondary or primary) represents that most aggressive variant in the myeloma spectrum. PCL is defined as secondary (with an antecedent medullary myeloma {40%}) or primary (no antecedent medullary disease {60%}). In either case, PCL is diagnosed with the identification of a population of clonal plasma cells representing greater then 20% of the total peripheral blood mononuclear cells or a an absolute plasma cell count of greater then 2000. Typically patients will present with a signs and symptoms of both AMM (CRAB-I) and leukemia (neutropenia, thrombocytopenia, and hepatosplenomegaly). This plasma cell dyscrasia has the worst prognosis with historical median overall survivals less than a year (Sher et al. 2010).

Genetic Contributions to Multiple Myeloma

Multiple myeloma lies along a continuum from the normal plasmablast to premalignant MGUS, to SMM, AMM, and to PCL in its most advanced form (Fig. 2). As with all cancers, multiple myeloma results from a sequence of genetic alterations within a malignant clone that facilitates an escape from the programmed constraints governing proliferation and death. It has been proposed that these initial requirements are at least partially fulfilled by the non-random translocation of a gene(s) that normally function as determinants of cell proliferation or cell survival to regions juxtaposed to active IgH enhancer elements located on the long arm of chromosome 14 (less frequently light chain enhancers; chromosome 2 and 22) that occur secondary to errors IgH switch recombination. These genetic alterations, in turn, facilitate the over-expression and activity of an oncogene (primarily MMSET, FGFR3, CCDN3, CCDN1, c-MAF, and MAFB) driving transformation and are observed in the premalignant condition MGUS (Table 1). Alternately, there exist a group of patients (approximately 50%) that do not present with these predictable translocations. Instead these patient's myeloma cells are characterized by increased genetic material

Table 1 Genetic anomalies defining High Risk Multiple Myeloma

High risk	Proposed gene target	Rate myeloma patients (%)
Deletion 13 or Aneuploidy by metaphase analysis*	Rb	50[a]
t(4;14) by FISH	MMSET/FGFR3	15
t(14;16) by FISH	c-maf	6
t(14;20) by FISH	MAFB	1–2
Deletion 17p by FISH	p53	8
Hypodiploidy		
Plasma cell labeling index >3.0		
Standard risk		
Hyperdiploidy by FISH		50–60
t(11;14)	Cyclin D1	20
t(6;14)	Cyclin D3	1

Patients with t(4;14), β2M<4 mg/l and Hb≥10g/dl may have *intermediate risk* disease

Patients with addition of chromosome 1q21 or deletion of 1p32 may have High Risk disease, occurring in 35 and 30% of patients, respectively

Treatment with Bortezomib and Dexamethasone containing regimens may overcome the High risk features of del 13 or t(4;14) having an *intermediate risk* disease

[a]50% of myeloma patients have monosomy or deletions of chromosome 13, not all are quantifiable by metaphase cytogenetic

*Deletion 13 by FISH only does not constitute High Risk

or hyperdiploidy (>2N DNA content). Thereby, facilitating a general dichotomous definition of myeloma patients in to groups of non-hyperdiploid multiple myeloma (NH-MM) or hyperdiploid multiple myeloma (H-MM- see below). It has been postulated that these two major divisions NH-MM (with associated translocations) and H-MM represent early transforming events in myeloma (Sawyer 2011).

Translocation (4;14)(p16;q32) –MMSET/FGFR3

Although not the first translocation identified t(4;14)(p16;q32) has been one of the most studied. It is identified in approximately 13–15% of patients and is associated with more aggressive disease and shorter disease free intervals (Rajkumar 2011). Like most of the translocations discussed it was initially identified in MM cell lines, but not by conventional karyotyping (Kyle and Rajkumar 2008), t(4;14) is cryptic in nature requiring FISH (or Polymerase chain reaction (PCR)) for identification. This translocation involves the juxtaposition of two genes of the telomeric side of 4p16, MMSET and FGFR3, with Ig enhancer elements. The translocation facilitates the increased production of Fibroblast Growth Factor Receptor 3 in myeloma cells and transformation. Interestingly, reports demonstrate that between 25 and 30% of patients harboring this translocation did not expression FGFR3. This observation lead to the identification of MMSET (multiple myeloma SET domain gene) as another gene product involved in the translocation. Although the true function of the MMSET-Igh chimeric protein has yet to be determined outright, it is present in all t(4;14) anomalies. To this end, it has been proposed to be the primary oncogenic event in this translocation. This translocation has been hypothesized to facilitate the transformation of plasma cells to MGUS, t(4;14) tends to be slightly underrepresented in MGUS relative to AMM; thereby may be a later or secondary transforming event hailing progression from MGUS to SMM or AMM. To this end, may not truly represent an initiating genomic event.

Translocation (6;14) and t(11;14)(q13;q32)-Cyclin D3 and Cyclin D1

These translocations involve cyclin D3 (t(6;14)) and cyclin D1 (t(11:14)(q13;q32)) to important determinants of cell cycle progression. Little is known about the clinical significance of the cyclin D3 translocation, but it has been identified in ~4% of myeloma patients. The translocation involving cyclin D1 was the first translocation identified in association with myeloma and is identical to that characterized in Mantle Cell Lymphoma (MCL). It is observed in approximately 15–20% of myeloma patients making it the most common translocation in myeloma. Throughout its history t(11;14) has been linked to both poorer and improved risk. Today, it is generally considered neutral and patients harboring this translocation are deemed standard risk. However, it is apparent that even within this subgroup of myeloma patients there is a large degree of heterogeneity in terms of risk. This is especially true when considering the high rate of t(11;14) observed in PCL.

Translocation (14;16)(q32; q23) and t(14;20)(q32;q11) –c-MAF and MAFB

These translocations are unique to myeloma. The t(14;16)(q32;q32) juxtaposes the v-maf muscolaponeuronic fibrosarcoma oncogene (MAF) to the IgH elements facilitating its overexpression. The MAF transcription factor has been demonstrated to regulate the expression of cyclin D2 (CCDN2) and beta 7 integrin (INTB7). This translocation is observed in approximately 6% of myeloma patients and is associated with a more aggressive clinical course and shorter disease free interval. The t(14;20)(q32;q11) is more rare representing less than 2% of myeloma patients and is also presumed to be associated with poorer prognosis. Interestingly, MAF is also overexpressed in a number of myeloma patients in the absence of translocation providing further evidence of the importance of this transcription factor in the pathogenesis of myeloma.

Hyperdiploidy and Trisomy

The most frequent abnormality in myeloma remains hyperdiploidy, observed in the majority of patients. As with the above listed predictable translocations, H-MM has been proposed to represent an early transforming event in myeloma. These patients contain myeloma cells with greater than 2N DNA content and is characterized by the trisomy in odd numbered chromosomes most frequently 9, 11, and 15. Trisomies of 3, 5, 7, 19, and 21 are also observed at lower frequencies. The manner in which trisomies contribute to myeloma remains undefined. However, has been hypothesized that trisomies contribute to myeloma cell proliferation and survival via a genome dosage effect.

Hyperdiploidy has generally been associated with better prognosis. However, as can be imagined in a group that makes up at least 50% of myeloma patients there is a significant degree of heterogeneity. A number of studies have looked into this population of patients (Chng et al. 2006). Most recently, gene expression profiling of H-MM revealed four separate subgroups. Chng et al. demonstrated that “cluster 1” was associated with poorer prognosis relative to the other three clusters and increased proliferative rate (Chng et al. 2006). Interestingly, this cluster was also characterized by a large number of genes localizing to chromosome 1. There are additional lesions that have been associated with poorer prognosis in H-MM patients including, but is not limited to IgH translocations with unknown partners and deletion 13 (Chng et al. 2006).

Loss of Tumor Suppressor Genes

Loss of tumor suppressors are also observed in patients with MGUS. These include loss of chromosome 13 (or long arm 13q) or deletion of the short arm of chromosome 17 (del17p). However, unlike losses of chromosome 13 (del13p), deletions of 17p are generally considered a later event with increasing rates with progression to AMM and PCL (Kyle and Rajkumar 2009).

Deletion of Chromosome 13

Loss of chromosome 13 was the first genetic event associated with myeloma by molecular profiling with metaphase cytogenetics. Deletion of chromosome 13 has been linked with more aggressive disease and a shorter progression free survival relative to patients without this anomaly (Tricot et al. 1995). However, there has been a subtle shift toward the hypothesis that the unfavorable prognostic relevance of monosomy 13 may be attributable to the associated molecular anomalies (*e.g.* t(4;14)). Chromosome 13 losses are observed in approximately 50% of myeloma patients. Monosomies are seen in at least 85% of these cases with deletions representing the other 15% (Sawyer 2011). As we have presumed in this chapter, the contribution of chromosome 13 losses to myeloma are related to the loss of the *RB1* (retinoblastoma 1) tumor suppressor. However, this has yet to be definitively elucidated and other candidate genes have been identified. For instance, the tripartite motif containing 13 (TRIM13) tumor suppressor may also be involved (Elnenaei et al. 2003). More recently, analysis of plasma cell dyscrasia patients with novel array comparative genomic hybridization (aCGH) strategy defined the 13q14 and 13q13 regions that affected the *RB1* and *NBEA* genes in patients with chromosome 13 deletions (each in 40% of the 20 myeloma and MGUS patients examined) (O'Neal et al. 2009). The gene product of *NBEA* has been implicated in membrane trafficking in neurons and PKA binding. Currently, no specific role for NBEA has been identified in cancer.

Deletion of Chromosome 17p

Loss of the short arm of chromosome 17 (17p13) represents the most clinically significant anomaly associated with multiple myeloma. Patients harboring deletion of 17p have short disease free survival, significantly foreshortened overall survival, more aggressive disease, and propensity for extramedullary disease (Kyle and Rajkumar 2008). Del17p is observed in 10% of myeloma patients. This event is linked to the loss of the tumor suppressor p53 (TP53) and presumed loss of heterogeneity. However, identification of TP53 mutations are rare in myeloma and remain almost exclusive to myeloma cell lines. As with loss of chromosome 13, additional genes may be important in the poor outcomes associated with del17p. Regardless of the gene(s) target, this chromosomal anomaly remains the most clinically significant issues and represents are group of patients that have a very limited life span.

Chromosome 1 Aberrations

Chromosome 1 anomalies represent the most common structural aberrations in myeloma. Changes in chromosome 1 primarily involve deletions of the short arm of chromosome 1 (1p) and amplification of the long arm (1q) (Sawyer 2011). Of primary importance is the unfavorable prognosis in individuals carrying deletion of 1p21 (Sawyer 2011). This anomaly is reported in approximately 30% of myeloma patients and is defined by varying interstitial deletions spanning the region from 1p13 to 1p31 (primarily 1p21). Of note, in the short arm of chromosome 1, a rare recurring reciprocal translocation has been reported involving *MYC* (Sawyer 2011). This t(1;8)(p12;q24) juxtaposes *MYC* to 1p12 representing a secondary aberration.

The long arm of chromosome 1 is also involved, characterized by the amplification of the 1q21 region in MM. This anomaly is has been demonstrated in between 35 and 40% of newly diagnosed myeloma patients by FISH. Consistent with the highly significant chromosome instability, amplifications in this region are seen in 70% of patients with relapsed disease (Hanamura et al. 2011). There remain a large number of potential gene targets within this region of 1q in myeloma, including MUC1, MCL1, PDZK1, IL6R, BCL9, CKS1B, PSMD4, UBAP2L, UBE2Q1, and ANP32E (Sawyer 2011). It is important to note that although chromosome 1 anomalies are hypothesized to correlate with unfavorable prognosis, there remains some controversy in using these abnormalities to define risk.

Secondary Mutations

Further progression form MGUS to AMM has been shown to correlate with activating mutations in RAS (NRAS & KRAS), FGFR3, and TRAF3 correlate with progression or relapse (Fig. 2) (Lonial 2010; Chapman 2011; Sawyer 2011). KRAS and NRAS mutations are seen in less than 5% of MGUS patients, but activating mutations targeting codons 12, 13, or 61 are seen in 40% of myeloma patients (60% NRAS and 40% KRAS) (Sawyer 2011). Activating mutations in FGFR3 are also observed in approximately 10% of patients with t(4;14). Importantly, the presence of a RAS or an FGFR3 mutation is mutually exclusive suggesting that they serve redundant pathways in the transformation from MGUS to AMM. More recently, activating mutations in BRAF have been identified via genome sequencing of myeloma patients (Chapman et al. 2011). These too may participate in late progression events, but this has yet to be demonstrated. The oncogene c-myc has long been understood to play a role in myeloma progression typified by secondary IgH-translocations with c-myc and is frequently associated with advanced disease. Myc anomalies are almost never seen MGUS or SMM and only 15% of AMM, unless associated with advanced disease- with ~50% of patients with advanced disease harboring c-myc anomalies (Sawyer 2011). Non-IgH myc translocations have also been noted including with chromosome 1 resulting in t(1;8)(p12;q24) (Sawyer 2011). These represent the most common known recurrent genetic abnormalitics. But the question remains: Are there other mutations to be found that may help us better understand the complex biology of myeloma? The use of genome sequencing and gene expression profiling (GEP) have provided novel tools to answer these questions.

Genome Sequencing

As discussed, genetic events are paramount in myeloma development and progression. With the high through-put genome sequencing, Chapman and colleagues utilized whole genome and exome sequencing of 38 myeloma genomes and compared them to normal plasma cell DNA (Chapman et al. 2011). The authors confirmed the existence of KRAS, NRAS, as well as mutations in TP53. Additionally, Chapman et al. identified mutations in previously unknown genes encompassing defined and undefined cancer related pathways in myeloma. These included RNA processing and protein homeostasis (*DIS3, FAM46C, XBP1, and LRRK2*), histone modification enzymes (*MLL, MLL2, MLL3, UTX, WHSC1, and WHSCH1L1*), NF-κB pathway (*BRTC, CARD11, CYCL:, IKBIP, IKBKB, MAP3K1, MAP3K14, RIPK4, TLR4, TNFRSF1A, and TRAF3*), cyclin D1, and BRAF mutations. The BRAF G469A activating mutation was identified in only 1 of the 38 patients sequenced. The authors subsequently genotyped an 161 additional myeloma specimens for activating mutations in BRAF. These results demonstrated activating mutations in 4% of myeloma patients (Chapman et al. 2011).

Gene Expression Profiling (GEP)

We have thus far discussed specific alterations in the myeloma genome. However, these known and unknown genomic aberrations translate to large-scale alterations in the transcriptome. Using microarray technology a number of groups have characterized myeloma patients by GEP using CD138 selected bone marrow myeloma cells (Zhan et al. 2007). Using a training set of 351 patients and a test set of 200 unsupervised hierarchical clustering identified seven reproducible classes

(Zhan et al. 2007). These molecular profiles were strongly influenced by GEP associated with known genetic lesions: Hyperdiploid (HY), MAF/MAFB (MF), MMSET (MS), cyclin D1/D3 (CD-1), Cyclin D1/D3/CD20 (CD-2), Low Bone disease(LB), and proliferation (PR). These GEP classes were also associated with different risk stratification. More recently, a 70 gene expression profile has been developed to assess risk independently of the know risk factors. Interestingly, this profile is enriched for gene form the short and long arm of chromosome 1, providing further evidence of the importance of this chromosome in myeloma.3 The 70 gene risk score GEP profiles was initially validated in patients receiving total therapy regimens and has since been validated in additional regimens including Bortezomib and high-dose Dexamethasone regimens (Shaughnessy et al. 2007). To this end, the 70 gene risk group has now been made available commercially as MyPRS™ (Myeloma personalized risks score; SignalGenetics) to assist in assessing risk in newly diagnosed and relapsed patients.

Therapeutic Implications

Currently, the utilization of the genetic information outlined above has been limited to prognosis. Therefore, the genetic aspects of myeloma are linked to clinical decision making. Risk assessment remains integral in choosing a patient-centric therapeutic regimens. Currently, risk is defined by metaphase cytogenetics and FISH studies with the potential of GEP in the near future. Based on these measures of the predicted myeloma genetic abnormalities patients have traditionally been divided in the Standard Risk and High Risk categories (Table 1) (Kyle and Rajkumar 2009). Standard risk is defined by the presence of hyperdiploidy, t(6;14), t(11:14), del13 by FISH, and the absence of high risk features. Standard risk patients are generally characterized by increased median overall survival and depending on competing factors including Durie-Salmon Stage, ISS, and age may expect to live a decade. High risk is defined by the presence of hypodiploidy, t(4:14), del13 by metaphase cytogenetics, t(14;16), del17, or t(14;20) (also include proliferative index >3). Chromosome 1 abnormalities have also been linked to high risk (addition of 1q21 and loss of 1p32) (Sawyer 2011). The integration of Lenalidomide and Bortezomib in myeloma therapy has made risk stratification a *moving target* to a degree. Studies looking at these agents have suggested that they may overcome the certain high risk genetic features including del13 and t(4;14) (Avet-Loiseau et al. 2010). The evidence is strongest for Bortezomib containing regimens and less so for Lenalidomide containing regimens.3 Patients with deletion 17p, however, remain a truly high risk population regardless of the therapeutic intervention. The potential of MyPRS in everyday clinical decision making remains in its infancy, but will be quite intriguing. However, questions remain as to the cost: benefit ratio and utility in the context of FISH and other hallmarks of cancer outside the setting of clinical trial patients.

Risk Adapted Therapy

Treatment of myeloma has evolved significantly in the nearly two centuries since the initial characterization of this plasma cell disease (Kyle and Rajkumar 2008). Over the last decade, equipped with traditional chemotherapeutics, novel agents (Lenalidomide and Bortezomib), and high-dose therapy and autologous stem cell transplant (HDT-ASCT), we have observed significant success in impacting overall survival (Kyle and Rajkumar 2009). So how does our increased comprehension of the genetic anomalies impact clinical decision making? One of the goals of myeloma therapy (and all of cancers) is that we can tailor our therapies to individuals. With the information outlined we are making strides in the personalizing myeloma therapy. Although more than risk stratification goes into clinical decision making, it should contribute to the foundation of induction therapy. Treatment plans are based on a number of factors including International Staging System (ISS), Durie-Salmon staging, age, performance status, co-morbidities and presenting end organ damage. Together, we can provide patients a "risk adapted therapeutic plan." We have made great strides in overall survival in myeloma in part do to the discovery and utilization of the novel agents Lenalidomide, Thalidomide, and Bortezomib, as well as high-dose therapy (HDT) and autologous stem cell transplant (ASCT) (Kyle and Rajkumar 2008). Risk adapted therapy is an attempt to utilize all of the available data to make reasonable treatment decisions for patients based risk stratification (Table 1) and transplant eligibility (Fig. 3) (Rajkumar 2011). In part, risk adapted therapy is also beholden to the hypothesis that better overcomes for high risk patients correlate with achieving complete responses (CR); whereas, standard risk patients are not fettered by the same criterion (Rajkumar 2011). As such, the schema outlined in Fig. 3 provides a guideline for therapy in newly diagnosed patients. Patients should initially be divided in to two groups based on their transplant eligibility (I generally err on the side of transplant if you believe that therapy may improve the patient's performance status).

In transplant eligible patients, risk stratification dictates therapy. In patients with high risk disease three drug regimens are recommended (VRd: Bortezomib, Lenalidomide, and Dex or CyBorD: Cyclophosphamide, Bortezomib, and Dex). Three agent combination therapies tend to have higher side effect rates, but also have higher rates of CR and VGPR relative to two agent combinations (Rajkumar 2011). Bortezomib backbones are preferred in high risk individuals do to data suggesting that proteosome inhibition at least partially overcome the high risk genetic lesions t(4;14) and del13.3 Furthermore, consider VRD for all patients with del17 and CyBorD for patients with other high risk features. Generally, 4–6 cycles of three agent therapy is recommended prior to stem cell harvest and HDT-ASCT. Maintenance therapy with either Bortezomib or Lenalidomide should be considered following a discussion about the potential risks and benefits of maintenance therapy. On the other hand, standard risk patients may be treated with a two agent combination (either Rd: Lenalidomide and low-dose Dex or Bortezomib and Dex). If Rd is used, 4 cycles is the optimal number of cycles to utilize prior to transplant to maximize stem cells yield. Standard risk patients may consider continuing with therapy until relapsed prior to HDT-ASCT or proceed to greater than 4 cycles (best response). Maintenance therapy with lenalidomide can also be considered in these patients.

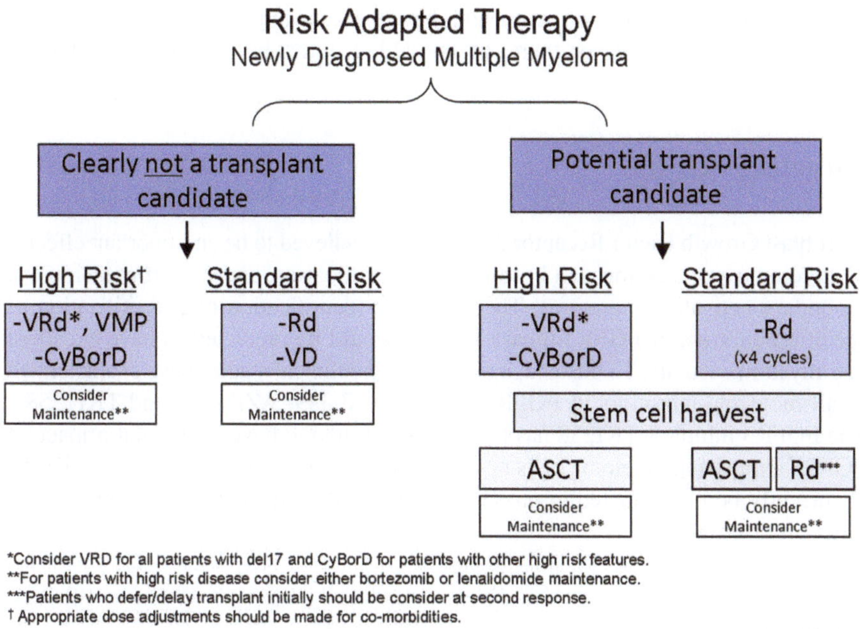

Fig. 3 Risk adapted therapy for newly diagnosed multiple myeloma. This schematic outlines a general approach towards patients with newly diagnosed myeloma based on (1) transplant eligibility and (2) risk stratification (Modified from Rajkumar (2011). VRd: Bortezomib, Lenalidomide, and Dexamethasone; VMP: Bortezomib, Melphalan, and Prednisone; CyBorD: Cyclophosphamide, Bortezomib, Dex; Rd: Lenalidomide and low-dose Dex; ASCT: autologous cell transplant (with high-dose therapy). Risk per Table 1)

In transplant ineligible patients risk also dictates therapy. However, in these cases transplant it is not an option. To this end, one may consider three agent therapies containing melphalan. Bortezomib containing regimens should still be considered for high risk patients. These may include VRd, VTD (Bortezomib, Thalidomide, and Dex), VMP (Bortezomib, Melphalan, and Prednisone) or CyBorD. Dose reductions should be utilized as appropriate based on pre-existing co-morbidities (Palumbo and Gay 2010). In standard risk transplant ineligible patients two agent combinations are recommended (either Rd: Lenalidomide and low-dose Dex or Bortezomib and Dex). Maintenance or continued therapy is recommended for patients tolerating therapy well.

Novel Therapies in Myeloma Targeting the Genome

We have made great strides over the last one to two decades in modifying disease. However, myeloma remains a mortal disease. Are there therapeutic targets myeloma equivalent to the BCR-Abl translocation in Chronic myelogenous leukemia (CML)? Although we have discussed an number of possible targets including RAS, FGFR3,

TRAF3, and BRAF, no known compounds have the same impact as the tyrosine kinase inhibitors in CML. However, there are a notable few that have clinical potential.

Targeting *FGFR3*

Fibroblast Growth Factor Receptor 3 (FGFR3) is believed to be an important effector of myeloma progression as (1) initiating translocation and (2) progression via secondary activating mutations. Therefore, the identification of and utilization of specific inhibitors of FGFR3 kinase activity would be predicted to have excellent anti-myeloma activity. To this end, a number of preclinical studies have characterized small molecule inhibitors of FGFR3 including NF449, AZD1480 and TKI1258.3 The FGFR3 inhibitor TKI258 has been demonstrated to have significant impact on FGFR3 expressing myeloma cells *in vitro* and in xenograft murine models. TKI258 is currently being evaluated in phase I and phase II clinical trials in myeloma and other disease states.3

Targeting *RAS*

Although this group of drugs has not been as beneficial as we had hoped, farnesyl-transferases inhibitors (FTIs) are group of compounds that were initially designed to inhibit the functional localization of Ras to the plasma membrane by altering the C-terminal prenylation of Ras and other GTPases. To this end, the use of FTIs would be able to preferentially treat patients with KRAS or NRAS mutations. However, although preclinical work was intriguing, the clinical translation was less then desired (Yanamandra et al. 2006; David et al. 2010; Armand et al. 2007; Alsina 2004). Tipifarnib remains a agent of interest in phase II trials with Bortezomib (NCT00972712).

Targeting *MYC*

As a master regulator cell proliferation, MYC plays an important role in development of many cancers. For years, MYC has been postulated to contribute as a late event in the progression of myeloma to advanced disease (Sawyer 2011). This is hypothesis is supported by the fact that MYC anomalies are rare in MGUS and SMM. Moreover, MYC aberrations are observed in only 15% of active myeloma patients. However, in advanced disease MYC anomalies can be seen in as many as 50% of patients. Furthermore, almost 90% of myeloma cells carry MYC mutations, suggesting that this is late event associated with aggressive disease. In addition to the known genetic alterations in MYC, additional GEP studies have identified MYC associated pathway activation can be observed in 60% of patient derived myeloma cells (Chng et al. 2011). These and other studies suggest that MYC represents a potential unifying mediated of myeloma disease progression (Chesi et al. 2008; Chng et al. 2011). Yet until recently a mechanism to target MYC related pathways had remained elusive. A relationship between MYC transcriptional pathways, histone acetylation templates, and the recruitment of transcriptional activators by the bromodomain and extraterminal (BET) family of proteins, specifically BRD4 has been demonstrated (Rahman et al. 2011; Dhalluin et al. 1999). Importantly, the small molecule inhibitor of BET bromodomains (BRD), JQ1, was recently characterized and demonstrated to pharmacologically modulate c-Myc transcriptional activity in multiple myeloma via an inhibition of chromatin dependent signaling by BRD proteins (Delmore et al. 2011). Delmore and colleagues went on to demonstrate that the global downregulation of c-Myc transcription was associated with cell cycle arrest and cellular senescence in *in vitro* and in *in vivo* models of myeloma (Delmore et al. 2011). Although the clinical benefits of this mode of therapy have not been established, these data provide important rationale for JQ1 or similar agents to target master regulators of myeloma oncogenesis.

Non-genetic Contributions to Myeloma

It is also important to recognize that although myeloma is a genetically based disease the bone marrow microenvironment plays a significant role in progression and therapeutic success. Recent clinical trials incorporating the novel compounds thalidomide, lenalidomide and bortezomib have provided a rationale for overcoming drug resistance (Gertz 2007; Mitsiades et al. 2007). The clinical success of these agents is attributable to both direct cytotoxicity and to negative regulation of pro-myeloma determinants of the microenvironment. Unfortunately, even with the clinical gains made from these agents, resistance to these compounds remains inevitable; as although highly effective, disease relapse remains unavoidable. This aspect of myeloma has fostered a great deal of research into delineating the mechanisms by which MM cells initially evade regimens of chemotherapy.

Microenvironmental Contributions to Multiple Myeloma

Within the backdrop of a genetic disease, it is appreciated that *both* genetic alterations and the microenvironmental effectors contribute to the deregulation of life and death of MM cells (Shain and Dalton 2009). The bone marrow niche provides a sanctuary to resident MM cells via both soluble and physical components (interactions between MM cells and specific extracellular matrix (ECM) components or other cellular components *e.g.* bone marrow stromal cells (BMSCs)). Soluble and physical determinants of the bone marrow are central in homing to and survival within the bone marrow microenvironment (Shain et al. 2009). Although discussed and examined as separate entities, resistance to therapy likely involves the "collaboration" between dynamic, *de novo* resistance modulated by the bone marrow microenvironment and heritable, acquired mechanisms of drug resistance selected by drug exposure. The former providing and initial protective effects and contributing to MRD and the later selected under chronic exposure to therapy ultimately resulting in the expansion of a MDR population (relapse). Based on this hypothesis elucidation of the environmental determinants that afford a protective advantage to resident myeloma cells, we may be able to overcome *de novo* and acquired MDR.

The Bone Marrow Niche

Normal plasma cell development culminates in homing of post-germinal centre (GC), long-lived, plasma cells to survival niches within the bone marrow microenvironment (Shapiro-Shelef et al 2005; Klein and Dalla-Favera 2008). Plasma cells are terminally differentiated, antibody producing, B lymphocytes that have undergone class-switch recombination, somatic hypermutation. Maturation of these cells has involved migration from the bone marrow, to secondary lymphoid organs (spleen or lymph nodes), and back to the bone marrow (arguments can be made that the bone marrow is also a secondary lymphoid organ and some plasma may not leave the marrow). Post-GC bone marrow homing correlates with the expression of BLIMP (B cell lymphocyte-induced maturation protein)-1 in secondary lymphoid organs modulating the repression of CXCR5 (CXC-chemokine receptor5) and the expression of CXCR4 and α4-integrins (Shapiro-Shelef et al. 2004; Shaffer et al. 2004, Sciammas et al. 2004). CXCR4 modulates migration to CXCL12 (CXC-ligand-12/stromal cell derived factor (SDF)-1)- rich regions of bone marrow niche rich and expression α4 integrins (Shapiro-Shelef et al. 2005; Azab et al. 2009; Azab et al. 2009). In turn, α4-containing integrin heterodimers facilitate adhesion to VCAM (vascular cell adhesion molecule)-1 as well as other homo- and hetero-typic adhesion molecules. Within the bone marrow plasma cell adherence to bone marrow stromal cells (BMSCs) and likely other juxtaposed cells leads to the production of IL (interleukin)-6, BAFF (B cell activating factor), bFGF (basic-fibroblast growth factor), and other soluble factors that provide crucial survival factors to these long-lived antibody producing cells concordant with bone marrow homeostasis.

It is not difficult to see how the pro-survival and proliferative effectors within the bone marrow microenvironment would also benefit the progression of a malignancy, if not its evolution. Multiple myeloma is a malignancy of clonal antibody producing plasma cells facilitated by errors in class-switch recombination and/or somatic hypermutation (Kyle and Rajkumar 2008). Like their normal counterparts, these mature B cells have migrated to the bone marrow via various chemo-attractants and adhesion molecules (Azab et al. 2009b). Unlike their normal counterparts, myeloma cells not only benefit from the normal effectors within the bone marrow niche, but also *hijack* and *contribute* to the environment in a manner to that promotes tumorigenesis, altered bone metabolism, neovascularisation, and drug resistance

(Kline et al. 2007; Pilarski et al. 2010). Myeloma cell adhesion induces bone marrow stroma cell secretion of chemokines, cytokines, and growth factors (Shain and Dalton 2009). IL-6 remains one of the most prominent myeloma growth factors; however, a growing list of soluble effectors are also induced by myeloma cell-stromal cell interactions. These factors include cytokines: IL-1β, IL-3, IL-15, IL-21, TNF-α, OSM, LIF; chemokines: IL-8, CXCL12, MIP1-α; growth factors: VEGF (vascular endothelial growth factor), FGF (family), IGF-1 (insulin-like GF-1), HGF (hepatocyte GF); and other pro-myeloma factors: Dkk-1 (Dickkopf-1), RANKL (Receptor activator of nuclear factor kappaB (NF-κB) ligand), BAFF (B cell activation growth factor), Wnt (family), and TGF-β (transforming GF) among others (Shain and Dalton 2009). In addition to direct proliferative and survival effects soluble factors also impart the adhesion of myeloma cells to adjacent cellular components (stromal cells, dendritic cells, mesenchymal stem cells (MSC), macrophages, and osteoclasts) and extracellular matrices (fibronectin, collagens, vitronectin, glucose-aminoglycans (GAG), hyaluronan, and laminins) via a host of adhesion molecules. The most prominent being VLA-4 (α4/β1/CD49d/CD29), VLA-5 (α5/β1/CD49e/CD29), CD44, VCAM-1, ICAM (CD54), NCAM (CD56), CD74, and CS-1 (CD2-subset-1) among others (Shain and Dalton 2009). Extracellular stimulation by soluble factors or direct adhesion facilitates networks of intracellular cascades. For most of the soluble and physical determinants listed above a number of signalling cascades have been delineated that translated these extracellular stimuli to biology function. Importantly, these pathways are typically investigated and discussed in a vacuum; however, it is evident that collaborative signalling alters that biologic outcome (Shain et al. 2009; Ishikawa 2006; McMillin 2010). To this end, it is important to account the network of effectors within the myeloma microenvironment when attempting to identify the most relevant targetable pathways.

It is important to remark that these myeloma cells have bearing on bone metabolism and angiogenesis within the bone marrow niche. Osteolytic disease is one of the more significant hallmarks of multiple myeloma occurring through alterations in osteoclast to osteoblast ratios and subsequent increase in focal bone catabolism (Yaccoby 2010). This is too mediated by the altered signalling between soluble and adherent stimuli between myeloma cells and osteoclasts, dendritic cells, activated T cells, and bone marrow stromal cells (Yaccoby 2010). Myeloma cell production of RANKL, TGF-β, HGF, IL-3, and Dkk-1 inhibit the growth of osteoblasts. Simultaneously, RANKL, SDF-1, MIP1-α, and TNF-α stimulate bone resorption through the positive regulation of osteoclasts (Yaccoby 2010; Roodman 2010). Together these effects result in the net resorption of bone in focal regions- lytic bone lesions. Focal lesions are sites of a number of heterotypic adhesive, paracrine, autocrine signalling between myeloma cells and adjacent cellularity. These lesions are likely relevant in drug resistance as these focal regions are potential sites for MRD. Evidence of this is suggested in studies demonstrating decreased EFS and OS in patients in CR with identifiable focal lesions on whole body-MRI or PET-CTs relative to patients with negative imaging (Bartel et al. 2009, Walker 2007).

A number of studies have implicated that increased microvessel density (MVD) correlates with disease state, suggesting that increased bone marrow angiogenesis is important to myeloma progression. This theorem originates from observations of the increased expression of proangiogenic factors including, VEGF, bFGF, angiopoietin (ang)-1, and ang-2 myeloma patients (Du et al. 2004). The most compelling evidence stems the examination of bone marrow MVD that revealed a greater density in patients with active disease relative to patients with MGUS (Du et al. 2004). Du et al. demonstrated that MVD increased with from normal bone marrow samples (normal stem cell donors) to MGUS to active disease (Du et al. 2004). The authors further demonstrated that treatment with thalidomide reduced MVD in 3 of 7 patients. Other evidence demonstrated that elevated MVD was a univariant predicted poorer overall survival in myeloma patients (Markovic et al. 2008). Studies characterizing the *multiple myeloma cancer stem cell* have suggested that two lineages are populated by these progenitor cells, (1) the prototypical *myeloma plasma cell* and (2) a *myeloma monocytoid cell* (Pilarski et al. 2010). With the use of patient specific FISH, further investigation demonstrated that the monocytoid lineage populated a *myeloma vascular endothelial cell* lineage (Pilarski et al. 2010). The authors suggest that not only do myeloma cells produced factors that promote angiogenesis, but may, in fact, directly contribute to increased MVD, nutrient delivery, and plasma cell homing to the marrow. Together, these data demonstrate that incredible influence the bone marrow microenvironment may have on myeloma pathobiology.

The Bone Marrow Niche and Drug Resistance

The unique dependence of myeloma on its niche has made this neoplasm a model for defining the influence of the tumor microenvironment network on disease. The cytokines, chemokines, growth factors, and adhesive matrices of the BM microenvironment facilitate MM cell homing and expansion (Shain and Dalton 2009). The biological mechanisms regulating MM cell homing to the BM involves a the regulation of soluble and physical determinants of the microenvironment. In turn, resident myeloma cells are afforded sanctuary from host immune surveillance as well as therapeutic agents. Therefore, elucidation of the mechanisms modulating homing of MM to the BM and *de novo* drug resistance will facilitate the rational development of novel pharmacological agents and/or regimens in MM by targeting MRD. Appreciation of this concept has spawned an arm of preclinical and clinical studies with the sole target of the bone marrow microenvironment.

Soluble Factors

Early observations of increased levels of cytokines and chemokines within the BM microenvironment of MM patients has long suggested a contribution of IL-6 and other soluble factors in MM pathogenesis. A number of these soluble factors, including interleukin (IL)-6, insulin-like growth factor (IGF)-1, interferon (IFN)-α, and fibroblast growth factor (FGF)-3 have subsequently been shown to confer resistance to cytotoxic insult (Pollet et al. 2002; Hideshima et al. 2007; Yasui et al. 2006). Other soluble factors are paramount in the homing to and maintenance of adherent myeloma cells within the BM. The chemokine SDF-1/CXCL12 is a critical regulator of myeloma-bone marrow niche via binding to CXC chemokine receptor (CXCR) -4/CD184 and the more recently identified CXCR7 (Alsayed et al. 2007; Burger and Kipps 2006; Burns et al. 2006). Signaling following CXCR4 (or CXCR7) ligation has been shown to activate JAK/STAT signalling, but primarily involves heterotrimeric G-proteins (Gα/Gβ/Gγ) (Teicher and Fricker 2010). Ligand-receptor binding facilitates the release of the trimeric protein and activation of PI3-kinase, Rac, Rho, Ras/Raf, and phospholipase C (PLC) (Azab et al. 2009a; Teicher and Fricker 2010). The SDF-1/CXCL12 and CXCR4 axis modulates adhesion and chemotaxis interactions with VCAM-1 and fibronectin (Sanz-Rodriguez et al. 2001). SDF-1 has additionally been demonstrated to promote cell survival (Lu et al. 2009).

Of the soluble determinants of the BM microenvironment, IL-6 remains the prototypical growth and survival factor in myeloma. Moreover, numerous additional factors contribute to myeloma by either directly or indirectly promoting the production/secretion of IL-6 by BM stromal cells (BMSCs), mesenchymal stem cells, osteoclasts, and malignant cells. TNF-α, IL-1β, FGF-3 TGF-β, and VEGF have been shown to modulated IL-6 expression in the context of the bone marrow niche (Mitsiades et al. 2007). These results highlight the prominent role of IL-6 in MM and propagated research characterizing the biology of IL-6 signaling. As such, IL-6 will be used an example of soluble factor signaling. Signal transduction follows binding of IL-6 to gp80/IL-6Ralpha and recruitment of gp130/CD130 (Heinrich et al. 2003; Li 2008). The induced dimerization of gp130 results in receptor phosphorylation by constitutively bound Janus kinase (JAK) family tyrosine kinases (JAK1, JAK2, and Tyk2) (Li 2008). IL-6 binding and receptor multimerization facilitates the phosphorylation of gp130 and initiation of three major signaling pathways in MM cells: (1) the Ras/Raf–mitogen-activated protein kinase kinase–extracellular signal-related kinase (ERK)1/2 pathway; (2) the phosphatidylinositol (PI) 3-kinase pathway; (3) the Janus kinase (JAK)/signal transducer and activator of transcription pathway. Src-family tyrosine kinases (SFKs) have also been identified as mediators of IL-6 signaling events (Zoubeidi et al. 2009). Whether these kinases (SFKs) act to enhance STAT3, PI3-kinase, and Ras signaling, or represent an independent pathway has yet to be completely determined. SFKs-dependent enhancement of STAT3 activation and proliferative signaling following IL-6 cross-linking has been observed (Zoubeidi et al. 2009). Each of these signal transduction pathways has been implicated in IL-6-mediated resistance to both physiological and chemotherapy-mediated apoptosis. The variety of downstream targets demonstrates that a single soluble factor has the potential to regulate drug sensitivity through multiple mechanisms. However, simultaneous targeting complementary signaling pathways will likely be necessary for improved clinical efficacy.

Direct Cell Contact

The physical components of the bone marrow niche have been demonstrated to play a prominent role in MM pathogenesis and MDR. Moreover, data suggests that the physical elements of the microenvironment evolve with progression. Alterations in expression of the ECM components fibronectin, collagen I and collagen IV have been shown to correlate with disease state (normal, MGUS vs. MM); (Tancred et al. 2009) thereby, providing further evidence that MM and the bone marrow microenvironment are intimately linked. Interactions between cells and their physical environment are mediated by the cell adhesion molecule (CAM) super-family. CAMs promote communication between the physical environment, cellular architecture, and intracellular signaling cascades. These cell surface molecules include: the Ig (immunoglobulin) family, cadherins, selectins, hyaluronate receptors, receptor tyrosine phosphatases, and integrins. These transmembrane receptors coordinate homing, lodging, and differentiation of MM cells in the marrow niche through specific homotypic and heterotypic interactions with environmental ligands (Shain and Dalton 2009). As with soluble factors associated with MM, interactions between MM cells and extracellular matrix (ECM) components or adjacent cells of the BM also confer *de novo* MDR (Shain and Dalton, McMillin et al. 2010). Although integrins will be primarily discussed below, a numerous other adhesion molecules involved in cell adhesion mediated drug resistance (CAM DR) including hyaluronan, LFA (lymphocyte associated function antigen –1), Notch-1/Jagged, Wnt/RhoA and MMSET gene products (Lauring et al. 2008; Ohwada et al. 2009; Nefedova et al. 2008). Moreover, these interactions are not limited to cell: ECM interactions. BMSCs, osteoclasts, bone marrow dendritic cells, vascular endothelial cells, and likely immune effectors also support myeloma drug resistance through direct contact (Shain and Dalton 2009; Nefedova et al. 2008).

When speaking about myeloma cell adhesion to ECM or adjacent cellular marrow components it is important to note that at least two levels of response can be separated temporally and by mechanism of activation. An *early phase*, characterized rapid dynamic biologic changes and a *delayed phase*, characterized by secondary signaling and transcriptional events (Meads et al. 2009). Within the early phase post-translational events result intracellular redistribution, degradation, and increased stability of critical proteins. What is unique about these early events is that they are reversible; within minutes of disassociating from the adherent matrix myeloma cells return to a drug sensitive phenotype. The rapid reversal of these events is important when considering adhesion as a target for therapy. As such, only transient alterations in the adhesive phenomenon may be necessary to sensitize myeloma cells to therapy. The delayed events are characterized by transcriptional activity and the production and secretion of soluble effectors such as IL-6, MIP-1α, VEGF, and others . In turn, these effectors participate in apparent feed-forward paracrine signaling between MM cells and the surrounding cellular milieu facilitating indirect EM-DR.

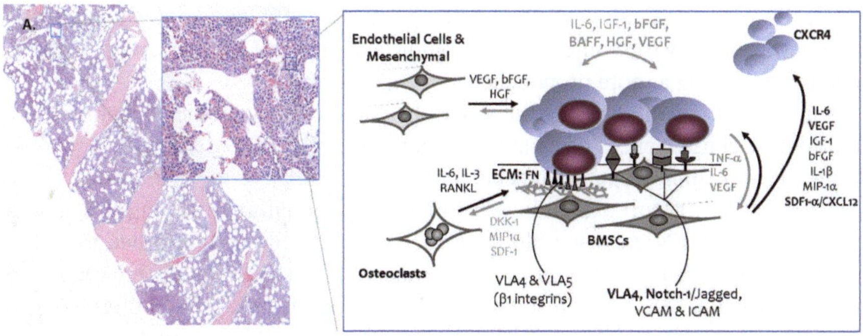

Fig. 4 The bone marrow niche is composed of a network of extracellular determinants. The BM microenvironment plays a key role in the disease pathology of MM. It is rich in cytokines, chemokines, growth factors, and adhesive matrices that facilitate MM cell homing, proliferation and survival. The protective determinants of the BM consist of both soluble factors and physical factors. These factors include cytokines: IL-1β, IL-3, IL-15, IL-21, TNF-α, OSM, LIF; chemokines: IL-8, CXCL12, MIP1-α; growth factors: VEGF (vascular endothelial growth factor), FGF (family), IGF-1 (insulin-like GF-1), HGF (hepatocyte GF); and other pro-myeloma factors: Dkk-1 (Dickkopf-1), RANKL (Receptor activator of nuclear factor kappaB (NF-κB) ligand), BAFF (B cell activation growth factor), Wnt (family), and TGF-β (transforming GF) among others.3 In addition to direct proliferative and survival effects soluble factors also impart the adhesion of myeloma cells to adjacent cellular components (stromal cells, dendritic cells, mesenchymal stem cells (MSC), macrophages, and osteoclasts) and extracellular matrices (fibronectin, collagens, vitronectin, glucose-aminoglycans (GAG), and laminins) via a host of adhesion molecules. The most prominent being VLA-4 (α4/β1/CD49d/CD29), VLA-5 (α5/β1/CD49e/CD29), CD44, VCAM-1, ICAM (CD54), NCAM (CD56), CD74, and CS-1 (CD2-subset-1) among others.3 Extracellular stimulation by soluble factors or direct adhesion facilitates networks of intracellular cascades

A Complex Network of Survival Signals

Soluble and physical determinants of the bone marrow microenvironment independently confer the EM-DR phenotype. However, logic dictates that these soluble and physical environmental effectors function in concert. This network is involves autocrine/paracrine signaling as well as regulatory events mediated by adhesion of myeloma cells to BMSCs, osteoclasts, osteoblasts, mesenchymal stem cells, immune effectors, and endothelial cells (Fig. 4) (Shain and Dalton 2009). These cellular interactions are important catalysts for the production of the numerous soluble factors (Shain and Dalton 2009). However, it is apparent that the maximal effect of EM-DR is not mediated by soluble or physical effectors alone. Instead, these factors cooperate in conferring EM-DR. Evidence for this has been provided in studies examining the anti-apoptotic nature of specific co-culturing conditions between myeloma cell and BMSCs (Nefedova et al. 2004). In these studies, the degree of chemotherapy-resistance was dependent upon the co-culture condition examined. Myeloma cells co-cultured in contact with BMSCs were protected to greater degree than cells co-cultured without contact. These results demonstrate the existence of at least two environmental networks elicited by interactions between

MM cells and BMSCs: a network involving anti-apoptotic paracrine signaling and a network involving the conjunction of soluble and physical effectors, with the combination providing a more pronounced survival advantage.

An explanation of these findings may be suggested by studies examining the intracellular consequences of co-stimulation (Shain et al. 2009). These data demonstrate that signaling events are specifically enhance relative to those elicited by a single effector. A report examining the intracellular signaling following stimulation with IL-6 alone, FN adhesion alone, or their combination demonstrated the complexity of converging extracellular determinants. The collaboration between IL-6 and FN adhesion resulted in a selective enhancement of JAK/STAT3 phosphorylation and activity, but not Akt or ERK1/2 phosphorylation. Co-stimulated cells were afforded a both a proliferative and MDR advantage. Further, this collaboration paralleled increased gp130 complex phosphorylation via a novel IL-6 independent pre-association of STAT3 (unphosphorylated) with gp130 when cells are adhered to FN. Together, these results suggest that examination of crosstalk between of intracellular signaling networks may identify specific components may contribute more (or less) significantly to MM therapy resistance and proliferation (*i.e.* more *in vivo*-like conditions; where cells are not modulated by a single determinant). To this end, identification of prominent signaling molecules under co-stimulatory conditions may direct us towards more appropriate drug targets.

Therapeutic Implications

As discussed above, EM-DR and MRD are prominent hurdles in the treatment failure. To this end, successful treatment may involve correctly identifying modalities that target the myeloma microenvironment and/or attendant signaling cascades. The novel agents bortezomib, thalidomide, and lenalidomide have had a significant impact on the treatment of this disease (Richardson et al. 2007). These therapeutic compounds have been demonstrated to function, at least in part, via a modulation of the microenvironment. Bortezomib is a direct inhibitor of the 26S proteosome and has number of *indirect* effects on the microenvironment through the down regulation of key paracrine effectors. Proteosome inhibition has been shown to overcome EM-DR in preclinical studies with cell lines and patient samples (Richardson et al. 2007). Bortezomib regulates the expression of numerous cytokines and growth factors. Bortezomib has been shown to inhibit angiogenesis via a modulation of VEGF and bFGF, as well as inhibit cellular adhesion via deregulation of vascular CAM (VCAM) and intracellular CAM (ICAM) (Richardson et al. 2007). Gene expression profiling of myeloma cells treated with bortezomib demonstrated decreased production of α4 integrin/CD49d consistent with a role in targeting EM-DR (Mitsiades et al. 2002). Thalidomide and the immune modulatory compound lenalidomide also have anti-EM-DR properties. The proposed action of thalidomide primarily involves the down regulation of VEGF and associated angiogenic factors (Richardson et al.

2007). Recent phase II studies using the VEGFR inhibitor pazopanib suggests that VEGF antagonism may not be sufficient as single agent therapy (Prince et al. 2009). Additionally, lenalidomide also has significant clinical benefit when coupled with dexamethasone (Richardson et al. 2007), likely secondary to its proposed activity of fortifying anti-tumor immune system. However, as with bortezomib and thalidomide, lenalidomide also demonstrates multi-targeted effects on the BM microenvironment (cytokine and growth factor expression, inhibition of angiogenesis and inhibition of cellular adhesion). Unfortunately, even with the wide spread utilization of these therapeutics myeloma remains incurable. As such, continued investigation into therapies to compliment their activity is required.

With our greater appreciation for the role of EM-DR, a large number of agents have been designed to targeting determinants of the bone marrow niche. Small molecule inhibitors of IL-6 and IL-6-dependent signaling have been identified. Early studies with Sant7 demonstrated positive preclinical results. More recently, monoclonal antibody therapy with the chimeric monoclonal antibody (CNTO 328) specific to IL-6 has shown preclinical success in myeloma cell lines and patient samples. A phase II study in previously treated patients demonstrated a 57% overall response rate following therapy with the combination of CNTO 328 and bortezomib suggesting that this anti-IL-6 chimeric immunoglobulin may have anti-myeloma activity (Rossi et al. 2008).

Drugs targeting other extracellular determinants are also being evaluated. Initially examined in the setting of HIV then stem cell harvesting, the CXCR4 inhibitor AMD3100/Plerixafor has been shown to attenuate myeloma-BMSC communication. These affects translated to enhance cytotoxicity of bortezomib (Devine et al. 2004; Alsayed et al. 2007; Azab et al. 2009b). AMD3100 acts to interfere with homing of cells to the bone marrow, but more importantly its effects may be to facilitate a deadhesion phenomenon; thereby attenuating the resistance conferred to resident myeloma cells by the physical microenvironment. AMD3100 has also been examined in other hematologic and non-hematologic malignancies (Wong and Korz 2008). A humanized monoclonal antibody to CXCR4 is also being evaluated in clinical and preclinical studies (BMS983564). The FGFR and VEGFR *specific* receptor tyrosine kinase inhibitor (BIBF100) was examined in MM cell lines and patient samples (Bisping et al. 2009). The cytotoxicity of bortezomib and dexamethasone was enhanced when combined with BIBF100 in poor risk patient samples (and cell lines) carrying t(4:14) (FGFR3/MMSET; IgH translocations) and t(14:16) (c-maf; IgH translocations-associated with c-maf -mediated VEGF signaling) (Bisping et al. 2009). Sorafenib, a putative VEGFR tyrosine kinase inhibitor has shown preclinical promise in myeloma (Ramakrishnan et al. 2010).

The physical microenvironment is also being targeted with novel therapeutics. The design of compounds to target $\beta 1$ integrins or other mediators of adhesion between tumor cells and the physical environment may provide therapeutic benefit. In AML VLA-4 ($\alpha 4\beta 1$) specific antibodies were shown to potentiate the anti-tumor effects of cytarabine *in vivo* by reducing MRD (Matsunaga et al. 2003). Volociximab, a chimeric monoclonal antibody to VLA-5 ($\alpha 5\beta 1$) was tested in solid tumors demonstrating mild clinical benefit and was well tolerated (Ricart et al. 2008). These early reports indicate that integrin-direct antibody therapy may be an important adjunct in targeting MRD. Integrin- ECM interactions involve the recognition of a specific three amino acid sequence, RGD (arginine, glycine, and aspartate). Linear peptides, cyclic peptides and peptidomimetics have been designed to mimic this peptide sequence to disrupt integrin-mediated adhesion (Kimura et al. 2009). Recently, a new group of RGD-blocking compounds has been identified. The cystine knot peptides or *knottins* have been engineered demonstrating nanomolar affinity for multiple integrin heterodimers (Kimura et al. 2009). Another example is the decapeptide HYD1 synthesized based on the RGD amino acid motif, has been shown to attenuate VLA-4 integrin- mediated adhesion to fibronectin in multiple cancer cell models. Interestingly, incubation of myeloma cell lines and patient samples with HYD1 not only abrogated adhesion to ECM, but HYD1 induced myeloma cell necrosis (caspase independent death) (Nair et al. 2009). These results demonstrate that inhibition of integrin-mediated adhesion may have direct (or indirect) anti-myeloma effects and reveal the potential of therapies designed to target the physical microenvironment in cancer therapy.

In the complex milieu of the bone marrow cell signaling cascades are also potential therapeutic targets (Fig. 4). Pyridone 6, a reversible ATPase inhibitor blocks JAK activity arresting growth of MM cells and patient samples with constitutive JAK/STAT3 activity (Pedranzini et al. 2006). Additional JAK inhibitors including INCB20, AZD1480, and INCB16562 have been shown to attenuate MM cell growth in response to IL-6 and BMSCs3 PI3-kinase, Akt, and mTOR are also promising targets for therapeutic intervention (Hoang et al. 2010; Richardson 2011; Podar et al. 2009). Ras/Raf-MEK-ERK1/2 signaling, especially MEK, is another pathway with a growing number of promising inhibitory compounds with anti-EM-DR properties (Annunziata et al. 2011; Popovic and Licht 2011). Lastly, SFK inhibitors are also being examined in MM. Preclinical data with the tyrosine kinase inhibitor dasatinib demonstrated growth arrest and synergism with conventional and novel therapeutics in MM cell lines (Coluccia et al. 2008). By no means is this a complete listing of novel anti-EM-DR compounds, but it demonstrates the principles behind drug design in the context of the microenvironment.

The inhibitors discussed above target both extracellular and intracellular determinants of the BM niche. The hope is that we can identify the appropriate factors within the complex network of the tumor cell microenvironment to target. To this end, we may divine therapies to overcome the coordinated effort between MM cells and the microenvironment. In so doing, we may be able interrupt the sequence of events (*de novo* and acquired) facilitating MRD culminating in therapy resistance. However, it is important to note that with the significant heterogeneity of signaling factors and transduction pathways within the bone marrow niche, we will need to design combination therapies with targeted agents. To this end, targeting of multiple pathways either simultaneously or in sequence may be that only measure by which to overcome the sanctuary of the bone marrow milieu.

Conclusion

As will most cancers, myeloma tumorigensis is a defined by both genetic transforming events and the growth and survival factors provided by the bone marrow microenvironment. Within this chapter we have discussed these two major forces affecting disease pathogenesis and progression. The genetic and environmental characteristics of myeloma are relatively well established And with increasing technology and understanding of myeloma (and bone marrow niche) biology we will continue to push for mechanisms to control, if not cure, this mortal disease. For the first time in decades we have novel therapeutic modalities at our disposal with significant clinical benefit (Thalidomide, Lenalidomide, and Bortezomib) due to the multi-targeted activities. As discussed, the ultimately refractory nature of this malignancy stems from the protective nature of bone marrow microenvironment and the underlying (and likely changing) genetic profile of the plasma cell clone(s). With whole genome sequencing, SNP analysis, GEP, and soon proteomics and phosphoproteomics we will continue to identify key modulators of myeloma biology and therapy resistance. In turn, this will facilitate improved patient specific or personalized care.

References

Alsayed Y, Ngo H, Runnels J, Leleu X, Singha UK, Pitsillides CM et al (2007) Mechanisms of regulation of CXCR4/SDF-1 (CXCL12)-dependent migration and homing in multiple myeloma Blood 109(7):2708-2717

Alsina M, Fonseca R, Wilson EF, Belle AN, Gerbino E, Price-Troska T, Overton RM, Ahmann G, Bruzek LM, Adjei AA, Kaufmann SH, Wright JJ, Sullivan D, Djulbegovic B, Cantor AB, Greipp PR, Dalton WS, Sebti SM (2004) Farnesyltransferase inhibitor tipifarnib is well tolerated, induces stabilization of disease, and inhibits farnesylation and oncogenic/tumor survival pathways in patients with advanced multiple myeloma. Blood 103(9):3271–3277 [Epub 2004 Jan 15]

Annunziata CM, Hernandez L, Davis RE, Zingone A, Lamy L, Lam LT et al (2011) A mechanistic rationale for MEK inhibitor therapy in myeloma based on blockade of MAF oncogene expression. Blood 117(8):2396–2404

Armand JP, Burnett AK, Drach J, Harousseau JL, Löwenberg B, San Miguel J (2007) The emerging role of targeted therapy for hematologic malignancies: update on bortezomib and tipifarnib. Oncologist 12(3):281–290. Review

Avet-Loiseau et al (2010) Bortezomib plus dexamethasone induction improves outcome of patients with t(4;14) myeloma but not outcome of patients with del(17p). Clin Oncol 28(30):4630–4634 [Epub 2010 Jul 19]

Azab AK, Azab F, Blotta S, Pitsillides CM, Thompson B, Runnels JM et al (2009) Rho-A and Rac-1 GTPases play major and differential roles in SDF1-induced cell adhesion and chemotaxis in multiple myeloma. Blood 114(3):619–629

Azab AK, Runnels JM, Pitsillides C, Moreau AS, Azab F, Leleu X et al (2009) CXCR4 inhibitor AMD3100 disrupts the interaction of multiple myeloma cells with the bone marrow microenvironment and enhances their sensitivity to therapy. Blood 113(18):4341–4351

Bartel TB, Haessler J, Brown TL, Shaughnessy JD Jr, van Rhee F, Anaissie E et al (2009) F18-fluorodeoxyglucose positron emission tomography in the context of other imaging techniques and prognostic factors in multiple myeloma. Blood 114(10):2068–2076

Bida JP, Kyle RA, Therneau TM, Melton LJ 3rd, Plevak MF, Larson DR et al (2009) Disease associations with monoclonal gammopathy of undetermined significance: a population-based study of 17,398 patients. Mayo Clin Proc 84(8):685–693

Bisping G, Wenning D, Kropff M, Gustavus D, Muller-Tidow C, Stelljes M et al (2009) Bortezomib, dexamethasone, and fibroblast growth factor receptor 3-specific tyrosine kinase inhibitor in t(4;14) myeloma. Clin Cancer Res 15(2):520–531

Burger JA, Kipps TJ (2006) CXCR4: a key receptor in the crosstalk between tumor cells and their microenvironment. Blood 107(5):1761–1767

Blade J, Dimopoulos M, Rosinol L, Rajkumar SV, Kyle RA (2010) Smoldering (asymptomatic) multiple myeloma: current diagnostic criteria, new predictors of outcome, and follow-up recommendations. J Clin Oncol 28(4):690–697

Burns JM, Summers BC, Wang Y, Melikian A, Berahovich R, Miao Z, et al (2006) A novel chemokine receptor for SDF-1 and I-TAC involved in cell survival, cell adhesion, and tumor development. J Exp Med 203(9):2201-2213

Chapman MA, Lawrence MS, Keats JJ, Cibulskis K, Sougnez C, Schinzel AC, Harview CL, Brunet JP, Ahmann GJ, Adli M, Anderson KC, Ardlie KG, Auclair D, Baker A, Bergsagel PL, Bernstein BE, Drier Y, Fonseca R, Gabriel SB, Hofmeister CC, Jagannath S, Jakubowiak AJ, Krishnan A, Levy J, Liefeld T, Lonial S, Mahan S, Mfuko B, Monti S, Perkins LM, Onofrio R, Pugh TJ, Rajkumar SV, Ramos AH, Siegel DS, Sivachenko A, Stewart AK, Trudel S, Vij R, Voet D, Winckler W, Zimmerman T, Carpten J, Trent J, Hahn WC, Garraway LA, Meyerson M, Lander ES, Getz G, Golub TR (2011) Initial genome sequencing and analysis of multiple myeloma. Nature 471(7339):467–472

Chesi M, Robbiani DF, Sebag M, Chng WJ, Affer M, Tiedemann R, Valdez R, Palmer SE, Haas SS, Stewart AK, Fonseca R, Kremer R, Cattoretti G, Bergsagel PL (2008) AID-dependent activation of a MYC transgene induces multiple myeloma in a conditional mouse model of post-germinal center malignancies. Cancer Cell 13(2):167–180

Chng WJ, Ketterling RP, Fonseca R (2006) Analysis of genetic abnormalities provides insights into genetic evolution of hyperdiploid myeloma. Genes Chromosom Cancer 45(12):1111–1120

Coluccia AM, Cirulli T, Neri P, Mangieri D, Colanardi MC, Gnoni A et al (2008) Validation of PDGFRbeta and c-Src tyrosine kinases as tumor/vessel targets in patients with multiple myeloma: preclinical efficacy of the novel, orally available inhibitor dasatinib. Blood 112(4):1346–1356

David E, Kaufman JL, Flowers CR, Schafer-Hales K, Torre C, Chen J, Marcus AI, Sun SY, Boise LH, Lonial S (2010) Tipifarnib sensitizes cells to proteasome inhibition by blocking degradation of bortezomib-induced aggresomes. Blood 116(24):5285–5288 [Epub 2010 Sep 15]

Delmore JE, Issa GC, Lemieux ME, Rahl PB, Shi J, Jacobs HM, Kastritis E, Gilpatrick T, Paranal RM, Qi J, Chesi M, Schinzel AC, McKeown MR, Heffernan TP, Vakoc CR, Bergsagel PL, Ghobrial IM, Richardson PG, Young RA, Hahn WC, Anderson KC, Kung AL, Bradner JE, Mitsiades CS (2011) BET bromodomain inhibition as a therapeutic strategy to target c-Myc. Cell 146(6):904–917 [Epub 2011 Sep 1]

Devine SM, Flomenberg N, Vesole DH, Liesveld J, Weisdorf D, Badel K et al (2004) Rapid mobilization of CD34+ cells following administration of the CXCR4 antagonist AMD3100 to patients with multiple myeloma and non-Hodgkin's lymphoma. J Clin Oncol 2004 22(6):1095–1102.

Dhalluin C, Carlson JE, Zeng L, He C, Aggarwal AK, Zhou MM (1999) Structure and ligand of a histone acetyltransferase bromodomain. Nature 399(6735):491–496

Dispenzieri A, Kyle RA, Katzmann JA, Therneau TM, Larson D, Benson J, Clark RJ, Melton LJ 3rd, Gertz MA, Kumar SK, Fonseca R, Jelinek DF, Rajkumar SV (2008) Immunoglobulin free light chain ratio is an independent risk factor for progression of smoldering (asymptomatic) multiple myeloma. Blood 111(2):785–789 [Epub 2007 Oct 17].

Du W, Hattori Y, Hashiguchi A, Kondoh K, Hozumi N, Ikeda Y et al (2004) Tumor angiogenesis in the bone marrow of multiple myeloma patients and its alteration by thalidomide treatment. Pathol Int 54(5):285–294

Elnenaei MO, Hamoudi RA, Swansbury J, Gruszka-Westwood AM, Brito-Babapulle V, Matutes E et al (2003) Delineation of the minimal region of loss at 13q14 in multiple myeloma. Genes Chromosom Cancer 36(1):99–106

Gertz MA (2008) New targets and treatments in multiple myeloma: Src family kinases as central regulators of disease progression. Leuk Lymphoma 49(12):2240–2245

Hanamura I, Stewart JP, Huang Y, Zhan F, Santra M, Sawyer JR, Hollmig K, Zangarri M, Pineda-Roman M, van Rhee F, Cavallo F, Burington B, Crowley J, Tricot G, Barlogie B, Shaughnessy JD Jr (2006) Frequent gain of chromosome band 1q21 in plasma-cell dyscrasias detected by fluorescence in situ hybridization: incidence increases from MGUS to relapsed myeloma and is related to prognosis and disease progression following tandem stem-cell transplantation. Blood 108(5):1724–1732 [Epub 2006 May 16]

Heinrich PC, Behrmann I, Haan S, Hermanns HM, Muller-Newen G, Schaper F (2003) Principles of interleukin (IL)-6-type cytokine signalling and its regulation. The Biochem J 374(Pt 1):1–20

Hideshima T, Mitsiades C, Tonon G, Richardson PG, Anderson KC (2007) Understanding multiple myeloma pathogenesis in the bone marrow to identify new therapeutic targets. Nat Rev 7(8):585-598

Hoang B, Frost P, Shi Y, Belanger E, Benavides A, Pezeshkpour G et al (2010) Targeting TORC2 in multiple myeloma with a new mTOR kinase inhibitor. Blood 116(22):4560–4568

Ishikawa H, Tsuyama N, Obata MMMK (2006) Mitogenic signals initiated via interleukin-6 receptor complexes in cooperation with other transmembrane molecules in myelomas. J Clin Exp Hematol 46(2):55–66

Kimura RH, Levin AM, Cochran FV, Cochran JR (2009) Engineered cystine knot peptides that bind alphavbeta3, alphavbeta5, and alpha5beta1 integrins with low-nanomolar affinity. Proteins 77(2):359–369

Klein U, Dalla-Favera R (2008) Germinal centres: role in B-cell physiology and malignancy. Nat Rev Immunol 8(1):22–33

Kline M, Donovan K, Wellik L, Lust C, Jin W, Moon-Tasson L et al (2007) Cytokine and chemokine profiles in multiple myeloma; significance of stromal interaction and correlation of IL-8 production with disease progression. Leuk Res 31(5):591–598

Kyle RA, Rajkumar SV (2008) Multiple myeloma. Blood 111(6):2962–2972

Kyle RA, Rajkumar SV (2009) Criteria for diagnosis, staging, risk stratification and response assessment of multiple myeloma. Leukemia 23(1):3–9

Kyle RA, Durie BG, Rajkumar SV, Landgren O, Blade J, Merlini G, Kröger N, Einsele H, Vesole DH, Dimopoulos M, San Miguel J, Avet-Loiseau H, Hajek R, Chen WM, Anderson KC, Ludwig H, Sonneveld P, Pavlovsky S, Palumbo A, Richardson PG, Barlogie B, Greipp P, Vescio R, Turesson I, Westin J, Boccadoro M, International Myeloma Working Group (2010) Monoclonal gammopathy of undetermined significance (MGUS) and smoldering (asymptomatic) multiple myeloma: IMWG consensus perspectives risk factors for progression and guidelines for monitoring and management. Leukemia 24(6):1121–1127 [Epub 2010 Apr 22] Review

Lauring J, Abukhdeir AM, Konishi H, Garay JP, Gustin JP, Wang Q et al (2008) The multiple myeloma associated MMSET gene contributes to cellular adhesion, clonogenic growth, and tumorigenicity. Blood 111(2):856–864

Li WX (2008) Canonical and non-canonical JAK-STAT signaling. Trends Cell Biol 18(11):545–551

Lonial S (2010) Presentation and risk stratification-improving prognosis for patients with multiple myeloma. Cancer Treat Rev 36 (Suppl 2):S12–S17 [Review]

Lu DY, Tang CH, Yeh WL, Wong KL, Lin CP, Chen YH et al (2009) SDF-1alpha up-regulates interleukin-6 through CXCR4, PI3K/Akt, ERK, and NF-kappaB-dependent pathway in microglia. Eur J Pharmacol 613(1–3):146–154

Markovic O, Marisavljevic D, Cemerikic V, Vidovic A, Perunicic M, Todorovic M et al (2008) Expression of VEGF and microvessel density in patients with multiple myeloma: clinical and prognostic significance. Med Oncol 25(4):451–457

Matsunaga T, Takemoto N, Sato T, Takimoto R, Tanaka I, Fujimi A et al (2003) Interaction between leukemic-cell VLA-4 and stromal fibronectin is a decisive factor for minimal residual disease of acute myelogenous leukemia. Nat Med 9(9):1158–1165

McMillin DW, Delmore J, Weisberg E, Negri JM, Geer DC, Klippel S et al (2010) Tumor cell-specific bioluminescence platform to identify stroma-induced changes to anticancer drug activity. Nat Med 16(4):483–489

Meads MB, Gatenby RA, Dalton WS (2009) Environment-mediated drug resistance: a major contributor to minimal residual disease. Nat Rev 9(9):665–674

Mitsiades N, Mitsiades CS, Poulaki V, Chauhan D, Fanourakis G, Gu X et al (2002) Molecular sequelae of proteasome inhibition in human multiple myeloma cells. Proceedings of the National Academy of Sciences of the United States of America. 99(22):14374–14379

Mitsiades CS, McMillin DW, Klippel S, Hideshima T, Chauhan D, Richardson PG, Munshi NC, Anderson KC (2007) The role of the bone marrow microenvironment in the pathophysiology of myeloma and its significance in the development of more effective therapies. Hematol Oncol Clin North Am 21(6):1007–1034, vii–viii

Moline JM, Herbert R, Crowley L, Troy K, Hodgman E, Shukla G et al (2009) Multiple myeloma in World Trade Center responders: a case series. J Occup Environ Med 51(8):896–902

Nair RR, Emmons MF, Cress AE, Argilagos RF, Lam K, Kerr WT et al (2009) HYD1-induced increase in reactive oxygen species leads to autophagy and necrotic cell death in multiple myeloma cells. Mol Cancer Ther 8(8):2441–2451

Nefedova Y, Cheng P, Alsina M, Dalton WS, Gabrilovich DI (2004) Involvement of Notch-1 signaling in bone marrow stroma-mediated de novo drug resistance of myeloma and other malignant lymphoid cell lines. Blood 103(9):3503–3510

Nefedova Y, Sullivan DM, Bolick SC, Dalton WS, Gabrilovich DI (2008) Inhibition of Notch signaling induces apoptosis of myeloma cells and enhances sensitivity to chemotherapy. Blood 111(4):2220–2229

O'Neal J, Gao F, Hassan A, Monahan R, Barrios S, Kilimann MW, Lee I, Chng WJ, Vij R, Tomasson MH (2009) Neurobeachin (NBEA) is a target of recurrent interstitial deletions at 13q13 in patients with MGUS and multiple myeloma. Exp Hematol 37(2):234–244. Erratum in: Exp Hematol. 2009 Apr; 37(4):532. Kilimann, Manfred W [added]

Ohwada C, Nakaseko C, Koizumi M, Takeuchi M, Ozawa S, Naito M et al (2008) CD44 and hyaluronan engagement promotes dexamethasone resistance in human myeloma cells. Eur J Haematol 80(3):245–250

Palumbo A, Gay F (2010) Towards a new standard of care for patients with myeloma? Lancet Oncol 11(1):3–4

Pilarski LM, Pilarski PM, Belch AR (2010) Multiple myeloma may include microvessel endothelial cells of malignant origin. Leuk Lymphoma 51(4):592–597

Popovic R, Licht JD (2011) MEK and MAF in myeloma therapy. Blood 117(8):2300–2302

Prince HM, Honemann D, Spencer A, Rizzieri DA, Stadtmauer EA, Roberts AW et al (2009) Vascular endothelial growth factor inhibition is not an effective therapeutic strategy for relapsed or refractory multiple myeloma: a phase 2 study of pazopanib (GW786034). Blood 113(19):4819-4820

Rahman S, Sowa ME, Ottinger M, Smith JA, Shi Y, Harper JW et al (2011) The Brd4 extraterminal domain confers transcription activation independent of pTEFb by recruiting multiple proteins, including NSD3. Mol Cell Biol 31(13):2641–2652

Rajkumar SV (2011) Multiple myeloma: 2011 update on diagnosis, risk-stratification, and management. Am J Hematol 86(1):57–65

Ramakrishnan V, Timm M, Haug JL, Kimlinger TK, Wellik LE, Witzig TE et al (2010) Sorafenib, a dual Raf kinase/vascular endothelial growth factor receptor inhibitor has significant anti-myeloma activity and synergizes with common anti-myeloma drugs. Oncogene 29(8):1190–1202

Ricart AD, Tolcher AW, Liu G, Holen K, Schwartz G, Albertini M et al (2008) Volociximab, a chimeric monoclonal antibody that specifically binds alpha5beta1 integrin: a phase I, pharmacokinetic, and biological correlative study. Clin Cancer Res. 14(23):7924–7929

Richardson PG, Hideshima T, Mitsiades C, Anderson KC (2007) The emerging role of novel therapies for the treatment of relapsed myeloma. J Natl Compr Canc Netw 5(2):149–162

Roodman GD (2010) Targeting the bone microenvironment in multiple myeloma. J Bone Miner Metab 28(3):244–250

Rossi J-F, Manges Robert F, Peter Voorhees, Sutherland Heather J, Orlowski Robert Z (2008) Preliminary results of CNTO 328, an anti-IL-6 monoclonal antibody, in combination with bortezomib in the treatment of relapsed or refractory multiple myeloma. Blood ASH Annu Meet Abstr 112:867

Sanz-Rodriguez F, Hidalgo A, Teixido J (2001) Chemokine stromal cell-derived factor-1alpha modulates VLA-4 integrin-mediated multiple myeloma cell adhesion to CS-1/fibronectin and VCAM-1. Blood 97(2):346-351

Sawyer JR (2011) The prognostic significance of cytogenetics and molecular profiling in multiple myeloma. Cancer Genet 204(1):3–12

Sciammas R, Davis MM (2004) Modular nature of Blimp-1 in the regulation of gene expression during B cell maturation. J Immunol 172(9):5427–5440

Shaffer AL, Shapiro-Shelef M, Iwakoshi NN, Lee AH, Qian SB, Zhao H et al (2004) XBP1, downstream of Blimp-1, expands the secretory apparatus and other organelles, and increases protein synthesis in plasma cell differentiation. Immunity 21(1):81–93

Shain KH, Dalton WS (2009) Environmental-mediated drug resistance: a target for multiple myeloma therapy. Expert Rev Hematol 2(6):649–662. Review

Shain KH, Yarde DN, Meads MB, Huang M, Jove R, Hazlehurst LA et al (2009) Beta1 integrin adhesion enhances IL-6-mediated STAT3 signaling in myeloma cells: implications for microenvironment influence on tumor survival and proliferation. Cancer Res 69(3):1009–1015

Shapiro-Shelef M, Calame K (2004) Plasma cell differentiation and multiple myeloma. Curr Opin Immunol 16(2):226–234

Shapiro-Shelef M, Calame K (2005) Regulation of plasma-cell development. Nat Rev Immunol 5(3):230–242

Shaughnessy JD Jr, Zhan F, Burington BE, Huang Y, Colla S, Hanamura I, Stewart JP, Kordsmeier B, Randolph C, Williams DR, Xiao Y, Xu H, Epstein J, Anaissie E, Krishna SG, Cottler-Fox M, Hollmig K, Mohiuddin A, Pineda-Roman M, Tricot G, van Rhee F, Sawyer J, Alsayed Y, Walker R, Zangari M, Crowley J, Barlogie B (2007) A validated gene expression model of high-risk multiple myeloma is defined by deregulated expression of genes mapping to chromosome 1. Blood 109(6):2276–2284 [Epub 2006 Nov 14]

Sher T, Miller KC, Deeb G, Lee K, Chanan-Khan A (2010) Plasma cell leukaemia and other aggressive plasma cell malignancies. Br J Haematol 150(4):418–427

Tancred TM, Belch AR, Reiman T, Pilarski LM, Kirshner J (2009) Altered expression of fibronectin and collagens I and IV in multiple myeloma and monoclonal gammopathy of undetermined significance. J Histochem Cytochem 57(3):239–247

Teicher BA, Fricker SP (2010) CXCL12 (SDF-1)/CXCR4 pathway in cancer. Clin Cancer Res 16(11):2927–2931

Tricot G, Barlogie B, Jagannath S, Bracy D, Mattox S, Vesole DH et al (1995) Poor prognosis in multiple myeloma is associated only with partial or complete deletions of chromosome 13 or abnormalities involving 11q and not with other karyotype abnormalities. Blood 86(11): 4250–4256

Walker R, Barlogie B, Haessler J, Tricot G, Anaissie E, Shaughnessy JD Jr et al (2007) Magnetic resonance imaging in multiple myeloma: diagnostic and clinical implications. J Clin Oncol 25(9):1121–1128

Weiss BM, Abadie J, Verma P, Howard RS, Kuehl WM (2009) A monoclonal gammopathy precedes multiple myeloma in most patients. Blood 113(22):5418–5422

Yaccoby S (2010) Advances in the understanding of myeloma bone disease and tumour growth. Br J Haematol 149(3):311–321
Zhan F, Barlogie B, Arzoumanian V, Huang Y, Williams DR, Hollmig K et al (2007) Gene-expression signature of benign monoclonal gammopathy evident in multiple myeloma is linked to good prognosis. Blood 109(4):1692–1700
Zoubeidi A, Rocha J, Zouanat FZ, Hamel L, Scarlata E, Aprikian AG et al (2009) The Fer tyrosine kinase cooperates with interleukin-6 to activate signal transducer and activator of transcription 3 and promote human prostate cancer cell growth. Mol Cancer Res 7(1):142–155

EBV-Positive Diffuse Large B-Cell Lymphoma of the Elderly

Jorge J. Castillo

The purpose of this chapter is to provide a thorough summary of the existing knowledge of EBV-positive DLBCL of the elderly. We will discuss the pathogenesis behind EBV-driven malignant transformation of B-cells, the different EBV latency patterns associated with DLBCL, the distinct pathological characteristics of EBV-positive DLBCL, the differential diagnosis of EBV-positive DLBCL, the clinical characteristics of patients with EBV-positive DLBCL of the elderly, the potential predictive and prognostic value of EBV tumoral status in patients with DLBCL, and potential strategies for the treatment of this rare entity.

Definitions

Epstein-Barr virus-positive diffuse large B-cell lymphoma (EBV-positive DLBCL) of the elderly is also known as age-related EBV-positive B-cell lymphoproliferative disorder (LPD) or senile EBV-associated B-cell LPD, and is among the newer inclusions in the 2008 WHO Classification of Tumors of Hematopoietic and Lymphoid Tissues as a provisional entity (Nakamura et al. 2008).

The proposed diagnostic criteria include an age older than 50 years, no underlying immunodeficiency and the expression of EBV by the malignant lymphomatous cells. Few cases have been described in apparently immunocompetent patients younger than 50 years (Oyama et al. 2003); however, immunodeficiency has to be

J.J. Castillo, M.D. (✉)
The Warren Alpert Medical School of Brown University, Providence, RI, USA

Division of Hematology and Oncology, The Miriam Hospital,
164 Summit Ave, Providence, RI 02906, USA
e-mail: jcastillo@lifespan.org

J. Tao and E. Sotomayor (eds.), *Hematologic Cancers: From Molecular Pathobiology to Targeted Therapeutics*, Cancer Growth and Progression 14,
DOI 10.1007/978-94-007-5028-9_5, © Springer Science+Business Media Dordrecht 2012

strongly considered in such patients, and extensive workups might be necessary to rule out the presence of an immunodeficient state. Other B-cell malignancies that are associated with EBV expression such as plasmablastic lymphoma, primary effusion lymphoma and lymphomatoid granulomatosis should be excluded (Nakamura et al. 2008).

Initial Descriptions

EBV-positive DLBCL of the elderly were initially described by Oyama and colleagues (2003). In this study, approximately 400 Japanese patients with a diagnosis of DLBCL or LPD were screened for the presence of EBV in the malignant cells using an *in situ* hybridization (ISH) technique to detect EBV-encoded RNA (EBER). Clinicopathological data on 22 patients with EBV-positive DLBCL or LDP were presented. In this cohort, three patients younger than 30 years were excluded form the analysis due to the suspicion of having an underlying immunodeficiency or chronic active EBV infection. The median age was 76 years with a slight male predominance (men 12, women 10). All patients tested were negative for HTLV-I and HIV infection; the patients that were not tested did not have history, signs or symptoms consistent with acquired or congenital immunodeficiency. The majority of patients (91%; n=20) had a good performance status, and 41% (n=9) had high-intermediate or high risk disease, calculated by the International Prognostic Index (IPI) score. Among the most striking clinical features of EBV-positive DLBCL of the elderly was the degree of extranodal involvement; 82% (n=18) of the patients presented with extranodal involvement. The most common extranodal sites affected were skin, stomach, respiratory tract and pancreas. Pathologically, patients were subdivided into two subtypes of EBV-positive DLBCL of the elderly. Nine patients (41%) showed a large B-cell morphology (DLBCL-like) and 13 patients (59%) showed a polymorphic appearance; the latter group was categorized as polymorphic LPD. Clinically, there was no difference between the clinical characteristics of these two groups of patients. However, the 2-year overall survival (OS) of patients with polymorphic LPD was significantly higher than for patients with DLBCL-like morphology. Approximately 40% of the patients died within 1 year of diagnosis.

This initial report established the platform to carry a larger study. Oyama and colleagues (2007) presented data on 96 patients with EBV-positive DLBCL found in a cohort of 1792 Japanese patients with LPDs with a large B-cell morphology (Oyama et al. 2007). The presence or absence of EBER was established by an ISH technique similar to the previous report, which was run in paraffin-embedded tissue. The researchers then proceeded to compare the characteristics of these 96 patients with 107 EBV-negative DLBCL controls. EBV-positive DLBCL patients were significantly older, had worse performance status, higher levels of lactate dehydrogenase (LDH) and higher rates of B symptoms (fevers, drenching night sweats and unintentional weight loss of >10% in 6 months). The rate of skin involvement was significantly higher in patients with EBV-positive DLBCL. Pathologically, EBV-positive DLBCL patients had lower expression of CD10 but higher expression of CD30. Based on their histological characteristics, 34 cases (35%) were categorized as DLBCL and 62 (65%) as polymorphic LPD. Strikingly, EBV-positive DLBCL cases had significantly lower complete response (CR) rates to chemotherapy (63% versus 91% in EBV-negative cases). Furthermore, the median OS for EBV-positive DLBCL patients was 24 months, while the median OS was not reached in EBV-negative patients. Using multivariate analytic models, the researchers were able to identify age older than 70 years and the presence of B symptoms as poor independent prognostic factors for survival in patients with EBV-positive DLBCL of the elderly, dividing patients in three prognostic groups. Patients with 0, 1 or 2 factors had a median OS of 56, 25 and 8.5 months, respectively.

The investigators concluded that patients meeting criteria for EBV-positive DLBCL of the elderly constitute a distinct group with its specific clinicopathological features and decreased response rates to chemotherapy and shorter survival (Oyama et al. 2007). It is important to note that none of these patients underwent therapy with the combination of rituximab and chemotherapy, which is considered the standard of care for the front line treatment of patients with EBV-negative DLBCL.

Pathogenesis

The pathogenetic mechanism behind EBV-positive DLBCL of the elderly is likely multifactorial (Fig. 1). In one hand, the lymphotropism and chronicity of EBV infection, and on the other, the immunological dysfunction associated to aging, would play a synergistic role in the development of this aggressive lymphoma. However, the participation of other more subtle and yet unknown mechanisms is also likely.

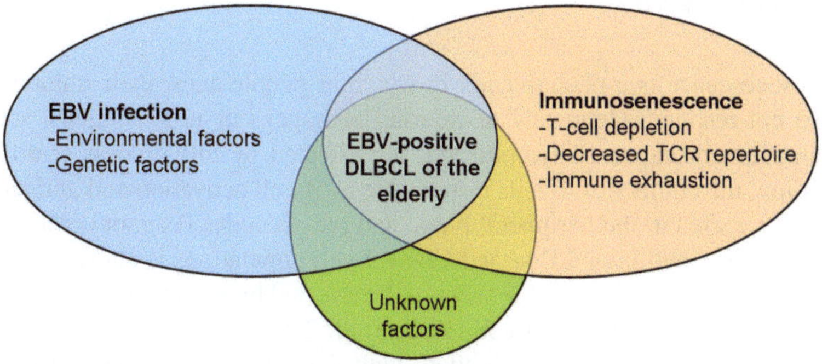

Fig. 1 Schematic pathophysiology of EBV-positive diffuse large B-cell lymphoma of the elderly. *EBV* Epstein Barr virus, *DLBCL* diffuse large B-cell lymphoma, *TCR* T-cell receptor

Epstein Barr Virus Infection

EBV was the first oncovirus ever described, and is a gamma-herpesvirus with a well recognized B-cell lymphotropism (Thompson and Kurzrock 2004). Once EBV has reached the bloodstream, EBV infection starts by the attachment of the virus to the CD21 antigen, which is located in the membrane of the B-lymphocyte; this initial step is characterized by increased production of IL-6 and mRNA producing a blastic transformation of the B-cell. EBV is then episomally inserted into the nucleus acquiring a circle-shaped configuration. EBV nuclear antigens (EBNA-LP, -1, -2 and –3) are the first EBV-related proteins to be produced after the infection; these products are essential for immortalization of the cell and upregulation of the expression of additional molecules such as the latent membrane proteins (LMP-1 and –2). EBNAs also upregulate C-MYC, a human oncogene associated with cell proliferation. LMPs increase expression of BCL-2 and drive the cell into a latent state, which is maintained by the production of EBV-encoded RNA (EBER-1 and –2). In this way, EBV-infected B-cells enter the resting phase avoiding immunosurveillance but, due to their activated phenotype, more prone to develop secondary oncogenic changes (Knecht et al. 2001).

B-cell differentiation represents a well-structured and intricate process that is initiated when a naïve B-cell recently released from the bone marrow encounters antigen within one of the follicles of a lymph node. This antigen-exposed B-cell undergoes a germinal centre (GC) reaction characterized by somatic hypermutation and class-switching, processes that will guarantee the production of antibodies specifically directed against the antigen that initiated the process. The final step of this process is the differentiation of post-germinal B-cells into either memory cells or plasma cells.

Interruption of this delicate process by a transforming event, such as EBV infection, may result in a clonal proliferation in which the differentiation of the cell is blocked, resulting potentially in the development of a lymphoid malignancy.

Immunosenescence

Immunosenescence is a phenomenon in which as people ages, their immune systems do not respond adequately to external pathogens or new antigens, such as immunizations or cancer. T-lymphocytes are affected by this process at different levels. First, the number of T-cells responsible of B-cell activation and antigen recognition decreases in the peripheral blood and lymph nodes (Fagnoni et al. 2000). Second, the distribution of CD4+ and CD8+ T-cell populations is altered; however, the changes in CD4+ T-cell profile seems to be affected by advancing age at a higher degree than the one of CD8+ T-cell (Koch et al. 2008). Third, the T-cell receptor repertoire (TCR) becomes more limited, with a TCR repertoire decreasing from 20,000,000 TCR beta-chains in younger individuals to less than 200,000 TCR beta-chains in individuals older than 60 years of age (Naylor et al. 2005).

The presence of persistent infections, such as EBV or other persistent antigens will also induce a phenomenon called immune exhaustion contributing to the loss of immunosurveillance (Zinkernagel et al. 1993). Other immunological changes associated with age include abnormalities in apoptotic mechanisms (Spaulding et al. 1999) and elevation of levels of proinflammatory cytokines (Franceschi et al. 2007). All the abovementioned phenomena could potentially explain the increased incidence of cancers in the elderly.

Patterns of EBV Latency in EBV-Positive DLBCL of the Elderly

Three different latency patterns have been described in EBV-related malignant disorders and EBV-derived cell lines (Table 1). These EBV latency patterns have been subdivided in type I, II and III. Different EBV latency patterns have been associated with different degrees of immunosuppression. For example, type I EBV latency pattern has been mostly described in malignancies that have developed in immunocompetent patients such as Burkitt lymphoma, and is characterized by the expression of EBNA2 and LMP2A. In type II EBV latency patterns, LMP1 and LMP2B are expressed in addition to EBNA2 and LMP2A suggesting a more advanced EBV infection or a mild underlying immunosuppression; important examples of LPDs with a type II EBV latency pattern are classical Hodgkin lymphoma, primary effusion lymphoma, angioimmunoblastic T-cell lymphoma and NK/T cell lymphoma. The mechanisms used by EBV to enter and infect T-cells, however, have not been elucidated, as they lack the expression of CD21 in their membranes. In type III EBV latency patterns, all EBV-associated proteins are expressed, including EBER-1 and –2. This is suggestive of a deeper degree of immunosuppression; in fact, posttransplant LPDs and AIDS-related immunoblastic lymphoma are characteristic examples of this type of EBV latency pattern (Knecht et al. 2001; Tan 2009).

Patients with EBV-positive DLBCL of the elderly show a type II or III EBV latency pattern. Using the abovementioned categorization of EBV latency patterns, Oyama and colleagues reported that 68% of the patients had a type II EBV latency pattern and 32% showed a type III latency pattern (Oyama et al. 2003). In a larger and more recent study (Oyama et al. 2007), Shimoyama and colleagues reported that 28% of the patients with EBV-positive DLBCL of the elderly had a type III EBV latency pattern. These results show the high level of immunodeficiency seen in patients who developed EBV-positive DLBCL of the elderly.

Pathological Features of EBV-Positive DLBCL of the Elderly

Although several pathological characteristics have been described in EBV-positive LDBCL of the elderly and will be discussed below, there are no histological, immunohistochemical or molecular features that are specific of this malignancy.

Table 1 EBV-related proteins expressed in various lymphoproliferative disorders according to the latency patterns

	EBV-associated antigens			
Latency pattern	EBNA1/LMP2A	LMP1/LMP2B	EBNA2/3A/3B/3C/LP	Lymphoproliferative disorders
I	+	–	–	Burkitt Lymphoma
II	+	+	–	Classical Hodgkin lymphoma
				Primary effusion lymphoma
				Angioimmunoblastic lymphoma
				NK/T-cell lymphoma
				EBV-positive DLBCL of the elderly
III	+	+	+	Post-transplant lymphoma
				AIDS-related lymphoma
				EBV-positive DLBCL of the elderly

EBV Epstein Barr virus

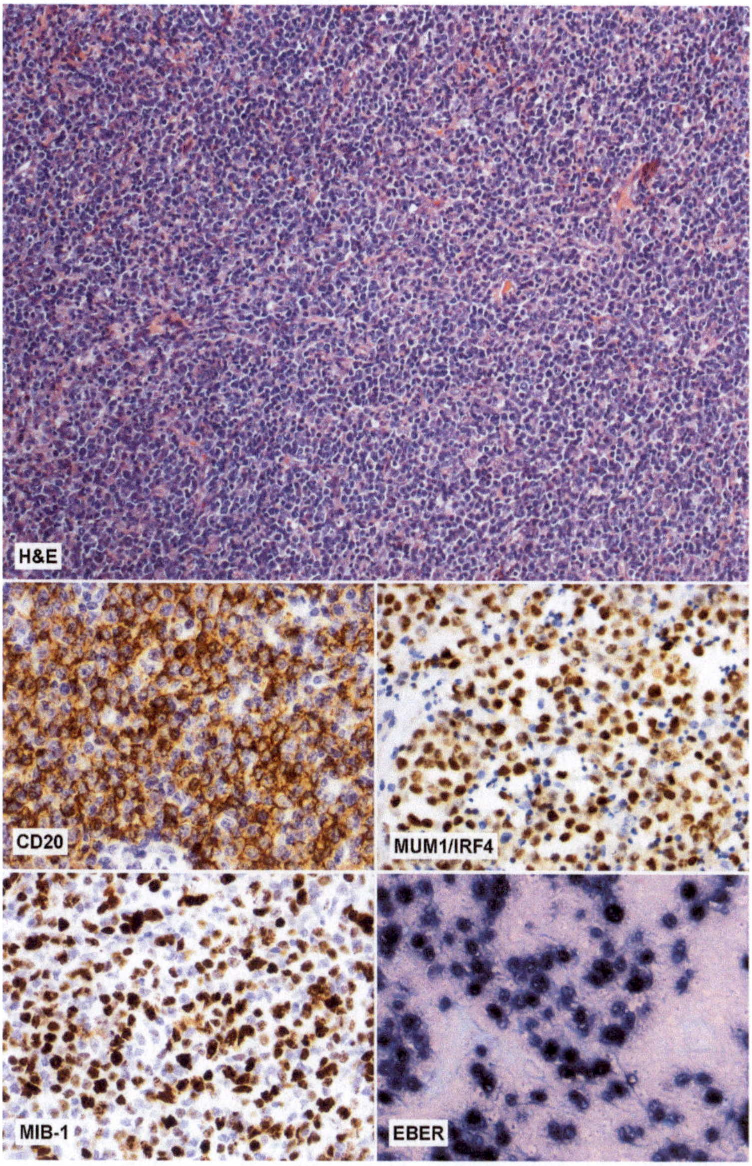

Fig. 2 Microscopic appearance and immunohistochemical profile of EBV-positive diffuse large B-cell lymphoma of the elderly. *Top*: Hematoxylin & eosin (100×). *Bottom*: CD20, MUM1/IRF4, MIB-1 and EBER (400×)

Histological Features

The most common pathological feature seen in EBV-positive DLBCL of the elderly is an effacement of the normal nodal architecture by malignant lymphocytes (Fig. 2, top), which is often accompanied by extensive necrosis or angiocentric destruction

(Oyama et al. 2003). Initial descriptions emphasized the presence of two separate subtypes, the polymorphous and the large cell subtypes (DLBCL-like). The polymorphous subtype shows variable amounts of reactive cells such as small lymphocytes, plasma cells, and histiocytes with the presence of scattered large cells. The large cell subtype refers to the presence of a rather monotonous population of large immunoblastic cells with minimal or no reactive component. Both variants may present with giant cells with a histological resemblance to Reed-Sternberg cells (Oyama et al. 2003). Most cases, however, commonly show both monomorphic and polymorphic patterns, suggesting the existence of a potential continuum between the two patterns (Wong and Wang 2009).

In the initial report, the monomorphic variant of EBV-positive DLBCL of the elderly correlated with a worse prognosis (Oyama et al. 2003). However, more recent data have not supported those initial findings (Shimoyama et al. 2009), and have shown that there are no clinical, therapeutic and/or survival differences between the two patterns. Hence, this histological subdivision has been abandoned and is no longer used (Wong and Wang 2009).

Immunohistochemical Features

Immunohistochemistry (IHC) has revolutionized the pathological classification of LPDs, allowing for a more precise definition of these malignancies. EBV-positive DLBCL of the elderly has a distinct but non-specific IHC profile. The neoplastic cells are usually positive for the leukocyte common antigen CD45 and for the B-cell markers CD20 (Fig. 2, bottom), CD19, CD79a and PAX-5, thus confirming the B-cell lineage of this malignancy. The GC markers CD10 and BCL-6 are usually negative, while IRF4/MUM-1 (Fig. 2, bottom) is commonly positive. This profile is consistent with a post-GC origin, according to the classification proposed by Hans and colleagues (Hans et al. 2004). A post-GC profile has been detected in normal B-lymphocytes that have acquired a certain level of plasmacytic differentiation and in LPDs arising from the malignant counterpart of this cell, such as plasmablastic lymphoma (Castillo et al. 2008). A few cases of EBV-positive DLBCL of the elderly with plasmablastic or immunoblastic features show a weak or absent expression of CD20 (Gibson and Hsi 2009). Several studies have shown that DLBCL with plasmacytic differentiation (post-GC) have a worse prognosis than DLBCL with a GC profile (Alizadeh et al. 2000; Rosenwald et al. 2002; Hans et al. 2004). Up to 50% of the cases of EBV-positive DLBCL of the elderly express CD30 but are negative for the expression of CD15, which may be helpful in differentiating these cases from Hodgkin lymphoma. Ki-67 (or MIB-1), a marker of cellular proliferation, usually is expressed in more than 70% of the malignant cells (Fig. 2, bottom), reflecting the aggressiveness of this lymphoma.

IHC is also helpful to classify the different EBV latency patterns as IHC methods for the detection of LMP1 and EBNA are commercially available. EBV-associated latent antigens such as LMP1 and EBNA-2 are positive in 94 and 28% of the cases of EBV-positive DLBCL of the elderly, respectively (Oyama et al. 2007).

Molecular Features

The diagnosis of EBV-positive DLBCL of the elderly is based on the detection of EBV genome within the tumoral cells. *In situ* hybridization (ISH) techniques, either chromogenic or fluorescent, are commercially available to detect EBV-encoded RNA (EBER). Per definition, EBER should be present in the neoplastic cells, although the expression of EBER can range from 10% to almost all of the tumor cells in the sample (Fig. 2, bottom). By means of double labeling techniques, it has been demonstrated that the large majority of the infected cells were of a B-cell lineage; however, rare T-cells were also EBER-positive [24]. Additionally, ISH studies have demonstrated a preferential localization of EBV in interfollicular or extrafollicular locations, which correlates with the lack of expression EBER by LPDs with a GC origin such as follicular lymphoma or nodular lymphocyte-predominant Hodgkin lymphoma (Tan 2009).

It is important to note that although EBER ISH is the test of choice for the pathological diagnosis of EBV-positive DLBCL of the elderly, it is not recommended that EBER testing should be routinely done in cases of DLBCL. EBER testing in DLBCL may only be justified after prospective clinical studies are performed to understand the prognostic and/or predictive role of EBER expression in DLBCL.

Differential Diagnosis

Infectious mononucleosis, chronic active EBV infection, and reactions to drugs, such as phenytoin and methotrexate, can produce pathological changes similar to EBV-positive DLBCL of the elderly. This reaction is characterized by an interfollicular expansion of cells ranging from small to intermediate and large lymphocytes with immunoblastic features, plasma cells and histiocytes with frequent residual and hyperplastic GCs. Infectious mononucleosis and drug reactions, however, are not associated with a clonal lymphoproliferative process; hence, immunoglobulin gene rearrangements are rarely seen in these conditions while in EBV-positive DLBCL of the elderly are observed in up to 90% of the cases. In chronic active EBV infection, few oligoclonal cells can be seen; however, most of the abnormal lymphocytes are of a T-cell lineage.

DLBCL is the most common lymphoma variant seen worldwide and accounts for up to 30% of cases of non-Hodgkin lymphoma. As with EBV-positive DLBCL of the elderly, certain categories of DLBCL have also been associated with expression of EBV in the malignant lymphocytes; these include lymphomatoid granulomatosis, post-transplant LPDs, plasmablastic lymphoma, primary effusion lymphoma and DLBCL associated with chronic inflammation. The clinical history is the most important criteria for the differentiation between EBV-positive DLBCL of the elderly and post-transplant LPDs. Patients with DLBCL associated with chronic inflammation usually have a history of pyothorax, chronic osteomyelitis or chronic suppurative skin ulcers. Plasmablastic lymphoma has been associated with EBV expression up to 75% of the cases but usually presents in the oral cavity of individuals infected with HIV and is almost exclusively CD20 negative with a variable expression of CD45 (Castillo et al. 2008). PEL has also been associated with co-infection by HIV and EBV; however, its association with the human herpesvirus 8 is almost 100% and is used for its diagnosis. Additionally, the malignant cells are usually negative for the expression of CD45 or CD20 (Brimo et al. 2007).

Similar to EBV-positive DLBCL of the elderly, classical Hodgkin lymphoma is characterized by an abundant reactive background, which includes small lymphocytes, plasma cells, eosinophils and histiocytes. One of the main differences is that in Hodgkin lymphoma the neoplastic cells usually represent approximately 1% of the cellular infiltrate, although occasional cases of Hodgkin lymphoma with sheets of large neoplastic cells have been described. Features that would raise the suspicion for Hodgkin lymphoma include the presence of sclerosis, the presence of retraction around neoplastic cells and the presence of CD30-positive large cells. These malignant cells frequently co-express CD15, which is usually negative in EBV-positive DLBCL of the elderly. Classical Hodgkin lymphoma can also be associated with EBV expression (Glaser et al. 1997), although this expression is highly variable depending on the specific histological variant of Hodgkin lymphoma. For example, the nodular sclerosis subtype shows 10–40% EBER positivity while the mixed cellularity subtype can be EBER-positive in up to 75% of cases (Chang et al. 2008).

The Predictive and Prognostic Role of EBV in DLBCL

In initial reports, the expression of EBV by the malignant DLBCL cells correlated with worse response to chemotherapy and shorter survival times. In the last few years, clinical evidence continues accumulating in support of these early findings. However, with the advent of chemoimmunotherapy as the new therapeutic standard for DLBCL, the role of EBV expression in survival (prognostic) or response to therapy (predictive) in patients with DLBCL is actively under investigation.

EBV as a Prognostic Factor in DLBCL

Few studies have evaluated the prognostic role that the EBV tumoral status plays on the survival of patients with DLBCL. In an earlier report, Oyama and colleagues evaluated 1792 patients with DLBCL for the expression of EBER within the tumoral cells (Oyama et al. 2007). The incidence of EBER positivity in this cohort was 8% (149 patients older than 50 years). The researcher then compared the characteristics of 96 patients with EBV-positive LPD/DLBCL with 107 EBV-negative DLBCL. In the survival analysis, EBER-positive patients had a statistically significant shorter survival than EBV-negative patients with a median OS of 24 months vs. not reached. In the subset analysis, EBER expression was an independent prognostic factor in

DLBCL patients regardless of the IPI score. The IPI score was developed in the early 1990s and is the most commonly used risk-stratification tool in patients with DLBCL (The International Non-Hodgkin's Lymphoma Prognostic Factors Project 1993). This score assigns one point per risk factor, namely age of 60 years or older, performance status of two or higher, elevated LDH levels, two or more extranodal sites and an advanced clinical stage (III or IV); hence, the importance of EBER positivity in DLBCL patients independently from the IPI scores. In a more recent Korean study, Park and colleagues evaluated the expression of EBER within the tumoral cells of patients with DLBCL treated with chemotherapy and chemoimmunotherapy (Park et al. 2007). The incidence of EBER positivity was 9% (34 out of 380 samples evaluated). EBER-positive DLBCL cases were associated with a worse OS (36 months vs. not reached) and a worse progression-free survival (13 vs. 36 months) than EBER-negative DLBCL patients. In the subset analysis, EBV-positivity was associated with a worse OS and PFS in patients with high IPI scores. In a smaller Peruvian study, Morales and colleagues evaluated the prognostic significance of EBER expression in a Peruvian population with DLBCL treated with chemotherapy. The incidence of EBER positivity was 15% (11 out of 74 DLBCL cases were EBER positive). EBER-positive DLBCL patients had a significantly shorter OS (7 vs. 47 months) than EBV-negative patents (Morales et al. 2010). In the multivariate analysis, EBER expression was a prognostic factor independent of the IPI score.

Scant data are available from Europe or the United States. More recently, Gibson and colleagues described the characteristics of five American patients and Hoeller and colleagues recently published data on eight European patients with EBV-positive DLBCL (Gibson and Hsi 2009; Hoeller et al. 2010). Due to the small numbers, the investigators in both studies did not perform a formal survival analysis; however, a poor outcome was observed in the EBV-positive DLBCL patients. The reported incidence of EBER positivity in both studies was less than 5%.

One small Japanese study did not show an association between EBER expression and survival in patients with DLBCL. Kuze and colleagues studied 114 DLBCL patients, from which 13 patients expressed EBER, accounting for an incidence of 11%. When comparing 58 EBER-negative patients with 12 EBER-positive patients, a statistical survival difference was not apparent (Kuze et al. 2000).

It is important to note that most of the studies on EBV-positive DLBCL of the elderly are largely limited to Asian or Hispanic populations, in which the incidence of this lymphoma seems higher than in the Caucasian population (Gibson and Hsi 2009; Hoeller et al. 2010). Of most importance is that the majority of patients received chemotherapy without the addition of rituximab; rituximab in combination with chemotherapy has been shown to increase survival in patients with DLBCL in several randomized controlled trials, and has become the standard of care for these patients (Feugier et al. 2005; Habermann et al. 2006; Pfreundschuh et al. 2006).

In summary, EBER positivity appears as a potential predictor of worse survival in patients with DLBCL treated with chemotherapy. However, several studies are ongoing evaluating the use of rituximab in combination with chemotherapy (chemoimmunotherapy) in patients with EBV-positive DLBCL of the elderly.

EBV as a Predictive Factor in DLBCL

The major goal of therapy in patients with DLBCL is the achievement of a complete response (CR), which is defined by the disappearance of any evidence of disease assessed by means of bone marrow biopsy, and imaging and laboratory studies. At least three studies have evaluated EBV expression as a predictive factor for response to therapy in DLBCL patients. In he biggest study, Oyama and colleagues evaluated 96 EBER-positive Japanese patients and compared their response to chemotherapy against 107 patients with EBV-negative DLBCL (Oyama et al. 2007). EBER-positive DLBCL patients treated with anthracycline-containing regimens showed an overall response rate (ORR) of 80% while EBER-negative DLBCL patients achieved an ORR of 99%. Furthermore, the CR rates were 66% and 91%, respectively. Park and colleagues showed similar results (Park et al. 2007). Among EBER-positive DLBCL Korean patients had an ORR of 72% while EBER-negative patients had an ORR of 92%. Unfortunately, no CR data was reported. A smaller Japanese study by Yoshino and colleagues found that 4 out of 50 patients with primary gastric DLBCL were EBER-positive (Yoshino et al. 2006). These patients had a poor response to radiochemotherapy; only one patient achieved a CR but relapsed shortly after therapy.

Based on these limited data, EBER-positive DLBCL patients seem to have worse ORR and CR rates to chemotherapy than EBV-negative DLBCL. It is important to emphasize that very few patients in these studies received rituximab as part of their therapeutic regimens. Hence, the predictive role of EBER expression in patients treated with immunochemotherapy is yet unknown.

Potential Therapeutic Approaches for EBV-Positive DLBCL

Currently, there is no uniformly accepted treatment for EBV-positive DLBCL of the elderly. Most of the published studies have evaluated chemotherapy alone with suboptimal results. The current standard treatment for EBV-negative DLBCL is the combination of rituximab and chemotherapy. In the 1980s, the CHOP regimen was compared against three combination chemotherapy regimens in a randomized controlled trial. More recently, rituximab in addition to CHOP (R-CHOP) has shown to prolong OS in patients with DLBCL in a series of randomized controlled trails (Feugier et al. 2005; Habermann et al. 2006; Pfreundschuh et al. 2006). Although it is likely that R-CHOP will also provide a survival benefit in EBV-positive

DLBCL of the elderly, the available published data is rather limited. In the largest report by Oyama and colleagues, none of the patients in their cohort received rituximab (Oyama et al. 2007). Park and colleagues reported that only 2 out of their 34 patients (6%) with EBV-positive DLBCL were treated with R-CHOP (Park et al. 2007). Finally, Gibson and colleagues reported four cases treated with R-CHOP, from which one patient obtained a CR relapsing shortly afterwards (Gibson and Hsi 2009). Large prospective studies will be necessary to evaluate the efficacy of chemoimmunotherapy in patients with EBV-positive DLBCL of the elderly. However, based on initial observations, it is likely that novel agents would be necessary to treat this poorly responsive lymphoma.

EBV-positive lymphomas such as post-transplant LPDs can be amenable to immunotherapy with EBV-specific cytotoxic T-cell lymphocytes (CTL). For instance, infusion of EBV-specific CTLs has achieved a moderate success on preventing and treating EBV-associated lymphomas in patients undergoing solid organ transplantation (Heslop et al. 2010). Unfortunately, this approach has not been used in EBV-positive DLBCL of the elderly. Arginine butyrate, which has histone deacetylating properties, in combination with ganciclovir, an antiviral, has been used in 15 patients with EBV-positive LPDs with encouraging results (Perrine et al. 2007). However, most of the patients had histologies other than DLBCL or had an underlying immunosuppressive state; hence the effectiveness in EBV-positive DLBCL is rather speculative.

Preclinically, the combination of valproic acid, with histone deacetylating properties, and ganciclovir have shown efficacy on depleting EBV-transformed cells (Jones et al. 2010). Bortezomib, a proteasome inhibitor approved for the treatment of plasma cell myeloma and mantle cell lymphoma, blocks nuclear factor kappa B, which has been associated with EBV-associated B-cell transformation (Zou et al. 2007). Bortezomib in combination with other molecularly targeted agents or chemotherapy may prove to be of value in the treatment of EBV-positive DLBCL of the elderly. Mammalian target of rapamycin inhibitors may also be of value on treating this malignancy (Vaysberg et al. 2007; Holtan et al. 2008).

Finally, there are several ongoing clinical trials focused on novel approaches to treat EBV-positive lymphomas. The Baylor College of Medicine in Texas and the Queensland Institute of Medical research in Australia are investigating the use of EBV-specific CTLs in patients with relapsed EBV-positive lymphomas (ClinicalTrials.gov. Epstein-Barr Virus (EBV)-Specific T Cells as Therapy for Relapsed/Refractory EBV-positive Lymphomas; ClinicalTrials.gov. Safety Study of EBV Specific Cytotoxic T-Cells to Treat Relapsed EBV-Positive Lymphoma). Boston University recently finished accruing patients with EBV-positive LPDs who have been treated with the combination of ganciclovir or valganciclovir and arginine butyrate (ClinicalTrials.gov. Study of Arginine Butyrate and Ganciclovir/Valganciclovir in EBV(+) Lymphoid Malignancies). Lastly, the University of California in Los Angeles in collaboration with the National Cancer Institute is evaluating the combination of bortezomib and ganciclovir in patients with EBV-positive lymphomas (ClinicalTrials.gov. Bortezomib and Ganciclovir in Treating Patients With Relapsed or Refractory Epstein Barr Virus-Positive Lymphoma).

Results from these trials are eagerly expected, as they could be of paramount importance in the treatment of EBV-positive DLBCL of the elderly. However, it is important to underline that none of these studies are specific for patients with EBV-positive DLBCL of the elderly.

Acknowledgements Dr. Castillo would like to thank Dr. Roberto N. Miranda from the Division of Hematopathology at MD Anderson Cancer Center in Houston, Texas for providing the pathological images included in this chapter.

References

Alizadeh AA, Eisen MB et al (2000) Distinct types of diffuse large B-cell lymphoma identified by gene expression profiling. Nature 403(6769):503–511

Brimo F, Michel RP et al (2007) Primary effusion lymphoma: a series of 4 cases and review of the literature with emphasis on cytomorphologic and immunocytochemical differential diagnosis. Cancer 111(4):224–233

Castillo J, Pantanowitz L et al (2008) HIV-associated plasmablastic lymphoma: lessons learned from 112 published cases. Am J Hematol 83(10):804–809

Chang KC, Chen PC et al (2008) Changing patterns in the frequency of Hodgkin lymphoma subtypes and Epstein-Barr virus association in Taiwan. Cancer Sci 99(2):345–349

ClinicalTrials.gov. Bortezomib and Ganciclovir in treating patients with relapsed or refractory Epstein Barr virus-positive lymphoma. Available at http://clinicaltrials.gov/ct2/show/NCT00093704. Accessed 30 Aug 2010

ClinicalTrials.gov. Epstein-Barr Virus (EBV)-specific T cells as therapy for relapsed/refractory EBV-positive lymphomas. Available at http://clinicaltrials.gov/ct2/show/NCT00779337. Accessed 30 Aug 2010

ClinicalTrials.gov. Safety study of EBV specific cytotoxic T-cells to treat relapsed EBV-positive lymphoma. Available at http://clinicaltrials.gov/ct2/show/NCT00058617. Accessed 30 Aug 2010

ClinicalTrials.gov. Study of arginine butyrate and ganciclovir/valganciclovir in EBV(+) lymphoid malignancies. Available at http://clinicaltrials.gov/ct2/show/NCT00917826. Accessed 30 Aug 2010

DLBCL (1993) A predictive model for aggressive non-Hodgkin's lymphoma. The international non-Hodgkin's lymphoma prognostic factors project. N Engl J Med 329(14):987–994

Fagnoni FF, Vescovini R et al (2000) Shortage of circulating naive CD8(+) T cells provides new insights on immunodeficiency in aging. Blood 95(9):2860–2868

Feugier P, Van Hoof A et al (2005) Long-term results of the R-CHOP study in the treatment of elderly patients with diffuse large B-cell lymphoma: a study by the Groupe d'Etude des Lymphomes de l'Adulte. J Clin Oncol 23(18):4117–4126

Franceschi C, Capri M et al (2007) Inflammaging and anti-inflammaging: a systemic perspective on aging and longevity emerged from studies in humans. Mech Ageing Dev 128(1):92–105

Gibson SE, Hsi ED (2009) Epstein-Barr virus-positive B-cell lymphoma of the elderly at a United States tertiary medical center: an uncommon aggressive lymphoma with a nongerminal center B-cell phenotype. Hum Pathol 40(5):653–661

Glaser SL, Lin RJ et al (1997) Epstein-Barr virus-associated Hodgkin's disease: epidemiologic characteristics in international data. Int J Cancer 70(4):375–382

Habermann TM, Weller EA et al (2006) Rituximab-CHOP versus CHOP alone or with maintenance rituximab in older patients with diffuse large B-cell lymphoma. J Clin Oncol 24(19):3121–3127

Hans CP, Weisenburger DD et al (2004) Confirmation of the molecular classification of diffuse large B-cell lymphoma by immunohistochemistry using a tissue microarray. Blood 103(1):275–282
Heslop HE, Slobod KS et al (2010) Long-term outcome of EBV-specific T-cell infusions to prevent or treat EBV-related lymphoproliferative disease in transplant recipients. Blood 115(5):925–935
Hoeller S, Tzankov A et al (2010) Epstein-Barr virus-positive diffuse large B-cell lymphoma in elderly patients is rare in Western populations. Hum Pathol 41(3):352–357
Holtan SG, Porrata LF et al (2008) mTOR inhibitor monotherapy is insufficient to suppress viremia and disease progression in Epstein-Barr virus-driven lymphoproliferative disorders (EBV-LPD). Am J Hematol 83(8):688–689
Jones K, Nourse J et al (2010) Sodium valproate in combination with ganciclovir induces lysis of EBV-infected lymphoma cells without impairing EBV-specific T-cell immunity. Int J Lab Hematol 32(1 Pt 1):e169–e174
Knecht H, Berger C et al (2001) The role of Epstein-Barr virus in neoplastic transformation. Oncology 60(4):289–302
Koch S, Larbi A et al (2008) Multiparameter flow cytometric analysis of CD4 and CD8 T cell subsets in young and old people. Immun Ageing 5:6
Kuze T, Nakamura N et al (2000) The characteristics of Epstein-Barr virus (EBV)-positive diffuse large B-cell lymphoma: comparison between EBV(+) and EBV(−) cases in Japanese population. Jpn J Cancer Res 91(12):1233–1240
Morales D, Beltran B et al (2010) Epstein-Barr virus as a prognostic factor in de novo nodal diffuse large B-cell lymphoma. Leuk Lymphoma 51(1):66–72
Nakamura S, Jaffe E et al (2008) EBV positive diffuse large B-cell of the elderly. In: Swerdlow S, Campo E, Harris N et al (eds) WHO classification of tumours of haematopoietic and lymphoid tissues. IARC, Lyon, pp 243–244
Naylor K, Li G et al (2005) The influence of age on T cell generation and TCR diversity. J Immunol 174(11):7446–7452
Oyama T, Ichimura K et al (2003) Senile EBV+B-cell lymphoproliferative disorders: a clinicopathologic study of 22 patients. Am J Surg Pathol 27(1):16–26
Oyama T, Yamamoto K et al (2007) Age-related EBV-associated B-cell lymphoproliferative disorders constitute a distinct clinicopathologic group: a study of 96 patients. Clin Cancer Res 13(17):5124–5132
Park S, Lee J et al (2007) The impact of Epstein-Barr virus status on clinical outcome in diffuse large B-cell lymphoma. Blood 110(3):972–978
Perrine SP, Hermine O et al (2007) A phase 1/2 trial of arginine butyrate and ganciclovir in patients with Epstein-Barr virus-associated lymphoid malignancies. Blood 109(6):2571–2578
Pfreundschuh M, Trumper L et al (2006) CHOP-like chemotherapy plus rituximab versus CHOP-like chemotherapy alone in young patients with good-prognosis diffuse large-B-cell lymphoma: a randomised controlled trial by the MabThera International Trial (MInT) Group. Lancet Oncol 7(5):379–391
Rosenwald A, Wright G et al (2002) The use of molecular profiling to predict survival after chemotherapy for diffuse large-B-cell lymphoma. N Engl J Med 346(25):1937–1947
Shimoyama Y, Asano N et al (2009) Age-related EBV-associated B-cell lymphoproliferative disorders: diagnostic approach to a newly recognized clinicopathological entity. Pathol Int 59(12):835–843
Spaulding C, Guo W et al (1999) Resistance to apoptosis in human CD8+ T cells that reach replicative senescence after multiple rounds of antigen-specific proliferation. Exp Gerontol 34(5):633–644
Tan LH (2009) A practical approach to the understanding and diagnosis of lymphoma: an assessment of the WHO classification based on immunoarchitecture and immuno-ontogenic principles. Pathology 41(4):305–326
Thompson MP, Kurzrock R (2004) Epstein-Barr virus and cancer. Clin Cancer Res 10(3):803–821

Vaysberg M, Balatoni CE et al (2007) Rapamycin inhibits proliferation of Epstein-Barr virus-positive B-cell lymphomas through modulation of cell-cycle protein expression. Transplantation 83(8):1114–1121

Wong HH, Wang J (2009) Epstein-Barr virus positive diffuse large B-cell lymphoma of the elderly. Leuk Lymphoma 50(3):335–340

Yoshino T, Nakamura S et al (2006) Epstein-Barr virus involvement is a predictive factor for the resistance to chemoradiotherapy of gastric diffuse large B-cell lymphoma. Cancer Sci 97(2):163–166

Zinkernagel RM, Moskophidis D et al (1993) Effector T-cell induction and T-cell memory versus peripheral deletion of T cells. Immunol Rev 133:199–223

Zou P, Kawada J et al (2007) Bortezomib induces apoptosis of Epstein-Barr virus (EBV)-transformed B cells and prolongs survival of mice inoculated with EBV-transformed B cells. J Virol 81(18):10029–10036

HIV and Lymphoma

Lubomir Sokol and Brady E. Betran

Introduction

It is well established that HIV-infected patients are at high risk for AIDS-associated lymphoma (ARL). Among people with AIDS before the advent of highly active antiretroviral therapy (HAART) NHL accounted for 3% of the AIDS-defining conditions while the relative risk of NHL was conservatively estimated 160-fold higher than the 3.1% expected in the non-AIDS population. The risk was higher for immunoblastic histology (630-fold) and Burkitt lymphoma (220-fold), intermediated for diffuse large cell (113-fold), and the lowest for low-grade (14-fold). Most NHL are B-cell type with 20% excess extranodal sites and 15% presenting as primary brain lymphoma (PBL) compared with the 1% expected for non-HIV individuals (Cote et al. 1997).

The first large retrospective series of patients with ARL was published in 1984 (Ziegler et al. 1984), followed by others (Levine et al. 1985; Kalter et al. 1985; Lowenthl et al. 1988; Knowles et al. 1988).

In 1985, the CDC incorporated immunoblastic lymphoma (or equivalent term), Burkitt's lymphoma (BL), and PBL as AIDS-defining conditions (United States, Centers for Disease Control, Department of Health and Human Services 1985). Other types of lymphoproliferative disorders also occurred in HIV-infected people, albeit with much less frequency, including primary effusion lymphoma (PEL) and its solid variants, plasmablastic lymphoma of the oral cavity, large B-cell lymphoma arising in KS-herpesvirus-associated multicentric Castleman disease (Carbone et al. 2008), and even Hodgkin lymphoma (Hessol et al. 1992).

L. Sokol, M.D., Ph.D. (✉)
Department of Malignant Hematology, H. Lee Moffitt Cancer Center & Research Institute, 12902 Magnolia Dr, 33612 Tampa, FL, USA
e-mail: lubomir.sokol@moffitt.org

B.E. Betran, M.D.
Hospital Nacional Edgardo Rebagliati Martins, Lima, Peru

J. Tao and E. Sotomayor (eds.), *Hematologic Cancers: From Molecular Pathobiology to Targeted Therapeutics*, Cancer Growth and Progression 14,
DOI 10.1007/978-94-007-5028-9_6, © Springer Science+Business Media Dordrecht 2012

Since the introduction of (HAART) in 1996, the epidemiology and demographic characteristics of HIV have change in United States. HIV and AIDS prevalence have increased among women, minority populations, and in those acquiring HIV through heterosexual relationships (Barclay et al. 2007)

Concomitantly, the incidence of ARL has fallen dramatically in over 50%, both for systemic NHL and PBL in patients receiving HAART with marked declined in morbidity and mortality (Bohlius et al. 2009; Palella et al. 1998).

Treatment

Pre-HAART Era

Prior to the widespread availability of HAART, the use of standard or high-dose chemotherapy regimens was associated with increased toxicity and high mortality without an increase in response rate or prolong survival (Gill et al. 1987; Kaplan et al. 1989; Bermudez et al. 1989; Gisselbrecht et al. 1993). Therefore, modification of the standard chemotherapy regimens was a logical step. These modifications included lower dose of the standard M-BACOD regimen or m-BACOD piloted by Levine et al. (2001) where a reduced dose of cyclosphosphamide and doxorubicin was implemented. The median survival of 42 patients so treated was only 5.6 months. Although, the hematologic toxicity was less yet the efficacy was similar in a randomized study of the AIDS Clinical Trial Group comparing low dose m-BACOD with the standard dose M-BACOD enrolling 198 patients. The median OS was only 35 and 31 weeks, respectively. This difference was no statically despite all patients receiving concomitant GM-CSF (Kaplan et al. 1997).

Post-HAART Era

The theoretical drug interactions between antineoplastic and antiretroviral therapies early raised concerns on the potential increase in toxicity and decreased efficacy of concurrent treatment with chemotherapy and HAART (Mounier et al. 2009).

Some studies were design including a sequential population of patients untreated with HAART (pre-HAART), followed by concomitant chemotherapy and HAART (Hoffmann et al. 2003; Sparano et al. 2004). Others, opted for withholding HAART during the chemotherapy administration to prevent adverse drug interactions that could reduce the lymphoma cure (Little et al. 2003) and yet many treated all patients with concomitant chemotherapy and HAART from the outset (Ratner et al. 2001; Little et al. 2003; Navarro et al. 2005; Levine et al. 2004). Finally, some investigators chose to stratified patients according with risk factors (Mounier et al. 2006; Mitrou et al. 2006) or utilizing dose- adjusted chemotherapy according with the hematologic toxicity (Ratner et al. 2001; Gea-Banacloche 2010). All patients included in the studies had a median CD4 cell count of>100/ul unlike most of the pre-HAART era that enrolled some patients with very low CD4 cell counts. See Table 1.

The lack of prospective randomized trials prevents to assess the merit of one chemotherapy regimen over the others and the true efficacy of HAART versus non-HAART treatment. However, general principles may be drawn.

The concomitant treatment of ARL patient with chemotherapy and HAART is safe and with higher efficacy.

The OS of patients on HAART seems to have improved significantly compared to historical pre-HAART series data (Vaccher et al. 2001; Antinori et al. 2001; Hoffmann et al. 2003; Sparano et al. 2004; Mounier et al. 2006), with less life-threatening toxicity (Sparano et al. 2004) and fewer opportunistic infections (OI) 21.

CR rates were significantly higher in HAART treated patients in some studies (Antinori et al. 2001; Hoffmann et al. 2003) but not in all trial (Vaccher et al. 2001). Those patients in CR and also HAART viral responders experienced an excellent OS of 83% at 39 months (Hoffmann et al. 2003).

Patient accrued in the only study withholding HAART during chemotherapy treatment achieved an excellent CR rate of 87% in those patients with CD4 >100/ul compared to 56% of those with CD4 <100, emphasizing that the achievement of CR may be highly related to the efficacy of the chemotherapy regimen selected as well as the immune-competence of the patients (Antinori et al. 2001).

Treatment of ARL Including Rituximab

The addition of rituximab to CHOP chemotherapy resulted in a significant improvement in clinical outcome for individuals with non-HIV-associated aggressive B-cell lymphoma (Coiffier et al. 2002).

The profound depletion of normal B cells that results after rituximab treatment and the potential viral reactivation that may occur with this drug (Gea-Banacloche 2010) grew concern among investigators about possible severe side effect of the concurrent treatment of chemotherapy, rituximab and HAART in ARL (Table 2).

Several early non-randomized studies in ARL have preliminary demonstrated not only the safety of combining chemotherapy with rituximab but also some benefit with R-CHOP or R-CDE treated patients (Spina et al. 2005b; Boue et al. 2006). Unexpectedly, the only randomized phase III study (AMC trial 010) that compared CHOP with and without rituximab in patients receiving HAART found only a trend towards greater efficacy in patients on the riruximab arm with CR of 47% vs. 57%, (p=NS) and in OS of 110 weeks compared to 139 weeks, (p=NS). Treatment-related infectious death occurred in 14% of patient receiving R-CHOP versus with 2% in the chemotherapy-alone group (p=.035). However, further analysis of the data showed that these infectious deaths occurred primarily among the 60% of the patients with CD 4 less than 50/ul. If these patients were excluded from the analysis, no significant difference in infectious death was seen.

Table 1 Studies including or comparing HAART without rituximab in the treatment of ARL

Author	Treatment	Patients	CD4/ul	CR %	OS	Comments
Ratner et al. (2001)	mCHOP	40	138	12 Mitrou et al. (2006)		CR duration 9 months
AMC	CHOP + G-CSF	25	122	11 Galicier et al. (2007)		Not reached
	All on HAART			NS		
Vaccher et al. (2001)	CHOP	80	146	36	7 months	52% OI vs.
Aviano, Italy	CHOP-HAART	24	190	50 (NS)	Not reached (p = 0.03)	18% (p = 0.005)
Antinori et al. (2001)	CHOP	34	144	71 on HAART responders	78%	Viral response to HAART was associated with better CR and prolonged survival
Rome, Italy Two Centers	Other	10		30 in HAART naïve/failed	18% at 1 year (p = 0.0001)	
Little et al. (2003)	DA-EPOCH	39	198	74%	60% at 53 months	CR 87% in CD4 >100. OS 87%
NCI, NIH	Infusion	HAART withheld				CR 56% in CD4 <100. OS 16%
Navarro et al.	CHOP	23 with HAART	134	74%	74% at 5 year	No difference compared with similar HIV-negative cohort
Barcelona, Spain Navarro et al. (2005)						
Spina et al. (2005a) Aviano, Italy	CHOP or CHOP-like	110 with 100 on HAART		51%	37% at 3 year	Better outcome for a similar HIV+cohort
Hoffmann et al. (2003)	CHOP-like	HAART 61	167	71	9.0 months	HAART responders had CR 77% vs. 50% for non-responders (p = 0.09)
5 German centers.		No HAART 93		48% (p = 0.006)		Pts in CR and HAART responders had an OS of 83% at 39 months

Sparano et al. (2004) ECOG 1494	CDE infusion	43 Didanosine 55 HAART	90 227	47% NS 44%	39% NS 45 at 2 year	HAART pts had less toxicity and improve survival
Levine et al. (2004) USC	CVP+liposomal doxo	24 all on HAART	112	75%	58% at 1 year	Liposomal doxo+CVP is active. CR are independent of HIV viral load control or MDR-1
Mounier et al. (2006)	Score 0: ACVBP or CHOP	485	129		GR 49%	HIV score, HAART and IPI affect survival but not intensity of CHOP-based chemotherapy
French/Italian	Score 1: CHOP or Ld-CHOP	HIV 0: 188			IR 26%	
GELA/GICAT Groups	Score 2, 3: Ld-CHOP or VS	HIV 1: 86 HIV 2, 3: 78 HAART 187 No HAART 298			PR l 7% at 6 year	
Weiss et al. (Mounier et al. 2006)	CHOP	77	175 SR	SR: 79%	HR 7.2 months	Risk-adapted treatment is effective and safe
German ARL Lymphoma Study Group	HR 75% dose in Cycle 1	HR: 24 SR: 48 All on HAART	223 HR 34	HR 29%	SR not reached at 26.2 months	Standard risk patients have a median OS rates similar to non-HIV NHL base on aaIPI.

Abbreviations: *AMC* AIDS Malignancy Consortium, *GELA* Adult Lymphoma Study Group, *GICAT* AIDS and Tumors Italian Cooperative Group, *Low-dose CHOP (mCHOP)* CTX 375, Doxo 25, prednisone 100 day 11–5, *ACVBP* Doxorubicin, cyclosphosphamide, vindesine, bleomycin and prednisolone, *VS* vincristine and prednisolone, HIV scores based on 3 independent risk factors: *PS 2–4, prior AIDS, and CD4 < 100/ul* HIV score 0 (good risk), HIV score 1 (intermediate risk) and HIV score 2–3 (poor risk), *SR* standard risk = 0–1 or 2–3 of risk factors: CD4 < 50/ul, prior OI and PS >3

Table 2 Treatment of ARL with regimens including rituximab

Authors	Treatment	Patients	CD4	CR	OS	Comments
Spina et al. (Mitrou et al. 2006)	R-CDE 76% on HAART	53 high-grade 21 BL	161	70%	64 at 2-year	Compared with CDE alone, R-CDE had higher G3/4 infections (31 vs. 20%) and lethal infection (2 vs. 0%) with similar G3/4 neutropenia
GICAT, Univ of Viena and Albert Einstein						
Kaplan et al. (Coiffier et al. 2002)	CHOP Vs R-CHOP	51	147 (<50 25%)	47	110 weeks	Treatment-related infection for R-CHOP 14% vs. 2% for CHOP (p=.035).
AMC	All on HAART	99	130 (<50 22%)	57.6 (NS)	139 weeks (NS)	If patients with CD4 <50 were excluded, no significant difference in infectious death
Boue et al. (Gea-Banacloche 2010)	R-CHOP	61	172	77	75 at 2-year	Best results obtained in patients with IPI score 0/1 and CD4 >100
20 French Centers	On HAART					
Ribera et al. (Spina et al. 2005b).	R-CHOP	86	<50=9	69	56 at 3-year	IPI score >1 associated with RR
PETHEMA, GELTAMO, GELCAB, GESIDA	On HAART	81 evaluable				Low and intermediate IPI score and virological response had a better OS
Sparano et al. (Kaplan et al. 2005)	Concurrent	51	181 (16<100)	73	70	50% of patients in sequential arm did not receive R.
AMC	R-EPOCH Vs Sequential R-EPOCH On HAART in 71% in each arm	56	194 (17 <100)	55	67 at 2-year	Treatment-associated death was 10% in concurrent arm with 38% occurring in patients with CD4<50.
Dunleavy et al. (Boue et al. 2006)	SC-EPOCH-RR x 3-6	33	<100 in 14	91	68 at 5-year	79% of patients receive only 3 cycles. PFS for GCB (n=21) was 95% vs. 44% for non-GCB (N=9) at 5-year
NCI, NIH	HAART withheld		>100 in 29			PET after cycle 2 not useful

Abbreviations: *R* rituximab, *CDE* infusional cyclosphosphamide, doxorubicin and etoposide
DA-EPOCH dose-adjusted infusional etoposide, doxorubicin, vincristine, cyclosphosphamide and prednisone
SC-EPOCH-RR short course EPOCH-dose dense, ritxumab on day 1 and 5

Later studies have confirmed the benefit of the addition of rituximab (Ribera et al. 2007; Sparano et al. 2010; Dunleavy et al. 2010). Ribera et al. (2007) and Boue et al. (2006) utilizing both R-CHOP regimen found a better outcome in viral load responders, lower IPI score and CD4 count of more than 100/ul (Boue et al. 2006). Continuing with previous the AMC experience with infusional EPOCH regimen, Sparano et al. reported the randomized result of concurrent versus sequential R-EPOCH x 4–6 cycles. The primary efficacy endpoint was met for the concurrent arm only. The toxicity was similar in the 2 arms despite the fact that the baseline CD4 cell count less than 50/ul had a high infectious death rate in the concurrent arm. Patients in the concurrent arm achieved a respectable CR rate of 75% with OS of 70% at 2 year.

Finally, Dunleavy et al. reported the NCI trial utilizing a short-course of EPOCH associated with a double dose of rituximab administered on days 1 and 5 (dose dense). Seventy nine percent of the patients receive only three cycles of treatment. The CR rate was 91 with 68% of the patients alive at 5 years. Interesting, the 21 patients with GCB tumor histogenesis had a PFS of 95% compared to 44% for 9 non-GBC patients. The inclusion of FDG-PET efficacy assessment after cycle 2 of treatment was considered problematic due to the difficult to differentiate inflammation from active lymphoma. Indeed, 65% of the PET scans were positive after cycle 2 and yet few patients relapsed. This abbreviated treatment seems highly effective and less immune-suppressive compared with standard strategies.

The lack of prospective randomized data comparing R-CHOP to R-EPOCH or SC-EPOCH-RR make it difficult for final recommendation of an optimal treatment regimen, although current phase 2 data positions infusional EPOCH as the regimen offering patients with ARL the most benefit.

HIGH- Dose Therapy and SCT in ARL

High dose therapy with hematopoietic peripheral blood stem cell transplant has been explored in several studies. See Table 3. The reports include single institution and cooperative trial. The study populations vary with inclusion of patients with high-risk in first remission and others with relapsed or refractory disease. All have included DLBL and other NHL histology, BL and HL.

Most of the trials include 20 or fewer patients except 2 multicenter studies with 50 and 68 patients with the longest follow-up (Re et al. 2009; Balsalobre et al. 2009). The median lymphoma-free survival for the these series was 48.9% at 45 months and 53% at 3 years, while the median OS was 49.8% at 45 months and 61% at 3 years (Re et al. 2009; Balsalobre et al. 2009).

Although no definitive conclusion about efficacy can be drawn from all these studies, sensitive disease to chemotherapy, low HIV viral load, good PS, higher CD4 cell count, DLBCL histology, and the achievement of CR, were all features associated with better outcomes.

In the first comparative analysis between HIV-positive and a matched cohort of HIV-negative lymphoma patients undergoing ASCT, Diez-Martin, et al. (2009)

Table 3 Autologous SCT studies in ARL

Authors	Pts	Disease	Median CD4	Conditioning regimen	PFS %	OS %	Comments
Spitzer et al. AMC 020	20	NHL 15 HL 5	203	BUCY	49.5 at 6 months	74.4 at 6 months	1 death
Krishnan et al.	20	NHL 12	175	BEC	85 at 32 months	85 at 32 months	All patients had low HIV viral load <10,000.
City of Hope		HL 2 BL 6	<100 first 5 pts excluded	FTBI + etoposide/ CY			1 death
Gabarre et al. Pitie-Salpetriere	14	NHL 6 BL 2 HL 6	209	BEAM CY/TBI BU/Ara-C, melphalan			10/14 achieved CR and 5 were alive at 2-year
Serrano et al. GELTAMO and GESIDA	11	NHL 8 BL 3 HL 1	186	BEAM BEAC	65 at 30 months	81 at 15 months	Similar efficacy to HIV + patients No deaths
Re et al. GICAT	50	NHL 29 BL 2 HL 19	190 11 <100	BEAM	48.9 at 45 months	49.8 at 45 months	CD4, marrow +, and poor PS independently affected survival
Balsalobre et al.	68	NHL 42	162	BEAM and variants 65	56 at 3-year	61% at 3-year	Mortality 7.5% at 12 months.
EGBMT		HD 18 BL 8	Viral load <200/ml in 80%	TBI 3			No CR, >2 prior Tx, and other than DBCL histology worse outcome.

Abbreviations: *BEAC* carmustine, etoposide, cytarabine, cyclosphosphamide, *BEAM* carmustine, etoposide, cytarabine, melphalan, *BEC* carmustine, etoposide, cyclosphosphamide, *BU* busulfan, *FTIB* fractionated total body irradiation
AMC AIDS Malignancies Consortium, *GELTAMO/GESIDA* Spanish Cooperative Group Lymphoma and Transplant and Spanish Group of AIDS, *GICAT* Italian Cooperative Group on AIDS and tumors, *EGBMT* European Group for Blood and Marrow Transplantation

reported a retrospective EGBMT study including 53 patients (66% NHL and 34% HL) within each cohort.

The incidence of relapse, OS and PFS were similar in both groups. At a median follow up of 30 months, the OS was 61.5 and 70% for HIV-positive and HIV-negative lymphoma, respectively (NS), and DFS was 61 and 56% (NS). A higher non-relapse mortality within the first year after AST was observed in the HIV-positive group (8% vs. 2%), predominantly because of early bacterial infections, although this was not statically significant and did not influence survival. This results suggest that that HIV-positive patients should be considered for AST according with the same indications as for HIV-negative lymphoma patients.

Similarly, in a single institution analysis Krishnan et al. (2010) compared 29 patients with HIV-positive NHL matched with HIV-negative-NHL patients. The non-relapse mortality was similar in HIV-positive-NHL (11%) and HIV-negative-NHL (4%) with a p=.18. Two-year DFS for HIV-positive-NHL was 76 and 56% for HIV-negative-NHL group (p=.33). OS was also similar with 2-year estimates 75 and 75%, respectively (p=.93).

Both studies strongly suggest that the HIV status alone should not longer exclude patients with ARL from transplant clinical trials.

Burkitt Lymphoma in HIV-Positive Patients

It has been recognized that the cure of Burkitt lymphoma is possible with appropriate intensive chemotherapy regimens in non-HIV-infected patients and similar approaches have been used for HIV-associated Burkitt lymphoma (Noy 2010).

The published studies (see Table 4) are single institutions trials (Cortes et al. 2001; Galicier et al. 2007; Spina et al. 1998) or cooperative multicenter efforts (Noy 2010; Montoto et al. 2010; Oriol et al. 2008) using different chemotherapy regimens. The CODOX-M/IVAC regimen has resulted in CR rates of 85% in the UK experience enrolling 30 patients with excellent OS of 52% at 3 years. A remarkable immunologic recovery after treatment was noted with 87% of patients had undetectable viral load and 80% had CD4 of more than 200/ul at 12 months post therapy (Spina et al. 1998). The preliminary AMC data revealed an OS of 85.7% at 1 year with the addition of rituximab to the CODX-M/IVAC combination. Likewise, the PETHEMA group reported a CR rate of 84% in 19 patients treated with an intensive regimen (GMALL German protocol) with 2 years OS of 73%. Interesting, this group noted that patients' responders to HAART had a better OS than non-responders (85% vs. 12%), respectively.

The Hyper-CVAD regimen widely use in non-HIV patients in USA seems an active combination in only 13 patients treated at MDA (Cortes et al. 2001). The CR rate of 92% and 2-year OS of 48% achieved in a population with low median CD4 of only 77/ul appears promising

Other less intensive treatment regimens such as standard CHOP or CHOP-like variations do not seems to achieved optimal outcomes as indicated by the OS of only 7 months and low CR rates of 40% in the Italian experience (Spina et al. 1998).

Table 4

Authors	Treat	Pts	CR	OS	CD4	Comments
Cortes et al. (2001)	Hyper-CVAD	13	92	48% at 2 year	77	6/7 pts on HAART alive in CR > 2 year
MDA						4 pts on no HAART died
Galicier et al. (2007)	LMB86	63	70	41.7 at 66 months	239	Pts with CD4 <200 and PS > 2 OS 12% at 2 year
St Louis Hospital, Paris		Stage IV				
Wang et al. (2003)	CODOX-M/IVAC	14	63	60	149	
MSKCC	7/8 stage IV		67	60 at 2 year		
	OTHER regimens 6					
Noy (2010)	CODOX-M/IVAC-R	22		85.7 at 1 year	290	No TRM
AMC 048					5 pts <100	
Montoto et al. (2010)	CODOX-M/IVAC	30	85	52 at 3 year	171	Viral load undetectable in 87% and CD4 >200 in 80% at 12 months
UK Hospitals						
Spina et al. (1998)	ACVBP	46	40	7 months	235	
Aviano, Italy	CHOP					
	mCHOP					
	CHVmP					
	Others					
Oriol et al. (2008)	GMALL protocol	19	84	73 at 2 year		HAART responders OS 85% at 2 year
PETHEMA						HAART no-responders OS 27% at 2 year

Abbreviations: *Hyper-CVAD* hyperfractionated cytoxan, doxorubicin, vincristine, dexamethasone (add-numbered courses x4), HD-MTX, and HD-ara-C (even-numbered courses × 4)

LMB86 cytoreductive: COP, Induction: COPADM (CHOP + HD-MTX) x 2 and consolidation: etoposide and H- ara-C, *CODOX-M* CHOP with HD-cytoxan and HD-MTX, *IVA* Ifosfamide, etoposide, HD-ara-C. *R* rituximab. *GMALL (GERMAN GROUP protocol)* prophase; cytoxan and prednisone, cycle A; rituximab, vincristine, MTX < ifosfamide, teniposide and ara-C; cycle B: rituximab, vincristine, MTX, cytoxan, dexa, doxorubicin; cycle C: rituximab, vindesine, MTX, dexa, etoposide and ara-C. *ACVBP* cytoxan, doxorubicin, vincristine, vindesine, bleomycin and prednisone. *CHVmP* cytoxan, doxorubicin, teniposide, prednisone, vincristine and bleomycin

References

Antinori A, Cingolani A, Alba L et al (2001) Better response to chemotherapy and prolonged survival in AIDS-related lymphomas responding to highly active antiretroviral therapy. AIDS 15:1483–1491

Balsalobre P, Diez-Martin JL, Re A et al (2009) Autologous stem cell transplantation in patients with HIV-related lymphoma. JCO 27:21922198

Barclay LR, Buskin SE, Kahle EM et al (2007) Clinical and immunologic profile of AIDS-related lymphoma in the era of highly active antiretroviral therapy. Clin Lymphoma Myeloma 7:272–279

Bermudez MA, Grant KM, Rodvien R et al (1989) Non-Hodgkin's lymphoma in a population with or at high risk for acquired immunodeficiency syndrome: indication for intensive chemotherapy. Am J Med 86:71–76

Bohlius J, Schmidlin K, Costagliola D et al (2009) Incidence and risk factors of HIV-related-Hodgkin's lymphoma in the era of combination antiretroviral therapy: a European multicohort study. Antivir Ther 14:1065–1074

Boue F, Gabarre J, Gisselbrecht C et al (2006) Phase II trial of CHOP plus rituximab in patients with HIV-associated non-Hodgkin's lymphoma. J Clin Oncol 24(25):4123–4128

Carbone A, Cesarman E, Spina M et al (2008) HIV-associated lymphomas and gamma-herpesviruses. Blood 113:1213–1224

Coiffier B, Lepage E, Briere J, Herbrecht R, Tilly H, Bouabdallah R, Morel P, Van Den Neste E, Salles G, Gaulard P, Reyes F, Lederlin P, Gisselbrecht C (2002) CHOP chemotherapy plus rituximab compared with CHOP alone in elderly patients with diffuse large B-cell lymphoma. N Engl J Med 346:235–242

Cortes J, Thomas D, Rios A et al (2001) Hyperfractionated cyclosphosphamide, vincristine, doxorubicin, and dexamethasone and highly active antiretroviral therapy for patients with acquired immunodeficiency syndrome-related Burkitt lymphoma/leukemia. Cancer 94:1492–1499

Cote TR, Biggar RJ, Rosenberg PS et al (1997) Non-Hodgkin's lymphoma among people with AIDS: incidence, presentation and public health burden. Int J Cancer 73:645–650

Diez-Martin JL, Balsalobre P, Re A et al (2009) Comparable survival between HIV+ and HIV- non-Hodgkin and Hodgkin lymphoma undergoing autologous peripheral blood stem cell transplantation. Blood 113:6011–6014

Dunleavy K, Little RF, Pittaluga S et al (2010) The role of tumor histogenesis, FDG-PET, and shot-course EPOCH with dose-dense rituximab (SC-EPOCH-RR) in HIV-associated diffuse large B-cell lymphoma. Blood 15:3017–3024

Gabarre J, Marcelin A-G, Azar N et al (2004) High-dose therapy plus autologous hematopoitic stem cell transplantation for human immunodeficiency virus (HIV)-related lymphoma: results and impact on HIV disease. Haematologica 89:1100–1108

Galicier L, Fieschi C, Borie R et al (2007) Intensive chemotherapy regimen (LMB86) for St Jude stage IV AIDS-related Burkitt lymphoma/leukemia: a prospective study. Blood 110:2846–2854

Gea-Banacloche JC (2010) Rituximab –associated infections. Semin Oncol 47:187–198

Gill PS, Levine AM, Krailo M et al (1987) AIDS-related malignant lymphoma: results of prospective treatment trials. J Clin Oncol 5:1322–1328

Gisselbrecht C, Oksenhendler E, Tirelli U et al (1993) Human immunodeficiency virus-related lymphoma treatment with intensive combination chemotherapy. Am J Med 95:188–196

Hessol NA, Katz MH, Liu JY, Buchbinder SP, Rubino CJ, Holmberg SD (1992) Increased incidence of Hodgkin disease in homosexual men with HIV infection. Ann Intern Med 117:309–311, Spectrum of AIDS-associated malignant disorders

Hoffmann C, Wolf E, Fatkenheuer G et al (2003) Response to highly active antiretroviral therapy strongly predicts outcome in patients with AIDS-related lymphoma. AIDS 17:1521–1529

Kalter SP, Riggs SA, Cabanillas F et al (1985) Aggressive non-Hodgkin's lymphoma in immunocompromised homosexual men. Blood 66:655–659

Kaplan LD, Abrams DI, Feigal E et al (1989) AIDS-associated non-Hodgkin's lymphoma in San Francisco. JAMA 3:719–724

Kaplan LD, Straus D, Testa MA et al (1997) Low-dose compared with standard-dose m-BACOD chemotherapy for non-Hodgkin's lymphoma associated with human immunodeficiency virus infection. The NEJM 336:1641–1648

Kaplan L, Lee J, Ambinder R et al (2005) Rituximab does not improve clinical outcome in randomized phase 3 trial of CHOP with or without rituximab in patients with HIV-associated non-Hodgkin lymphoma: AIDS—malignancies consortium trial 010. Blood 106(15):1538–1543

Knowles DM, Chamulak GA, Subar M et al (1988) Lymphoid neoplasia associated with the acquired immunodeficiency syndrome (AIDS). Ann Intern Med 108:744–753

Krishnan A, Molina A, Zaia J et al (2005) Durable remissions with autologous stem cell transplantation for high-risk HIV-associated lymphomas. Blood 105:874–878

Krishnan A, Palmer JM, Zaia JA et al (2010) HIV status does not affect the outcome of autologous stem cell transplantation (ASCT) for non-Hodgkin lymphoma (NHL). Biol Blood Marrow Transplant:1–7

Levine AM, Gill PS, Meyer PR et al (1985) Retrovirus and malignant lymphoma in homosexual men. JAMA 254:1921–1925

Levine AM, Wernz JC, Kaplan L et al (2001) Low-dose chemotherapy with central nervous system prophylaxis and zidovudine maintenance in AIDS-related lymphoma. A prospective multi-institutional trial. JAMA 266:84–88

Levine SM, Tulpule A, Espina B et al (2004) Liposome-encapsulated doxorubicin in combination with standard agents (cyclosphosphamide, vincristine, prednisone) in patients with newly diagnosed AIDS-related non-Hodgkin's lymphoma: results of therapy and correlates of response. JCO 22:2662–2670

Little RF, Pittaluga S, Grant N et al (2003) Highly effective treatment of acquired immunodeficiency syndrome-related lymphoma with dose-adjusted EPOCH: impact of antiretroviral therapy suspension and tumor biology. Blood 101:4653–4659

Lowenthl DA, Straus DJ, Wise-Campbell S et al (1988) AIDS-related lymphoid neoplasia. The memorial experience. Cancer 61:2325–2337

Mitrou P, Arasteh J, Weiss R et al (2006) Acquired immunodeficiency syndrome-related lymphoma. Simultaneous treatment with combined cyclosphophamide, doxorubicin, vincristine, and prednisone chemotherapy and highly active antiretroviral therapy is safe and improves survival-results of the German multicenter trial. Cancer 106:1560–1568

Montoto S, Wilson J, Shaw K et al (2010) Excellent immunological recovery following CODOX-M/IVAC, an effective intensive chemotherapy for HIV-associated Burkitt's lymphoma. AIDS 24:851–856

Mounier N, Spina M, Gabarre J et al (2006) AIDS-related non-Hodgkin lymphoma: final analysis of 485 patients treated with risk-adapted intensive chemotherapy. Blood 107:3832–3840

Mounier N, Katlama C, Costagliola D et al (2009) Drug interactions between antineoplastic and antiretroviral therapies: implications and management for clinical practice. Crit Rev Oncol Hematol 72:10–20

Navarro JT, Lloveras N, Ribera JM et al (2005) The prognosis of HIV-infected patients with diffuse large B-cell lymphoma treated with chemotherapy and highly active antiretroviral therapy is similar to that of HIV-negative patients receiving chemotherapy. Hematologica 90:704–705

Noy A (2010) Controversies in the treatment of Burkitt lymphoma in AIDS. Curr Opin Oncol 22:443–448

Oriol A, Ribera J-M, Bergua J, Mesa EG, Grande C, Esteve J, Brunet S, Moreno M-J, Escoda L, Hernandez-Rivas J-M, Hoelzer D (2008) High-dose chemotherapy and immunotherapy in adult Burkitt lymphoma. Comparison of results in human immunodeficiency virus-infected and noninfected patients. Cancer 113:117–125

Palella FJ, Delaney KM, Moorman AC et al (1998) Declining morbidity and mortality among patients with advanced human immunodeficiency virus infection. The New Engl J Med 338:853–860

Ratner K, Lee J, Tang S et al (2001) Chemotherapy for human immunodeficiency virus-associated non-Hodgkin's lymphoma in combination with highly active antiretroviral therapy. J Clin Oncol 19:2171–2178

Re A, Michieli M, Casari S et al (2009) High-dose therapy and autologous peripheral blood stem cell transplantation as salvage treatment for AIDS-related lymphoma: long term results of the Italian cooperative group on AIDS and tumors (GICAT) study with analysis of prognostic factors. Blood 114:1306–1313

Ribera MJ, Oriols A, Morgade M et al (2007) Safety and efficacy of cyclosphosphamide, adriamycin, vincristine, prednisone and rituximab in patients with human immunodeficiency virus-associated diffuse B-cell lymphoma: results of a phase II trial. Br J Haematol 140:411–419

Serrano D, Carrion R, Balsalobre P et al (2005) HIV-associated lymphoma successfully treated with peripheral stem cell transplantation. Exp Hematol 33:487–494

Sparano JA, Lee S, Chen MG et al (2004) Phase II trial of infusional cyclosphosphamide, doxorubicin, and etoposide in patients with HIV-associated non-Hodgkin's lymphoma: an Eastern Cooperative Oncology Group trial (E1494). JCO 22:1491–1500

Sparano JA, Lee JY, Kaplan LD et al (2010) Rituximab plus concurrent infusional EPOCH chemotherapy is highly effective in HIV-associated B-cell non-Hodgkin lymphoma. Blood 115:3008–3016

Spina M, Tirelli U, Zagonet V et al (1998) Burkitt's lymphoma in adults with and without human immunodeficiency virus infection. Cancer 82:766–774

Spina M, Carbone A, Vaccher E et al (2005a) Outcome in patients with non-Hodgkin lymphoma and with or without human immunodeficiency virus infection. HIV/AIDS 38:142–144

Spina M, Jaeger U, Sparano J et al (2005b) Rituximab plus infusional cyclosphosphamide, doxorubicin, and etoposide in HIV-associated non-Hodgkin lymphoma: pooled results from 3 phase 2 trials. Blood 105(5):1891–1897

Spitzer TR, Ambinder DM, Lee JY et al (2008) Dose-reduce busulfan, cyclosphosphamide, and autologous stem cell transplantation for human immunodeficiency virus-associated lymphoma: AIDS malignant consortium study 020. Biol Blood Marrow Transplant 14:59–66

Thomas DA, Faderl S, O'Brien S et al (2006) Chemoimmunotherapy with hyper-CVAD plus riruximab for the treatment of adult Burkitt and Burkitttype lymphoma or acute lymphoblastic leukemia. Cancer 106:1569–1580

United States, Centers for Disease Control, Department of Health and Human Services (1985) Revision of the case definition of acquired immunodeficiency syndrome for national reporting. Ann Intern Med 103:402–403

Vaccher E, Spina M, di Gennaro G et al (2001) Concomitant cyclosphosphamide, doxorubicin, vincristine, and prednisone chemotherapy plus highly active antiretroviral therapy in patients with human immunodeficiency virus-related, non-Hodgkin lymphoma. Cancer 91:155–163

Wang ES, Straus FJ, Teruya-Feldstein J et al (2003) Intensive chemotherapy with cyclosphosphamide, doxorubicin, high-dose metotrexate/ifosfamide, etopsodise, and high-dose cytarabine (CODX-M/IVAC) for human immunodeficiency virus-associated Burkitt lymphoma. Cancer 98:1196–1205

Ziegler JL, Beckstead JA, Volberding PA et al (1984) Non-Hodgkin's lymphoma in 90 homosexual men. N Engl J Med 311:565–570

Molecular Biology of Mantle Cell Lymphoma

Bijal Shah, Peter Martin, Jianguo Tao, and Eduardo M. Sotomayor

Introduction

Mantle cell lymphoma (MCL) is an aggressive Non-Hodgkin's lymphoma with nearly uniform tendency for recurrence and untimely demise despite intensive chemotherapeutic approaches. This has underscored a need for more targeted approaches to improve remission duration, salvage relapsed disease, and ultimately cure this malignancy. The following chapter will highlight progress to date in unraveling the complex biology of this disease with an emphasis on pathways amenable to targeted intervention.

Epidemiology

MCL accounts for approximately 2–10% of NHL, with a predilection for elderly (median age 68) and male individuals (~2:1)(Leux et al. 2011; Morton et al. 2006; Sant et al. 2010; Smedby and Hjalgrim 2011; The Non-Hodgkin's Lymphoma Classification Project 1997). A family history of lymphoma, exposure to European

B. Shah (✉) • E.M. Sotomayor
Departments of Malignant Hematology, H. Lee Moffitt Cancer Center & Research Institute, Magnolia Drive, FOB-3 Room 5.3113, Tampa, FL 12902, USA
e-mail: Bijal.Shah@moffitt.org

P. Martin
Division of Hematology/Oncology, Weill-Cornell Medical School, New York, NY, USA

J. Tao
Departments of Hematopathology, H. Lee Moffitt Cancer Center & Research Institute, Tampa, FL, USA

J. Tao and E. Sotomayor (eds.), *Hematologic Cancers: From Molecular Pathobiology to Targeted Therapeutics*, Cancer Growth and Progression 14, DOI 10.1007/978-94-007-5028-9_7, © Springer Science+Business Media Dordrecht 2012

strains of Borellia Burgdefori , and genetic polymorphisms in the pro-inflammatory cytokine IL-10 have all been weakly associated with development of this disease (Wang et al. 2007; Schöllkopf et al. 2008; Skibola et al. 2010). However, there is no uniform exposure or inherited abnormality sufficient to explain the majority of cases. Survival following diagnosis is improving, though remains inferior to other low and intermediate grade B-cell NHL, with current data suggesting median overall survival of approximately 5 years (Herrmann et al. 2008). It is not clear whether survival improvements reflect improvements in therapy or an increased perception for earlier, more indolent stages of disease.

Biology

MCL is characterized by the translocation of cyclin D1 on 11q13 to the immunoglobin heavy-chain joining region on 14q32. This leads to aberrant production of cyclinD1, a protein critical for regulating transition from G1 to S phase. Interestingly, rare cases of MCL have been identified which lack this translocation, but, instead are accompanied by dysregulated production of cyclinD2 or cyclinD3, highlighting the importance of proliferative drive (Rosenwald et al. 2003).

The immunophenotype of MCL includes CD5, CD19, CD20, and CD22 (Klapper 2011). Although initial reports suggested a pregerminal center origin, subsequent data showing biased IGVH use and somatic hypermutation suggest that at least a subset derive from antigenically exposed B-cells (Camacho et al. 2003; Hadzidimitriou et al. 2011; Kienle et al. 2003; Thelander and Rosenquist 2008). Indeed Kolar et al. have identified a possible population of tonsillar B-cells with a similar immunophenotype which could reflect the cell of origin for MCL. These cells appear to be intermediate between naïve and germinal center cells, with expression of activation induced cytidine deaminase (AID), a low burden of somatic hypermutation, and in the G1 phase of cell division (Kolar et al. 2007). The expression of chemokine receptor genes involved in lymphocyte trafficking in MCL may similarly support this concept that this lymphoma is derived from an "early" antigen exposed B-cell (Kurtova et al. 2009).

MCL is classically described as being slightly larger than normal lymphocytes with a thin rim of cytoplasm, and an indented nucleus with slight relaxation of chromatin, but without prominent nucleoli. Morphologic variants include the small cell type (closely resembling CLL), and the blastoid/pleomorphic subtype which can resemble a large cell lymphoma or lymphoblastic leukemia. As might be anticipated, those MCL with small cell histology tend to behave more indolently, while those with blastoid/pleomorphic histology have aggressive patterns of growth with generally poorer response to therapy (Tiemann et al. 2005).

Among classical morphologic variants, three growth patterns have been recognized: mantle zone, nodular, diffuse. The mantle zone pattern is uncommon (~2%), and reflects malignant expansion of the mantle region of the follicle with impingement upon small, "punched-out" residual germinal centers (Tiemann et al. 2005). A rare, but distinct, variant of the mantle zone pattern includes "in situ" MCL, in which scattered lymphocytes possessing the cyclin D1 translocation can be identified in the mantle zone without accompanying expansion of this region (Aqel et al. 2007; Carvajal-Cuenca et al. 2011; Richard et al. 2006). Ongoing expansion of the mantle zone is later accompanied by loss of residual germinal centers, producing nodular variant, which can be seen in ~18% of cases (Tiemann et al. 2005). Ultimately, progressive growth leads to complete loss of the nodal architecture, producing the more commonly observed diffuse pattern.

Thus, these patterns of growth likely reflect a continuum, progressing from an 'in situ', to a mantle zone, to a nodular, and finally to a diffuse pattern of growth. Indeed, a similar progression from a small cell variant to classical variant, and ultimately blastoid/pleomorphic disease may also occur in parallel with changes in the histological pattern of growth. Together, these changes appear to reflect an increasing proliferative drive, and highlight an ongoing process of unregulated genetic damage proceeding from the fundamental cyclin D1 translocation.

This notion of superimposed genomic injury is especially important when attempting to reconcile failed attempts to develop mouse models of MCL via overexpression of cyclinD1. Success was ultimately obtained in two contexts. The first relied on mitogenic stimulation (using pristane) in aged Eμ-cyclinD1 transgenic mice. In the second approach, Eμ is coupled to a mutated isoform of cyclinD1 which lacks the phosphorylation site (Threonine 268) important in nuclear export (Eμ—D1/T286A). Interestingly, the age of the mouse was important in both experiments. In the first model, pristane was incabable of generating lymphoma in mice less than 10 months old, suggesting that an aged immunological background is necessary for lymphomagenesis. In Em-D1/T286A model, mice had to be followed for 13–14 months for lymphomagenesis (Gladden et al. 2005; Smith et al. 2006).

Secondary Genomic Changes

Comparative genomic hybridization (CGH) can be helpful to identify unbalanced chromosomal gains or losses. Several CGH studies in MCL have been conducted, and these have shown a characteristic profile of genomic changes (Beà et al. 1999). Regions most frequently showing losses include 1p, 3p, 6q, 8p, 9p, 9q, 10p14-15, 11q14-23, 13q, and 17p. Alternatively, those most commonly showing gains include 3q, 7p, 7q, 8q, 10p13, 12q, 13q, 15q, and 18q. In some cases, alterations may cluster, including simultaneous loss of 8p and gain 8q; while in many other cases, specific alterations were uncommonly observed together, such as loss of 1p or gain of 3q. Blastoid/pleomorphic cases were more likely to have gains of 3q, 7p, and 12q, and losses of 9p21 and 17p. Similarly, gain of 3q, as well as losses of 8p, 9p, and 9q tended to be associated with inferior survival. Those cases with three or more chromosomal imbalances were also more likely to share an inferior survival (Allen et al. 2002; Beà et al. 1999; Bentz et al. 2000; Flordal et al. 2002; Hutter et al. 2009; Jarosová et al. 2004; Martinez-Climent et al. 2001; Monni et al. 1998; Salaverria et al. 2007, 2008; Vizcarra et al. 2001).

The use of higher resolution platforms, such as array CGH and SNP-chip microarrays have facilitated the identification of additional genomic alterations, and have similarly helped to refine those areas identified in cytogenetic and metaphase CGH studies identified above. Coupling of these technologies with gene expression profiling has helped to identify candidate genes which may be involved in chromosomal losses/gains. Identification of biallelic deletions frequently corresponded with poor prognosis. These deletions include 9p21.3, which may reflect loss of CDKN2A, CDKN2B, and MTAP; while those involving 1p32.3/33, 2q13, and 2q37 may involve FAF1 & CDKN2C, BCL2L11, and SP100, respectively. Monoallelic losses also appear to contribute to pathogenesis, as suggested by losses at 11q (ATM), 17p (p53), 13q14.2 (RB1), 6q23.3-q24.1 (TNFAIP3/A20, an inhibitor of NF-KB), 6q25.1 (LATS1), 9p21.2 (MOBKL2B), 19p13.3 (MOBKL2A), 1q32 (PROX1), 8p21.3 (MCPH1), 13q33-34 (ING1), and 13q34 (CUL4A). Amplified regions include, necessarily, 11q13 (CCND1 & MAP6), as well as 8q24.21 (MYC), 13q31.3 (MIR17HG, encoding the mir-17-92 cluster), 12q14 (CDK4, MDM2, CENTG1), 10p12.2-12.31 (BMI1), 18q21.33 (BCL2). Amplified gene(s) associated with gains of 3q26 have not been identified. One likely candidate is PIK3CA, which encodes the catalytic subunit of PI3K, important in activation of the AKT pathway (Beà et al. 2009; Flordal Thelander et al. 2007; Halldórsdóttir et al. 2011; Hartmann et al. 2010; Kawamata et al. 2009; Kohlhammer et al. 2004; Rubio-Moscardo et al. 2005; Schraders et al. 2005, 2008; Tagawa et al. 2005; Takeuchi et al. 2009; Vater et al. 2009).

This mutation pattern suggests that MCL pathogenesis integrates cell cycle dysregulation, alterations in DNA damage response, and anti-apoptotic pathways, as described in Table 1.

Cell Cycle Dysregulation

While cyclin D1 is not sufficient by itself to reproduce the phenotype of MCL, it clearly plays a major role in the pathogenesis of this disease. Perhaps the best known function of cyclin D1 (as well as D2 and D3) is to promote cell cycle proliferation via binding to cyclin dependent kinases 4 and 6 (cdk4/6). This binding promotes phosphorylation of RB1, leading to the release of E2F transcription factors capable of driving G1/S phase transition. Amplification of proliferative signals include E2F mediated accumulation of cyclin E, cyclin D1/CKD4 binding of p27kip (an inhibitor of cyclin E/CKD2), and cyclin E/CDK2 mediated degradation of p27kip and further phosphorylation of RB1.

The cyclin D1/CDK4 complex also acts to inhibit degradation of CDT1, a rate limiting factor in DNA replication (Aggarwal et al. 2007). Interestingly, CDT1 when active during S-phase, can lead to an increase in double strand breaks (Lovejoy et al. 2006). Emerging data have shown numerous other substrates for cyclin D1/CDK4, with a potential impact on differentiation (GATA4, MEF2), survival (RUNX), DNA repair (BRCA1), DNA replication/segregation, mitochondrial function, and protein synthesis (tuberous sclerosis complex/mTOR), ribosome biogenesis

Table 1 A summary of commonly observed genomic losses and gains in MCL (Royo et al. 2011)

Loss	1p32.3-p33a	CDKN2C, FAF1	Cell cycle/cell survival
Loss	1q32	PROX1	Proliferation
Loss	2q13a	BCL2L11	Cell survival
Loss	2q37.1	SP100, SP140	DNA Damage
Loss	6q23.3	TNFAIP3	NF-KB inhibitor
Loss	6q25	LATS1	Hippo signaling pathway
Loss	8p21.3	MCPH1	DNA Damage
Loss	9p21.2	MOBKL2B	Hippo signaling pathway
Loss	9p21.3a	CDKN2A/B, MTAP	Cell Cycle
Loss	11q22.3	ATM	DNA Damage
Loss	13q14.2	RB1	Cell Cycle
Loss	13q34	CUL4A, ING1	Cell cycle/DNA damage
Loss	17p13	TP53	Cell cycle/DNA damage
Loss	19p13.3	MOBKL2A	Hippo signaling pathway
Gain	3q26.1-q26.32	?PIK3CA	Proliferation
Gain	7p	IGF2BP3	Proliferation
Gain	8q24.21	MYC	Proliferation
Gain	10p12.2-12.31	BMI1	Cell Cycle
Gain	11q13.3-q21	CCND1, MAP6	Cell cycle/microtubule dynamics
Gain	12q14	CDK4, MDM2, CENTG1/PIKE	Cell cycle, apoptosis, DNA Damage, proliferation
Gain	13q31.3	MIR17HG (miR-17-92)	Cell cycle, apoptosis
Gain	18q21.33	BCL2	Apoptosis

(MEP50/PRMT5) (Aggarwal et al. 2010; Kehn et al. 2007; Lazaro et al. 2002; Nakajima et al. 2011; Ren et al. 2010; Shen et al. 2006; Zacharek et al. 2005; Zhang et al. 2008). Cyclin D1 similarly has several non-catalytic, CDK independent roles as well, influencing migration (p27/RHOA/smathmin), DNA damage response (BRCA2/RAD51, p21), cell cycle progression (STAT3/TAF1), as well as pleotropic influences on transcription through the binding of HDACs/HATs and nuclear hormone receptors (Besson et al. 2004; Bienvenu 2001; Coqueret 2002; Fu et al. 2004, 2005; Jirawatnotai et al. 2011; Li et al. 2006, 2010; McMahon et al. 1999; Musgrove et al. 2011; Reutens et al. 2001; Sherr and Roberts 1999; Siegert et al. 2000).

Additional factors which may cooperate with the cyclin D1 translocation to increase functional levels of cyclin D1 include activation of signaling through receptor tyrosine kinases/RAS , WNT/beta-catenin, NF-KappaB, PI3K/mTOR pathways (Albanese et al. 1995; Boyd et al. 2009; Cecconi et al. 2008; Gladden et al. 2005; Musgrove et al. 2011; Musgrove 2006; Pérez-Galán et al. 2011a, b; Pighi et al. 2011; Psyrri et al. 2009; Wander et al. 2011; Wang et al. 2011a). Indeed, attempts to target many of these pathways have been met with efficacy in both pre-clinical and clinical models. Additional data have shown that progressive accumulation of a truncated form of cyclin D1, lacking the long 3′ untranslated region is correlated with blastoid phenotype and inferior prognosis (Wiestner et al. 2007). Loss of this region contributes to increased stability of the cyclin D1 transcript, likely due to loss of micro RNA binding sites (including mir-15/16, mir-503, mir-34a, mir-195, mir-424, and the mir-17-92 cluster), as well as other mRNA destabilizing elements (Deshpande et al. 2009; Jiang et al. 2009; Zhao et al. 2010).

It is clear that additional events leading progressive accumulation of CDK4/6 are likely also occurring (reviewed in Navarro et al. 2011). Of particular relevance are losses at 9p21, where the CDKN2A locus resides. CDKN2A encodes two critical elements: p16INK4A, a CDK inhibitor, and ARF, a positive regulator of p53. Similarly, amplification of BMI1 at 10p12.2 may bypass this mechanism, acting to transcriptionally repress the CDKN2A locus. CUL4A and p33ING1, located at 13q34 also contribute to this pathway. CUL4A is an E3 ubiquitin ligase important in mediating DNA damage responses through its binding and epigenetic activation of the CDKN2A locus (Hartmann et al. 2010; Kotake et al. 2009). Interestingly CUL4A is also repressed by cyclin D1 (Aggarwal et al. 2010). Other CDK inhibitors are also frequently influenced by loss, including CDKN2B (9p21) and CDKN2C (1p32.3). Finally RB1 loss (13q14) itself may also occur, though, this has not been a consistent finding (Pinyol et al. 2007; Dreyling et al. 1997).

DNA Damage Response

MCL is among the most genomically instable of B-cell non-Hodgkins lymphomas, with 90% of cases showing at least one alteration. Additionally, the total number of alterations per case is also higher in MCL (Beà et al. 2009). These data suggest significant abnormalities both in DNA repair and sensing of DNA damage. Indeed, as noted above, cyclinD1/CDK4 may directly inhibit DNA damage response elements, including BRCA1, as well potentiate DNA damage through overexpression of CDT1. Genomic instability may be further potentiated by mutations involving the microtubule associated protein MAP2 (2q34), and in particular, loss of ATM (11q22-23), which is seen in up to 50% of cases (Camacho et al. 2002; Schraders et al. 2008; Stilgenbauer et al. 2000; Vater et al. 2009). ATM is typically recruited to sites of double strand DNA breaks via the MRN complex (Mre11, RAD50, NBS1), leading to phosphorylation of CHK2, degradation of the CDK2-Cylin E activating phosphatase (CDC25A), and subsequent cell cycle arrest. Failure to repair the double strand break leads to subsequent phosphorylation of p53 and MDM2 by ATM/CHK2, leading to stabilization of p53 and downstream activation of CDK inhibitors, including CDKN1A to potentiate cell cycle arrest, and even apoptosis. Inactivation of CHK2 (and CHK1), which are downstream of ATM, has also been described in MCL (Tort et al. 2002, 2005). Amplification of 13q31.3 (mir-17-92 cluster) may similarly be associated with decreased expression of CDKN1A (Inomata et al. 2009). P53 stability may be further compromised by amplification of MDM2 (12p13), loss of ARF (9p21, CDNK2A locus), p33/ING1 (13q33-34), as well as mutation or loss of p53 itself (17p13). Interestingly, a commonly deleted region at 8p21 may lead to loss of the BRIT1/MCPH1 gene, which normally binds to E2F1 to drive transcription of CHK1, BRCA1, RAD51, DDB2, TOPBP1, and p73 – and additionally functions to recruit these and other DNA damage response elements to sites of damage (Rai et al. 2006).

Loss of the nuclear body components, SP100/SP140 (2q37.1) have also been described as recurring event in MCL (Beà et al. 2009). SP100 has been shown in fibroblasts to synergize with p53 to activate CDKN1A; and its loss has been associated with rapid transformation to a stem cell phenotype, followed by activation of KRAS, MYC, and TERT, sequestration of p53, and significant genomic instability (Negorev et al. 2010). Downstream consequences of SP140 loss are less clear. While it appears to share significant homology with SP100, it appears to be uniformly expressed in lymphocytes with nuclear membrane pattern that is only partially overlapping with SP100 (Bloch et al. 1996, 1999; Zong et al. 2000). Interestingly, SNPs have been frequently identified among patients with the closely related syndrome, chronic lymphocytic leukemia, suggesting that this protein may play a central, though as yet undefined role lymphomagenesis (Bernardo et al. 2008; Sillé et al. 2012).

Apoptosis

The evolution of MCL from inception through its indolent and ultimately aggressive phases suggests an ability to ignore, or potentially evade immune regulation. This may relate to a loss of normal response to extrinsic cues, such as tumor necrosis factor, as well as perturbed downstream signaling from intrinsic and extrinsic apoptotic cues. Amplification of the antiapoptotic gene BCL2 (18q21), as well as loss of pro-apoptotic genes, such as BCL2L11/BIM (2q13), have both been observed in this regards (Kawamata et al. 2009; Vater et al. 2009). Amplification of the mir-17-92 cluster (13q31.3) can also lead to inhibition of apoptosis through down-regulation of BCL2L11/BIM (Inomata et al. 2009; Molitoris et al. 2011). Similarly, recurrent genomic losses at 1p32.3 (FAF1) and 6q23 (TNFAIP3/A20) leading to constitutive activation of the NF-KB pathway and overexpression of downstream anti-apoptotic targets, including cFLIP, BCL-2, BCL-XL, and XIAP (Hartmann et al. 2010; Honma et al. 2009).

Proliferation

Cell cycle dysregulation, as described above, may be accompanied by additional mutations which influence proliferation. Amplifications involving SYK (9q22), CENTG1/PIKE (12q14), and likely PIK3CA (3q26) highlight the importance of aberrant cell signaling in MCL (Hartmann et al. 2010; Psyrri et al. 2009; Rinaldi et al. 2006). Indeed, all these mutations overlap with regards to activation of PI3K and downstream phosphorylation of AKT. AKT in turn leads to phosphorylation of GSK3B, a kinase important in facilitating nuclear export of cyclin D1 as well as promoting destruction of the pro-survival protein, beta-catenin. AKT also functions to activate the mTOR pathway, leading to activation of S6K1 and 4E-BP1 – proteins important in cell growth and proliferation – as well as amplification of cell survival/proliferation signals via the RAS/RAF/MAPK pathway. Amplification at 7p (IGF2BP3), which leads to increased translation of IGF-2, similarly serves to activate both the PI3K and RAS/RAF/MAPK pathways (Suvasini et al. 2011).

The RAS/RAF/MAPK pathway is particularly important in activating C-Myc (Kapeli and Hurlin 2011). C-Myc has numerous effects, including activation of cell proliferation (via upregulation cyclins and downregulation of CDKN1A), inhibition of apoptosis (via downregulation of bcl2), protein/ribosomal production, and DNA replication(reviewed in Eilers and Eisenman 2008). Indeed, amplification of c-Myc (8q24) is associated with poor prognosis, not only in mantle cell lymphoma, but several other lymphoma subtypes (Reddy et al. 2008; Smith et al. 2010).

Deletions involving members of the Hippo signaling pathway, including MOBKL2A (19p13.3), MOBKL2B (9p21.2), LATS1 (6q25), and likely LATS2 (13q11-q12), have surprisingly been found to be a rather common event in MCL, occurring in ~40% of evaluated samples. Mutations in this pathway have been associated with high genomic complexity and shortened survival (Hartmann et al. 2010). The Hippo pathway contributes to cell proliferation, and potentially lymphomagenesis (Mauviel et al. 2011; Saucedo and Edgar 2007).

Translation to the Clinic

Gene and protein expression profiling have been explored to better understand the upstream and downstream consequences of this broad constellation of recurring genomic gains and losses. Although such explorations have been plagued by disease heterogeneity and reproducibility, several relevant pathways have emerged which are being translated into highly active therapies. Perhaps not surprisingly, genes associated with proliferation and survival appear to have the clearest relationship to overall survival in MCL (Hartmann et al. 2008; Martínez et al. 2003; Rosenwald et al. 2003). Interestingly, more recent gene expression data suggest a unique profile for indolent variants of MCL, an important subset for which treatment may often be delayed (Fernàndez et al. 2010; Royo et al. 2012).

Putting Current Clinical Approaches in Context

The addition of anthracyclines to combination chemotherapy has become standard for MCL, however their contribution to remission and survival is less clear (Bosch et al. 1998; Fisher et al. 1995; Meusers et al. 1989; Teodorovic et al. 1995; Zucca et al. 1995). This may be a reflection of topoisomerase II alpha (Topo IIa) expression, a known target of the anthracycline class. Topo IIa plays an important role in repairing double stranded supercoils, which may form during DNA replication/transcription. Topo IIa expression was explored in cohort of non-anthracycline treated MCL, which ranged from 0.7 to 57.8%, with a mean of 12.7%. The mean expression was lower in classic vs. blastoid variant (10.7% vs. 21.9%), and this expression correlated with Ki-67. In this group, those with ≤10% expression demonstrated a median overall survival of 49 months, compared to those with higher expression, who had a median overall survival of 17 months. Multi-variable analysis found that only topoisomerase IIa, IPI, and LDH were prognostic for survival in this study, with Ki-67 falling out of model (Schrader et al. 2004).

Subsequent studies of anthracycline treated patients could no longer show independence of topoisomerase IIa in predicting survival, instead, Ki-67 and other genes involved in proliferation, now retained influence (Hartmann et al. 2008; Galimberti et al. 2007). Interestingly, one of the five genes validated in these studies was SLC29A2, which is part of the SLC29 family of nucleoside carriers important in purine and pyrimidine transport (Baldwin et al. 2004). This could explain the benefit of purine analogues, a prominent component of the HyperCVAD regimen, as well as the autologous transplant conditioning regimen BEAM (Endo et al. 2007).

Getting Around Cyclin D1

Attempts to directly target cyclin D1 in preclinical models has been complicated by aberrant upregulation of cyclin D2 and D3, suggesting that it will be necessary to address either upstream influences on cyclin D expression, or, alternatively, block some of its functions by targeting its catalytic partners, the CDK's (Tchakarska et al. 2009).

Flavopiridol, the first-in-class pan-CDK inhibitor, is currently under study in MCL. Although initial studies suggested little clinical activity, subsequent modification of the dosing regimen has facilitated a marked improvement in response (Holkova et al. 2011; Kouroukis 2003; Lin et al. 2009). An alternative approach using a PD0332991, a selective inhibitor of CDK4/6, is being explored in MCL. Single agent data show a significant potential for abrogation of cell cycling, with extended duration of stable disease noted in 50% of refractory patients (Leonard et al. 2008, 2012). Interestingly, withdrawal of his compound appears to induce "coordination" of cell cycling, which could facilitate enhanced response rates by sequencing PD0332991 with other active, and potentially synergistic, agents. In fact, this approach is currently under study using the proteasome inhibitor bortezomib (NCT01111188).

Two important interconnected pathways sit upstream of cyclin D1, namely the B-cell receptor (BCR) and the PI3K/AKT/MTOR pathways. Upregulation of the BCR pathway has been observed in MCL by both microarray and phosphoproteomic analyses, possibly reflecting antigenic drive either by foreign or self-peptides (Hadzidimitriou et al. 2011; Pighi et al. 2011; Rinaldi et al. 2006). The BCR is composed of two antigen responsive membrane immunoglobulin molecules coupled to CD79a/

Table 2 Novel agents in MCL (Friedberg et al. 2010; Hess et al. 2009; Kahl et al. 2010; Kirschbaum et al. 2011; Kouroukis 2003; Leonard et al. 2012; Morschhauser et al. 2008; Renner et al. 2012; Rizzieri et al. 2008; Wang et al. 2011b)

	n	ORR
BCR inhibition		
Enzastaurin	60	0% (10% SD at 6 month)
Fostamatinib	9	11%
CAL-101	16	62
PCI-32765	24	67%
mTOR inhibition		
Temsirolimus	162	22%
Everolimus	36	20%
Deferolimus	9	33%
Cell cycle inhibition		
Flavopiridol	30	11%
PD-0332991	17	18%
HDAC inhibition		
Vorinostat	9	0% (11% with SD × 26month)

C79b heterodimers, which transduce intracellular signaling via the Src family kinases (Lyn, Blk, Fyn), as well as Syk and Btk, which in turn ultimately activate PI3k/AKT/mTOR via CD19, Ras/Raf/Mek/Erk, and PKC-beta/NF-KB/NFAT. Negative regulation occurs via a feedback loop involving Lyn-CD22-Shp1. Significant complexity is introduced by other signaling molecules, such as CD40, which can modulate intracellular signals through the BCR (reviewed in (Dal Porto et al. 2004; Kurosaki 2011; Werner et al. 2010). This may explain lackluster responses following targeted inhibition of Syk with Fostamatinib (ORR 11%), and the PKC-Beta inhibitor, Enzastaurin, which elicited no objective responses (Friedberg et al. 2010; Morschhauser et al. 2008). Novel approaches targeting Btk, on the other hand, are showing significant promise, with an overall response rate of 67% in highly refractory patients (Wang et al. 2011b) (Table 2).

Inhibition of the PI3K/AKT pathway represents another approach for inhibiting tonic pro-survival/proliferation signaling. In addition to aberrant signaling through the BCR, it is likely that amplification in PI3KCA, PDK1, AKT1, as well as inactivation of the PI3K pathway regulating protein, PTEN, also contribute to the malignant phenotype (Rizzatti et al. 2005; Rudelius et al. 2006). Signaling through the PI3K pathway may lead to constitutive activation of AKT, as confirmed by GEP in classic and blastoid MCL. This is associated with phosphorylation of several downstream targets including MDM2 (leading to degradation of p53), BAD (anti-apoptotic), p27 (proliferation), and, importantly, mTOR. mTOR is a highly conserved kinase which regulates translation of proteins important in cell proliferation and growth. mTOR exists as two separate complexes, mTORC1 and mTORC2. mTORC1 classically sits downstream of AKT signaling, and functions as a critical component in bridging PI3K/AKT signals with replicative/proliferative drive (reviewed in Oh and Jacinto

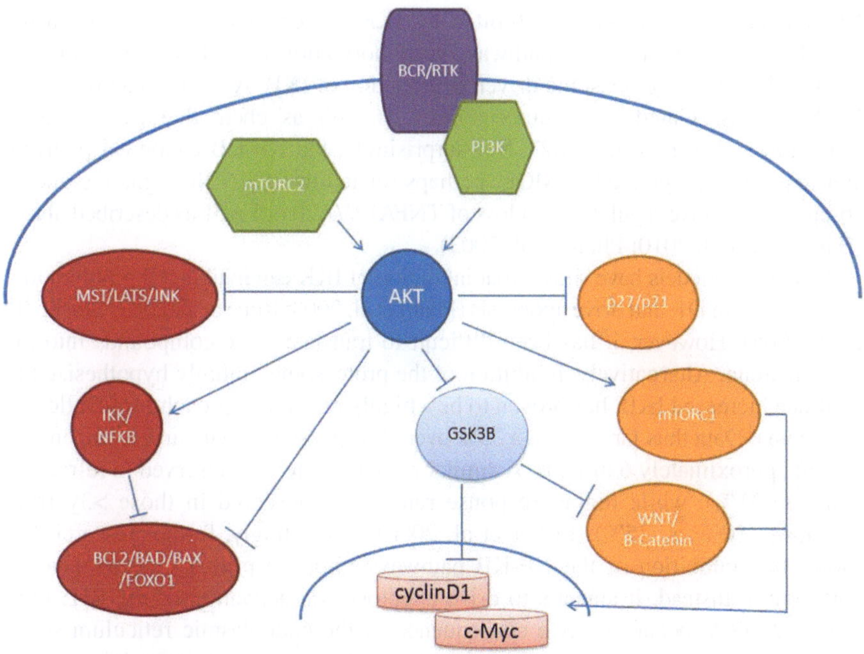

Fig. 1 Aberrant signaling via AKT influences multiple downstream pathways in MCL, leading to inhibition of apoptosis (*red*) and activation of proliferation (*orange*)

2011; Rosner et al. 2008; Wander et al. 2011). Attempts to inhibit mTORC1 using compounds such as Temsirolimus, and its derivatives, Everolimus and Deferolimus have shown therapeutic benefit in MCL (Hess et al. 2009; Renner et al. 2010; Rizzieri et al. 2008). However, the magnitude of benefit has proven to be modest driving further investigation into mechanisms of resistance. Indeed, inhibition of mTORC1 led to decrease in SB6 and EIF4E. Surprisingly, however, inhibition led to increased levels of phospho-AKT driven by mTORC2. While the driving stimuli for mTORC2 expression/activity have not been fully elucidated, it appears that the mTORC1 pathway plays a central role in regulating mTORC2. Dual mTORC inhibitors, as well as mTORC/PI3K inhibitors have been developed and are currently under study, and may provide a novel means for enhancing response and survival (Bhagwat et al. 2011; Gupta et al. 2010, 2011) (Fig. 1).

NF-KB and Beyond

NF-KB is a ubiquitously expressed family of several different transcription factors, including p65 (RelA), p50/105, p52/p100, c-Rel, and RelB. Each of these factors possesses the ability to bind to DNA either as homo- or heterodimers where they regulate the transcription of genes involved in cell growth, survival, and adhesion.

NF-KB function is normally inhibited by IKB, which bind these factors in the cytoplasm. In the canonical pathway, phosphorylation of IKB by IKK releases NF-KB, leading to proteasome driven degradation of IKB. Aberrant inactivation of NF-KB is associated with tumorigenesis as well as chemotherapy resistance (reviewed in Karin et al. 2002). Not surprisingly, the NF-KB canonical pathway appears to be upregulated in MCL, perhaps related to BCR/TNF signaling and/or absence of negative regulation via loss of TNFAIP3/A20 or FAF1 as described above (Hartmann et al. 2010; Pham et al. 2003).

Preclinical models have shown that inhibition of IKK can inhibit cell proliferation, decrease cyclin D1, and drive apoptosis (Pham et al. 2003; Roué et al. 2007; Shishodia et al. 2005). However, it has been difficult to translate these compounds into the clinical arena. Alternatively, inhibition of the proteasome, initially hypothesized to facilitate increased IKB, has proven to be a highly relevant approach for mantle cell lymphoma. Data thus far suggest a 33% overall response rate with an accompanying PFS of approximately 6 months. A similar response rate was observed in refractory patients (31%), while higher response rates were observed in those >3y from diagnosis (50% vs. 25%) (Fisher et al. 2006). Interestingly, further research has shown that inhibition of the NF-KB pathway is not the primary mechanism for bortezomib. Instead, it appears to drive apoptosis via activation of the BH3-only protein NOXA, occurring as a consequence of the endoplasmic reticulum stress response to excessive polyubiquitinated proteins (Hideshima et al. 2009; Pérez-Galán et al. 2006; Rizzatti et al. 2008; Yang et al. 2008). Loss of response to bortezomib is associated with plasmacytic differentiation, as suggested by increases in BLIMP & IRF4, as well as increased surface expression of CD38 and CD138, and corresponding decreases in expression of CD19, CD24, and CD52 (Pérez-Galán et al. 2011b). Interestingly, in cells which have not undergone plasmacytic differentiation, BLIMP is critical to the activation of NOXA (Desai et al. 2010). This may suggest increased tolerance to ER stress related to the unfolded protein response pathway. Specifically, it was found that resistant cases had increased levels of the heat shock protein, BiP/Grp78, an endoplasmic pro-survival chaperone. BiP/Grp78, as well as several other pro-survival proteins including cyclinD1, p53, BID and AKT, appears to be stabilized by a separate heat shock protein, HSP90. Indeed, pre-clinical models in MCL show strong synergy between bortezomib, and HSP90 inhibitors, suggesting a strong rationale for further study (Roué et al. 2011).

Inhibition of Histone Deactylases

The methylation of genes and the acetylation of core histones represent major mechanism for gene silencing in the absence of genomic loss. However, MCL is characterized by both hypomethylation and hypermethylation of relevant gene promoters, which may prove difficult to target, and, further the impact of hypermethylation may not necessarily prove equivalent to gene loss (Chim et al. 2007; Hutter et al. 2006; Leshchenko et al. 2010). Alternatively, histone deacetylases

(HDACs) may prove a more facile target, as specific HDAC subtypes may generate a more predictable influence on gene expression. A recent study by Tao and colleagues uncovered HDAC3/c-Myc interactions were critical in facilitating changes in gene expression signatures driven by c-Myc (Zhang et al. 2011). Competition between different histone deacetylases may provide an additional mechanism for controlling distinct differentiation pathways. This was recently explored in novel study by Sotomayor and colleagues who found that HDAC6 and HDAC11 likely compete for binding to the IL-10 promoter, with opposite effects on transcription of this gene (Villagra et al. 2008a).

These data may help to explain the modest benefits seen with pan-HDAC inhibitors, as such approaches are unlikely to address competition between HDAC's, or may fail to fully inhibit a relevant HDAC prior to the onset of excessive systemic toxicity (Kirschbaum et al. 2011). This suggests a strong therapeutic rationale for exploring targeted HDAC inhibition in MCL (Cheng et al. 2010, 2011; Shah et al. 2010; Villagra et al. 2008b).

Conclusions

The heterogenous clinical and biological profile of MCL continues to pose challenges in defining optimal therapy. The application of modern molecular approaches has helped to underscore the overlapping roles of the pro-proliferative & anti-apoptotic pathways which occur on a background of cell cycle dysregulation and diminished DNA damage response. Decoding aberrantly regulated pathways occurring in parallel and upstream of these responses may provide a unique opportunity to truly "target" MCL, and break the cycle of inevitable relapse and untimely death.

References

Aggarwal P, Lessie MD, Lin DI, Pontano L, Gladden AB, Nuskey B, Goradia A et al (2007) Nuclear accumulation of cyclin D1 during S phase inhibits Cul4-dependent Cdt1 proteolysis and triggers p53-dependent DNA rereplication. Genes Dev 21(22):2908–2922. doi:10.1101/gad.1586007

Aggarwal P, Vaites LP, Kim JK, Mellert H, Gurung B, Nakagawa H, Herlyn M et al (2010) Nuclear cyclin D1/CDK4 kinase regulates CUL4 expression and triggers neoplastic growth via activation of the PRMT5 methyltransferase. Cancer Cell 18(4):329–340. doi:10.1016/j.ccr.2010.08.012

Albanese C, Johnson J, Watanabe G, Eklund N, Vu D, Arnold A, Pestell RG (1995) Transforming p21ras mutants and c-Ets-2 activate the cyclin D1 promoter through distinguishable regions. J Biol Chem 270(40):23589–23597

Allen JE, Hough RE, Goepel JR, Bottomley S, Wilson GA, Alcock HE, Baird M et al (2002) Identification of novel regions of amplification and deletion within mantle cell lymphoma DNA by comparative genomic hybridization. Br J Haematol 116(2):291–298

Aqel N, Barker F, Patel K, Naresh KN (2007) In-situ mantle cell lymphoma-a report of two cases. Histopathology 52(2):256–260. doi:10.1111/j.1365-2559.2007.02906.x

Baldwin SA, Beal PR, Yao SYM, King AE, Cass CE, Young JD (2004) The equilibrative nucleoside transporter family, SLC29. Pflügers Archiv Eur J Physiol 447(5):735–743. doi:10.1007/s00424-003-1103-2

Beà S, Ribas M, Hernández JM, Bosch F, Pinyol M, Hernández L, García JL et al (1999) Increased number of chromosomal imbalances and high-level DNA amplifications in mantle cell lymphoma are associated with blastoid variants. Blood 93(12):4365–4374

Beà S, Salaverria I, Armengol L, Pinyol M, Fernández V, Hartmann EM, Jares P et al (2009) Uniparental disomies, homozygous deletions, amplifications, and target genes in mantle cell lymphoma revealed by integrative high-resolution whole-genome profiling. Blood 113(13):3059–3069. doi:10.1182/blood-2008-07-170183

Bentz M, Plesch A, Bullinger L, Stilgenbauer S, Ott G, Müller-Hermelink HK, Baudis M et al (2000) t(11;14)-positive mantle cell lymphomas exhibit complex karyotypes and share similarities with B-cell chronic lymphocytic leukemia. Gene Chromosom Cancer 27(3):285–294

Bernardo MCD, Crowther-Swanepoel D, Broderick P, Webb E, Sellick G, Wild R, Sullivan K et al (2008) A genome-wide association study identifies six susceptibility loci for chronic lymphocytic leukemia. Nat Genet 40(10):1204–1210. doi:10.1038/ng.219

Besson A, Assoian RK, Roberts JM (2004) Regulation of the cytoskeleton: an oncogenic function for CDK inhibitors? Nat Rev Cancer 4(12):948–955. doi:10.1038/nrc1501

Bhagwat SV, Gokhale PC, Crew AP, Cooke A, Yao Y, Mantis C, Kahler J et al (2011) Preclinical characterization of OSI-027, a potent and selective inhibitor of mTORC1 and mTORC2: distinct from rapamycin. Mol Cancer Ther 10(8):1394–1406. doi:10.1158/1535-7163.MCT-10-1099

Bienvenu F (2001) Cyclin D1 represses STAT3 activation through a Cdk4-independent mechanism. J Biol Chem 276(20):16840–16847. doi:10.1074/jbc.M100795200

Bloch DB, de la Monte SM, Guigaouri P, Filippov A, Bloch KD (1996) Identification and characterization of a leukocyte-specific component of the nuclear body. J Biol Chem 271(46):29198–29204. doi:10.1074/jbc.271.46.29198

Bloch DB, Chiche J-D, Orth D, de la Monte SM, Rosenzweig A, Bloch KD (1999) Structural and functional heterogeneity of nuclear bodies. Mol Cell Biol 19(6):4423–4430

Bosch F, López-Guillermo A, Campo E, Ribera JM, Conde E, Piris MA, Vallespí T (1998) Mantle cell lymphoma: presenting features, response to therapy, and prognostic factors. Cancer 82(3):567–575. doi:10.1002/(SICI)1097-0142(19980201)82:3<567::AID-CNCR20>3.0.CO;2-Z

Boyd RS, Jukes-Jones R, Walewska R, Brown D, Dyer MJS, Cain K (2009) Protein profiling of plasma membranes defines aberrant signaling pathways in mantle cell lymphoma. Mol Cell Proteomics 8(7):1501–1515

Camacho E, Hernández L, Hernández S, Tort F, Bellosillo B, Beà S, Bosch F et al (2002) ATM gene inactivation in mantle cell lymphoma mainly occurs by truncating mutations and missense mutations involving the phosphatidylinositol-3 kinase domain and is associated with increasing numbers of chromosomal imbalances. Blood 99(1):238–244

Camacho FI, Algara P, Rodrı-guez A, Ruíz-Ballesteros E, Mollejo M, Martínez N, Martínez-Climent JA et al (2003) Molecular heterogeneity in MCL defined by the Use of specific VH genes and the frequency of somatic mutations. Blood 101(10):4042–4046. doi:10.1182/blood-2002-11-3456

Carvajal-Cuenca A, Sua LF, Silva NM, Pittaluga S, Royo C, Song JY, Sargent RL et al (2011) In situ mantle cell lymphoma: clinical implications of an incidental finding with indolent clinical behavior. Haematologica. doi:10.3324/haematol.2011.052621

Cecconi D, Zamò A, Bianchi E, Parisi A, Barbi S, Milli A, Rinalducci S et al (2008) Signal transduction pathways of mantle cell lymphoma: a phosphoproteome-based study. Proteomics 8(21):4495–4506

Cheng F, Wang H, Woan K, Rock-Klotz J, Wang Z, Merino O, Gill A et al (2010) Targeting histone deacetylase 6 (HDAC6) as a novel approach to overcome T-cell anergy. ASH Annu Meet Abstr 116(21):1483

Cheng F, Wang Z, Wang H, Woan KV, Sahakian E, Klotz JR, Wang L et al (2011) Epigenetic modulation of STAT3 by histone deacetylase 6 (HDAC6) regulates IL-10 gene expression and immune tolerance mediated by antigen-presenting cells (APCs). ASH Annu Meet Abstr 118(21):519

Chim CS, Wong KY, Loong F, Lam WW, Srivastava G (2007) Frequent epigenetic inactivation of Rb1 in addition to p15 and p16 in mantle cell and follicular lymphoma. Hum Pathol 38(12):1849–1857. doi:10.1016/j.humpath.2007.05.009

Coqueret O (2002) Linking cyclins to transcriptional control. Gene 299(1–2):35–55

Dal Porto JM, Gauld SB, Merrell KT, Mills D, Pugh-Bernard AE, Cambier J (2004) B cell antigen receptor signaling 101. Mol Immunol 41(6–7):599–613. doi:10.1016/j.molimm.2004.04.008

Desai S, Maurin M, Smith MA, Bolick SCE, Dessureault S, Tao J, Sotomayor E et al (2010) PRDM1 is required for mantle cell lymphoma response to bortezomib. Mol Cancer Res MCR 8(6):907–918. doi:10.1158/1541-7786.MCR-10-0131

Deshpande A, Pastore A, Deshpande AJ, Zimmermann Y, Hutter G, Weinkauf M, Buske C et al (2009) 3′ UTR mediated regulation of the cyclin D1 proto-oncogene. Cell Cycle (Georgetown, Tex) 8(21):3584–3592

Dreyling MH, Bullinger L, Ott G, Stilgenbauer S, Müller-Hermelink HK, Bentz M, Hiddemann W et al (1997) Alterations of the cyclin D1/p16-pRB pathway in mantle cell lymphoma. Cancer Res 57(20):4608–4614

Eilers M, Eisenman RN (2008) Myc's broad reach. Genes Dev 22(20):2755–2766. doi:10.1101/gad.1712408

Endo Y, Obata T, Murata D, Ito M, Sakamoto K, Fukushima M, Yamasaki Y et al (2007) Cellular localization and functional characterization of the equilibrative nucleoside transporters of antitumor nucleosides. Cancer Sci 98(10):1633–1637. doi:10.1111/j.1349-7006.2007.00581.x

Fernàndez V, Salamero O, Espinet B, Solé F, Royo C, Navarro A, Camacho F et al (2010) Genomic and gene expression profiling defines indolent forms of mantle cell lymphoma. Cancer Res 70(4):1408–1418. doi:10.1158/0008-5472.CAN-09-3419

Fisher RI, Dahlberg S, Nathwani B, Banks P, Miller T, Grogan T (1995) A clinical analysis of two indolent lymphoma entities: mantle cell lymphoma and marginal zone lymphoma (including the mucosa-associated lymphoid tissue and monocytoid B-cell subcategories): a Southwest Oncology Group study. Blood 85(4):1075

Fisher RI, Bernstein SH, Kahl BS, Djulbegovic B, Robertson MJ, de Vos S, Epner E et al (2006) Multicenter phase II study of bortezomib in patients with relapsed or refractory mantle cell lymphoma. J Clin Oncol 24(30):4867–4874. doi:10.1200/JCO.2006.07.9665

Flordal E, Berglund M, Rosenquist R, Erlanson M, Enblad G, Roos G, Larsson C et al (2002) Clonal development of a blastoid mantle cell lymphoma studied with comparative genomic hybridization. Cancer Genet Cytogenet 139(1):38–43

Flordal Thelander E, Ichimura K, Collins VP, Walsh SH, Barbany G, Hagberg A, Laurell A et al (2007) Detailed assessment of copy number alterations revealing homozygous deletions in 1p and 13q in mantle cell lymphoma. Leuk Res 31(9):1219–1230. doi:10.1016/j.leukres.2006.10.022

Friedberg JW, Sharman J, Sweetenham J, Johnston PB, Vose JM, LaCasce A, Schaefer-Cutillo J et al (2010) Inhibition of Syk with fostamatinib disodium has significant clinical activity in non-Hodgkin lymphoma and chronic lymphocytic leukemia. Blood 115(13):2578–2585. doi:10.1182/blood-2009-08-236471

Fu M, Wang C, Li Z, Sakamaki T, Pestell RG (2004) Minireview: cyclin D1: normal and abnormal functions. Endocrinology 145(12):5439–5447. doi:10.1210/en.2004-0959

Fu M, Rao M, Bouras T, Wang C, Wu K, Zhang X, Li Z et al (2005) Cyclin D1 inhibits peroxisome proliferator-activated receptor gamma-mediated adipogenesis through histone deacetylase recruitment. J Biol Chem 280(17):16934–16941. doi:10.1074/jbc.M500403200

Galimberti S, Nagy B, Benedetti E, Pacini S, Brizzi S, Caracciolo F, Papineschi F et al (2007) Evaluation of the MDR1, ABCG2, Topoisomerases IIalpha and GSTpi gene expression in patients affected by aggressive mantle cell lymphoma treated by the R-Hyper-CVAD regimen. Leuk Lymphoma 48(8):1502–1509. doi:10.1080/10428190701402895

Gladden AB, Woolery R, Aggarwal P, Wasik M, Diehl J (2005) Expression of constitutively nuclear cyclin D1 in murine lymphocytes induces B-cell lymphoma. Oncogene 25(7):998–1007

Gupta M, Hendrickson AW, Han JJ, Stenson M, Wellik L, Barr S, Kaufmann SH et al. (2010) Dual inhibition of mTORC1/mTORC2 induces apoptosis of mantle cell lymphoma by Preventing Rictor Mediated AKTS473 phosphorylation by potentiating AKT2-PHLPP1 association. ASH Annu Meet Abstr 116(21):772

Gupta M, Wahner Hendrickson AE, Yun SS, Han JJ, Schneider PA, Koh BD, Stenson MJ et al (2011) Dual mTORC1/mTORC2 inhibition diminishes Akt activation and induces Puma-dependent apoptosis in lymphoid malignancies. Blood. doi:10.1182/blood-2011-04-346601

Hadzidimitriou A, Agathangelidis A, Darzentas N, Murray F, Delfau-Larue M-H, Pedersen LB, Lopez AN et al (2011) Is there a role for antigen selection in mantle cell lymphoma? Immunogenetic support from a series of 807 cases. Blood 118(11):3088–3095. doi:10.1182/blood-2011-03-343434

Halldórsdóttir AM, Sander B, Göransson H, Isaksson A, Kimby E, Mansouri M, Rosenquist R et al (2011) High-resolution genomic screening in mantle cell lymphoma–specific changes correlate with genomic complexity, the proliferation signature and survival. Gene Chromosom Cancer 50(2):113–121. doi:10.1002/gcc.20836

Hartmann E, Fernandez V, Moreno V, Valls J, Hernandez L, Bosch F, Abrisqueta P et al (2008) Five-gene model to predict survival in mantle-cell lymphoma using frozen or formalin-fixed, paraffin-embedded tissue. J Clin Oncol 26:4966–4972. doi:10.1200/JCO.2007.12.0410

Hartmann EM, Campo E, Wright G, Lenz G, Salaverria I, Jares P, Xiao W et al (2010) Pathway discovery in mantle cell lymphoma by integrated analysis of high-resolution gene expression and copy number profiling. Blood 116(6):953–961. doi:10.1182/blood-2010-01-263806

Herrmann A, Hoster E, Zwingers T, Brittinger G, Engelhard M, Meusers P, Reiser M et al (2008) Improvement of overall survival in advanced stage mantle cell lymphoma. J Clin Oncol 27:511–518. doi:10.1200/JCO.2008.16.8435

Hess G, Herbrecht R, Romaguera J, Verhoef G, Crump M, Gisselbrecht C, Laurell A et al (2009) Phase III study to evaluate temsirolimus compared with Investigator's choice therapy for the treatment of relapsed or refractory mantle cell lymphoma. J Clin Oncol 27(23):3822–3829. doi:10.1200/JCO.2008.20.7977

Hideshima T, Ikeda H, Chauhan D, Okawa Y, Raje N, Podar K, Mitsiades C et al (2009) Bortezomib induces canonical nuclear factor-kappaB activation in multiple myeloma cells. Blood 114(5):1046–1052. doi:10.1182/blood-2009-01-199604

Holkova B, Perkins EB, Ramakrishnan V, Tombes MB, Shrader E, Talreja N, Wellons MD et al (2011) Phase I trial of bortezomib (PS-341; NSC 681239) and alvocidib (flavopiridol; NSC 649890) in patients with recurrent or refractory B-cell neoplasms. Clin Cancer Res Off J Am Assoc Cancer Res 17(10):3388–3397. doi:10.1158/1078-0432.CCR-10-2876

Honma K, Tsuzuki S, Nakagawa M, Tagawa H, Nakamura S, Morishima Y, Seto M (2009) TNFAIP3/A20 functions as a novel tumor suppressor gene in several subtypes of non-Hodgkin lymphomas. Blood 114(12):2467–2475. doi:10.1182/blood-2008-12-194852

Hutter G, Scheubner M, Zimmermann Y, Kalla J, Katzenberger T, Hübler K, Roth S et al (2006) Differential effect of epigenetic alterations and genomic deletions of CDK inhibitors [p16(INK4a), p15(INK4b), p14(ARF)] in mantle cell lymphoma. Gene Chromosom Cancer 45(2):203–210. doi:10.1002/gcc.20277

Hutter G, Scheubner M, Ott G, Zimmermann Y, Hübler K, Roth S, Stilgenbauer S et al (2009) Allelic genotyping reveals a hierarchy of genomic alterations in mantle cell lymphoma associated to cell proliferation. Ann Hematol 88(9):821–828. doi:10.1007/s00277-008-0686-2

Inomata M, Tagawa H, Guo YM, Kameoka Y, Takahashi N, Sawada K (2009) MicroRNA-17-92 down-regulates expression of distinct targets in different B-cell lymphoma subtypes. Blood 113(2):396

Jarosová M, Papajík T, Holzerová M, Dusek L, Pikalová Z, Lakomá I, Raida L et al (2004) High incidence of unbalanced chromosomal changes in mantle cell lymphoma detected by comparative genomic hybridization. Leuk Lymphoma 45(9):1835–1846. doi:10.1080/10428190410001683796

Jiang Q, Feng M-G, Mo Y-Y (2009) Systematic validation of predicted microRNAs for cyclin D1. BMC Cancer 9:194. doi:10.1186/1471-2407-9-194

Jirawatnotai S, Hu Y, Michowski W, Elias JE, Becks L, Bienvenu F, Zagozdzon A et al (2011) A function for cyclin D1 in DNA repair uncovered by protein interactome analyses in human cancers. Nature 474(7350):230–234. doi:10.1038/nature10155

Kahl B, Byrd JC, Flinn IW, Wagner-Johnston N, Spurgeon S, Benson DM, Furman RR et al (2010) Clinical safety and activity in a phase 1 study of CAL-101, an isoform-selective inhibitor of phosphatidylinositol 3-kinase P110{delta}, in patients with relapsed or refractory non-Hodgkin lymphoma. ASH Annu Meet Abstr 116(21):1777

Kapeli K, Hurlin PJ (2011) Differential regulation of N-Myc and c-Myc synthesis, degradation, and transcriptional activity by the Ras/mitogen-activated protein kinase pathway. J Biol Chem 286(44):38498–38508. doi:10.1074/jbc.M111.276675

Karin M, Cao Y, Greten FR, Li Z-W (2002) NF-kappaB in cancer: from innocent bystander to major culprit. Nat Rev Cancer 2(4):301–310. doi:10.1038/nrc780

Kawamata N, Ogawa S, Gueller S, Ross SH, Huynh T, Chen J, Chang A et al (2009) Identified hidden genomic changes in mantle cell lymphoma using high-resolution single nucleotide polymorphism genomic array. Exp Hematol 37(8):937–946. doi:10.1016/j.exphem.2009.04.012

Kehn K, Berro R, Alhaj A, Bottazzi ME, Yeh W-I, Klase Z, Van Duyne R et al (2007) Functional consequences of cyclin D1/BRCA1 interaction in breast cancer cells. Oncogene 26(35):5060–5069. doi:10.1038/sj.onc.1210319

Kienle D, Kröber A, Katzenberger T, Ott G, Leupolt E, Barth TFE, Möller P et al (2003) VH mutation status and VDJ rearrangement structure in mantle cell lymphoma: correlation with genomic aberrations, clinical characteristics, and outcome. Blood 102(8):3003–3009. doi:10.1182/blood-2003-05-1383

Kirschbaum M, Frankel P, Popplewell L, Zain J, Delioukina M, Pullarkat V, Matsuoka D et al (2011) Phase II study of vorinostat for treatment of relapsed or refractory indolent non-Hodgkin's lymphoma and mantle cell lymphoma. J Clin Oncol 29(9):1198–1203. doi:10.1200/JCO.2010.32.1398

Klapper W (2011) Histopathology of mantle cell lymphoma. Semin Hematol 48(3):148–154. doi:10.1053/j.seminhematol.2011.03.006

Kohlhammer H, Schwaenen C, Wessendorf S, Holzmann K, Kestler HA, Kienle D, Barth TFE et al (2004) Genomic DNA-chip hybridization in t(11;14)-positive mantle cell lymphomas shows a high frequency of aberrations and allows a refined characterization of consensus regions. Blood 104(3):795–801. doi:10.1182/blood-2003-12-4175

Kolar GR, Mehta D, Pelayo R, Capra JD (2007) A novel human B cell subpopulation representing the initial germinal center population to express AID. Blood 109(6):2545–2552

Kotake Y, Zeng Y, Xiong Y (2009) DDB1-CUL4 and MLL1 mediate oncogene-induced p16INK4a activation. Cancer Res 69(5):1809–1814. doi:10.1158/0008-5472.CAN-08-2739

Kouroukis CT (2003) Flavopiridol in untreated or relapsed mantle-cell lymphoma: results of a phase II study of the national cancer institute of Canada clinical trials group. J Clin Oncol 21(9):1740–1745. doi:10.1200/jco.2003.09.057

Kurosaki T (2011) Regulation of BCR signaling. Mol Immunol 48(11):1287–1291. doi:10.1016/j.molimm.2010.12.007

Kurtova AV, Tamayo AT, Ford RJ, Burger JA (2009) Mantle cell lymphoma cells express high levels of CXCR4, CXCR5, and VLA-4 (CD49d): importance for interactions with the stromal microenvironment and specific targeting. Blood 113(19):4604–4613. doi:10.1182/blood-2008-10-185827

Lazaro J-B, Bailey PJ, Lassar AB (2002) Cyclin D-cdk4 activity modulates the subnuclear localization and interaction of MEF2 with SRC-family coactivators during skeletal muscle differentiation. Genes Dev 16(14):1792–1805. doi:10.1101/gad.U-9988R

Leonard JP, LaCasce A, Smith MR, Noy A, Yap JT, Van den Abbeele AD, Yu JQ et al. (2008) Cdk4/6 Inhibitor PD 0332991 demonstrates cell cycle inhibition via FLT-PET imaging and tissue analysis in patients with recurrent mantle cell lymphoma. ASH Annu Meet Abstr 112(11):264

Leonard JP, Lacasce AS, Smith MR, Noy A, Chirieac LR, Rodig SJ, Yu JQ et al (2012) Selective CDK4/6 inhibition with tumor responses by PD0332991 in patients with mantle cell lymphoma. Blood. doi:10.1182/blood-2011-10-388298

Leshchenko VV, Kuo P-Y, Shaknovich R, Yang DT, Gellen T, Petrich A, Yu Y et al (2010) Genomewide DNA methylation analysis reveals novel targets for drug development in mantle cell lymphoma. Blood 116(7):1025–1034. doi:10.1182/blood-2009-12-257485

Leux C, Herry A, Maynadie M, Cabrera Q, Troussard X, Le Guyader-Peyrou S, Le Gouill S et al (2011) Epidemiology of mantle cell lymphoma: a population-based study in France. ASH Annu Meet Abstr 118(21):5222

Li Z, Wang C, Jiao X, Lu Y, Fu M, Quong AA, Dye C et al (2006) Cyclin D1 regulates cellular migration through the inhibition of thrombospondin 1 and ROCK signaling. Mol Cell Biol 26(11):4240–4256. doi:10.1128/MCB.02124-05

Li Z, Jiao X, Wang C, Shirley LA, Elsaleh H, Dahl O, Wang M et al (2010) Alternative cyclin D1 splice forms differentially regulate the DNA damage response. Cancer Res 70(21):8802–8811. doi:10.1158/0008-5472.CAN-10-0312

Lin TS, Blum KA, Fischer DB, Mitchell SM, Ruppert AS, Porcu P, Kraut EH et al (2009) Flavopiridol, fludarabine, and rituximab in mantle cell lymphoma and indolent B-cell lymphoproliferative disorders. J Clin Oncol 28(3):418–423. doi:10.1200/JCO.2009.24.1570

Lovejoy CA, Lock K, Yenamandra A, Cortez D (2006) DDB1 maintains genome integrity through regulation of Cdt1. Mol Cell Biol 26(21):7977–7990. doi:10.1128/MCB.00819-06

Martínez N, Camacho FI, Algara P, Rodríguez A, Dopazo A, Ruíz-Ballesteros E, Martín P et al (2003) The molecular signature of mantle cell lymphoma reveals multiple signals favoring cell survival. Cancer Res 63(23):8226–8232

Martinez-Climent JA, Vizcarra E, Sanchez D, Blesa D, Marugan I, Benet I, Sole F et al (2001) Loss of a novel tumor suppressor gene locus at chromosome 8p is associated with leukemic mantle cell lymphoma. Blood 98(12):3479–3482

Mauviel A, Nallet-Staub F, Varelas X (2011) Integrating developmental signals: a Hippo in the (path)way. Oncogene. doi:10.1038/onc.2011.363

McMahon C, Suthiphongchai T, DiRenzo J, Ewen ME (1999) P/CAF associates with cyclin D1 and potentiates its activation of the estrogen receptor. Proc Natl Acad Sci U S A 96(10):5382–5387

Meusers P, Engelhard M, Bartels H, Binder T, Fülle HH, Görg K, Gunzer U et al (1989) Multicentre randomized therapeutic trial for advanced centrocytic lymphoma: anthracycline does not improve the prognosis. Hematol Oncol 7(5):365–380

Molitoris JK, McColl KS, Distelhorst CW (2011) Glucocorticoid-mediated repression of the oncogenic microRNA cluster miR-17~92 contributes to the induction of Bim and initiation of apoptosis. Mol Endocrinol (Baltimore, Md) 25(3):409–420. doi:10.1210/me.2010-0402

Monni O, Oinonen R, Elonen E, Franssila K, Teerenhovi L, Joensuu H, Knuutila S (1998) Gain of 3q and deletion of 11q22 are frequent aberrations in mantle cell lymphoma. Gene Chromosom Cancer 21(4):298–307

Morschhauser F, Seymour JF, Kluin-Nelemans HC, Grigg A, Wolf M, Pfreundschuh M, Tilly H et al (2008) A phase II study of enzastaurin, a protein kinase C beta inhibitor, in patients with relapsed or refractory mantle cell lymphoma. Ann Oncol 19(2):247–253

Morton LM, Wang SS, Devesa SS, Hartge P, Weisenburger DD, Linet MS (2006) Lymphoma incidence patterns by WHO subtype in the United States, 1992–2001. Blood 107(1):265–276

Musgrove EA (2006) Cyclins: roles in mitogenic signaling and oncogenic transformation. Growth Fact (Chur, Switzerland) 24(1):13–19. doi:10.1080/08977190500361812

Musgrove EA, Caldon CE, Barraclough J, Stone A, Sutherland RL (2011) Cyclin D as a therapeutic target in cancer. Nat Rev Cancer 11(8):558–572

Nakajima K, Inagawa M, Uchida C, Okada K, Tane S, Kojima M, Kubota M et al (2011) Coordinated regulation of differentiation and proliferation of embryonic cardiomyocytes by a jumonji (Jarid2)-cyclin D1 pathway. Development (Cambridge, England) 138(9):1771–1782. doi:10.1242/dev.059295

Navarro A, Royo C, Hernández L, Jares P, Campo E (2011) Molecular pathogenesis of mantle cell lymphoma: new perspectives and challenges with clinical implications. Semin Hematol 48(3):155–165. doi:10.1053/j.seminhematol.2011.04.001

Negorev DG, Vladimirova OV, Kossenkov AV, Nikonova EV, Demarest RM, Capobianco AJ, Showe MK et al (2010) Sp100 as a potent tumor suppressor: accelerated senescence and rapid malignant transformation of human fibroblasts through modulation of an embryonic stem cell program. Cancer Res 70(23):9991–10001. doi:10.1158/0008-5472.CAN-10-1483

Oh W, Jacinto E (2011) mTOR complex 2 signaling and functions. Cell Cycle (Georgetown, Tex) 10(14):2305–2316

Pérez-Galán P, Roué G, Villamor N, Montserrat E, Campo E, Colomer D (2006) The proteasome inhibitor bortezomib induces apoptosis in mantle-cell lymphoma through generation of ROS and Noxa activation independent of p53 status. Blood 107(1):257–264. doi:10.1182/blood-2005-05-2091

Pérez-Galán P, Dreyling M, Wiestner A (2011a) Mantle cell lymphoma: biology, pathogenesis, and the molecular basis of treatment in the genomic era. Blood 117(1):26–38. doi:10.1182/blood-2010-04-189977

Pérez-Galán P, Mora-Jensen H, Weniger MA, Shaffer AL 3rd, Rizzatti EG, Chapman CM, Mo CC et al (2011b) Bortezomib resistance in mantle cell lymphoma is associated with plasmacytic differentiation. Blood 117(2):542–552. doi:10.1182/blood-2010-02-269514

Pham LV, Tamayo AT, Yoshimura LC, Lo P, Ford RJ (2003) Inhibition of constitutive NF-kappa B activation in mantle cell lymphoma B cells leads to induction of cell cycle arrest and apoptosis. J Immunol (Baltimore, Md: 1950) 171(1):88–95

Pighi C, Gu T-L, Dalai I, Barbi S, Parolini C, Bertolaso A, Pedron S et al (2011) Phospho-proteomic analysis of mantle cell lymphoma cells suggests a pro-survival role of B-cell receptor signaling. Cell Oncol 34(2):141–153. doi:10.1007/s13402-011-0019-7

Pinyol M, Bea S, Plà L, Ribrag V, Bosq J, Rosenwald A, Campo E et al (2007) Inactivation of RB1 in mantle-cell lymphoma detected by nonsense-mediated mRNA decay pathway inhibition and microarray analysis. Blood 109(12):5422–5429. doi:10.1182/blood-2006-11-057208

Psyrri A, Papageorgiou S, Liakata E, Scorilas A, Rontogianni D, Kontos CK, Argyriou P et al (2009) Phosphatidylinositol 3′-kinase catalytic subunit alpha gene amplification contributes to the pathogenesis of mantle cell lymphoma. Clin Cancer Res Off J Am Assoc Cancer Res 15(18):5724–5732. doi:10.1158/1078-0432.CCR-08-3215

Rai R, Dai H, Multani AS, Li K, Chin K, Gray J, Lahad JP et al (2006) BRIT1 regulates early DNA damage response, chromosomal integrity, and cancer. Cancer Cell 10(2):145–157. doi:10.1016/j.ccr.2006.07.002

Reddy K, Ansari-Lari M, Dipasquale B (2008) Blastic mantle cell lymphoma with a Burkitt translocation. Leuk Lymphoma 49(4):740–750. doi:10.1080/10428190701852024

Ren J, Wang Y, Liang Y, Zhang Y, Bao S, Xu Z (2010) Methylation of ribosomal protein S10 by protein-arginine methyltransferase 5 regulates ribosome biogenesis. J Biol Chem 285(17):12695–12705. doi:10.1074/jbc.M110.103911

Renner C, Zinzani PL, Gressin R, Klingbiel D, Favet L, Hitz F, Bargetzi M et al (2010). A multicenter phase II study (SAKK 36/06) of single agent everolimus (RAD001) in patients with relapsed or refractory mantle cell lymphoma. ASH Annu Meet Abstr 116(21):2803

Renner C, Zinzani P, Gressin R, Klingbiel D, Dietrich P-Y, Hitz F, Bargetzi M et al (2012) A multicenter phase II trial (SAKK 36/06) of single-agent Everolimus (RAD001) in patients with relapsed or refractory mantle cell lymphoma. Haematologica. doi:10.3324/haematol.2011.053173

Reutens AT, Fu M, Wang C, Albanese C, McPhaul MJ, Sun Z, Balk SP et al (2001) Cyclin D1 binds the androgen receptor and regulates hormone-dependent signaling in a p300/CBP-associated factor (P/CAF)-dependent manner. Mol Endocrinol (Baltimore, Md) 15(5):797–811

Richard P, Vassallo J, Valmary S, Missoury R, Delsol G, Brousset P (2006) "In situ-like" mantle cell lymphoma: a report of two cases. J Clin Pathol 59(9):995–996. doi:10.1136/jcp. 2005.030783

Rinaldi A, Kwee I, Taborelli M, Largo C, Uccella S, Martin V, Poretti G et al (2006) Genomic and expression profiling identifies the B-cell associated tyrosine kinase Syk as a possible therapeutic target in mantle cell lymphoma. Br J Haematol 132(3):303–316

Rizzatti EG, Falcão RP, Panepucci RA, Proto-Siqueira R, Anselmo-Lima WT, Okamoto OK, Zago MA (2005) Gene expression profiling of mantle cell lymphoma cells reveals aberrant expression of genes from the PI3K-AKT, WNT and TGFβ signalling pathways. Br J Haematol 130(4):516–526. doi:10.1111/j.1365-2141.2005.05630

Rizzatti EG, Mora-Jensen H, Weniger MA, Gibellini F, Lee E, Daibata M, Lai R et al (2008) Noxa mediates bortezomib induced apoptosis in both sensitive and intrinsically resistant mantle cell lymphoma cells and this effect is independent of constitutive activity of the AKT and NF-kappaB pathways. Leuk Lymphoma 49(4):798–808. doi:10.1080/10428190801910912

Rizzieri DA, Feldman E, DiPersio JF, Gabrail N, Stock W, Strair R, Rivera VM et al (2008) A phase 2 clinical trial of deforolimus (AP23573, MK-8669), a novel mammalian target of rapamycin inhibitor, in patients with relapsed or refractory hematologic malignancies. Clin Cancer Res 14(9):2756

Rosenwald A, Wright G, Wiestner A, Chan WC, Connors JM, Campo E, Gascoyne RD et al (2003) The proliferation gene expression signature is a quantitative integrator of oncogenic events that predicts survival in mantle cell lymphoma. Cancer Cell 3(2):185–197

Rosner M, Hanneder M, Siegel N, Valli A, Fuchs C, Hengstschläger M (2008) The mTOR pathway and its role in human genetic diseases. Mutat Res 659(3):284–292. doi:10.1016/j.mrrev.2008.06.001

Roué G, Pérez-Galán P, López-Guerra M, Villamor N, Campo E, Colomer D (2007) Selective inhibition of IkappaB kinase sensitizes mantle cell lymphoma B cells to TRAIL by decreasing cellular FLIP level. J Immunol (Baltimore, Md: 1950) 178(3):1923–1930

Roué G, Pérez-Galán P, Mozos A, López-Guerra M, Xargay-Torrent S, Rosich L, Saborit-Villarroya I et al (2011) The Hsp90 inhibitor IPI-504 overcomes bortezomib resistance in mantle cell lymphoma in vitro and in vivo by down-regulation of the prosurvival ER chaperone BiP/Grp78. Blood 117(4):1270–1279. doi:10.1182/blood-2010-04-278853

Royo C, Salaverria I, Hartmann EM, Rosenwald A, Campo E, Beà S (2011) The complex landscape of genetic alterations in mantle cell lymphoma. Semin Cancer Biol 21(5):322–334. doi:10.1016/j.semcancer.2011.09.007

Royo C, Navarro A, Clot G, Salaverria I, Giné E, Jares P, Colomer D et al (2012). Non-nodal type of mantle cell lymphoma is a specific biological and clinical subgroup of the disease. Leuk: Off J Leuk Soc Am, Leuk Res Fund U K doi:10.1038/leu.2012.72

Rubio-Moscardo F, Climent J, Siebert R, Piris MA, Martín-Subero JI, Nieländer I, Garcia-Conde J et al (2005) Mantle-cell lymphoma genotypes identified with CGH to BAC microarrays define a leukemic subgroup of disease and predict patient outcome. Blood 105(11):4445–4454. doi:10.1182/blood-2004-10-3907

Rudelius M, Pittaluga S, Nishizuka S, Pham TH-T, Fend F, Jaffe ES, Quintanilla-Martinez L et al (2006) Constitutive activation of Akt contributes to the pathogenesis and survival of mantle cell lymphoma. Blood 108(5):1668–1676. doi:10.1182/blood-2006-04-015586

Salaverria I, Zettl A, Beà S, Moreno V, Valls J, Hartmann E, Ott G et al (2007) Specific secondary genetic alterations in mantle cell lymphoma provide prognostic information independent of the gene expression–based proliferation signature. J Clin Oncol 25(10):1216–1222. doi:10.1200/JCO.2006.08.4251

Salaverria I, Espinet B, Carrió A, Costa D, Astier L, Slotta-Huspenina J, Quintanilla-Martinez L et al (2008) Multiple recurrent chromosomal breakpoints in mantle cell lymphoma revealed by a combination of molecular cytogenetic techniques. Genes Chromosom Cancer 47(12):1086–1097. doi:10.1002/gcc.20609

Sant M, Allemani C, Tereanu C, De Angelis R, Capocaccia R, Visser O, Marcos-Gragera R et al (2010) Incidence of hematologic malignancies in Europe by morphologic subtype: results of the HAEMACARE project. Blood 116(19):3724–3734. doi:10.1182/blood-2010-05-282632

Saucedo LJ, Edgar BA (2007) Filling out the Hippo pathway. Nat Rev Mol Cell Biol 8(8):613–621. doi:10.1038/nrm2221

Schöllkopf C, Melbye M, Munksgaard L, Smedby KE, Rostgaard K, Glimelius B, Chang ET et al (2008) Borrelia infection and risk of non-Hodgkin lymphoma. Blood 111(12):5524–5529

Schrader C, Meusers P, Brittinger G, Teymoortash A, Siebmann J-U, Janssen D, Parwaresch R et al (2004) Topoisomerase IIα expression in mantle cell lymphoma: a marker of cell proliferation and a prognostic factor for clinical outcome. Leukemia 18(7):1200–1206. doi:10.1038/sj.leu.2403387

Schraders M, Pfundt R, Straatman HMP, Janssen IM, van Kessel AG, Schoenmakers EFPM, van Krieken JHJM et al (2005) Novel chromosomal imbalances in mantle cell lymphoma detected by genome-wide array-based comparative genomic hybridization. Blood 105(4):1686–1693. doi:10.1182/blood-2004-07-2730

Schraders M, Jares P, Bea S, Schoenmakers EFPM, van Krieken JHJM, Campo E, Groenen PJTA (2008) Integrated genomic and expression profiling in mantle cell lymphoma: identification of gene-dosage regulated candidate genes. Br J Haematol 143(2):210–221. doi:10.1111/j.1365-2141.2008.07334.x

Shah BD, Villagra A, Merino O, Rock-Klotz J, Woan K, Seto E, Martin P et al (2010). HDAC profiling in mantle cell lymphoma unveils HDAC11 and HDAC10 as potential molecular targets. ASH Annu Meet Abstr 116(21):2506

Shen R, Wang X, Drissi H, Liu F, O'Keefe RJ, Chen D (2006) Cyclin D1-cdk4 induce runx2 ubiquitination and degradation. J Biol Chem 281(24):16347–16353. doi:10.1074/jbc.M603439200

Sherr CJ, Roberts JM (1999) CDK inhibitors: positive and negative regulators of G1-phase progression. Genes Dev 13(12):1501–1512

Shishodia S, Amin HM, Lai R, Aggarwal BB (2005) Curcumin (diferuloylmethane) inhibits constitutive NF-kappaB activation, induces G1/S arrest, suppresses proliferation, and induces apoptosis in mantle cell lymphoma. Biochem Pharmacol 70(5):700–713. doi:10.1016/j.bcp.2005.04.043

Siegert JL, Rushton JJ, Sellers WR, Kaelin WG Jr, Robbins PD (2000) Cyclin D1 suppresses retinoblastoma protein-mediated inhibition of TAFII250 kinase activity. Oncogene 19(50):5703–5711. doi:10.1038/sj.onc.1203966

Sillé FCM, Thomas R, Smith MT, Conde L, Skibola CF (2012) Post-GWAS functional characterization of susceptibility variants for chronic lymphocytic leukemia. PLoS One 7(1):e29632. doi:10.1371/journal.pone.0029632

Skibola CF, Bracci PM, Nieters A, Brooks-Wilson A, De Sanjosé S, Hughes AM, Cerhan JR et al (2010) Tumor necrosis factor (TNF) and lymphotoxin-alpha (LTA) polymorphisms and risk of Non-Hodgkin lymphoma in the InterLymph consortium. Am J Epidemiol 171(3):267

Smedby KE, Hjalgrim H (2011) Epidemiology and etiology of mantle cell lymphoma and other non-Hodgkin lymphoma subtypes. Semin Cancer Biol 21(5):293–298. doi:10.1016/j.semcancer.2011.09.010

Smith M, Joshi I, Jin F, Al-Saleem T (2006) Murine model for mantle cell lymphoma. Leukemia 20(5):891–893

Smith SM, Anastasi J, Cohen KS, Godley LA (2010) The impact of MYC expression in lymphoma biology: beyond Burkitt lymphoma. Blood Cell Mol Dis 45(4):317–323. doi:10.1016/j.bcmd.2010.08.002

Stilgenbauer S, Schaffner C, Winkler D, Ott G, Leupolt E, Bentz M, Möller P et al (2000) The ATM gene in the pathogenesis of mantle-cell lymphoma. Ann Oncol Off J Eur Soc Med Oncol/ESMO 11(Suppl 1):127–130

Suvasini R, Shruti B, Thota B, Shinde SV, Friedmann-Morvinski D, Nawaz Z, Prasanna KV et al (2011) Insulin growth factor-2 binding protein 3 (IGF2BP3) is a glioblastoma-specific marker that activates phosphatidylinositol 3-kinase/mitogen-activated protein kinase (PI3K/MAPK) pathways by modulating IGF-2. J Biol Chem 286(29):25882–25890. doi:10.1074/jbc.M110.178012

Tagawa H, Karnan S, Suzuki R, Matsuo K, Zhang X, Ota A, Morishima Y et al (2005) Genome-wide array-based CGH for mantle cell lymphoma: identification of homozygous deletions of the proapoptotic gene BIM. Oncogene 24(8):1348–1358. doi:10.1038/sj.onc.1208300

Takeuchi I, Tagawa H, Tsujikawa A, Nakagawa M, Katayama-Suguro M, Guo Y, Seto M (2009) The potential of copy number gains and losses, detected by array-based comparative genomic hybridization, for computational differential diagnosis of B-cell lymphomas and genetic regions involved in lymphomagenesis. Haematologica 94(1):61–69. doi:10.3324/haematol.12986

Tchakarska G, Le Lan-Leguen A, Roth L, Sola B (2009) The targeting of the sole cyclin D1 is not adequate for mantle cell lymphoma and myeloma therapies. Haematologica 94((12):1781–1782

Teodorovic I, Pittaluga S, Kluin-Nelemans J, Meerwaldt J, Hagenbeek A, Van Glabbeke M, Somers R et al (1995) Efficacy of four different regimens in 64 mantle-cell lymphoma cases: clinicopathologic comparison with 498 other non-Hodgkin's lymphoma subtypes. European Organization for the Research and Treatment of Cancer Lymphoma Cooperative Group. J Clin Oncol 13(11):2819–2826

The Non-Hodgkin's Lymphoma Classification Project (1997) A clinical evaluation of the international lymphoma study group classification of non-Hodgkin's lymphoma. The non-Hodgkin's Lymphoma Classification Project. Blood 89(11):3909–3918

Thelander EF, Rosenquist R (2008) Molecular genetic characterization reveals new subsets of mantle cell lymphoma. Leuk Lymphoma 49(6):1042–1049. doi:10.1080/10428190801947559

Tiemann M, Schrader C, Klapper W, Dreyling MH, Campo E, Norton A, Berger F et al (2005) Histopathology, cell proliferation indices and clinical outcome in 304 patients with mantle cell lymphoma (MCL): a clinicopathological study from the European MCL Network. Br J Haematol 131(1):29–38

Tort F, Hernàndez S, Beà S, Martínez A, Esteller M, Herman JG, Puig X et al (2002) CHK2-decreased protein expression and infrequent genetic alterations mainly occur in aggressive types of non-Hodgkin lymphomas. Blood 100(13):4602–4608. doi:10.1182/blood-2002-04-1078

Tort F, Hernández S, Beà S, Camacho E, Fernández V, Esteller M, Fraga MF et al (2005) Checkpoint kinase 1 (CHK1) protein and mRNA expression is downregulated in aggressive variants of human lymphoid neoplasms. Leuk Off J Leuk Soc Am Leuk Res Fund UK 19((1):112–117

Vater I, Wagner F, Kreuz M, Berger H, Martín-Subero JI, Pott C, Martinez-Climent JA et al (2009) GeneChip analyses point to novel pathogenetic mechanisms in mantle cell lymphoma. Br J Haematol 144(3):317–331. doi:10.1111/j.1365-2141.2008.07443.x

Villagra A, Cheng F, Wang H-W, Suarez I, Glozak M, Maurin M, Nguyen D et al (2008) The histone deacetylase HDAC11 regulates the expression of interleukin 10 and immune tolerance. Nat Immunol 10:92–100. doi:10.1038/ni.1673

Vizcarra E, Martínez-Climent JA, Benet I, Marugan I, Terol MJ, Prosper F, Marco J et al (2001) Identification of two subgroups of mantle cell leukemia with distinct clinical and biological features. Hematol J Off J Eur Haematol Assoc/EHA 2(4):234–241. doi:10.1038/sj/thj/6200111

Wander SA, Hennessy BT, Slingerland JM (2011) Next-generation mTOR inhibitors in clinical oncology: how pathway complexity informs therapeutic strategy. J Clin Invest 121(4):1231–1241. doi:10.1172/JCI44145

Wang S, Slager S, Brennan P, Holly E, De Sanjose S, Bernstein L, Boffetta P et al (2007) Family history of hematopoietic malignancies and risk of non-Hodgkin lymphoma (NHL): a pooled analysis of 10 211 cases and 11 905 controls from the International Lymphoma Epidemiology Consortium (InterLymph). Blood 109(8):3479–3488

Wang C, Lisanti MP, Liao DJ (2011a) Reviewing once more the c-myc and Ras collaboration. Cell Cycle 10(1):57–67. doi:10.4161/cc.10.1.14449

Wang L, Martin P, Blum KA, Kahl BS, Maeda LS, Advani R, Williams ME et al (2011b). The Bruton's tyrosine kinase inhibitor PCI-32765 is highly active as single-agent therapy in previously-treated Mantle Cell Lymphoma (MCL): preliminary results of a phase II trial. ASH Annu Meet Abstr 118(21):442

Werner M, Hobeika E, Jumaa H (2010) Role of PI3K in the generation and survival of B cells. Immunol Rev 237(1):55–71. doi:10.1111/j.1600-065X.2010.00934.x

Wiestner A, Tehrani M, Chiorazzi M, Wright G, Gibellini F, Nakayama K, Liu H et al (2007) Point mutations and genomic deletions in CCND1 create stable truncated cyclin D1 mRNAs that are associated with increased proliferation rate and shorter survival. Blood 109(11):4599–4606. doi:10.1182/blood-2006-08-039859

Yang DT, Young KH, Kahl BS, Markovina S, Miyamoto S (2008) Prevalence of bortezomib-resistant constitutive NF-kappaB activity in mantle cell lymphoma. Mol Cancer 7:40. doi:10.1186/1476-4598-7-40

Zacharek SJ, Xiong Y, Shumway SD (2005) Negative regulation of TSC1-TSC2 by mammalian D-type cyclins. Cancer Res 65(24):11354–11360. doi:10.1158/0008-5472.CAN-05-2236

Zhang L, Fried FB, Guo H, Friedman AD (2008) Cyclin-dependent kinase phosphorylation of RUNX1/AML1 on 3 sites increases transactivation potency and stimulates cell proliferation. Blood 111(3):1193–1200. doi:10.1182/blood-2007-08-109702

Zhang X, Chen X, Lin J, Lwin T, Wright G, Moscinski LC, Dalton WS et al (2011) Myc represses miR-15a/miR-16-1 expression through recruitment of HDAC3 in mantle cell and other non-Hodgkin B-cell lymphomas. Oncogene. Retrieved from http://dx.doi.org/10.1038/onc.2011.470
Zhao JJ, Lin J, Lwin T, Yang H, Guo J, Kong W, Dessureault S et al (2010) microRNA expression profile and identification of miR-29 as a prognostic marker and pathogenetic factor by targeting CDK6 in mantle cell lymphoma. Blood 115(13):2630–2639
Zong R-T, Das C, Tucker PW (2000) Regulation of matrix attachment region-dependent, lymphocyte-restricted transcriptio n through differential localization within promyelocytic leukemia nuclear bodies. EMBO J 19(15):4123–4133. doi:10.1093/emboj/19.15.4123
Zucca E, Roggero E, Pinotti G, Pedrinis E, Cappella C, Venco A, Cavalli F (1995) Patterns of survival in mantle cell lymphoma. Annu Oncol Off J Eur Soc Med Oncol/ESMO 6(3):257–262

Pathogenesis of Non-Hodgkin Lymphoma Derived from Inflammatory, Autoimmune or Immunologic Disorders

Ling Zhang and Jianguo Tao

The risk of lymphoproliferative disorders has been reported to be increased in autoimmune rheumatic disease and chronic inflammatory diseases. The autoimmune rheumatic diseases, also called autoimmune disorders, that are frequently associated with lymphomas are namely rheumatoid arthritis, Sjörgen syndrome, systemic lupus erythematous, celiac disease, Crohn disease, and Hashimoto thyroiditis (Dias and Isenberg 2011; Tvarůzková et al. 2011; Maverakis et al. 2012; Mellemkjaer et al. 2008).

Studies have demonstrated that there are various kinds of lymphomas that develop in patients with systemic autoimmune disorders (Suvajdzic et al. 2011; Dias and Isenberg 2011). The most frequent encountered lymphoma type was non-Hodgkin lymphoma, accounting for 88% (35/40) patients, as reported in one study (Suvajdzic et al. 2011). Of them, diffuse large B-cell lymphoma was the most common subtype, comprising 34% of patients in the group (Suvajdzic et al. 2011). Links to autoimmune/inflammatory diseases have also been described in chronic lymphocytic leukemia (Maverakis et al. 2012; Tvarůzková et al. 2011), as well as in Hodgkin lymphoma (Váróczy et al. 2012). In Váróczy's retrospective study, it reported that 11.5% (14/121) Hodgkin lymphoma patients had associated autoimmune disease (Váróczy et al. 2012). Pediatric Hodgkin lymphomas have also been found in many autoimmunity states such as lupus syndrome, antiphospholipid syndrome with transient ischemic attack, Evans syndrome, leukocytoclastic vasculitis, autoimmune hemolytic anemia, autoimmune thyroiditis, and juvenile idiopathic arthritis (Jarrassé et al. 2011). Patients with these autoimmune rheumatic diseases, particularly those with detectable autoantibodies and systemic involvement, are at increased risk of

L. Zhang, M.D. (✉) • J. Tao, M.D., Ph.D.
Hematopathology and Laboratory Medicine, Hematologic Malignancies,
H. Lee Moffitt Cancer Center & Research Institute,
12902 Magnolia Drive, Tampa, FL 33612, USA
e-mail: ling.zhang@moffitt.org

J. Tao and E. Sotomayor (eds.), *Hematologic Cancers: From Molecular Pathobiology to Targeted Therapeutics*, Cancer Growth and Progression 14,
DOI 10.1007/978-94-007-5028-9_8,

developing B-cell lymphoma, especially diffuse large B-cell lymphoma and marginal zone lymphoma (Dias and Isenberg 2011). The other risk factors for increased lymphoma in these patients are male sex, advanced age, prolonged disease course, and increased disease severity, but not family history of autoimmune conditions (Dias and Isenberg 2011).

The exact pathoetiology of autoimmune disease is still under investigation. Infection (viral or bacterial), chemical or toxin exposure, and genetic susceptibility could attribute to its development. Chronic immune stimulation, genetic and other environmental factors, and some immunosuppressive drugs might also be involved in lymphomagenesis in these patients (Dias and Isenberg 2011; Tvarůzková et al. 2011). It is believed that antigens or immunological elements can trigger transformation of polyclonal population of normal lymphocyte into monoclonal neoplastic process--lymphoproliferative disease (Tvarůzková et al. 2011). Alteration of gene expression is one of the significant factors in lymphoma formation. Genetic susceptibility (for example, HLA typing) plays a considerable role in concurrence of autoimmune disorder and non-Hodgkin lymphoma based on the data from genome-wide association analyses. In Conde's study, it was shown that non-Hodgkin lymphoma and several autoimmune disorders shared common genomic regions and susceptibility loci, suggestive of the overlapping risk for both diseases (Conde et al. 2011). Certainly, superimposed infections in patients with autoimmune rheumatic disease might increase the risks to developing lymphoproliferative disorder or lymphoma. The infectious agents commonly consist of viruses [e.g. Epstein-Barr virus (EBV), human immunodeficiency virus (HIV) or hepatitis C virus], and bacteria (e.g. Helicobacter pylori, Borrelia burgdorferi). EBV is one of the most plausible candidates for playing a role in the pathophysiology of lymphomas in autoimmune diseases. High titers of anti-EBV antibodies and impaired T-cell responses to EBV antigens have been found in patients with systemic lupus erythematous (SLE), rheumatoid arthritis, and primary Sjörgen syndrome versus levels shown in normal subjects (Toussirot and Roudier 2008). In addition, administration of some immunomodulators or immune suppression of autoimmune rheumatic disease has been proved to be akin to lymphoproliferative disorders/lymphoma.

In light of the numerous kinds of autoimmune disorders and the various levels of understanding of the pathogenesis of these diseases, the following sections will more specifically discuss the relationship between several common autoimmune disorders and lymphoproliferative disease.

Sjögren Syndrome

Primary Sjögren syndrome is an autoimmune syndrome characterized by chronic dry month (xerostomia) and dry eyes (xerophthalmia) resulting from dysregulated immune cells attacking and destroying the exocrine glands and corresponding organ tissue.(Delaleu et al. 2008; Jonsson et al. 2005; Skopouli et al. 2000; Theander et al. 2005). Up to 20% of patients with this disease will display systemic or extraglandular features (Jonsson et al. 2005; Skopouli et al. 2000; Theander et al. 2005).The incidence rate of Sjögren syndrome is 4/1,000,000 people in the United States with female prevalence (estimated female-to-male ratio of 9:1 or even higher) (Westhoff and Zink 2010). Most patients with Sjögren syndrome are more than 40 years old (Westhoff and Zink 2010).

Non-Hodgkin lymphomas are frequently encountered in patients with Sjögren syndrome when compared to patients with other autoimmune diseases or healthy individuals (Sugai et al. 1987; Voulgarelis and Skopouli 2007; Isaacson et al. 2008b; Tzioufas and Voulgarelis 2007). The estimated occurrence rate of lymphoid malignancy in patients with Sjögren syndrome is 5% (Tzioufas and Voulgarelis 2007). A 10-year follow-up study in 66 patients with primary Sjögren syndrome showed 4 (6.7%) developed non-Hodgkin lymphoma (Krylova and Isenberg 2010; Westhoff and Zink 2010) while an incidence rate of lymphoma in the general population is reportedly 0.03% (ISD Cancer Information Programme 2009). A single-center review of 352 patients with lymphoma found that Sjögren syndrome (8 cases, 3.4%) was the leading cause versus the other autoimmune disorders: undifferentiated connective tissue disease (3%) thyroiditis (2.6%) rheumatoid arthritis (1.7%), and systemic vasculitis (1.7%) (Váróczy et al. 2012).

Patients with severe Sjögren syndrome are much more likely to develop lymphomas than those with mild or moderate diseases (Smedby et al. 2006). Studies have shown that patients with systemic involvement often display palpable purpura, low serum levels of the complement component C4 and mixed type II cryoglobulins and secondary lymphoma (Voulgarelis and Skopouli 2007; Voulgarelis and Tzioufas 2010). In addition, lymphadenopathy, splenomegaly, and neutropenia are predicting factors for systemic disease (Le Guern and Mouthon 2011). Thus, systemic involvement beyond epithelial lesions reflect an overall worse outcome than those with localized presentation (Voulgarelis and Tzioufas 2010). Of note, most lymphomas derived from primary Sjögren syndrome are of B-cell lineage and often low grade. B-cell non-Hodgkin lymphoma has been reported in 82.3% (14 of 17) of patients with primary Sjögren syndrome (Wang et al. 2010). An extranodal marginal zone B cell lymphoma (MALT lymphoma) in the salivary glands with or without transforming to diffuse large B-cell lymphoma is the most common one (Smedby et al. 2006; Voulgarelis and Skopouli 2007; Isaacson 2008). MALT lymphomas could also be found in the stomach and lung or adjacent lymph nodes in these patients (Mariette 1999; Fox and Kang 1992).

The transition from a chronic inflammatory condition or autoimmune reaction to malignant lymphoma takes multistep processes and the exact pathogenesis is yet poorly understood. There is more than one mechanism has been proposed in primary Sjögren syndrome and its secondary lymphoproliferative disorder/lymphoma. The factors initiating and driving autoimmunity in this entity are of importance. Infection e.g. viral infection could be predisposing factor, where a consequence of disturbances of the immune system e.g. B-cell hyperreactivity and enhanced levels of B-cell-activating factor/B-lymphocyte stimulator play a key role in disease development (Hansen et al. 2005). In addition, genetic features, including certain HLA phenotypes and polymorphisms in genes encoding cytokines or factors implicated in cytokine signaling and hormonal factors are also thought to participate in disease pathogenesis (Tzioufas et al. 2012; Voulgarelis and Skopouli 2007). And certainly, additional molecular oncogenic abnormalities e.g. chromosomal translocation, gene deletion or mutation, microsatellite instability, deregulation of B cell cycle, and overproduction of specific B cell biologic stimulators also contribute to the emergence and progression of the neoplastic process (Voulgarelis and Skopouli 2007).

People were thought that primary Sjögren syndrome is due to the combined effects of chronic EBV infection and autoimmunity (Whittingham et al. 1987). It was documented that presence of EBV DNA copy found in patients was higher that that noted in control subjects (Whittingham et al. 1987; Mariette 1999). Emergent data showed hepatitis C infection might also be the triggers for abnormal immune response and B-cell lymphoproliferative disorder in primary Sjögren syndrome (Ramos-Casals et al. 2007).

It has been well known that genetic factors (for example, HLA-DR3) can predispose patients to primary Sjögren syndrome (Fox and Kang 1992). Overexpression of HLA-DR antigens in epithelial cells in these patients suggested a high susceptibility to autoimmune disease (Fox and Stern 2002). Growing evidence is also suggesting the important role of epigenetics in primary Sjögren syndrome (Dantec et al. 2012). This study demonstrated an association between DNA methylation and autoantibody production in primary Sjögren syndrome (Dantec et al. 2012).

Sustainably abnormal lymphoid proliferation consisting of T cells and B cells has long been demonstrated in a subpopulation of patients with Sjögren syndrome. Increased B-lymphocyte activity in primary Sjögren syndrome is manifested by the presence of a number of autoantibodies, such as anti-SS-A and anti-SS-B antibodies, rheumatoid factor, type 2 cryoglobulins, and hypergammaglobulinemia (Kallenberg et al. 2011). B cells within the salivary glands produce refractory factor-like autoantibodies with clonal B cell expansions. The latter can be detected by molecular diagnostic tool such as PCR for immunoglobin gene rearrangement (Kallenberg et al. 2011). BAFF, B cell activating factor belonging to the Tumor Necrosis Factor family, has been observed to be overexpressed in primary Sjögren syndrome and is associated with expansion of transitional type 2 (T2) cells and marginal-zone -like B cells in the salivary glands (Varim et al. 2010). Prolonged B-cell survival and excessive B-cell activity is probably related to increased production of BAFF, which may eventually lead to MALT lymphomas in these patients (Varim et al. 2010; Baimpa et al. 2009; Kassan et al. 1978). BAFF plays a key role in B-cell repopulation after B-cell depletion therapy in primary Sjögren syndrome (Varim et al. 2010). A recent study illustrated that a toll-like-receptor (TLR9) in primary Sjögren's syndrome, is not only able to keep B cells alive by delivering sufficiency of tonic signaling, but also able to confer autoreactive B cells with an marginal-zone-like phenotype (Guerrier et al. 2012).

T cells are implicated in both local and systemic pathophysiology of primary Sjögren syndrome. Data have shown that lymphocytic infiltrates in exocrine glands are dominated by CD4+ T cells, which produce cytokines, including IL-2 and interferon-gamma, and exhibit cytotoxic potential (Fox and Stern 2002). A recent study showed that oligoclonal expansion of T cells is found locally and systemically in these patients (Busch et al. 2012) in response to inflammation and autoimmune dysregulation. The self-perpetuating T cell dependent autoimmune is triggered by autoantigen present in Sjögren syndrome (Katsifis et al. 2007). This immune response drive proinflammatory (Th1 and Th17) and limited inhibitory (Treg) pathways (Katsifis et al. 2007, 2009). As squeals, persistent tissue damage and compromised function of salivary and lacrimal glands would occur (Katsifis et al. 2007). The roles of the lymphoid structure and lymphoid chemokines are important in extranodal B-cell lymphomagenesis (Bombardieri and Pitzalis 2012). Abnormal proliferation of T cells and cytokine release could support a B-cell lymphoma formation. *In vitro* and *in vivo* studies of Sjögren syndrome have shown that the formation and growth of the lymphoid structures in the salivary gland critically rely on the ectopic expression of lymphoid chemokines (for example, CXCL13, CCL19, CCL21, and CXCL12), which are likely produced by the background stromal cells, as well as different subsets of infiltrating immune cells during chronic inflammatory processes in Sjögren syndrome (Salomonsson et al. 2002; Barone et al. 2005; Bombardieri and Pitzalis 2012). It has been noted that increased infiltration by IL-18(+) cells and C4-hypocomplementemia are adverse prognostic factors for lymphoma development (Tzioufas et al. 2012).

Taken together, secondary lymphoma occurs as the results of multifactorial interaction in primary Sjogren syndrome. More studies are necessary to explore the networks of inflammatory and immunological processes.

Systemic Lupus Erythematosus

Systemic lupus erythematosus (SLE) is an autoimmune disease with very heterogeneous clinical manifestations and various immune abnormalities, producing a plethora of autoantibodies with deposition of immune complexes in various organs, ultimately ending in organ damage or secondary malignancies (Anolik 2007). The ratio of women to men with the disease is 10:1 (Veeranki and Choubey 2010). As with other autoimmune disorders, genetic predisposition, chronic antigen stimulus, disproportional immune responses, and chronic administration of immunosuppressive medications also contribute to the development of malignancies in lupus (Kiss et al. 2010).

It is not uncommon for patients with SLE to develop malignant tumors. One study revealed that cervical cancer, non-Hodgkin lymphoma, and bladder cancer were three common malignancies that developed in female SLE patients, as shown by high standardized incidence ratios of 3.42 (CI 0.00–7.26), 15.37 (CI 2.90–37.68), and 43.55 (CI 8.21–106.78), respectively (Kang et al. 2010). SLE patients also had a three- to four-fold increased risk of developing non-Hodgkin lymphoma and Hodgkin lymphoma versus the general population (Veeranki and Choubey 2010; Rossi et al. 2011; Bernatsky et al. 2009), with a preference to diffuse large B-cell lymphomas (DLBCL) (Rossi et al. 2011; Bernatsky et al. 2009). There was also an increased trend toward Hodgkin lymphoma (Bernatsky et al. 2009). In an investigation of 258 patients with SLE at a hospital between 1982 and 2009, six patients developed lymphomas, including four developing DLCBL, one Hodgkin lymphoma, and one indolent lymphocytic lymphoma (Rossi et al. 2011).

Even though the exact pathogenesis of SLE and related lymphoma has not been fully revealed, several aspects have been delineated. The combination of germline and somatic mutations, persistent immune overstimulation, and the impairment of immune surveillance facilitated by immunosuppressive drugs is thought to be at the origin of the increased lymphomagenesis (Rossi et al. 2011). Defect of tumor suppressors could be one of the triggers in development of secondary malignancies, including lymphoma. For example, microtubule-associated tumor suppressor 1(MTUS-1) is involved in the angiotensin II type 2 receptor (AT2R)-mediated growth inhibitory signaling cascade (Zuern et al. 2012). In a mouse model, MTUS-1 deficiency caused spontaneous heart hypertrophy and SLE-like lymphoproliferative disease (Zuern et al. 2012).

B cells serve a key role in immune regulation in the healthy population. The role of B cells in development of SLE and related immunologic abnormalities is also very crucial. Alteration of B-cell signaling cascades, B-cell tolerance, B-cell longevity, and microenvironments contribute to pathogenesis in SLE (Anolik 2007). Clinically, efficacy of anti-B cell antibodies (rituxan, epratuzumab) in SLE patients also point to B cells being involved in the pathogenesis of this disease (Anolik 2007; Dorner et al. 2006; Daridon et al. 2010). However, the molecular mechanism involved in regulation of B cell homeostasis is not yet clear in SLE patients, especially in those with B-cell malignancies.

A balanced cytokine release is the key step in maintaining immune system function. Therefore, the type or level of cytokine release could play a pivotal role in autoimmune disorders. A study reported that cytokine A proliferating-inducing Ligand (APRIL) was overexpressed in patients with DLBCL and the cytokine ligand showed a high concentrations in sera from a subset of patients with autoimmune disorder including rheumatoid arthritis and SLE (Löfström et al. 2011). Therefore, dysregulation of cytokine A proliferating-inducing Ligand (APRIL) could be associated with increased risk of lymphoproliferative disorder in SLE and rheumatoid arthritis.

A cohort of patients with SLE demonstrated a high potentiality of virus-induced malignancies (standardized incidence ratio of 2.9 (95% CI 2.0–4.1)) in systemic lupus erythematosus, suggestive of some oncogenic virus infection in the pathogenesis of SLE-related lymphoproliferative disorders and carcinoma (Dreyer et al. 2011).

An advanced molecular profile study also showed the importance of some genes in SLE and B-cell lymphoma. In a microarray study of marrow mononuclear cells from 20 SLE patients (11 with active disease), a significant similarity was shown in gene profiles between patients with active SLE and non-Hodgkin's lymphoma, with overlapping main pathways, including kinases (MAPK, ERK, JNK, PKC), transcription factors (NF-kappa B, STAT3), and insulin (Nakou et al. 2010).

Regarding immunosuppressant use in SLE, someone believed that a high-dose cyclophosphamide therapy seemed to contribute to the increased risk of cancer in patients with SLE (Kang et al. 2010). In addition to the use of cyclophosphamide, the use of anti-TNF-α agent was also reported to be associated with non-Hodgkin lymphoma in SLE patients (Aringer et al. 2009). A long-term follow up of 13 patients with SLE who were administrated TNF-alpha blockade with infliximab showed that two patients had fatal CNS lymphoma (Aringer et al. 2009). The drug-related lymphoma may be partially attributed to immunocompromised state with superimposing lymphotrophic virus infection e.g. EBV. The detailed mechanism is to be explored.

Together, there is a close clinical and molecular relationship between SLE and secondary lymphoproliferative disorders or lymphoma. Alterations of normal immune response and gene profile are considered the main pathway toward abnormal lymphoproliferation and development of lymphoma.

Rheumatoid Arthritis

Rheumatoid arthritis is defined as a chronic, systemic inflammatory disorder that may affect many tissues and organs, but that particularly attacks flexible (synovial) joints. The disease process often ends with the destruction of articular cartilage and ankylosis of the joints (Harris and Firestein 2008). However, the chronic inflammation and autoimmune process may also involve the lungs, pleura, heart, sclera, and subcutaneous tissue.

Rheumatoid arthritis was is a fairly common disease, with occurrence of 0.5–1% of the World's population and female predominance (Alamanos et al. 2006; Carmona et al. 2010). The male-to-female ratio is 3–5:1 (Alamanos et al. 2006). Onset of the disease most frequently occurs between the ages of 40 and 50 years; however, rheumatoid arthritis can affect people of any age (Alamanos et al. 2006; Carmona et al. 2010). Rheumatoid arthritis is defined as an immune-mediated disease because of detectable autoantibodies, B-cell proliferation, accumulation of immunoglobulins and other inflammatory products such as complements, and positive response to B-cell depletion therapy (Rantapaa-Dahlqvist 2005; Weyand et al. 2005; Low and Moore 2005; Perosa et al. 2005). Rheumatoid arthritis could be triggered by an environmental (microbial) factor in a genetically susceptible individual (Ebringer and Rashid 2006). It is widely accepted that autoimmunity plays a key role in both the disease chronicity and progression. The mechanism of development of rheumatoid arthritis is unclear. Nevertheless, the interaction of genetic admixture, environmental factors, and socio-demographic background might be the most important underlying causes of rheumatoid arthritis (Carmona et al. 2010).

Lymphoma risks in patients with rheumatoid arthritis are heterogeneous between studies, ranging from no increase to borderline significantly increase (Ekström Smedby et al. 2008). The heterogenesity could be attributed to the prevalence of reported rheumatoid arthritis among controls in different studies. (Ekström Smedby et al. 2008). Dias's group reported an increased risk of lymphoma in rheumatoid arthritis, especially diffuse large B-cell lymphoma and marginal zone lymphoma. (Dias and Isenberg 2011).

The pathogenesis has never been completely understood. A recent study pointed out there is a direct causal relationship between autoimmune and chronic inflammatory disorders including rheumatoid arthritis and the various lymphoproliferative malignancies (van de Schans et al. 2011). Of many other inflammatory diseases, rheumatoid arthritis was more often present in newly diagnosed patients with most lymphomas (van de Schans et al. 2011). Genetic and environmental factors might also independently or synthetically play pivotal roles in lymphomagenesis in these patients (Dias and Isenberg 2011). Some date suggested that the overall risk of cancer and lymphoma in patients with rheumatoid arthritis did not appear to differ greatly from those recorded among untreated patients; rather, it was associated with the degree of disease activity (Rosenblum and Amital 2011).

Infection is one of etiologies. Numerous evidences showed EBV infection may contribute indirectly to the pathophysiology of rheumatoid arthritis, namely impaired lymphocyte response to EBV, detectable EBV antigens and genetic material within the rheumatoid synovium, presence of cell-mediated responses to EBV replicative cycle proteins and to glycoprotein 110 similar epitope of EBV antigens to some self antigens, increased virus load in blood than in controls (Toussirot and Roudier 2007). Impairing immune control of EBV replication, increased exposure to EBV antigens and, and chronic inflammation might also trigger lymphogenesis, and cause sustained chronic inflammation or disease progression in this group of patients (Toussirot and Roudier 2007).

Whether anti-inflammatory drugs and/or immunomodulants could induce lymphoproliferative disorders in patients with rheumatoid arthritis is still debatable. Numerous drugs have been developed in an attempt to control this disease. Disease-modifying anti-rheumatic drugs are the most popular ones, which commonly include azathioprine, cyclosporin A, D-penicillamine, gold salts, hydroxychloroquine, leflunomide, methotrexate, sulfasalazine, and cyclophosphamide. These medications can be used individually or in combination. Although the use of combinations could minimize the toxicity of drugs, long-term use of some of these medications (for example, methotrexate) has been associated with lymphoproliferative disorders, either "pseudolymphoma" or true malignant lymphoma. Besides diffuse large B-cell lymphoma (the most common one), follicular lymphoma was also reported in a patient with rheumatoid arthritis receiving methotrexate treatment (Nakanishi et al. 2011). Lymphomatoid granulomatosis is another rare scenario reported in the same setting (Inaba et al. 2011).

Biological agents have also been approved for treatment of rheumatoid arthritis. Introduction of three main anti-TNF-α inhibitory agents (infliximab, adalimumab, and etanercept) have been recently applied in the control of rheumatoid arthritis, especially in those RA patients resistant to routine therapy (Rosenblum and Amital 2011). A large study cohort including 2,306 control patients and 5,179 patients treated with anti-TNFα therapy was conducted. Four of 2,306 (0.17%) patients from control group developed hematolymphoid neoplasms while 11 (11/5,179, 0.21%) patients receiving anti-TNFα therapy developed lymphomas. The results of statistical analysis suggested an increase in lymphomas in the TNF alpha group, with predominance of B-cell lymphomas but do not reach statistical significance ($p > 0.05$) (Wong et al. 2012). The other study data has shown that short- to intermediate-term treatment with biologics (for example, up to 4 years), especially anti-TNF-α agents, should be secured with respect to lymphoma risk (Dommasch and Gelfand 2009).

Interestingly, most lymphomas that occurred in rheumatoid arthritis were of B-cell origin; a rare extranodal NK/T-cell lymphoma, nasal type, was also found in a patient with rheumatoid arthritis receiving anti-tumor necrosis factor (TNF) therapy (Arai et al. 2011). In the reported case, discontinuation of methotrexate and anti-TNF agent ("etanercept" or Enbrel) did not stop growth of the lymphoma (Arai et al. 2011). Another group suggested that attention should be paid to the high risk of hepatosplenic T-cell lymphoma in young male patients treated with anti-TNF-α agents together with thiopurines (Ye 2011). The pathoetiology here is unclear.

In conclusion, secondary lymphomas have been reported in patients with rheumatoid arthritis with not yet determined risk factors. Mechanisms of the link between the two entities are worthy of further exploration.

Hashimoto Thyroiditis

Hashimoto thyroiditis, also called chronic lymphocytic thyroiditis, is an autoimmune disease. The disease has been earlier described by Japanese scholar Hashimoto in 1912 and is now recognized as an antibody-mediated immune response toward thyroid glands, resulting in damaged glandular structures and chronic lymphocytic infiltrate (Nakazawa 2008). It affects middle-aged adults, with a significant female predominance (female to male ratio of 15;1) (Brent et al. 2008) The annual incidence of Hashimoto thyroiditis worldwide is estimated to be 0.3–1.5 cases per 1,000 persons (Vanderpump et al. 1995 and Vanderpump et al. 1998). In a large study cohort of 917 patients with surgical resection of thyroid gland(s) for thyroid diseases, 77 patients (~8.4%) showed with histologic features compatible with chronic lymphocytic thyroditis. Hashimoto thyroiditis was identified in only 20.8% of the 77 patients with chronic lymphocytic thyroditis, corresponding to 1.7% of all patients experienced thyroidectomy (Büyükaşık et al. 2011).

Lymphoma of the thyroid is quite a rare entity, comprising approximately 3% of all non-Hodgkin lymphoma and 5% of all thyroid tumor patients (Ruggiero et al. 2005; Ha et al. 2001; Pasieka 1998). Although rare, thyroid lymphomas occur more frequently in patients with a history of Hashimoto thyroiditis (Ruggiero et al. 2005; Ha et al. 2001; Holm et al. 1985; Mellemkjaer et al. 2008). Primary thyroid lymphomas essentially belong to the non-Hodgkin lymphoma subcategory and can be further divided into two clinical settings: indolent or aggressive. Primary thyroid mucosa-associated lymphoid tissue (MALT) lymphoma, usually having indolent behavior, constitutes about 6–28% of all primary thyroid lymphomas (Harris et al. 1994; Jaffe et al. 1999) with one study reporting up to 46% of lymphomas (Watanabe et al. 2011). An aggressive variant, diffuse large B-cell lymphoma, *de novo* or transformed, has been shown to account for 51% (81 of 171) of patients with primary thyroid lymphoma (Watanabe et al. 2011). Extrathyroidal MALT lymphoma has been observed in patients with Hashimoto thyroiditis (Troch et al. 2008). Despite the majority of non-Hodgkin lymphoma in thyroid being of B-cell lineage, T-cell lymphoma associated with Hashimoto thyroiditis has also been occasionally reported (Abdul-Rahman et al. 1996; Kim et al. 2010). Coexistence of thyroid carcinoma with primary thyroid lymphoma in a background of Hashimoto thyroiditis has been documented (Vassilatou et al. 2011; Li et al. 2010; Panayiotides et al. 2010).

The pathoetiology of Hashimoto thyroiditis and related thyroid lymphoma is still under investigation. The mechanisms implicated in development of Hashimoto thyroiditis might also play a pivotal role in the development of lymphoma in the same patient group. Genetic predisposition with HLA-DR5 carrier seems to be an important factor implicated in Hashimoto thyroiditis. The production of autoantibodies against thyroglobubin, thyroid-stimulating hormone (TSH), and thyroid peroxidase, so as to destroy thyroid cells, is the major concern. Studies demonstrated that cytotoxic T-cell lymphocyte associated-4 (CTLA-4) gene polymorphisms are also associated with the autoimmune process in development of Hashimoto thyroiditis (Kavvoura et al. 2007) and possibly effect on secondary lymphoma formation. It is also recognized that increased CD8-positive cytotoxic T-cells in response to the cell-mediated immune response and cytokine release might expedite the disease course.

An evolution from Hashimoto thyroiditis to lymphoma is not well understood. Genetic predisposition, chronic inflammatory processes, abnormal cytokine release, increased CD8-positive cytotoxic T-cells, and tissue-specific antibody in thyroid tissue are the imperative factors that contribute to lymphoma development. A case study of peripheral T-cell lymphoma developed in Hashimoto thyroiditis found the neoplastic T-cells with phenotype (CD3+, CD4+, CD8-, TCR+, CXCR3+, CCR5+ and ST2(L)-) corresponding to a Th1 cell origin (Koida et al. 2007). The study assumed that Th1 activation induced by chronic inflammation could lead to peripheral T-cell lymphoma as well as possible MALT-lymphoma of B cells in the setting (Koida et al. 2007). As B-cells are increased in Hashimoto thyroiditis, the accumulation of intraepithelial B-cells were considered the probable promoters of the autoimmune process and when clonal evolution and immortalization was occurred, it could be regard as causative agent in the development of B-cell lymphoma (Tiemann et al. 1996). The clonal relationship of B-cells in Hashimoto disease and secondary lymphoma has been illustrated by a recent study conducted by a Canadian group (Moshynska and Saxena 2008). This study illustrated reproducible clonal IgH peaks with high sequence homology between 60 and 100% in 3 of 4 patients with Hashimoto thyroiditis who subsequently developed lymphoma (Moshynska and Saxena 2008). These molecular findings suggested a likelihood of clonal evolution from Hashimoto thyroiditis to primary thyroid lymphoma.

Inflammatory Bowel Disease

Inflammatory diseases include ulcerative colitis and Crohn disease. In general, the risk of lymphoproliferative disorders in patients with inflammatory bowel disease (IBD) exists, especially Crohn disease. In a recent retrospective study of 8,780 records of patients with IBD, 14 patients had a diagnosis of lymphoma (0.15%) (Sultan et al. 2012). In another large study, the incidence rate of lymphoma in IBD was reported to be 0.28% (46 of 16,023 patients) (Herrinton et al. 2011). A large study cohort that included 1,374 patients diagnosed with IBD revealed that, when compared with the same disease in adults, the overall risk of lymphoma in children with IBD is lower (Ashworth et al. 2012).

It is still uncertain whether the development of lymphoproliferative disorders in these patients is associated with the use of anti-inflammatory regimens or immunomodulants in that the pathoetiology could be multifactorial. Use of immunosuppressive agents was thought to be one of the major contributors to secondary lymphoma in IBD. Although there were conflicting results from hospital and population-based studies, purine analog-based therapies demonstrated a 4-folder higher risk of lymphoma. The contribution of anti-TNF-α agents is difficult to determine (Jones and Loftus 2007). However, a combination of thiopurin and anti-TNF-α therapy in IBD has been shown to increase the risk for lymphoma, commonly B-cell non-Hodgkin lymphomas, including diffuse large B-cell lymphoma (44%), follicular lymphoma (14%), and Hodgkin lymphoma (12%) (Jones and Loftus 2007; Herrinton et al. 2011). Increased risk of lymphoma was reported in children who had suffered from IBD and were treated with thiopurines, but the standardized incidence ratio was similar to that in adults with the same disease (Ashworth et al. 2012). A single case of hepatosplenic T-cell lymphoma was reported in a young male patient with IBD (Ye et al. 2011). A pseudo-Hodgkin-lymphoma that occurs after TNF-α treatment in Crohn disease has been occasionally reported (Cassaday et al. 2011). Withdrawal of immunosuppression caused tumor regression (Cassaday et al. 2011).

In addition to some immunosuppressants, EBV infection or reactivation appears to be a trigger in lymphoproliferative disorders/lymphoma in IBD. A Dutch group reported that 11 of 12 IBD patients who were treated with azathioprine/mercaptopurine developed EBV-positive lymphoma (Vos et al. 2011). It is not yet proven whether there is an association between azathioprine/mercaptopurine therapy and the development of lymphoma in IBD (van Hogezand et al. 2002). A review article suggested the risks associated with azathioprine and mercaptopurine use are myelosuppression and infection, whereas lymphoma was of minimal clinical significance (Kwon and Farrell 2005). It seems more data needed to understand the relationship and mechanism of lymphoma in IBD.

Celiac Disease

Celiac disease is regarded as an autoimmune disorder of the small intestine that occurs in genetically predisposed people of all ages from middle infancy onward. Clinically the disease is characterized with chronic diarrhoea, failure to thrive, especially in children, and fatigue. It can be asymptomatic or accompanied by numerous systemic disorders, including nutritional deficiencies, osteoporosis, hypofertility, cardiovascular disorder, venous thrombosis, peripheral neuropathy, and cirrhosis

(Cosnes and Nion-Larmurier 2011) An association with an increased risk of autoimmune diseases such as type 1 diabetes, thyroiditis, and gastrointestinal carcinoma, hepatocellular carcinoma and lymphoma has also been reported (Cosnes and Nion-Larmurier 2011). There are two main subtypes of non-Hodgkin lymphomas that occur in celiac disease: enteropathy-associated T-cell lymphoma and hepatosplenic T-cell lymphoma.

Enteropathy-associated T-cell lymphoma is a rare lymphoma subtype of mature alpha-beta T-cell lymphoma that is strongly associated with celiac disease, an autoimmune disease triggered by the ingestion of gluten with increasing incidence (Isaacson et al. 2008a; Sharaiha et al. 2012). The overall age-adjusted and sex-adjusted annual incidence for all bowel lymphomas has increased from 0.006 to 0.024 per 100,000 populations with predominance in the Hispanic population (Sharaiha et al. 2012). The Sharaiha study indicated that the significant increase in the incidence of enteropathy-associated T-cell lymphoma in the United States may reflect the increasing seroprevalence of celiac disease and better recognition of rare types of T-cell lymphomas.

There is limited information regarding the molecular mechanisms linking celiac disease and enteropathic T-cell lymphoma. Studies focused on a population of intraepithelial intestinal lymphocytes from refractory celiac disease show that the cells expressed clonal V gamma, (V)γ, gene rearrangement, which was identical to variable region (V)γ of T-cell receptor in enteropathic T-cell lymphoma. (Perfetti et al. 2012).

Intestinal T-cell lymphoma, as a complication of celiac disease, can be involved in liver or spleen as a metastatic lesion. However, celiac disease could also be accompanied by a primary hepatic or hepatosplenic T-cell lymphoma, although rare (Farcet et al. 1990; Harris et al. 1994; Prasad et al. 2011). Farcet's study demonstrated that hepatic T-cell lymphoma displays T-cell receptor gene rearrangement involving a gamma-delta zone (Farcet et al. 1990). The exact mechanisms of hepatic T-cell lymphoma involvement in celiac disease are unclear. It has been hypothesized that the celiac disease shares genetic or immunopathogenetic factors with primary biliary cirrhosis, primary sclerosing cholangitis, and autoimmune hepatitis (Freeman 2010). The same mechanisms may apply to primary hepatic T-cell lymphoma in this setting.

Autoimmune Lymphoproliferative Syndrome

Autoimmune lymphoproliferative syndrome (ALPS) is an inherited disease of the immune system with an earlier onset, chronic, nonmalignant lymphoproliferation, and autoimmune manifestation characterized by lymphadenopathy, hepatosplenomegaly, autoimmune disorder, multilineage cytopenia, polyclonal hypergammaglobulinemia, and life-long high risk to develop lymphoma. The syndrome was defined in 1990s, with nearly 500 patient families investigated. It commonly affects both children and adults. Immunophenotyping performed on peripheral blood revealed an elevation of the characteristic double negative CD4 and CD8 T-cells (CD4-/CD8-, double negative T (DNT) cells), which is one of pathognomonic features of ALPS (Rao and Oliveira 2011).

ALPS is recognized as a genetic and family disorder (Neven et al. 2011). In ALPS, a genetic defect in programmed cell death, or apoptosis, leads to breakdown of lymphocyte homeostasis and normal immunologic tolerance. A cell surface receptor (FAS), also named CD95 or APO-1, is a member of the tumor necrosis factor receptor (TNF-R) superfamily of proteins and considered a putative tumor suppressor. It functions as a direct apoptosis trigger in order to maintain lymphocyte homeostasis and peripheral immune tolerance and prevent autoimmunity (Oehm et al. 1992; Cohen and Eisenberg 1991). Patients with ALIPS usually carry heterozygous germline or somatic mutations involving the TNFSF6 gene, previously known as APT1, which codes for the cell surface receptor FAS (Neven et al. 2011). According to Neven's report, heterozygous germline mutations of TNFRSF6 were identified in 83% of probands, whereas somatic TNFRSF6 mutations (located within intracellular domain of FAS) were found in 17% of index cases (Neven et al. 2011). Because of various mutations involved in ALPS, the syndrome is subdivided into three types: type Ia with mutant Fas, type Ib with mutations in the ligand for Fas as well as systemic lupus erythematosus with lymphadenopathy, type II with mutant caspase 10, and type III without any defined genetic cause (Bleesing et al. 2011; Madkaikar et al. 2011). Although majority of patients with ALPS harbor heterozygous germline mutations in Fas, random somatic Fas mutations are also noted, and become the second most common genetic etiology of ALPS. There is a subpopulation of patients with ALPS exhibited additionally gene mutations involving Fas-ligand (FASLG), caspase 10 (CASP10) and caspase 8 (CASP8), NRAS and KRAS (Madkaikar et al. 2011). In the newly proposed consensus several gene abnormalities (Somatic or germline pathogenic mutation in FAS, FASLG, *or* CASP10) are listed primary diagnosis criteria for ALPS (Oliveira et al. 2010)

There is a clear link between ALPS and lymphoma, including both Hodgkin and non-Hodgkin lymphomas. The estimated risk of Hodgkin lymphoma and non-Hodgkin lymphoma is 50-fold and 14-fold higher than risks shown in the general population, respectively (Straus et al. 2001). As observed, lymphomas occurring in ALPS typically originate in B cells. T-cell lymphomas were found as well, although much less frequently (Bleesing et al. 2011). All aforementioned gene mutations resulted in reduced lymphomcyte apoptosis are primary pathogenesis for abnormal lymphoproliferation and lymphomas in ALPS patients and their families (Oliveira et al. 2010; Hsu et al. 2012). Interestingly, unlike other immunocomprised scenario lymphoma in ALPS is not related to EBV infection (Bleesing et al. 2011). The critical role of FAS in lymphoma surveillance cannot be ignored. Mutations involved in FAS share the same pathogenesis for both diseases. Somatic mutations in exon 9 of the FAS gene, coding for the intracellular death domain of the protein, were detected in 20% of follicular center cell origin B-cell lymphomas (Müschen et al. 2002). Somatic CASP10 gene mutations were noted in 14.5% of non-Hodgkin lymphomas (Poppema et al. 2004).

In general, autoimmune disorders link with lymphoproliferative disorders because both disorders share genetic and molecular backgrounds; however, the more biologic and molecular mechanisms are needed for further investigation. More efforts should be made to shed light in the field of autoimmunity and lymphomagenesis.

References

Abdul-Rahman ZH, Gogas HJ, Tooze JA, Anderson B, Mansi J, Sacks NP, Finlayson CJ (1996) T-cell lymphoma in Hashimoto's thyroiditis. Histopathology 29(5):455–459

Alamanos Y, Voulgari PV, Drosos AA (2006) Incidence and prevalence of rheumatoid arthritis, based on the 1987 American College of Rheumatology criteria: a systematic review. Semin Arthritis Rheum 36(3):182–188

Anolik JH (2007) B cell biology and dysfunction in SLE. Bull NYU Hosp Jt Dis 65(3):182–186

Arai H, Maki K, Tadokoro J, Nakamura Y, Sasaki K, Mitani K (2011) Extranodal NK/T-cell lymphoma, nasal type, developed in a patient with rheumatoid arthritis. Rinsho Ketsueki 52(7):551–555

Aringer M, Houssiau F, Gordon C, Graninger WB, Voll RE, Rath E, Steiner G, Smolen JS (2009) Adverse events and efficacy of TNF-alpha blockade with infliximab in patients with systemic lupus erythematosus: long-term follow-up of 13 patients. Rheumatology (Oxford) 48(11):1451–1454

Ashworth LA, Billett A, Mitchell P, Nuti F, Siegel C, Bousvaros A (2012) Lymphoma risk in children and young adults with inflammatory bowel disease: analysis of a large single-center cohort. Inflamm Bowel Dis 18(5):838–843, Epub 2011 Sep 1

Baimpa E, Dahabreh U, Voulgarelis M, Moutsopoulos HM (2009) Hematologic manifestations and predictors of lymphoma development in primary Sjogren's syndrome: clinical and pathophysiological aspects. Medicine 88:284–293

Barone F, Bombardieri M, Manzo A, Blades MC, Morgan PR, Challacombe SJ, Valesini G, Pitzalis C (2005) Association of CXCL13 and CCL21 expression with the progressive organization of lymphoid-like structures in Sjögren's syndrome. Arthritis Rheum 52(6):1773–1784

Bernatsky S, Ramsey-Goldman R, Clarke AE (2009) Malignancy in systemic lupus erythematosus: what have we learned? Best Pract Res Clin Rheumatol 23(4):539–547

Bleesing JJH, Johnson J, Zhang K (2011) Autoimmune lymphoproliferative syndrome. In: Pagon RA, Bird TD, Dolan CR, Stephens K (eds) GeneReviews [Internet]. University of Washington, Seattle, 1993-.2006 Sep 14 [updated 2011 Sep 08]

Bombardieri M, Pitzalis C (2012) Ectopic lymphoid neogenesis and lymphoid chemokines in Sjogren's syndrome: at the interplay between chronic inflammation, autoimmunity and lymphomagenesis. Curr Pharm Biotechnol Jan 2. [Epub ahead of print]

Brent GA, Larsen PR, Davies TF (2008) Hypothyroidism and thyroiditis. In: Kronenberg HM, Melmed S, Polonsky KS, Larsen PR (eds) Williams textbook of endocrinology, 11th edn. Saunders Elsevier, Philadelphia, chap 12

Busch R, Hadjinicolaou AV, Hall FC (2012) Local activation and systemic dysregulation of T Lymphocytes in Sjögren's Syndrome. Curr Pharm Biotechnol Jan 2. [Epub ahead of print]

Büyükaşık O, Hasdemir AO, Yalçın E, Celep B, Sengül S, Yandakçı K, Tunç G, Küçükpınar T, Alkoy S, Cöl C (2011) The association between thyroid malignancy and chronic lymphocytic thyroiditis: should it alter the surgical approach? Endokrynol Pol 62(4):303–308

Carmona L, Cross M, Williams B, Lassere M, March L (2010) Rheumatoid arthritis. Best Pract Res Clin Rheumatol 24(6):733–745

Cassaday RD, Malik JT, Chang JE (2011) Regression of Hodgkin lymphoma after discontinuation of a tumor necrosis factor inhibitor for Crohn's disease: a case report and review of the literature. Clin Lymphoma Myeloma Leuk 11(3):289–292

Cohen PL, Eisenberg RA (1991) Lpr and gld: single gene models of systemic autoimmunity and lymphoproliferative disease. Annu Rev Immunol 9:243–269
Conde L, Bracci PM, Halperin E, Skibola CF (2011) A search for overlapping genetic susceptibility loci between non-Hodgkin lymphoma and autoimmune diseases. Genomics 98(1):9–14
Cosnes J, Nion-Larmurier I (2011) Complications of celiac disease. Pathol Biol (Paris) May 26. [Epub ahead of print] [Article in French]
Dantec CL, Varin MM, Brooks WH, Pers JO, Youinou P, Renaudineau Y (2012) Epigenetics and Sjögren's syndrome. Curr Pharm Biotechnol Jan 2. [Epub ahead of print]
Daridon C, Blassfeld D, Reiter K, Mei HE, Giesecke C, Goldenberg DM, Hansen A, Hostmann A, Frölich D, Dörner T (2010) Epratuzumab targeting of CD22 affects adhesion molecule expression and migration of B-cells in systemic lupus erythematosus. Arthritis Res Ther 12(6):R204, Epub 2010 Nov 4
Delaleu N, Immervoll H, Cornelius J, Jonsson R (2008) Biomarker profiles in serum and saliva of experimental Sjögren's syndrome: associations with specific autoimmune manifestations. Arthritis Res Ther 10(1):R22
Dias C, Isenberg DA (2011) Susceptibility of patients with rheumatic diseases to B-cell non-Hodgkin lymphoma. Nat Rev Rheumatol 7(6):360–368
Dommasch E, Gelfand JM (2009) Is there truly a risk of lymphoma from biologic therapies? Dermatol Ther 22(5):418–430
Dorner T, Kaufmann J, Wegener WA, Teoh N, Goldenberg DM, Burmester GR (2006) Initial clinical trial of epratuzumab (humanized anti-CD22 antibody) for immunotherapy of systemic lupus erythematosus. Arthritis Res Ther 8(3):R74, Epub 2006 Apr 21
Dreyer L, Faurschou M, Mogensen M, Jacobsen S (2011) High incidence of potentially virus-induced malignancies in systemic lupus erythematosus: a long-term followup study in a Danish cohort. Arthritis Rheum 63(10):3032–3037
Ebringer A, Rashid T (2006) Rheumatoid arthritis is an autoimmune disease triggered by Proteus urinary tract infection. Clin Dev Immunol 13(1):41–48
Ekström Smedby K, Vajdic CM, Falster M, Engels EA, Martínez-Maza O, Turner J, Hjalgrim H, Vineis P, Seniori Costantini A, Bracci PM, Holly EA, Willett E, Spinelli JJ, La Vecchia C, Zheng T, Becker N, De Sanjosé S, Chiu BC, Dal Maso L, Cocco P, Maynadié M, Foretova L, Staines A, Brennan P, Davis S, Severson R, Cerhan JR, Breen EC, Birmann B, Grulich AE, Cozen W (2008) Autoimmune disorders and risk of non-Hodgkin lymphoma subtypes: a pooled analysis within the InterLymph Consortium. Blood 111(8):4029–4038, Epub 2008 Feb 8
Farcet J, Gaulard P, Marolleau J et al (1990) Hepatosplenic T-cell lymphoma: sinusal/sinusoidal localization of malignant cells expressing the T-cell receptor gamma-delta. Blood 75(11):2213–2219
Fox RI, Kang HI (1992) Pathogenesis of Sjögren's syndrome. Rheum Dis Clin North Am 18(3):517–538
Fox RI, Stern M (2002) Sjögren's syndrome: mechanisms of pathogenesis involve interaction of immune and neurosecretory systems. Scand J Rheumatol 116(Suppl):3–13, Review
Freeman HJ (2010) Hepatic manifestations of celiac disease. Clin Exp Gastroenterol 3:33–39
Guerrier T, Le Pottier L, Devauchelle V, Pers JO, Jamin C, Youinou P (2012) Role of toll-like receptors in primary Sjögren's syndrome with a special emphasis on B-cell maturation within exocrine tissues. J Autoimmun 39(1–2):69–76, Epub 2012 Feb 21
Ha CS, Shadle KM, Medeiros LJ, Wilder RB, Hess MA, Cabanillas F, Cox JD (2001) Localized non-Hodgkin lymphoma involving the thyroid gland. Cancer 91:629–635
Hansen A, Lipsky PE, Dörner T (2005) Immunopathogenesis of primary Sjögren's syndrome: implications for disease management and therapy. Curr Opin Rheumatol 17(5):558–565
Harris ED Jr, Firestein GS (2008) Clinical features of rheumatoid arthritis. In: Firestein GS, Budd RC, Harris ED Jr (eds) Kelley's textbook of rheumatology, 8th edn. Saunders Elsevier, Philadelphia, chapter 66
Harris NL, Jaffe ES, Stein H et al (1994) A revised European-American classification of lymphoid neoplasms: a proposal from the International Lymphoma Study Group. Blood 84(5):1361–1392
Herrinton LJ, Liu L, Weng X, Lewis JD, Hutfless S, Allison JE (2011) Role of thiopurine and anti-TNF therapy in lymphoma in inflammatory bowel disease. Am J Gastroenterol 106(12):2146–2153

Holm LE, Blomgren H, Lowhagen T (1985) Cancer risks in patients with chronic lymphocytic thyroiditis. N Engl J Med 312(10):601–604

Hsu AP, Dowdell KC, Davis J, Niemela JE, Anderson SM, Shaw PA, Rao VK, Puck JM (2012) Autoimmune lymphoproliferative syndrome due to FAS mutations outside the signal-transducing death domain: molecular mechanisms and clinical penetrance. Genet Med 14(1):81–89, Epub 2011 Oct 7

Inaba M, Ushijim S, Hirata N, Saisyoji T, Kitaoka M, Yoshinaga T (2011) Methotrexate-related lymphomatoid granulomatosis in a patient with rheumatoid arthritis. Nihon Kokyuki Gakkai Zasshi 49(8):597–601 [Article in Japanese]

Isaacson PG, Chott A, Ott G, Stein H (2008a) Enteropathy-associated T-cell lymphoma. In: Swerdlow S (ed) WHO classification of tumors of hematopoictic and lymphoid tissue. IARC, 69008, Lyon, France pp 289–291

Isaacson PG, Chott A, Nakamura S, Mueller-Hermelink HK, Harris NL, Swerdlow SH (2008b) Extradonal marginal zone lymphoma of mucosa-associated lymphoid tissue (MALT lymphoma). In: Swerdlow SH (ed) WHO classification of tumors of hematopoietic and lymphoid tissues. IARC, 69008, Lyon, France pp 214–219

ISD Cancer Information Programme, Cancer statistics (2009) Non-Hodgkin's lymphoma: ICD-10 C82-C85. http://www.isdscotland.org/isd/1483.html Cancer prevalence

Jaffe ES, Harris NL, Diebold J, Muller-Hermelink HK (1999) World Health Organization classification of neoplastic diseases of the hematopoietic and lymphoid tissues. A progress report. Am J Clin Pathol 111(1 Suppl 1):S8–S12

Jarrassé C, Pagnier A, Edan C, Landman-Parker J, Mazingue F, Mansuy L, Bertrand Y, Paillard C, Pellier I, Margueritte G, Plantaz D (2011) Hodgkin disease and autoimmunity in children: 11 case reports. Arch Pediatr 18(4):376–382 [Article in French]

Jones JL, Loftus EV Jr (2007) Lymphoma risk in inflammatory bowel disease: is it the disease or its treatment? Inflamm Bowel Dis 13(10):1299–1307

Jonsson R, Bowman SJ, Gordon TP (2005) Sjögren's syndrome. Ch. 78. In: Koopman WJ, Moreland LW (eds) Arthritis and allied conditions, 15th edn. Lippincott Williams and Wilkins, Philadelphia, Pennsylvania, USA, pp 1681–1705

Kallenberg CG, Vissink A, Kroese FG, Abdulahad WH, Bootsma H (2011) What have we learned from clinical trials in primary Sjögren's syndrome about pathogenesis? Arthritis Res Ther 13(1):205

Kang KY, Kim HO, Yoon HS, Lee J, Lee WC, Ko HJ, Ju JH, Cho CS, Kim HY, Park SH (2010) Incidence of cancer among female patients with systemic lupus erythematosus in Korea. Clin Rheumatol 29(4):381–388

Kassan SS, Thomas TL, Moutsopoulos HM, Hoover R, Kimberly RP, Budman DR, Costa J, Decker JL, Chused TM (1978) Increased risk of lymphoma in sicca syndrome. Ann Intern Med 88:888–892

Katsifis GE, Moutsopoulos NM, Wahl SM (2007) T lymphocytes in Sjögren's syndrome: contributors to and regulators of pathophysiology. Clin Rev Allergy Immunol 32(3):252–264

Katsifis GE, Rekka S, Moutsopoulos NM, Pillemer S, Wahl SM (2009) Systemic and local interleukin-17 and linked cytokines associated with Sjögren's syndrome immunopathogenesis. Am J Pathol 175(3):1167–1177, Epub 2009 Aug 21

Kavvoura FK, Akamizu T, Awata T et al (2007) Cytotoxic T-lymphocyte associated antigen 4 gene polymorphisms and autoimmune thyroid disease; a meta-analysis. J Clin Endocrinol Metab 92(92):3162–3170

Kim NR, Ko YH, Lee YD (2010) Primary T-cell lymphoma of the thyroid associated with Hashimoto's thyroiditis, histologically mimicking MALT-lymphoma. J Korean Med Sci 25(3):481–484

Kiss E, Kovacs L, Szodoray P (2010) Malignancies in systemic lupus erythematosus. Autoimmun Rev 9(4):195–199

Koida S, Tsukasaki K, Tsuchiya T, Harasawa H, Fukushima T, Yamada Y, Ohshima K, Kamihira S, Kikuchi M, Tomonaga M (2007) Primary T-cell lymphoma of the thyroid gland with chemokine receptors of Th1 phenotype complicating autoimmune thyroiditis. Haematologica 92(3):e37–e40

Krylova L, Isenberg D (2010) Assessment of patients with primary Sjögren's syndrome--outcome over 10 years using the Sjögren's Syndrome Damage Index. Rheumatology (Oxford) 49(8):1559–1562

Kwon JH, Farrell RJ (2005) The risk of lymphoma in the treatment of inflammatory bowel disease with immunosuppressive agents. Crit Rev Oncol Hematol 56(1):169–178

Le Guern V, Mouthon L (2011) Sjögren's syndrome and non-Hodgkin's lymphoma: from pathophysiology to treatment. Presse Med 40(12 Pt 1):1113–1119, Epub 2011 Oct 2. [Article in French]

Li B, Tang Y, Xiao D, Zhong M (2010) Coexistence of thyroid metastasis carcinoma in the background of HT and primary thyroid mucosa-associated lymphoid tissue B cell lymphoma in a thyroid gland. Ann Hematol 89(10):1053–1056

Löfström B, Backlin C, Pettersson T, Lundberg IE, Baecklund E (2011) Expression of APRIL in diffuse large B cell lymphomas from patients with systemic lupus erythematosus and rheumatoid arthritis. J Rheumatol 38(9):1891–1897

Low JM, Moore TL (2005) A role for the complement system in rheumatoid arthritis. Curr Pharm Des 11:655–670

Madkaikar M, Mhatre S, Gupta M, Ghosh K (2011) Advances in autoimmune lympho-proliferative syndromes. Eur J Haematol 87(1):1–9

Mariette X (1999) Lymphomas in patients with Sjögren's syndrome: review of the literature and physiopathologic hypothesis. Leuk Lymphoma 33(1–2):93–99

Maverakis E, Goodarzi H, Wehrli LN, Ono Y, Garcia MS (2012) The etiology of paraneoplastic autoimmunity. Clin Rev Allergy Immunol 42(2):135–144

Mellemkjaer L, Pfeiffer RM, Engels EA, Gridley G, Wheeler W, Hemminki K, Olsen JH, Dreyer L, Linet MS, Goldin LR, Landgren O (2008) Autoimmune disease in individuals and close family members and susceptibility to non-Hodgkin's lymphoma. Arthritis Rheum 58(3):657–666

Moshynska OV, Saxena A (2008) Clonal relationship between Hashimoto thyroiditis and thyroid lymphoma. J Clin Pathol 61(4):438–444, Epub 2007 Nov 15

Müschen M, Rajewsky K, Krönke M, Küppers R (2002) The origin of CD95-gene mutations in B-cell lymphoma. Trends Immunol 23(2):75–80

Nakanishi H, Muta T, Oshiro Y, Hida A, Mizuno Y, Yasunaga M, Ueda Y, Fujisaki T (2011) B-cell lymphoma possessing t(14;18)(q32;q21)in a patient with rheumatoid arthritis receiving methotrexate treatment. Gan To Kagaku Ryoho 38(8):1365–1369

Nakazawa D (2008) The autoimmune epidemic. Simon & Schuster, New York, pp 32–35

Nakou M, Bertsias G, Stagakis I, Centola M, Tassiulas I, Hatziapostolou M, Kritikos I, Goulielmos G, Boumpas DT, Iliopoulos D (2010) Gene network analysis of bone marrow mononuclear cells reveals activation of multiple kinase pathways in human systemic lupus erythematosus. PLoS One 5(10):e13351

Neven B, Magerus-Chatinet A, Florkin B, Gobert D, Lambotte O, De Somer L, Lanzarotti N, Stolzenberg MC, Bader-Meunier B, Aladjidi N, Chantrain C, Bertrand Y, Jeziorski E, Leverger G, Michel G, Suarez F, Oksenhendler E, Hermine O, Blanche S, Picard C, Fischer A, Rieux-Laucat F (2011) A survey of 90 patients with autoimmune lymphoproliferative syndrome related to TNFRSF6 mutation. Blood 118(18):4798–4807, Epub 2011 Sep 1

Oehm A, Behrmann I, Falk W et al (1992) Purification and molecular cloning of the APO-1 cell surface antigen, a member of the tumor necrosis factor/nerve growth factor receptor superfamily: sequence identity with the Fas antigen. J Biol Chem 267(15):10709–10715

Oliveira JB, Bleesing JJ, Dianzani U, Fleisher TA, Jaffe ES, Lenardo MJ, Rieux-Laucat F, Siegel RM, Su HC, Teachey DT, Rao VK (2010) Revised diagnostic criteria and classification for the autoimmune lymphoproliferative syndrome (ALPS): report from the 2009 NIH International Workshop. Blood 116(14):e35–e40, Epub 2010 Jun 10

Panayiotides IG, Foukas PG, Meristoudis C, Zourla AP, Peros G, Karakitsos P (2010) Simultaneous occurrence of Warthin-like papillary carcinoma and lymphoma of the mucosa associated lymphoid tissue in Hashimoto thyroiditis. J Clin Pathol 63(7):662–663

Pasieka JL (1998) Anaplastic cancer, lymphoma, and metastases of the thyroid gland. Surg Oncol Clin N Am 7(4):707–720

Perfetti V, Brunetti L, Biagi F, Ciccocioppo R, Bianchi PI, Corazza GR (2012) TCRβ clonality improves diagnostic yield of TCRγ clonality in refractory celiac disease. J Clin Gastroenterol Jan 30. [Epub ahead of print]
Perosa F, Favoino E, CaragnanoMA Prete M, Dammacco F (2005) A target antigen for immunotherapy of autoimmune diseases. Autoimmu Rev 4:526–531
Poppema S, Maggio E, van den Berg A (2004) Development of lymphoma in autoimmune lymphoproliferative syndrome (ALPS) and its relationship to Fas gene mutations. Leuk Lymphoma 45(3):423–431
Prasad KK, Debi U, Sinha SK, Nain CK, Singh K (2011) Hepatobiliary disorders in celiac disease: an update. Int J Hepatol 2011:438184, Epub 2010 Nov 14
Ramos-Casals M, la Civita L, de Vita S, Solans R, Luppi M, Medina F, Caramaschi P, Fadda P, de Marchi G, Lopez-Guillermo A, Font J (2007) SS-HCV study group. Characterization of B cell lymphoma in patients with Sjögren's syndrome and hepatitis C virus infection. Arthritis Rheum 57(1):161–170
Rantapaa-Dahlqvist S (2005) Diagnostic and prognostic significance of autoantibodies in early rheumatoid arthritis. Scand J Rheumatol 34:83–96
Rao VK, Oliveira JB (2011) How I treat autoimmune lymphoproliferative syndrome. Blood 118(22):5741–5751
Rosenblum H, Amital H (2011) Anti-TNF therapy: safety aspects of taking the risk. Autoimmun Rev 10(9):563–568, Epub 2011 May 5
Rossi E, Catania G, Truini M, Ravetti GL, Grassia L, Marmont AM (2011) Patients with systemic lupus erythematosus (SLE) having developed malignant lymphomas. Complete remission of lymphoma following high-dose chemotherapy, but not of SLE. Clin Exp Rheumatol 29(3):555–559, Epub 2011 Jun 30
Ruggiero FP, Frauenhoffer E, Stack BC Jr (2005) Thyroid lymphoma: a single institution's experience. Otolaryngol Head Neck Surg 133:888–896
Salomonsson S, Larsson P, Tengnér P, Mellquist E, Hjelmström P, Wahren-Herlenius M (2002) Expression of the B cell-attracting chemokine CXCL13 in the target organ and autoantibody production in ectopic lymphoid tissue in the chronic inflammatory disease Sjögren's syndrome. Scand J Immunol 55(4):336–342
Sharaiha RZ, Lebwohl B, Reimers L, Bhagat G, Green PH, Neugut AI (2012) Increasing incidence of enteropathy-associated T-cell lymphoma in the United States, 1973–2008. Cancer 118(15):3786–3792, Epub 2011 Dec 13
Skopouli FN, Urania D, Ioannidis JPA (2000) Clinical evolution and morbidity and mortality of primary Sjögren's syndrome. Semin Arthritis Rheum 29:296–304
Smedby KE, Baecklund E, Askling J (2006) Malignant lymphomas in autoimmunity and inflammation: a review of risks, risk factors, and lymphoma characteristics. Cancer Epidemiol Biomarkers Prev 15(11):2069–2077
Straus SE, Jaffe ES, Puck JM et al (2001) The development of lymphomas in families with autoimmune lymphoproliferative syndrome with germline Fas mutations and defective lymphocyte apoptosis. Blood 98(1):194–200
Sugai S, Tachibana J, Sawada M, Shimizu S, Hirose Y, Takiguchi T, Konda S (1987) Malignant lymphomas in patients with autoimmune diseases: a report of 6 cases and a review of the Japanese literature. Jpn J Med 26(3):339–347
Sultan K, Korelitz BI, Present D, Katz S, Sunday S, Shapira I (2012) Prognosis of lymphoma in patients following treatment with 6-mercaptopurine/azathioprine for inflammatory bowel disease. Inflamm Bowel Dis Jan 12. [Epub ahead of print]
Suvajdzic N, Djurdjevic P, Todorovic M, Perunicic M, Stojanović R, Novkovic A, Mihaljevic B (2011) Clinical characteristics of patients with lymphoproliferative neoplasms in the setting of systemic autoimmune diseases. Med Oncol Jul 14. [Epub ahead of print]
Theander E, Andersson SI, Manthorpe R, Jacobsson LH (2005) Proposed core set of outcome measures in patients with primary Sjögren's syndrome: 5 year follow-up. Rheumatology 32:109–116

Tiemann M, Asbeck R, Wacker HH (1996) Clonal B-cell reaction in Sjögren disease and Hashimoto autoimmune thyroiditis. Pathologe 17(4):289–295 [Article in German]
Toussirot E, Roudier J (2007) Pathophysiological links between rheumatoid arthritis and the Epstein-Barr virus: an update. Joint Bone Spine 74(5):418–426
Toussirot E, Roudier J (2008) Epstein-Barr virus in autoimmune diseases. Best Pract Res Clin Rheumatol 22(5):883–896
Troch M, Woehrer S, Streubel B, Weissel M, Hoffmann M, Müllauer L, Chott A, Raderer M (2008) Chronic autoimmune thyroiditis (Hashimoto's thyroiditis) in patients with MALT lymphoma. Ann Oncol 19(7):1336–1339, Epub 2008 Mar 11
Tvarůzková Z, Pavlová S, Doubek M, Mayer J, Pospísilová S (2011) Lymphoproliferative disease in patients with autoimmune and inflammatory diseases: significance of antigenic stimulation and inflammatory processes. Cas Lek Cesk 150(3):161–168 [Article in Czech]
Tzioufas AG, Voulgarelis M (2007) Update on Sjögren's syndrome autoimmune epithelitis: from classification to increased neoplasias. Best Pract Res Clin Rheumatol 21(6):989–1010
Tzioufas AG, Kapsogeorgou EK, Moutsopoulos HM (2012) Pathogenesis of Sjögren's syndrome: what we know and what we should learn. J Autoimmun 39(1–2):4–8, Epub 2012 Feb 10
van de Schans SA, van Spronsen DJ, Hooijkaas H, Janssen-Heijnen ML, Coebergh JW (2011) Excess of autoimmune and chronic inflammatory disorders in patients with lymphoma compared with all cancer patients: a cancer registry-based analysis in the south of the Netherlands. Autoimmun Rev 10(4):228–234
van Hogezand RA, Eichhorn RF, Choudry A, Veenendaal RA, Lamers CB (2002) Malignancies in inflammatory bowel disease: fact or fiction? Scand J Gastroenterol 236:48–53
Vanderpump MP, Tunbridge WM, French JM (1995) The incidence of thyroid disorders in the community: a twenty-year follow-up of the Whickham Survey. Clin Endocrinol (Oxf) 43(1):55–68
Vanderpump MP, French JM, Appleton D (1998) The prevalence of hyperprolactinaemia and association with markers of autoimmune thyroid disease in survivors of the Whickham Survey cohort. Clin Endocrinol (Oxf) 48(1):39–44
Varim MM, Le Pottier L, Youinou P, Saulep D, Mackay F, Pers JO (2010) B-cell tolerance breakdown in Sjogren's syndrome: focus on BAFF. Autoimmun Rev 9:604–608
Váróczy L, Páyer E, Kádár Z, Gergely L, Miltényi Z, Magyari F, Szodoray P, Illés A (2012) Malignant lymphomas and autoimmunity-a single center experience from Hungary. Clin Rheumatol 31(2):219–224, Epub 2011 Jul 7
Vassilatou E, Economopoulos T, Tzanela M, Rontogianni D, Karakitsos P, Magou C, Hadjidakis D, Dimitriadis G (2011) Coexistence of differentiated thyroid carcinoma with primary thyroid lymphoma in a background of Hashimoto's thyroiditis. J Clin Oncol 29(25):e709–e712
Veeranki S, Choubey D (2010) Systemic lupus erythematosus and increased risk to develop B cell malignancies: role of the p200-family proteins. Immunol Lett 133(1):1–5
Vos AC, Bakkal N, Minnee RC, Casparie MK, de Jong DJ, Dijkstra G, Stokkers P, van Bodegraven AA, Pierik M, van der Woude CJ, Oldenburg B, Hommes DW, Initiative on Crohn's and Colitis (ICC) (2011) Risk of malignant lymphoma in patients with inflammatory bowel diseases: a Dutch nationwide study. Inflamm Bowel Dis 17(9):1837–1845, Epub 2010 Dec 22
Voulgarelis M, Skopouli FN (2007) Clinical, immunologic, and molecular factors predicting lymphoma development in Sjogren's syndrome patients. Clin Rev Allergy Immunol 32(3):265–274
Voulgarelis M, Tzioufas AG (2010) Pathogenetic mechanisms in the initiation and perpetuation of Sjögren's syndrome. Nat Rev Rheumatol 6(9):529–537
Wang L, Zhao Y, Zhang FC (2010) Malignant lymphoma associated with primary Sjögren's syndrome. Zhonghua Yi Xue Za Zhi 90(39):2773–2775 [Article in Chinese]
Watanabe N, Noh JY, Narimatsu H, Takeuchi K, Yamaguchi T, Kameyama K, Kobayashi K, Kami M, Kubo A, Kunii Y, Shimizu T, Mukasa K, Otsuka F, Miyara A, Minagawa A, Ito K (2011) Clinicopathological features of 171 cases of primary thyroid lymphoma: a long-term study involving 24553 patients with Hashimoto's disease. Br J Haematol 153(2):236–243, Epub 2011 Mar 4
Westhoff G, Zink A (2010) Epidemiology of primary Sjörgren's syndrome. Z Rheumatol 69(1):41–49 [Article in German]

Weyand CM, Seyler TM, Goronzy JJ (2005) B cells in rheumatoid synovitis. Arthritis Res Ther 7(suppl 3):S9–S12
Whittingham S, McNeilage LJ, Mackay IR (1987) Epstein-Barr virus as an etiological agent in primary Sjögren's syndrome. Med Hypotheses 22(4):373–386
Wong AK, Kerkoutian S, Said J, Rashidi H, Pullarkat ST (2012) Risk of lymphoma in patients receiving antitumor necrosis factor therapy: a meta-analysis of published randomized controlled studies. Clin Rheumatol 31(4):631–636, Epub 2011 Nov 18
Ye BD (2011) Inflammatory bowel disease and lymphoproliferative disorders. Korean J Gastroenterol 58(4):171–177 [Article in Korean]
Zuern C, Krenacs L, Starke S, Heimrich J, Palmetshofer A, Holtmann B, Sendtner M, Fischer T, Galle J, Wanner C, Seibold S (2012) Microtubule associated tumor suppressor 1 deficient mice develop spontaneous heart hypertrophy and SLE-like lymphoproliferative disease. Int J Oncol 40:1079–1088

Pathogenesis of Non-Hodgkin Lymphoma Derived from Infection Diseases

Ling Zhang and Roger Klein

Recent improvements in diagnostic techniques and technology have resulted in genome-based studies in lymphoma and new insights into the novel pathway regarding pathoetiology of subclinical and clinical lymphoproliferative disorders.

It is well recognized that there is a strong link between some infections (viral and bacterial) and lymphoproliferative disorders or lymphomas; HIV/AIDS (human immunodeficiency virus/acquired immunodeficiency syndrome) is the best example of lymphoma resulting from virus-induced immunoincompetency that predisposes toward lymphoproliferative disorders and lymphoma (Roman and Smith 2011). In addition to HIV, there are a number of other viruses that may increase lymphoma risk, through other mechanisms, e.g., human T-cell leukemia virus type , Epstein-Barr-virus (EBV), human herpesvirus-8 (HHV-8) (Kaposi's sarcoma herpesvirus), and hepatitis C (Roman and Smith 2011). Helicobacter pyroli (H pyroli) infection is another clinical scenario representative of a potential evolution from bacterium induced lymphoproliferation to extradonal marginal zone lymphoma of mucosa associated lymphoid tissue (MALT lymphoma) (Wotherspoon et al. 1991; Issacson et al. 2008). As reported, the infection of H. pyroli results in an increased lymphoma risk of approximately sixfold greater than the normal population (Swerdlow et al. 2008; Parsonnet et al. 1994; Parsonnet and Isaacson 2004; Sagaert et al. 2010).

However, the underlying molecular mechanisms that contribute to development of lymphoproliferative disorders/lymphomas might be different and complicated in light of different infectious etiologies. Despite many efforts made in the fields since the last decade, it still remains a broad unknown zone for further exploration.

L. Zhang, M.D. (✉)
Hematopathology and Laboratory Medicine, Hematologic Malignancies, H. Lee Moffitt Cancer Center & Research Institute, 12902 Magnolia Drive, Tampa, FL 33612, USA
e-mail: ling.zhang@moffitt.org

R. Klein, M.D., Ph.D.
Department of Molecular Pathology, Cleveland Clinic, Cleveland, OH, USA

J. Tao and E. Sotomayor (eds.), *Hematologic Cancers: From Molecular Pathobiology to Targeted Therapeutics*, Cancer Growth and Progression 14,
DOI 10.1007/978-94-007-5028-9_9, © Springer Science+Business Media Dordrecht 2012

Bacteria Infection

Evidence has been increasing regarding a clear pathogenic link between some bacteria and non-Hodgkin lymphoma (Groves et al. 2000). Many studies have established the association between Helicobacter pylori, Borrelia burgdorferi, and Chlamydia psittaci and Campylobacter jejuni and organ-specific infection, as well as the corresponding lymphomas including extranodal marginal zone lymphomas of mucosa-associated lymphoid tissue (MALT) of the stomach, MALT of the skin, MALT of the ocular adnexa and immunoproliferative small intestinal disease (IPSID)/alpha heavy-chain disease, respectively (Du 2007; Ferreri and Zucca 2007; Wyatt and Rathbone 1988; Slater 2001).

It is no doubt that bacteria can be triggers for organ-specific chronic disorders. In addition to acute infection, many bacteria perpetuate chronic inflammatory disease and abnormal proliferation of epithelial or lymphoid tissue (Marteau and Chaput 2011). The exact molecular mechanisms involved in irreversible lymphoid proliferation or lymphoma in bacteria-infected organs and tissues are still unknown. Chronic antigenic stimulation of infectious agents has been considered to elicit host immune responses that could induce sustained clonal B-cell expansion (Wyatt and Rathbone 1988). A novel study disclosed that dysregulations of a cluster of B-cel homeostatic and activation-dependent chemokine receptors (e.g. decreased expression of CCR8 and CXCR1, loss of XCR1, and increased expression of CXCR1 and CXCR2) played a pivotal role in the site of origin, development and progression of extragastic MALT lymphoma rather than gastric MALT (Deutsch et al. 2008).

Acquisition of genetic anomalies and dysregulation of the some key signaling pathways during infectious and inflammatory processes also play an active role in development of lymphoma (Marteau and Chaput 2011). As reported, translocation involving chromosome 14 or 18 (e.g., t(11;18)/API2-MALT1, t(1;14)/IGH-BCL10, t(14;18)/IGH-MALT1, and t(3;14)/IGH-FOXP1) occurs in MALT lymphomas of different sites (Du 2007). The all four chromosome aberrations are specifically linked to the MALT lymphoma entity and uncommonly seen with the other type of B-cell lymphoma. As a consequence of translation, the oncogenic products have been shown to target a common nuclear factor-κB pathway (Du 2007). It seems that no direct relationship exists between bacterial infections and acquired cytogenetic aberrations. Nevertheless, these indirect associations of the infections with chronic antigen elicitation and altered immune responses in abnormal lymphoid proliferation might exist and they are required further exploration.

Immunoglobulin gene rearrangement could be also one of the key steps in the initiation of immune response as well as non-Hodgkin lymphoma. It is known that marginal-zone B-cells are functionally heterogeneous and may differ with respect to the pattern of somatic hypermutation in their immunoglobulin variable genes. Sequence study of the mutation pattern of immunoglobulin heavy chain variable genes suggested that during the B-cell lymphoma formation, marginal zone lymphoma may arise from different subsets of marginal-zone B-cells (Ferreri and Zucca 2007).

Although several therapeutic strategies have been implicated in treatment of bacterial infection-related lymphoma, target antibiotics are the initial approach. Surgical resection, radiotherapy, and alkylating agent-based chemotherapy are also considered standard management for refractory lymphomas. Anti-CD20 therapy and other novel forms of immunotherapy are experimental or alternative choices.

In the section, we will focus on some bacteria-associated lymphomas arising from mucosa-associated lymphoid tissue.

Helicobacter pylori

The gastrointestinal tract is one of the common locations of extranodal lymphomas, comprising 10–15% of all non-Hodgkin lymphoma cases and approximately 30–40% of extranodal lymphomas (Arora et al. 2011). Nearly 80% of gastrointestinal MALT lymphomas are found in the stomach (Isaacson and Spencer 1987; Issacson et al. 2008). It is well known that MALT lymphomas of the stomach are causatively linked to Helicobacter pylori infection (Issacson et al. 2008), and this organism is present in over 90% of gastric MALT lymphoma cases (Wotherspoon et al. 1991). In addition, epidemiological data have shown that MALT lymphomas of the stomach constitute nearly 3% of all gastric malignancies and 10% of all lymphomas (Kamangar et al. 2011). Helicobacter pylori also induces gastritis, ulcer disease, and gastric cancer (Naumov and Fenjvesi 2011; Adamu et al. 2012).

H. pylori, a type of spiral-shaped bacterium residing in the stomach of adults, have been widely accepted as the major etiological factor in gastric MALT. The accurate pathogenicity of the organism has been in investigation and partially illustrated. H. pylori secrete urease, which is able to convert urea to ammonia so as to buffer hydrogen iron and allow H. pylori to survive in the harsh, low-pH environment of the stomach (Collins 1992; Atherton 2006; Kusters et al. 2006). H. pyroli also produce lectin to bind to gastric mucosa and epithelial cells (Collins 1992; Atherton 2006; Kusters et al. 2006). These bacteria release harmful toxins and enzymes into the gastric cells, resulting in cellular damage and eliciting an immune response (Collins 1992; Atherton 2006; Kusters et al. 2006). During infection, H. pylori have developed their own way to interfere with body immune responses, making the body ineffective in eliminating the bacteria (Atherton 2006; Kusters et al. 2006).

Besides these specific functions and the survival ability of H. pyroli in the stomach, host factors such as genetic predisposition, blood group, stress, drugs, and smoking may also affect disease process. Studies have shown that host genetic polymorphisms produced a high-level pro inflammatory cytokine release in response to infection so as to increase risk of malignant transformation (Atherton 2006). NADPH oxidases (NOX) could induce cell damage by generating reactive oxygen species (ROS) and possibly lead to carcinogenesis in H. pylori infections. A novel study demonstrated that NOX expression was increased in a series of well-characterized gastric MALT lymphoma patients but stayed normal in healthy patient controls (Bessède et al. 2012). Ample evidence has shown that nearly all patients with gastric MALT lymphoma exhibited symptom and signs of H. pylori infection with elevated serum anti-H. Pylori antibody and the risk of developing MALT lymphoma is sixfold higher in infected people than in uninfected people (Parsonnet et al. 1994; Sagaert et al. 2010). Normally, the human stomach is devoid of mucosal or submucosal lymphoid tissue. However, after colonization of the lining by H. pyroli on the stomach, mucosal lymphoid tissue can develop in response to bacterial stimulation in approximately 25% of H. pylori-positive patients (Kusters et al. 2006; Wotherspoon et al. 1991). As far as the pathogenesis of the two subtypes of H. pylori is concerned, CagA-positive strains show higher risk of MALT lymphoma than CagA-negative strains (Eck et al. 1997).

A recent study has also demonstrated that the presence of T-regulator cells in gastric lesions, the number of CD4 (+), CD25 (+), and Foxp3 (+) cells, and the ratio of CD25 (+)/CD4 (+) to Foxp3 (+)/CD4 (+) cells, were correlated with the degree of inflammation (acute or chronic), lymphoid follicle number, and H. pylori infection (Cheng et al. 2012). However, the exact role of T lymphocytes in development of MALT lymphomas of the stomach has not been elucidated. The interaction of T- and B-cells might be considered one important step in immune modulation and greatly influences the development of B-cell lymphoma.

Treatment data have been very convincing in proving the role of H. pylori in causing MALT lymphoma (Wotherspoon et al. 1993). Administration of antibiotics specific to H. pylori will eradicate the bacterium and achieve a long-term complete remission of the MALT lymphoma in approximately 20% of patients or a partial regression (Du 2007; Roggero et al. 1995; Steinbach et al. 1999; Fischbach et al. 2004).

Borrelia burgdorferi

Borrelia burgdorferi is a spirochete and causes the most common tick-borne infection in both Europe and the United States, called Lyme disease (Stanek and Strle 2003; Steere 2006). The most common manifestation of Lyme disease to skin is erythema migrans (EM). The B. burgdorferi infection could have a variety of symptoms as it could affect different organs, including joints, heart muscle, or nervous system (Christen et al. 1993).

There is an association between B. burgdorferi and primary cutaneous B cell lymphomas (Kütting et al. 1997; Goodlad et al. 2000a, b; Slater 2001). Primary cutaneous marginal zone lymphoma, a cutaneous counterpart of extranodal marginal zone B-cell lymphoma, is a predominant subtype, whereas cutaneous B-cell chronic leukemia or cutaneous follicle center cell lymphoma has also been reported to be associated with B. burgdorferi infection (Goodlad et al. 2000a, b; Fühler et al. 2010; Kash et al. 2011). However, the prevalence and incidence rates of B. burgdorferi-derived cutaneous B-cell lymphoma varied as a result of geographic variability or heterogeneity in the molecular techniques that have been applied (Ponzoni et al. 2011a; Goodlad et al. 2000a). An Italian group showed that B burgdorferi played no significant role in primary cutaneous B- or T-cell lymphoma according to PCR tests performed on 98 patients with cutaneous lymphoma and 19 normal skin biopsies

(Ponzoni et al. 2011a) despite many reports from Europe showing an etiological link (Takino et al. 2008). The study from Italy used PCR primers targeting the hbb, flagellin, and Osp-A genes of B. burgdorferi and revealed no targeted DNA products in control samples as well as in the majority of biopsies from patients with cutaneous lymphomas, including follicle center lymphoma (3 cases), marginal zone lymphoma (31 cases), and mycosis fungoides (30 cases) (Ponzoni et al. 2011a). However, Goodlad's study in United Kingdom ensured that B. burgdorferi-specific DNA was detected in a subpopulation of cutaneous B-cell lymphoma (35%, 7/20 patients). By lymphoma subtyping, bacterial DNA was found in 41.7% of marginal zone lymphomas (5/12) , 20% of primary cutaneous follicle center cell lymphomas (1/5), and 33% of diffuse, large B-cell lymphomas of the leg (1/3) (Goodlad et al. 2000a).

Interestingly, only one study indicated that a history of B. burgdorferi infection and the presence of serum anti-Borrelia antibodies were both associated with risk of mantle cell lymphoma, with odds ratio (OR) of 2.5 [1.2-5.1] and 3.6 [1.8-7.4], respectively (Schöllkopf et al. 2008). No further investigations of the association between B. burgdorferi and other systemic or non-cutaneous lymphomas have been conducted.

Primary cutaneous marginal zone lymphoma could arise from B. burgdorferi infection but also originates from unknown causes. The subtype of cutaneous lymphoma did not conform to the general characteristics of non-cutaneous extranodal marginal zone B-cell lymphoma. The majority of which exhibites class-switched immunoglobulins and developed in a T-helper type 2 inflammatory environment (van Maldegem et al. 2008). They commonly express IgG, IgA, and IgE, instead of IgM and do not have distinct immunoglobulin repertoire bias (van Maldegem et al. 2008). Also, unlike most noncutaneous marginal zone B-cell lymphomas that express CXCR3 (a receptor for interferon-gamma-induced chemokines upregulated under the chronic inflammatory media), primary cutaneous marginal zone lymphoma in majority have shown absent expression of CXCR3 (van Maldegem et al. 2008). The above findings suggest that primary cutaneous marginal zone lymphoma could arise in an inflammatory microenvironment that is distinct from other non-cutaneous marginal zone lymphomas (van Maldegem et al. 2008).

For skin involvement of lyme borreliosis, antibiotic treatment is the primary choice in all patients to eliminate the spirochete, cure current disease, and prevent late sequelae. First-line antibiotics include oral doxycycline and topical application of Anaplasma phagocytophilum (Müllegger and Glatz 2008). It has been earlier noted that cutaneous marginal zone lymphoma has responded to antibiotic therapy and intralesional injection of interferon alfa-2a (Kütting et al. 1997).

Chlamydophila psittaci

Chlamydophila psittaci is Gram-negative, nonmotile, intracellular bacterium and grows in wide variety of mammalian hosts, including humans (Longbottom and Coulter 2003; Vanrompay 2007). It is transmitted by inhalation, contact, or ingestion among birds to mammals (Vanrompay 2007) and has a tendency to cause persistent infections (Lietman et al. 1998). The virulence of C. psittaci differs among hosts and infected strains. An in vitro study in mice showed that C. psittaci could induce different host immunities when exposed to attenuated or virulent strain (Miyairi et al. 2011). C. psittaci–related follicular conjunctivitis is often observed (Lietman et al. 1998), which may eventually develop into ocular adnexal lymphomas via a chronic antigenic stimulation (Ferreri et al. 2004, 2009; Guidoboni et al. 2006).

C. psittaci infection has been known to be a link with lymphomas of the ocular adnexa marginal zone B-cell lymphomas (OAMZL) in some geographical areas, suggesting chronic stimulation by persistent infections of C. psittaci cast an important role in OAMZL development. C. psittaci infection is also a considered etiology in the development of nongastrointestinal MALT lymphomas and their precursor lesions (Pannekoek and van der Ende 2011; Ponzoni et al. 2008). A study detected C. psittaci DNA in tissue biopsy from 80% of ocular adnexal lymphomas (conjunctiva, lachrymal gland, and orbital soft tissues) and 40% of the peripheral blood mononuclear cells (PBMCs) of 40% of these patients (Ferreri et al. 2004). Ponzoni' study demonstrated that monocytes/macrophages are the carriers of C. psittaci (Ponzoni et al. 2008). C. psittaci is transferred by peripheral blood monocytes to tissue, and then cause systemic infection (Ferreri et al. 2009). The prevalence of C. psittaci was equal in non-ocular adnexal lymphomas including both nodal and extradonal types (Ponzoni et al. 2008).

Despite advanced messages of this bacterium–lymphoma association, the actual pathogenetics of C. psittaci infection is not clear. Because of geographic variation areas, the related diagnostic and therapeutic implications are still a major unsolved issue. A molecular study of immunoglobin in 44 patients with OAMZL (40% positive for C. psittaci infection) showed the lymphoma cells express a distinctive immunoglobin repertoire, along with several electropositive antigen (Ag)-binding sites, allowing autoantibodies (auto-Abs) to recognize antigen stimulation (Dagklis et al. 2011).

Some cases of OAMZL (11%) expressed autoreactive immunoglobulins resembling those in patients with chronic lymphocytic leukemia (Dagklis et al. 2011). The finding suggested a pathogenic mechanism of B-cell autoreactivity in OAMZL. It is assumed that C. psittaci acts indirectly on the malignant B cells to result in an abnormal proliferation of B-cells through creating an inflammatory milieu and production of auto-antigens (Dagklis et al. 2011). Antibiotics are the choice for infection related lymphoma exclusively. A study reported that in nine patients with ocular adnexal lymphoma, administration of doxycycline, an antibiotic therapy specific for C. psittaci, two achieved complete remission and two achieved partial remission, leading to consideration of this drug as a first-line therapy for this disease (Ferreri et al. 2005).

Besides OAMZL, C. Psittaci infection is also associated with other MALT lymphomas of different location. The detection rate of C. Psittaci is variable according to the affecting sites: lung, 100% (5/5); thyroid gland, 30% (3/10); salivary gland, 13% (2/15); and skin, 25% (1/4) (Aigelsreiter et al. 2011). The study also demonstrated that C. psitassi was frequently noted in autoimmune related disorders e.g. Hashimoto thyroiditis and Sjögren syndrome (11 of total 27 cases, 41%). The overall findings suggest C. Psittaci is a potential pathogen driving autoimmune and lymphoproliferative disorders (Aigelsreiter et al. 2011) although the actual pathogenesis is not yet well understood. Rarely, C. psittaci infection is associated with primary central nervous system marginal zone B-cell lymphoma involving the choroid plexus (Ponzoni et al. 2011b).

Campylobacter jejuni

Campylobacter species are gram-negative bacilli that have a curved or spiral shape. Campylobater jejuni belongs to Campylobater species and is considered one of the most common causes of gastroenteritis worldwide (Allos 2001).

Clinically, infection of C. jejuni may present with diarrhea and malabsoption. Gastrointestinal spread of Campylobacter infections results in cholecystitis, pancreatitis, peritonitis, and massive gastrointestinal hemorrhage. Rare extraintestinal manifestations of Campylobacter infection may be encountered in patients with meningitis, endocarditis, septic arthritis, osteomyelitis, and neonatal sepsis (Allos 2001). A serious bacteremic situation might occur in AIDS or immunocomprised patients (Tee and Mijch 1998; Pacanowski et al. 2008). Recurrence of C. jejuni infection has been frequently observed in some children and adults with immunoglobin deficiencies. Infected patients would develop specific IgA antibodies in intestinal secretions in addition to IgG, IgM, and IgA antibodies in serum (Allos 2001).

Immunoproliferative small intestinal disease (IPSID) is a new entity, with molecular and immunohistochemical studies proving an association with C. jejuni (Lecuit et al. 2004; Al-Saleem and Al-Mondhiry 2005), that belongs to a subcategory of infectious pathogen-associated human lymphomas. IPSID involves mainly the proximal small intestine, resulting in malabsorption, diarrhea, and abdominal pain. Its pathologic and molecular nature is a variant of the B-cell lymphoma of MALT with hybridized features of MALT lymphoma, lymphoplasmacytic lymphoma, and plasma cell neoplasms. (Al-Saleem and Al-Mondhiry 2005). IPSID is also called Mediterranean lymphoma because of its geographic prevalence, mainly affecting Israel, Egypt, Saudi Arabia, and North Africa (Al-Saleem and Al-Mondhiry 2005; Harris et al. 2008). It was previously thought that the disease in its early stages "does not appear to be truly malignant lymphoma" (Bull 1976). In the recent WHO classification of hematopoietic and lymphoid tissue, IPSID is subclassified under heavy-chain diseases as a special variant of extranodal marginal zone B-cell (MALT) lymphoma (Harris et al. 2008).

Morphologically, IPSID exhibits excessive plasma cell differentiation. Large cell transformation or extraenteric involvement is uncommonly observed in the disease (Al-Saleem and Zardawi 1979; Rogers et al. 1988; Chang et al. 1992). Molecularly, the encoding gene sequence of immunoglobin in patients with this disease has revealed a deletion of V region and parts of C (H) 1 domain, truncated alpha heavy-chain proteins, and lack of the light chains as well as the first constant domain (Al-Saleem and Al-Mondhiry 2005). In the corresponding mRNA level, multiple defects have also been detected (e.g., lacking variable heavy chain and the constant heavy chain 1 sequences and deletions or insertions of unknown origin) (Al-Saleem and Al-Mondhiry 2005). In the earlier phase of IPSID, immunoglobulin heavy-chain or light-chain gene rearrangements have been detected (Smith et al. 1987), suggestive of possible clonal evolution for this entity. Chromosomal analysis has shown clonal rearrangements involving predominantly the heavy- and light-chain genes, such as t(9;14)/PAX5-IgH (Al-Saleem and Al-Mondhiry 2005).

Recently, the complete genome sequence of C. jejuni was characterized. Hypervariable regions by genome sequence might be important in the survival of the organism (Parkhill et al. 2000). Chronic antigen stimulation is one of the pathologic mechanisms leading to abnormal immune responses to treatment. Consequent immunoglobulin gene rearrangement or deletion contributes to further development of IPSID. However, the mechanisms involved the interaction between bacterium and host in development of IPSID remain to be elucidated.

Eradication of this bacterium would cure IPSID at its earlier stage (Marteau and Chaput 2011). Studies have shown that approximately 30–70% of patients with early-stage IPSID responded to antibiotics and achieved complete remission (Al-Saleem and Al-Mondhiry 2005). However, chemotherapy is necessary for patients with advanced disease at presentation or those refractory to antibiotics. The most common chemotherapeutic regimens are the CHOP (cyclophosphamide, vincristine, doxorubicin, and prednisone) regimen (Mrabti et al. 2011).

Viral Infections

The main consideration of infection-related lymphoma is virus-driven atypical lymphoproliferation proceeding to lymphoma although the relevance of some viral infection to human lymphoma development has often been debated. There are at least six oncogenic human viruses, namely Epstein-Barr virus (EBV), hepatitis B virus, hepatitis C virus (HCV), human papilloma virus (HPV), human T-cell lymphotropic virus (HTLV-1), and Kaposi-associated herpesvirus (KSHV). These six viruses contribute to 10–15% of the cancers worldwide (Martin and Gutkind 2008). Of them, EBV, HCV, and HTLV are the main three causative etiologies of lymphoproliferative disorders or lymphomas (Martin and Gutkind 2008) in addition to, human immunodeficiency virus (HIV). In the chapter, we would only focus on HCV, HTLV, HHV6 and HHV8 viruses induced lymphoproliferative disorders/lymphomas. HIV and EBV associated lymphomas will be discussed in separate chapters.

Hepatitis C Virus (HCV)

Hepatitis C virus (HCV) belongs to family Flavivridae and is a small, enveloped, single-stranded RNA virus (Suzuki et al. 2007). There is a high global prevalence of HCV infection, affecting about 3% of world population (about 180 million people) and becoming the second most common chronic viral infection in the world (Lauer and Walker 2001). HCV is both a hepatotropic and a lymphotropic virus. Chronic infection could result in chronic hepatitis, cirrhosis, and hepatocellular carcinoma, as well as several extrahepatic diseases, including mixed cryoglobulinemia and lymphoma as the leading one (Craxì et al. 2008). The evidence of the association between hepatic (hepatocellular carcinoma) and extrahepatic malignancies (lymphoma and thyroid cancer) have justified HCV as being listed as a human oncogenic virus (Craxì et al. 2008). A causative association between HCV and non-Hodgkin lymphoma (NHL) was postulated recently and has been the subject of intense investigation, as well as some debate.

Increased risk of lymphoma in patients with HCV is reported. In a large cohort of 8,234 patients with HCV, six developed non-Hodgkin lymphoma. The overall adjusted standardized incidence ratio of development of lymphoma in the study was calculated at 3.42 (CI: 1.25-7.45) (Proby et al. 2011). In the other aspect, increased HCV infection is also found in immunocompromised patients with lymphoma post therapy. A study of 207 consecutive NHL patients who received chemotherapy found a higher prevalence of HCV infection (9.2%) than that shown in the general population in Italy (3%) (Nosotti et al. 2011). Whether HCV infection preceded lymphoma in these patients was unclear. A recent large-scale epidemiologic study was able to demonstrate the marked differences in geographic distribution in the prevalence of HCV in non-Hodgkin lymphomas (Nicolosi-Guidicelli et al. 2011). HCV infection is reported to be associated with NHL in Europe, North America, and Australia but not in Northern Europe, the United Kingdom, and Canada (Nicolosi-Guidicelli et al. 2011), Geographic differences in HCV genotype could be the reason for these discrepancies (Nieters et al. 2006).

HCV infection is often associated with a non-malignant B-cell lymphoproliferation as well as overt malignant B-cell lymphomas. Studies report that there is an incremental increased development of B-cell lymphomas in patients with HCV versus that shown in the general population (Turner et al. 2003; de Sanjose et al. 2008; Nicolosi-Guidicelli et al. 2011; Proby et al. 2011; Hartridge-Lambert et al. 2011) HCV-associated B-cell lymphomas represent a variety of histological subtypes, including marginal zone lymphoma (splenic, nodal and extranodal), small lymphocytic lymphoma/chronic lymphocytic leukemia, lymphoplasmacytic lymphoma, diffuse large B-cell lymphoma and occasionally plasmablastic lymphoma (de Sanjose et al. 2008; Pellicelli et al. 2011; Plaza et al. 2011; Viswanatha and Dogan 2007). Many studies have shown that HCV-associated lymphomas had a high response rate to antiviral therapy, thus providing a clinical evidence for the relationship between HCV and lymphoma (Viswanatha and Dogan 2007). Emerging biological investigations and clinical observations suggest that HCV is one of the growing pathogens associated with the development of lymphomas (Suarez et al. 2007). However, it is still elusive whether HCV-associated antigen-driven lymphomagenesis was the sole mechanism in lymphoma occurred in the group of patients.

Mixed cryglobulinemia is commonly associated with HCV. HCV-associated lymphatic malignancies are observed during the course of mixed cryoglobulinemia or they may be idiopathic forms. The more and more evidences indicated that presence of cryoglobulinemia is an independent risk factor for the development of lymphoma in HCV patients. Several investigations reported that HCV patients with mixed cryoglobulinemia had a 35-fold higher risk of non-Hodgkin lymphoma than the general population (Pozzato et al. 1994; Ferri et al. 1994; Monti et al. 2005). Case study of HCV-associated splenic lymphomas with villous lymphocytes also demonstrated a consistent association with mixed cryoglobulinemia (Suarez et al. 2007). In the same study, the HCV viral load strongly correlated with the tumor burden in patients, also suggestive of HCV-associated antigen-driven B-cell proliferation and neoplastic transformation (Suarez et al. 2007) beyond cryglobulinemia.

Infection of HCV itself could be the main trigger to initiate lymphoma regarding either the inflammatory or oncologic process. HCV does not only incorporate in hepatocytes but is also found in peripheral blood mononuclear cells, in particular B-lymphocytes (Zignego et al. 1992; Ito et al. 2011). Researchers considered that peripheral blood mononuclear cells served as HCV reservoirs. Nevertheless, how HCV impairs the innate immune response in infected B-cells is still unclear. HCV is highly variable, divided into at least 6 major genotypes and more than 50 subtypes based on a nucleotide diversity study within core, E1, and NS5 genes (Zein 2000). Replication of HCV virus is occurred in CD19 expressing peripheral blood mononuclear cells (Ito et al. 2010). In infected peripheral blood mononuclear cells including B-cells, HCV virus not only replicates its RNA but not produces HCV-encoding protein (NS5) in circulation (Gong et al. 2003). The virus carrying B-cells may stay in nodal or extranodal tissue. A virus antigen-driven indirect stimulation leads to monoclonal expansion of these B cells and in some circumstances to malignancy (Landau et al. 2007). HCV-E2 antigen is one of key candidate antigens for the development of HCV associated lymphoma (Landau et al. 2007). Binding of the B-cell receptor (CD81) by viral antigens HCV-E2 has revealed a strong signal of B-cell proliferation through communicating directly with immunoglobulin superfamily member 8 and CD36, and forming a signal transduction complex with CD19, CD21 and CD225 (Bartosch et al. 2003; Landau et al. 2007). Additional regulatory elements are also affected in HCV-related B-cell clonal expansion, including the Fas and BLyS signaling mechanisms (Raghuraman et al. 2005). Similar to HCVE2, NS3 protein of HCV is here also critical in antigen-related B-lymphomagenesis. A study demonstrated HCV-NS3 and IgG-Fc crossreactive IgM in patients with type II mixed cryoglobulinemia and B-cell clonal proliferations (De Re et al. 2006) Finally, genetic events such as bcl-2 rearrangement may also be involved in clonal expansion (Landau et al. 2007). Lymphoma development could also relate to the genome subtyping of hepatitis C virus. A study reported that there was a high prevalence of HCV genotype 1 observed in patients who had DLBCL (13 of 24 cases), whereas HCV genotype 2 was more frequently detected in patients who had indolent lymphoma (Pellicelli et al. 2011). The key role of the genome type of HCV in regard to indolent or aggressive B-cell lymphoproliferative disorders requires further exploration.

HCV induced gene dysregulation or mutation in B-cells is the mainstream pathogenesis for lymphoma formation (Ito 2011, Durand et al. 2010). There are several signaling pathways engaged in B-cell immune response (e.g., interferon, toll-like-receptor and retinoid-acid-inducible-gene-I), which are also crucial in lymphomagenesis (Vilcek 2006; Akira et al. 2006; Yoneyama et al. 2004). Clinically, interferon treatment has not only been used for eradication of HCV virus but also for treatment of B-cell lymphomas (Mazzaro et al. 1996; Agnello et al. 2002; Hermine et al. 2002). HCV core protein could induce production of IL-6 via Toll-like receptor 2 (TLR2) and lead to increased B-cell proliferation in vitro. Both TLR2 expression and IL-6 serum concentrations were increased in HCV-infected patients, especially in those with mixed cryglobulinemia, in comparison to healthy controls ($P<0.05$) (Feldmann et al. 2006). A study demonstrated that constitutive phosphorylation of the downstream genes of insulin growth receptor-3 (IGR-3) pathway; including TANK-binding kinase-1 (TBK1) and IκB kinase-ε (IKKε) was present in chronic hepatitis C B-cells. The feedback of activation of TBK1 and IKKε is to down-regulate IRF-3 and reduce interferon production in B-cells, further allowing a persistence of HCV infection (Schr¨oder et al. 2008). In addition, increased expression of oncogenesis-related genes such as cyclin D1, cyclin D2, BAL, STK15, and galectin-3 was noted in chronic hepatitis C B-cells (Ito et al. 2010). Of these oncogenes, both STK15 and galectin-3 have been known to be highly expressed in high-grade B-cell non-Hodgkin lymphoma, DLBCL, and other aggressive types (Hamada et al. 2003; Hoyer et al. 2004). Alteration of cyclin D1 and cyclin D2 expression is also know to change cell cycle and induce abnormal proliferation of lymphocytes, forming lymphoma (Aguiar et al. 2000). Moreover, the BAL (B-aggressive lymphoma) gene map to chromosome 3q21 is a novel oncogene involved in B-cell lymphoma and DLBCL subtype (Aguiar et al. 2000). Expression of BAL at significantly higher levels in fatal high-risk DLBCLs has been reported (Aguiar et al. 2000).

Cytokines released from infected B-cells also play a key role in chronic inflammation and immunomodulation, further resulting in lymphoid malignancy. As mentioned above, elevated IL6 level is commonly present in HCV patients, resulting in TLK2 mediated lymphoproliferation (Feldmann et al. 2006) A transgenic mouse model that expressed full HCV genome in B-cells showed both substantially elevated serum-soluble IL-2 receptor alpha and development of B-cell lymphoma. The mouse with increased IL-2 also exhibited a high level of BCL-2 expression (Tsukiyama-Kohara et al. 2011). BCL-2 was clearly related to lymphomagenesis in HCV patients (Landau et al. 2007; Tsujimoto et al. 1985).

Recent studies have also indicated that HCV infects not only B-cell lines, but also T-cell lines, especially naïve CD4+ T-cells (Kondo et al. 2010). That study also pointed out that HCV replication in T-cells suppressed T-cell function (Kondo et al. 2010). The investigators speculated that dysfunction of T-cells also plays a critical role in B-cell lymphoma (Kondo et al. 2010).

Patients having indolent HCV-related lymphomas are benefited by antiviral therapy with complete hematologic response rate of 55%, whereas patients with DLBCL responded to immunochemotherapy (Pellicelli et al. 2011). It has been suggested ultilizing anti-B-cell antibodies (for example, rituximab), along with antiviral regimens (peginterferon and ribavirin), to get rid of HCV-infected peripheral B cells in patients and prevent from their development of B-cell lymphoma (Ito et al. 2011).

However, a long term use of anti-HCV regimens. e.g. immunomodulants or anti-TNF therapy might raise a concern about secondary malignancies. Further investigations are needed before making a definitive conclusion.

Human Herpesvirus-6

Human herpesvirus-6 (HHV-6) is the smallest of the herpesviruses, with a genome size of 170 kb. HHV-6 infection occurs very early in life; after the acute phase, the virus becomes dormant and then reactivates at a later time during infectious diseases, proliferative disorders, and immune deficiencies (Pellett et al. 1992; Cone et al. 1993; Drobyski et al. 1993). The functions of HHV-6 virus include inducing and upregulating CD4 receptors and cytokine expression, as well as enhancing release of lytic virus, e.g., HIV, from the infected cells (Flamand et al. 1998; Ensoli et al. 1989). HHV-6 has been associated with some characteristic T-cell lymphoproliferative diseases and is a causative agent for the childhood disease Roseola infantum (exanthem subitum or sixth disease) (Pruksananonda et al. 1992), febrile illness in young children (Caserta et al. 2001), and EBV- and CMV-negative cases of mononucleosis in young adults as well as B-cell lymphoma (Pellett et al. 1992; Caserta et al. 2001; Zhou et al. 2007). However, its role in lymphomagenesis is still uncertain.

Activation of HHV-6 has been noted in Hodgkin and non-Hodgkin lymphomas (Liedtke et al. 1995; Loutfy et al. 2010). HHV6 was noted in Hodgkin lymphoma, more frequent in the nodular sclerosis variant, 34% (25 of 73), than in mixed cellularity subtype, 10% (1 of 10) (Lacroix et al. 2007). HHV6 is also found in non-Hodgkin lymphoma, with most common reports in angioimmunoblastic T-cell lymphoma (AILT) (Luppi et al. 1993).

However, HHV-6 being an independent factor triggering lymphoma formation is not well illustrated. At most circumstances, coinfection of HHV with the other versus e.g. HIV, CMV and EBV is present. In a study of 50 pediatric lymphomas, 23 children (46%) showed detectable herpes virus (HHV-6 and CMV) DNA in their white blood cells or plasma, which was much higher than that noted in control groups (2/20, 10%) ($P=0.0005$) (Loutfy et al. 2010). HHV-6 is also known to be associated with angioimmuneblastic T cell lymphoma (AITL) with or without coexistence of EBV. A PCR study demonstrated molecular evidence of EBV and HHV-6 infection in 36/42 cases (85%) and 19/42 cases (45%) of angioimmuneblastic T cell lymphoma, respectively (Zhou et al. 2007). Coinfection of EBV and HHV8 was found in 17/49 specimens (15/42 cases) (Zhou et al. 2007). In an earlier study, the sole viral genome of HHV-6 was detected in 4 of 18 lymph node biopsies (22%) from patients with angioimmuneblastic T cell lymphoma; all were TCR gamma clonal (Vrsalovic et al. 2004). In Zhou's study, it also demonstrated a high EBV load being associated with B-cell monoclonality (Zhou et al. 2007). Therefore, presence of HHV6 plays an important role in the pathogenesis of abnormal T-cell proliferation/lymphoma and concurrence of HHV6 and EBV might promote disease severity and aggression (Zhou et al. 2007). Increased viral load could also be considered as the consequence of increasing dysfunction of the immune system during lymphoma progression (Zhou et al. 2007).

The incidence of CMV and HHV-6 infection after hematopoietic stem cell transplant (HSCT) is substantial: 38.5% (20/52) and 26.9% (14/52) in all HSCT patients, respectively (Lv et al. 2010). Co-infection occurred in 6 of 20 patients (30%) (Lv et al. 2010). Genotyping revealed that HHV-6 infection was all type B (Lv et al. 2010). Different mechanisms of HHV-6 infection from EBV or CMV in transplant patients have been hypothesized (Jaskula et al. 2010) by analyzing these three viruses as posttransplant complications. Reactivation of HHV-6 after hematopoietic stem cell transplant has been suggested to be associated with increased mortality and severe clinical manifestations, including graft versus host disease (de Pagter et al. 2008). However, whether HHV-6 could drive transplant-related lymphoproliferative processes toward malignant clonal lymphoproliferative disorders, as EBV does, is unclear.

Taken together, HHV6 could be reactivated at a time during other infectious diseases, proliferative disorders, and immune deficient state. The exact role or mechanism of HHV6 infection in lymphomagenesis is uncertain.

Human Herpesvirus-8

Kaposi sarcoma-associated herpesvirus (KSHV) or human herpesvirus type 8 (HHV-8) is newly identified herpes virus, in 1995 from AIDS patients (Carbone and Gloghini 2008). HHV-8 has been consistently detected in HIV-associated lymphoproliferative disorders, one of which is primary effusion lymphoma (PEL), whereas the other one is multicentric Castleman disease (Carbone and Gloghini 2008; Platt et al. 2000). Primary effusion lymphoma is lymphoma localized in body cavities and presenting as pleural, peritoneal, and pericardial lymphomatous effusions (Platt et al. 2000). Recently, the concept of KSHV/HHV-8-associated lymphomas has also been expanded to extracavitary solid lymphomas without serous effusions, regardless of HIV status (Zhang et al. 2010). KSHV/HHV-8 infection has been reported in multicentric Castleman disease as well as in multicentric Castleman disease-associated plasmablastic lymphoma (Du et al. 2007; Carbone and Gloghini 2008). A study including 12 cases of HIV-infected patients with serous effusions showed all patients had large HHV-8(+) lymphomatous cells and associated visceral involvement (Boulanger et al. 2001). A monoclonal rearrangement of IgH genes was demonstrated in 6 of 12 cases, Cytogenetic analysis by array comparative genomic hybridization (aCGH) analysis demonstrated a complex karyogeneic abnormality in all cases without recurrent characteristics (Boulanger et al. 2001). Because of broad diversification in the clinical presentation of KSHV/HHV-8-associated lymphomas, the identification of the similar histology, immunophenotype, and KSHV/HHV-8 virus in these disorders is necessary for diagnosis. The KSHV/HHV-8-associated lymphomas show adverse clinical outcomes (Boulanger et al. 2001).

The exact oncogenic mechanisms of KSHV/HHV-8 have not been clearly defined. The presence of an HHV-8 oncogene plays an important role in the development of lymphoma, in addition to cytokine-induced tumor growth and immunocompromised state. Several virally encoded genes (for example, bcl-2, interleukin 6, cyclin D, G-protein-coupled receptor, and interferon regulatory factor) provide key functions regarding cellular proliferation and survival (Hengge et al. 2002). Most studies have indicated that HHV-8 infection of tumor cells is predominately latent, with reactivation resulting in release of viral gene productions during lytic phase and causing consequential re-infection (Hengge et al. 2002). HHV-8 has recently been found to escape HLA-class-I-restricted antigen presentation to CD8+ cytotoxic T lymphocytes, which is the key factor for HHV-8 latent infection and immune escape from primary and chronic infection (Hengge et al. 2002). In addition, there are a few latent viral gene products, including ORF73, ORF72, and ORF71, which could increase proliferation and impair apoptosis through variant regulatory processes (Zhang et al. 2010). As hypothesized in infection-related lymphomas, the role of HHV-8 is similar to EBV and HCV by mimicking B-cell antigen receptor signaling (Rui and Goodnow 2006). Thus, modulation of B-cell antigen receptor signaling is potentially a powerful strategy for HHV-8-associated lymphoma. There are several viral genes that may be attributed to the particular plasmacytic features of primary effusion lymphoma and Castleman disease. Of these, a viral homologue of interleukin 6 (vIL6) has been revealed to promote the growth of plasma cells. The D-type cyclin homologue in HHV-8 genome, the latent nuclear antigen LANA, and viral homologues of interferon regulatory factors are critical for lymphoplasmacytic proliferation (Schulz 2001).

It is uncertain how the immune system controls replication of HHV-8 virus so as to inhibit abnormal lymphoplasmacytic proliferation in Castleman disease, primary effusion lymphoma and plasmablastic lymphoma. HHV-8 infection often presents in HIV-infected patients, whereas these patients have an impaired immune modulation. When human dendritic cells are exposed to HHV virus, their ability to stimulate immune response is reduced by mixed lymphocyte culture assay (Cirone et al. 2007). The surface CD1a expression and costimulatory molecular CD80 are also markedly decreased (Cirone et al. 2007). Impairment of dendritic cell development and function would help HHV-8 to escape from the host defense mechanisms (Cirone 2007). Human natural killer cells play a critical role in immune surveillance and tumor cell death. Decreased NK cell activity is noted in HIV and/or HHV-8 Infection (Sirianni et al. 2007). In addition, some specific cytotoxic T lymphocytes may provide protection from persistent HHV-8 infection (Lambert et al. 2006). Albeit the mechanism of HHV-8 driven lymphoma is not clear, antiviral agents target DNA synthesis in herpes viruses can inhibit HHV-8 lytic replication so as to prevent HHV-8-driven lymphoma and Kaposi sarcoma (Gantt and Casper 2011). In addition to antiviral therapy, conventional chemotherapy with CHOP (cyclophosphamide, doxorubicin, prednisone, vincristine) has also been used to treat HHV-8-associated lymphoma. Novel therapy including some inhibition of specific cellular targets will be pursued in order to cure this type of aggressive lymphoma.

Adult T Cell Leukemia Virus Type I (HTLV-1)

Human T-lymphotropic virus type-1 (HTLV-1) belongs to the Deltaretrovirus genera of the Orthoretrovirinae subfamily and is the first discovered human retrovirus, isolated in the early 1980s from peripheral blood samples of a patient with cutaneous T-cell lymphoma. HTLV virus can be subcategorized into two subtypes: HTLV 1 and HTLV-2 (Poiesz et al. 1980; Jeang 2010).

Adult T-cell leukemia/lymphoma (ATLL) is a rare aggressive form of T-cell lymphoma/leukemia associated with HTLV-1 (Poiesz et al. 1980; Hinuma et al. 1981; Takatsuki 2005; Matsuoka and Jeang 2007). Although the disease was originally reported in Japan, adult T cell lymphoma/leukemia cases are also found in Western countries (Tajima and Cartier 1995; Farcet et al. 1983). It is characterized as a malignant proliferation of CD4 T-cells infected by HTLV-1 showing "flower-shaped" nuclear features in tissue or blood. Besides adult T cell lymphoma/leukemia, HTLV-1 is also the causative agent of inflammatory disorders, including tropical spastic paraparesis/HTLV-associated myelopathy (TSP/HAM), arthritis, uveitis, dermatitis, lymphadenitis and Sjögren syndrome (Proietti et al. 2005). Hypothetically, these HTLV-related inflammatory diseases must eventually cause lymphoproliferative disorders or lymphoma. However, only a small population of HTLV-1 carriers (infection lasting life long without clinical manifestation) will progress to adult T cell lymphoma/leukemia (2.1% of females and 6.6% of males) (Boxus and Willems 2009; Tajima and Cartier 1995). Adult T cell lymphoma/leukemia cells are consistently monoclonal with respect to proviral integration and originate from initial polyclonal/oligoclonal expansion of HTLV-1-infected cells (Boxus and Willems 2009). Clinically, adult T cell lymphoma/leukemia behaves variably and could present with acute, chronic, and smoldering phase (Ohshima et al. 2008). Patients with acute disease could die within weeks, whereas those with chronic course demonstrate much longer survival unless converting to acute phase under certain circumstances (Ohshima et al. 2008).

Recent research has revealed that HLTV-1 TAX1, a regulatory factor, plays a key role in HTLV-1-specific pathogenesis such as cell survival, cell proliferation, and persistent infection (Matsuoka and Jeang 2007; Higuchi and Fujii 2009). TAX and another regulatory factor (HBZ) allow favored proliferation of infected cells. HBZ (HTLV-1 bZIP factor), p12, and p30 (accessory proteins encoded by different pX mRNAs formed by alternative splicing events) either reduce viral expression or inhibit immune recognition. Permanent TAX-induced proliferation and abnormal expansion of infected cells generate DNA lesions characteristic of adult T cell lymphoma. Inhibition of host checkpoint machinery allows further proliferation of infected cells harboring DNA damage. Progressive stabilization of these abnormalities provides an increased proliferative capacity to the infected cells and ultimately leads to adult T cell lymphoma/leukemia (Boxus and Willems 2009). The role of TAX and HBZ in viral persistence and the development of adult T cell lymphoma/leukemia have been well studied. The virally encoded oncoprotein TAX activates the transcription of HTLV-1 and cellular genes by altering cellular signaling pathways (Higuchi and Fujii 2009). In asymptomatic HTLV carriers, TAX stimulates both viral and cellular gene expression to allow viral replication as well as T-cell proliferation. Because the host immune system is able to efficiently get rid of lymphocytes expressing viral antigens, TAX. Infected cell proliferation relied on another viral factor, HBZ, encoded by the complementary strand. A sustained cell replication regulated by viral TAX causes DNA damage that activates checkpoints, a second barrier to transformation (Higuchi and Fujii 2009; Shoji et al. 2009). There are evidences of higher expression levels of cyclin D1 and D2 mRNA in HTLV-I-infected T-cell lines, which is partially mediated by the viral transforming protein Tax (Mori et al. 2002). In vitro study revealed that HLTV-1 TAX1 binds the cyclin D1 promoter-proximal cyclic AMP response element (CRE) in the presence of phosphorylated CREB (pCREB). In an in vivo study also demonstrated that cyclin D1 transcript levels were elevated in HTLV-1 infected cells (Kim et al. 2010). TAX-induced accumulations of cyclin D1 make the G1 phase of cell cycle shortened, mitotic replication of the virus increased, and malignant T-cells expanded (Kim et al. 2010). Thus, HTLV-I infection had changed cellular D-type cyclin gene expression, resulting in T-cell transformation and subsequent development of T-cell leukemia (Mori et al. 2002). A recent study showed that activating transcription factor 3 (ATF3), an HBZ-interacting protein, is upregulated in adult T cell lymphoma/leukemia (Hagiya et al. 2011). Inhibition of ATF3 expression reduced cell proliferation, causing death of adult T cell lymphoma/leukemia cells. Although ATF3 enhanced p53 transcriptional activity, this activity was suppressed by HBZ. When ATF3 and HBZ interact, tumor cells are amplified in adult T cell lymphoma/leukemia (Hagiya et al. 2011). Moreover, both HBZ and ATF3 suppress TAX expression, enabling infected cells to escape the host immune system (Hagiya et al. 2011). Gene dysregulation not only affect the localized cell growth of adult T cell lymphoma/leukemia but also systemic dissemination. A novel surface marker, Cell adhesion molecule 1 (CADM1/TSLC1), which encodes a multifunctional immunoglobulin-like cell adhesion molecule, has been identified for adult T-cell leukemia/lymphoma (Matsuoka et al. 2010). In conjunction with other signals (Tiam1 and RAC), CADM1 plays a role in tissue infiltration of leukemic cells in adult T cell lymphoma/leukemia patients (Matsuoka et al. 2010), resulting in disease dissemination.

Combination conventional chemotherapy has little impact on the long-term survival of these patients. Blockage of CD25 by daclizumab (Zenapax®), a humanized murine anti-CD 25 monoclonal antibody, on adult T-cell lymphoma/leukemia cells is considered a novel way to cure the disease. On-going phase II clinical trial using CHOP-Zenapax to treat 15 patients with aggressive adult T cell lymphoma/leukemia demonstrated that 33% patients (5 patients) achieved complete response (CR) lasting for 2 months or more, and 20% (3 patients) having partial response (Ceesay et al. 2011). Understanding molecular pathogenesis of the lymphoma would benefit patient care and extend survival.

References

Adamu MA, Schöttker B, Weck MN, Brenner H (2012) Helicobacter pylori infection is strongly associated with gastric and duodenal ulcers in a large prospective study. Clin Gastroenterol Hepatol 10(5):487–493.e1, Epub 2012 Jan 8

Agnello V, Mecucci C, Casato M et al (2002) Regression of splenic lymphoma after treatment of hepatitis C virus infection. N Engl J Med 347(26):2168–2170

Aguiar RC, Yakushijin Y, Kharbanda S, Salgia R, Fletcher JA, Shipp MA (2000) BAL is a novel risk-related gene in diffuse large B-cell lymphomas that enhances cellularmigration. Blood 96(13):4328–4334

Aigelsreiter A, Gerlza T, Deutsch AJ, Leitner E, Beham-Schmid C, Beham A, Popper H, Borel N, Pospischil A, Raderer M, Kessler HH, Neumeister P (2011) Chlamydia psittaci infection in nongastrointestinal extranodal MALT lymphomas and their precursor lesions. Am J Clin Pathol 135(1):70–75

Akira S, Uematsu S, Takeuchi O (2006) Pathogen recognition and innate immunity. Cell 124(4):783–801

Allos BM (2001) Campylobacter jejuni infections: update on emerging issues and trends. Clin Infect Dis 32(8):1201–1206

Al-Saleem T, Al-Mondhiry H (2005) Immunoproliferative small intestinal disease (IPSID): a model for mature B-cell neoplasms. Blood 105(6):2274–2280

Al-Saleem T, Zardawi IM (1979) Primary lymphomas of the small intestine in Iraq: a pathological study of 145 cases. Histopathology 3:89–106

Arora N, Manipadam MT, Pulimood A, Ramakrishna BS, Chacko A, Kurian SS, Nair S (2011) Gastrointestinal lymphomas: pattern of distribution and histological subtypes: 10 years experience in a tertiary centre in South India. Indian J Pathol Microbiol 54(4):712–719

Atherton JC (2006) The pathogenesis of Helicobacter pylori-induced gastro-duodenal diseases. Annu Rev Pathol 1:63–96

Bartosch B, Vitelli A, Granier C, Goujon C, Dubuisson J, Pascale S, Scarselli E, Cortese R, Nicosia A, Cosset FL (2003) Cell entry of hepatitis C virus requires a set of co-receptors that include the CD81 tetraspanin and the SR-B1 scavenger receptor. J Biol Chem 278(43):41624–41630

Bessède E, Copie-Bergman C, Lehours P, Lévy M, Leroy K, Baia M, Riou A, Mégraud F, Delchier JC, Salles N (2012) Is elevated gastric tissue NOX2 associated with lymphoma of mucosa-associated lymphoid tissue? Antioxid Redox Signal 16(11):1205–1211, Epub 2012 Mar 15

Boulanger E, Agbalika F, Maarek O, Daniel MT, Grollet L, Molina JM, Sigaux F, Oksenhendler E (2001) A clinical, molecular and cytogenetic study of 12 cases of human herpesvirus 8 associated primary effusion lymphoma in HIV-infected patients. Hematol J 2(3):172–179

Boxus M, Willems L (2009) Mechanisms of HTLV-1 persistence and transformation. Br J Cancer 101:1497–1501

Bull D (1976) Alpha-chain disease and related small intestinal lymphoma: a memorandum. (Geneva Switzerland) World Health Organ 54:615–624

Carbone A, Gloghini A (2008) KSHV/HHV8-associated lymphomas. Br J Haematol 140(1):13–24, Epub 2007 Nov 7

Caserta MT, Mock DJ, Dewhurst S (2001) Human herpesvirus 6. Clin Infect Dis 33(6):829–833, Epub 2001 Aug 10

Ceesay MM, Matutes E, Taylor GP, Fields P, Cavenagh J, Simpson S, Ho A, Devereux S, Mufti GJ, Pagliuca A (2012) Phase II study on combination therapy with CHOP-Zenapax for HTLV-I associated adult T-cell leukaemia/lymphoma (ATLL). Leuk Res 36(7):857–861, Epub 2011 Dec 29

Chang CS, Lin SF, Chen TP, Liu HW, Liu TC, Li CY, Chao MC, Wu PL (1992) Leukemic manifestation in a case of alpha-chain disease with multiple polypoid intestinal lymphocytic lymphoma. Am J Hematol 41:209–214

Cheng HH, Tseng GY, Yang HB, Wang HJ, Lin HJ, Wang WC (2012) Increased numbers of Foxp3-positive regulatory T cells in gastritis, peptic ulcer and gastric adenocarcinoma. World J Gastroenterol 18(1):34–43

Christen HJ, Hanefeld F, Eiffert H, Thomsen R (1993) Epidemiology and clinical manifestations of Lyme borreliosis in childhood. A prospective multicentre study with special regard to neuroborreliosis. Acta Paediatr 386(Suppl):1–75

Cirone M, Lucania G, Bergamo P, Trivedi P, Frati L, Faggioni A (2007) Human herpesvirus 8 (HHV-8) inhibits monocyte differentiation into dendritic cells and impairs their immunostimulatory activity. Immunol Lett 113(1):40–46, Epub 2007 Aug 21

Collins JS (1992) Role of Helicobacter pylori in gastritis and duodenitis in man. Agents Actions Spec No:C47-9

Cone RW, Hackman RC, Huang ML, Bowden RA, Meyers JD, Metcalf M, Zeh J, Ashley R, Corey L (1993) Human herpesvirus 6 in lung tissue from patients with pneumonitis after bone marrow transplantation. N Engl J Med 329(3):156–161

Craxì A, Laffi G, Zignego AL (2008) Hepatitis C virus (HCV) infection: a systemic disease. Mol Aspects Med 29(1–2):85–95, Epub 2007 Nov 21

Dagklis A, Ponzoni M, Govi S, Cangi MG, Pasini E, Charlotte F, Vino A, Doglioni C, Davì F, Lossos IS, Ntountas I, Papadaki T, Dolcetti R, Ferreri AJ, Stamatopoulos K, Ghia P (2012) Immunoglobulin gene repertoire in ocular adnexal lymphomas: hints on the nature of the antigenic stimulation. Leukemia 26(4):814–821, Epub 2011 Oct 25

de Pagter PJ, Schuurman R, Meijer E, van Baarle D, Sanders EA, Boelens JJ (2008) Human herpesvirus type 6 reactivation after haematopoietic stem cell transplantation. Clin Virol 43(4):361–366, Epub 2008 Oct 1

De Re V, Sansonno D, Simula MP, Caggiari L, Gasparotto D, Fabris M, Tucci FA, Racanelli V, Talamini R, Campagnolo M, Geremia S, Dammacco F, De Vita S (2006) HCV-NS3 and IgG-Fc crossreactive IgM in patients with type II mixed cryoglobulinemia and B-cell clonal proliferations. Leukemia 20(6):1145–1154

de Sanjose S, Benavente Y, Vajdic CM et al (2008) Hepatitis C and non-Hodgkin lymphoma among 4784 cases and 6269 controls from the international lymphoma epidemiology consortium. Clin Gastroenterol Hepatol 6(4):451–458

Deutsch AJ, Aigelsreiter A, Steinbauer E, Frühwirth M, Kerl H, Beham-Schmid C, Schaider H, Neumeister P (2008) Distinct signatures of B-cell homeostatic and activation-dependent chemokine receptors in the development and progression of extragastric MALT lymphomas. J Pathol 215(4):431–444

Drobyski WR, Dunne WM, Burd EM, Knox KK, Ash RC, Horowitz MM, Flomenberg N, Carrigan DR (1993) Human herpesvirus-6 (HHV-6) infection in allogeneic bone marrow transplant recipients: evidence of a marrow-suppressive role for HHV-6 in vivo. J Infect Dis 167(3):735–739

Du MQ (2007) MALT lymphoma: recent advances in aetiology and molecular genetics. J Clin Exp Hematop 47(2):31–42

Du MQ, Bacon CM, Isaacson PG (2007) Kaposi sarcoma-associated herpesvirus/human herpesvirus 8 and lymphoproliferative disorders. J Clin Pathol 60(12):1350–1357

Durand T, Di Liberto G, Colman H et al (2010) Occult infection of peripheral B cells by hepatitis C variants which have low translational efficiency in cultured hepatocytes. Gut 59(7):934–942

Eck M, Schmausser B, Haas R, Greiner A, Czub S, Muller-Hermelink HK (1997) MALT-type lymphoma of the stomach is associated with Helicobacter pylori strains expressing the CagA protein. Gastroenterology 112:1482–1486

Ensoli B, Lusso P, Schachter F, Josephs SF, Rappaport J, Negro F, Gallo RC, Wong-Staal F (1989) Human herpes virus-6 increases HIV-1 expression in co-infected T cells via nuclear factors binding to the HIV-1 enhancer. EMBO J 8(10):3019–3027

Farcet JP, Kuentz M, Andre C, Darves JM, Reyes F, Dreyfus B, Rochant H (1983) Adult T-cell lymphoma leukemia in Western countries. Am J Hematol 15(4):403–411

Feldmann G, Nischalke HD, Nattermann J, Banas B, Berg T, Teschendorf C, Schmiegel W, Dührsen U, Halangk J, Iwan A, Sauerbruch T, Caselmann WH, Spengler U (2006) Induction of interleukin-6 by hepatitis C virus core protein in hepatitis C-associated mixed cryoglobulinemia and B-cell non-Hodgkin's lymphoma. Clin Cancer Res 12(15):4491–4498

Ferreri AJ, Zucca E (2007) Marginal-zone lymphoma. Crit Rev Oncol Hematol 63(3):245–256, Epub 2007 Jun 20. Reviews

Ferreri AJ, Guidoboni M, Ponzoni M, De Conciliis C, Dell'Oro S, Fleischhauer K, Caggiari L, Lettini AA, Dal Cin E, Ieri R, Freschi M, Villa E, Boiocchi M, Dolcetti R (2004) Evidence for an association between Chlamydia psittaci and ocular adnexal lymphomas. J Natl Cancer Inst 96:586–594

Ferreri AJ, Ponzoni M, Guidoboni M, De Conciliis C, Resti AG, Mazzi B, Lettini AA, Demeter J, Dell'Oro S, Doglioni C, Villa E, Boiocchi M, Dolcetti R (2005) Regression of ocular adnexal lymphoma after Chlamydia psittaci-eradicating antibiotic therapy. J Clin Oncol 23(22):5067–5073, Epub 2005 Jun 20

Ferreri AJ, Dolceltti R, Magnino S, Doglioni C, Ponzoni M (2009) Chlamydial infection: the link with ocular adnexal lymphomas. Nat Rev Clin Oncol 6(11):658–669, Epub 2009 Oct 6

Ferri C, Monti M, La Civita L, Careccia G, Mazzaro C, Longombardo G, Lombardini F, Greco F, Pasero G, Bombardieri S (1994) Hepatitis C virus infection in non-Hodgkin's B-cell lymphoma complicating mixed cryoglobulinaemia. Eur J Clin Invest 24:781–784

Fischbach W, Goebeler-Kolve ME, Dragosics B, Greiner A, Stolte M (2004) Long-term outcome of patients with gastric marginal zone B cell lymphoma of mucosa associated lymphoid tissue (MALT) following exclusive Helicobacter pylori eradication therapy: experience from a large prospective series. Gut 53:34–37

Flamand L, Romerio F, Reitz MS, Gallo RC (1998) CD4 promoter transactivation by human herpesvirus 6. J Virol 72(11):8797–8805

Fühler M, Ottmann KW, Tronnier M (2010) Cutaneous marginal zone lymphoma (SALT) and infection with Borrelia burgdorferi. Hautarzt 61(2):145–147

Gantt S, Casper C (2011) Human herpesvirus 8-associated neoplasms: the roles of viral replication and antiviral treatment. Curr Opin Infect Dis 24(4):295–301

Gong GZ, Lai LY, Jiang YF, He Y, Su XS (2003) HCV replication in PBMC and its influence on interferon therapy. World J Gastroenterol 9(2):291–294

Goodlad JR, Davidson MM, Hollowood K, Ling C, MacKenzie C, Christie I, Batstone PJ, Ho-Yen DO (2000a) Primary cutaneous B-cell lymphoma and Borrelia burgdorferi infection in patients from the Highlands of Scotland. Am J Surg Pathol 24(9):1279–1285

Goodlad JR, Davidson MM, Hollowood K, Batstone P, Ho-Yen DO (2000b) Borrelia burgdorferi-associated cutaneous marginal zone lymphoma: a clinicopathological study of two cases illustrating the temporal progression of B. burgdorferi-associated B-cell proliferation in the skin. Histopathology 37(6):501–508

Groves FD, Linet MS, Travis LB, Devesa SS (2000) Cancer surveillance series: non-Hodgkin's lymphoma incidence by histologic subtype in the United States from 1978 through 1995. J Natl Cancer Inst 92:1240–1251

Guidoboni M, Ferreri AJ, Ponzoni M, Doglioni C, Dolcetti R (2006) Infectious agents in mucosa-associated lymphoid tissue-type lymphomas: pathogenic role and therapeutic perspectives. Clin Lymphoma Myeloma 6(4):289–300

Hagiya K, Yasunaga J, Satou Y, Ohshima K, Matsuoka M (2011) ATF3, an HTLV-1 bZip factor binding protein, promotes proliferation of adult T-cell leukemia cells. Retrovirology 8:19

Hamada M, Yakushijin Y, Ohtsuka M, Kakimoto M, Yasukawa M, Fujita S (2003) Aurora2/BTAK/STK15 is involved in cell cycle checkpoint and cell survival of aggressive non-Hodgkin's lymphoma. Br J Haematol 121(3):439–447

Harris NL, Isaacson PG, Grogan TM, Jaffe ES (2008) Heavy chain disease. In: Swerdlow SH (ed) WHO classification of tumors of haematopoietic and lymphoid tissues. IARC, 69008 Lyon, France pp 196–199

Hartridge-Lambert SK, Stein EM, Markowitz AJ, Portlock CS (2012) Hepatitis C and non-Hodgkin lymphoma: the clinical perspective. Hepatology 55(2):634–641

Hengge UR, Ruzicka T, Tyring SK, Stuschke M, Roggendorf M, Schwartz RA, Seeber S (2002) Update on Kaposi's sarcoma and other HHV8 associated diseases. Part 2: pathogenesis, Castleman's disease, and pleural effusion lymphoma. Lancet Infect Dis 2(6):344–352

Hermine O, Lefr`ere F, Bronowicki JP et al (2002) Regression of splenic lymphoma with villous lymphocytes after treatment of hepatitis C virus infection. N Engl J Med 347(2):89–94

Higuchi M, Fujii M (2009) Distinct functions of HTLV-1 Tax1 from HTLV-2 Tax2 contribute key roles to viral pathogenesis. Retrovirology 6:117

Hinuma Y, Nagata K, Hanaoka M, Nakai M, Matsumoto T, Kinoshita KI, Shirakawa S, Miyoshi I (1981) Adult T-cell leukemia: antigen in an ATL cell line and detection of antibodies to the antigen in human sera. Proc Natl Acad Sci USA 78:6476–6480

Hoyer KK, Pang M, Gui D et al (2004) An anti-apoptotic role for galectin-3 in diffuse large B-cell lymphomas. Am J Pathol 164(3):893–902

Isaacson PG, Spencer J (1987) Malignant lymphoma of mucosa-associated lymphoid tissue. Histopathology 11(5):445–462

Issacson PG, Chott A, Nakamura S, Nueler-Hermelink HK, Harris NL, Swerdlow SH (2008) Extranodal marginal zone lymphoma of mucosa-associated lymphoid tissue (MALT lymphoma). In: Swerdlow SH (ed) WHO classification of tumours of haematopoietic and lymphoid tissues. IARC, 69008 Lyon, France pp 214–219

Ito M, Murakami K, Suzuki T et al (2010) Enhanced expression of lymphomagenesis-related genes in peripheral blood B cells of chronic hepatitis C patients. Clin Immunol 135(3):459–465

Ito M, Kusunoki H, Mochida K, Yamaguchi K, Mizuochi T (2011) Adv Hematol 2011:835314, Epub 2011 Jul 20

Jaskula E, Dlubek D, Sedzimirska M, Duda D, Tarnowska A, Lange A (2010) Reactivations of cytomegalovirus, human herpes virus 6, and Epstein-Barr virus differ with respect to risk factors and clinical outcome after hematopoietic stem cell transplantation. Transplant Proc 42(8):3273–3276

Jeang KT (2010) HTLV-1 and adult T-cell leukemia: insights into viral transformation of cells 30 years after virus discovery. J Formos Med Assoc 109(10):688–693

Kamangar F, Sheikhattari P, Mohebtash M (2011) Helicobacter pylori and its effects on human health and disease. Arch Iran Med 14(3):192–199

Kash N, Fink-Puches R, Cerroni L (2011) Cutaneous manifestations of B-cell chronic lymphocytic leukemia associated with Borrelia burgdorferi infection showing a marginal zone B-cell lymphoma-like infiltrate. Am J Dermatopathol 33(7):712–715

Kim YM, Geiger TR, Egan DI, Sharma N, Nyborg JK (2010) The HTLV-1 tax protein cooperates with phosphorylated CREB, TORC2 and p300 to activate CRE-dependent cyclin D1 transcription. Oncogene 29(14):2142–2152, Epub 2010 Jan 25

Kondo Y, Ueno Y, Kakazu E et al (2010) Lymphotropic HCV strain can infect human primary na¨ıve CD4+ cells and affect their proliferation and IFN-γ secretion activity. J Gastroenterol 46(2):232–241

Kusters JG, van Vliet AH, Kuipers EJ (2006) Pathogenesis of Helicobacter pylori infection. Clin Microbiol Rev 19(3):449–490

Kütting B, Bonsmann G, Metze D, Luger TA, Cerroni L (1997) Borrelia burgdorferi-associated primary cutaneous B cell lymphoma: complete clearing of skin lesions after antibiotic pulse therapy or intralesional injection of interferon alfa-2a. J Am Acad Dermatol 36(2 Pt 2):311–314

Lacroix A, Jaccard A, Rouzioux C, Piguet C, Petit B, Bordessoule D, Ranger-Rogez S (2007) HHV-6 and EBV DNA quantization in lymph nodes of 86 patients with Hodgkin's lymphoma. J Med Virol 79(9):1349–1356

Lambert M, Gannagé M, Karras A, Abel M, Legendre C, Kerob D, Agbalika F, Girard PM, Lebbe C, Caillat-Zucman S (2006) Differences in the frequency and function of HHV8-specific CD8 T cells between asymptomatic HHV8 infection and Kaposi sarcoma. Blood 108(12):3871–3880, Epub 2006 Aug 22

Landau DA, Saadoun D, Calabrese LH, Cacoub P (2007) The pathophysiology of HCV induced B-cell clonal disorders. Autoimmun Rev 6(8):581–587, Epub 2007 Apr 19

Lauer GM, Walker BD (2001) Hepatitis C virus infection. N Engl J Med 345(1):41–52

Lecuit M, Abachin E, Martin A (2004) Immunoproliferative small intestinal disease associated with Campylobacter jejuni. N Engl J Med 350(3):239–248

Liedtke W, Trübner K, Schwechheimer K (1995) On the role of human herpesvirus 6 in viral latency in nervous tissue and in cerebral lymphoma. J Neurol Sci 134(1–2):184–188

Lietman T, Brooks D, Moncada J, Schachter J, Dawson C, Dean D (1998) Chronic follicular conjunctivitis associated with Chlamydia psittaci or Chlamydia pneumoniae. Clin Infect Dis 26:1335–1340

Longbottom D, Coulter LJ (2003) Animal chlamydioses and zoonotic implications. J Comp Pathol 128:217–244

Loutfy SA, Fawzy M, El-Wakil M, Moneer MM (2010) Presence of human herpes virus 6 (HHV6) in pediatric lymphomas: impact on clinical course and association with cytomegalovirus infection. Virol J 7:287

Luppi M, Marasca R, Barozzi P, Artusi T, Torelli G (1993) Frequent detection of human herpesvirus-6 sequences by polymerase chain reaction in paraffin-embedded lymph nodes from patients with angioimmunoblastic lymphadenopathy and angioimmunoblastic lymphadenopathy-like lymphoma. Leuk Res 17(11):1003–1011

Lv DY, Qin MQ, Wang Y, Gao LW, Xie ZD (2010) Incidence and relationship of human cytomegalovirus infection and human herpesvirus 6 infection in pediatric patients after hemopoietic stem cell transplantation. Zhonghua Shi Yan He Lin Chuang Bing Du Xue Za Zhi 24(6):455–457 [Article in Chinese]

Marteau P, Chaput U (2011) Bacteria as trigger for chronic gastrointestinal disorders. Dig Dis 29(2):166–171

Martin D, Gutkind JS (2008) Human tumor-associated viruses and new insights into the molecular mechanisms of cancer. Oncogene 27(supple 2):S31–S42

Masuda M, Maruyama T, Ohta T, Ito A, Hayashi T, Tsukasaki K, Kamihira S, Yamaoka S, Hoshino H, Yoshida T, Watanabe T, Stanbridge EJ, Murakami Y (2010) CADM1 interacts with Tiam1 and promotes invasive phenotype of human T-cell leukemia virus type I-transformed cells and adult T-cell leukemia cells. J Biol Chem 285(20):15511–15522, Epub 2010 Mar 9

Matsuoka M, Jeang KT (2007) Human T-cell leukaemia virus type 1 (HTLV-1) infectivity and cellular transformation. Nat Rev Cancer 7:270–280

Mazzaro C, Franzin F, Tulissi P et al (1996) Regression of monoclonal B-cell expansion in patients affected by mixed cryoglobulinemia responsive to α-interferon therapy. Cancer 77(12):2604–2613

Miyairi I, Laxton JD, Wang X, Obert CA, Arva Tatireddigari VR, van Rooijen N, Hatch TP, Byrne GI (2011) Chlamydia psittaci genetic variants differ in virulence by modulation of host immunity. J Infect Dis 204(4):654–663

Monti G, Pioltelli P, Saccardo F, Campanini M, Candela M, Cavallero G, De Vita S, Ferri C, Mazzaro C, Migliaresi S, Ossi E, Pietrogrande M, Gabrielli A, Galli M, Invernizzi F (2005) Incidence and characteristics of non-Hodgkin lymphomas in a multicenter case file of patients with hepatitis C virus-related symptomatic mixed cryoglobulinemias. Arch Intern Med 165:101–105

Mori N, Fujii M, Hinz M, Nakayama K, Yamada Y, Ikeda S, Yamasaki Y, Kashanchi F, Tanaka Y, Tomonaga M, Yamamoto N (2002) Activation of cyclin D1 and D2 promoters by human T-cell leukemia virus type I tax protein is associated with IL-2-independent growth of T cells. Int J Cancer 99(3):378–385

Mrabti H, Raiss G, Raissouni S, Tazi EM, Inghaouen H, El Ghissassi I, Errihani H (2011) Intestinal non-Hodgkin lymphoma. Immunoproliferative small intestinal disease [Article in French]. Presse Med 40(11):995–1000

Müllegger RR, Glatz M (2008) Skin manifestations of lyme borreliosis: diagnosis and management. Am J Clin Dermatol 9(6):355–368

Naumov I, Fenjvesi A (2011) Correlation between rapid urease test and pathohistological gastrobiopsy finding with positive immunological test in detecting Helicobacter pylori infection [Article in Serbian]. Med Pregl 64(7–8):413–417

Nicolosi-Guidicelli S, Lopez-Guillermo A, Falcone U, Conconi A, Christinat A, Rodriguez-Abreu D, Grisanti S, Lobetti-Bodoni C, Piffaretti JC, Johnson PW, Mombelli G, Cerny A, Montserrat E, Cavalli F, Zucca E (2011) Hepatitis C virus and GBV-C virus prevalence among patients with B-cell lymphoma in different European regions: a case–control study of the International Extranodal Lymphoma Study Group. Hematol Oncol 21. doi:10.1002/hon.1015

Nieters A, Kallinowski B, Brennan P et al (2006) Hepatitis C and risk of lymphoma: results of the European multicenter case–control study EPILYMPH. Gastroenterology 131(6):1879–1886

Nosotti L, D'Andrea M, Pitidis A, Pimpinelli F, Dessanti ML, Pisani F, Vignally P, Petti MC (2012) Hepatitis C virus infection prevalence and liver dysfunction in a cohort of B-cell non-Hodgkin's lymphoma patients treated with immunochemotherapy. Scand J Infect Dis 44(1):70–73, Epub 2011 Sep 12

Ohshima K, Jaffe ES, Kikuchi M (2008) Adult T-cell leukemia/lymphoma. In: Swerdlow S (ed) WHO classification of tumors of hematopoieitc and lymphoid tissues. IARC, 69008 Lyon, France pp 281–284

Pacanowski J, Lalande V, Lacombe K, Boudraa C, Lesprit P, Legrand P, Trystram D, Kassis N, Arlet G, Mainardi JL, Doucet-Populaire F, Girard PM, Meynard JL, CAMPYL Study Group (2008) Campylobacter bacteremia: clinical features and factors associated with fatal outcome. Clin Infect Dis 47(6):790–796

Pannekoek Y, van der Ende A (2011) Chlamydia psittaci infection in nongastrointestinal MALT lymphomas and their precursor lesions. Am J Clin Pathol 136(3):480

Parkhill J, Wren BW, Mungall K, Ketley JM, Churcher C, Basham D, Chillingworth T, Davies RM, Feltwell T, Holroyd S, Jagels K, Karlyshev AV, Moule S, Pallen MJ, Penn CW, Quail MA, Rajandream MA, Rutherford KM, van Vliet AH, Whitehead S, Barrell BG (2000) The genome sequence of the foodborne pathogen Campylobacter jejuni reveals hypervariable sequences. Nature 403:66–68

Parsonnet J, Isaacson PG (2004) Bacterial infection and MALT lymphoma. N Engl J Med 350(3):213–215

Parsonnet J, Hansen S, Rodriguez L, Gelb AB, Warnke RA, Jellum E, Orentreich N, Vogelman JH, Friedman GD (1994) Helicobacter pylori infection and gastric lymphoma. N Engl J Med 330:1267–1271

Pellett PE, Black JB, Yamamoto M (1992) Human herpesvirus 6: the virus and the search for its role as a human pathogen. Adv Virus Res 41:1–52

Pellicelli AM, Marignani M, Zoli V, Romano M, Morrone A, Nosotti L, Barbaro G, Picardi A, Gentilucci UV, Remotti D, D'Ambrosio C, Furlan C, Mecenate F, Mazzoni E, Majolino I, Villani R, Andreoli A, Barbarini G (2011) Hepatitis C virus-related B cell subtypes in non Hodgkin's lymphoma. World J Hepatol 3(11):278–284

Platt GM, Cannell E, Cuomo ME, Singh S, Mittnacht S (2000) Detection of the human herpesvirus 8-encoded cyclin protein in primary effusion lymphoma-derived cell lines. Virology 272(2):257–266

Plaza R, Ponferrada A, Benito DM, Arevalo N, Foncillas MA, de Fuenmayor ML, Aldeguer M (2011) Plasmablastic lymphoma associated to Crohn's disease and hepatitis C virus chronic infection. J Crohns Colitis 5(6):628–632, Epub 2011 Aug 15

Poiesz BJ, Ruscetti FW, Mier JW, Woods AM, Gallo RC (1980) T-cell lines established from human T-lymphocytic neoplasias by direct response to T-cell growth factor. Proc Natl Acad Sci U S A 77(11):6815–6819

Ponzoni M, Ferreri AJ, Guidoboni M, Lettini AA, Cangi MG, Pasini E, Sacchi L, Pecciarini L, Grassi S, Dal Cin E, Stefano R, Magnio S, Dolcetti R, Doglioni C (2008) Chlamydia infection and lymphomas: association beyond ocular adnexal lymphomas highlighted by multiple detection methods. Clin Cancer Res 14(18):5794–5800

Ponzoni M, Ferreri AJ, Mappa S, Pasini E, Govi S, Facchetti F, Fanoni D, Tucci A, Vino A, Doglioni C, Berti E, Dolcetti R (2011a) Prevalence of borrelia burgdorferi infection in a series of 98 primary cutaneous lymphomas. Oncologist 16(11):1582–1588

Ponzoni M, Bonetti F, Poliani PL, Vermi W, Bottelli C, Dolcetti R, Cangi MG, Ferreri AJ, Cin ED, Pasini E, Liserre R, Doglioni C, Rossi G, Facchetti F (2011b) Central nervous system marginal zone B-cell lymphoma associated with Chlamydophila psittaci infection. Hum Pathol 42(5):738–742, Epub 2011 Jan 15

Pozzato G, Mazzaro C, Crovatto M, Modolo ML, Ceselli S, Mazzi G, Sulfaro S, Franzin F, Tulissi P, Moretti M (1994) Low-grade malignant lymphoma, hepatitis C virus infection, and mixed cryoglobulinemia. Blood 84:3047–3053

Proby C, Minello A, Quantin C, Jooste V, Binquet C, Hägi M, Mounier M, Benzenine E, Hillon P, Maynadié M, Tsukiyama-Kohara K, Sekiguchi S, Kasama Y, Salem NE, Machida K, Kohara M (2012) Chronic viral hepatitis and risk of lymphoid malignancies: a retrospective twelve-year population-based cohort study in Côte d'Or, France. Dig Liver Dis 44(2):160–165, Epub 2011 Oct 4

Proietti FA, Carneiro-Proietti AB, Catalan-Soares BC, Murphy EL (2005) Global epidemiology of HTLV-I infection and associated diseases. Oncogene 24:6058–6068

Pruksananonda P, Hall CB, Insel RA, McIntyre K, Pellett PE, Long CE, Schnabel KC, Pincus PH, Stamey FR, Dambaugh TR et al (1992) Primary human herpesvirus 6 infection in young children. N Engl J Med 326(22):1445–1450

Raghuraman S, Abraham P, Daniel HD, Ramakrishna BS, Sridharan G (2005) Characterization of soluble FAS, FAS ligand and tumour necrosis factor-alpha in patients with chronic HCV infection. J Clin Virol 34(1):63–70

Rogers P, Hill I, Sinclair-Smith C, Bowie M (1988) Clinical and pathological evolution of alpha chain disease to immunoblastic lymphoma and response to COMP chemotherapy. Med Pediatr Oncol 16:128–131

Roggero E, Zucca E, Pinotti G, Pascarella A, Capella C, Savio A, Pedrinis E, Paterlini A, Venco A, Cavalli F (1995) Eradication of Helicobacter pylori infection in primary low-grade gastric lymphoma of mucosa-associated lymphoid tissue. Ann Intern Med 122:767–769

Roman E, Smith AG (2011) Epidemiology of lymphomas. Histopathology 58:4–14

Rui L, Goodnow CC (2006) Lymphoma and the control of B cell growth and differentiation. Curr Mol Med 6(3):291–308

Sagaert X, Van Cutsem E, De Hertogh G, Geboes K, Tousseyn T (2010) Gastric MALT lymphoma: a model of chronic inflammation-induced tumor development. Nat Rev Gastroenterol Hepatol 7(6):336–346

Schöllkopf C, Melbye M, Munksgaard L, Smedby KE, Rostgaard K, Glimelius B, Chang ET, Roos G, Hansen M, Adami HO, Hjalgrim H (2008) Borrelia infection and risk of non-Hodgkin lymphoma. Blood 111(12):5524–5529

Schr¨oder M, Baran M, Bowie AG (2008) Viral targeting of DEAD box protein 3 reveals its role in TBK1/IKKε- mediated IRF activation. EMBO J 27(15):2147–2157

Schulz TF (2001) KSHV/HHV8-associated lymphoproliferations in the AIDS setting. Eur J Cancer 37(10):1217–1226

Shoji T, Higuchi M, Kondo R, Takahashi M, Oie M, Tanaka Y, Aoyagi Y, Fujii M (2009) Identification of a novel motif responsible for the distinctive transforming activity of human T-cell leukemia virus (HTLV) type 1 Tax1 protein from HTLV-2 Tax2. Retrovirology 6:83

Sirianni MC, Campagna M, Scaramuzzi D, Carbonari M, Toschi E, Bacigalupo I, Monini P, Ensoli B (2007) Control of human herpes virus type 8-associated diseases by NK cells. Ann N Y Acad Sci 1096:37–43

Slater DN (2001) Borrelia burgdorferi–associated primary cutaneous B-cell lymphoma. Histopathology 38:73–77

Smith WJ, Price SK, Isaacson PG (1987) Immunoglobulin gene rearrangement in immunoproliferative small intestinal disease (IPSID). J Clin Pathol 40:1291–1297

Stanek G, Strle F (2003) Lyme borreliosis. Lancet 362(9396):1639–1647

Steere AC (2006) Lyme borreliosis in 2005, 30 years after initial observations in Lyme Connecticut. Wien Klin Wochenschr 118(21–22):625–633

Steinbach G, Ford R, Glober G, Sample D, Hagemeister FB, Lynch PM, McLaughlin PW, Rodriguez MA, Romaguera JE, Sarris AH, Younes A, Luthra R, Manning JT, Johnson CM, Lahoti S, Shen Y, Lee JE, Winn RJ, Genta RM, Graham DY, Cabanillas FF (1999) Antibiotic treatment of gastric lymphoma of mucosa-associated lymphoid tissue. An uncontrolled trial. Ann Intern Med 131:88–95

Suarez F, Lefrere F, Besson C, Hermine O (2007) Splenic lymphoma with villous lymphocytes, mixed cryoglobulinemia and HCV infection: deciphering the role of HCV in B-cell lymphomagenesis. Dig Liver Dis 39(Suppl 1):S32–S37

Suzuki T, Ishii K, Aizaki H, Wakita T (2007) Hepatitis C viral life cycle. Adv Drug Deliv Rev 59(12):1200–1212

Swerdlow S, International Agency for Research on Cancer, World Health Organization (2008) WHO classification of tumours of haematopoietic and lymphoid tissues, 4th edn. IARC, Lyon

Tajima K, Cartier L (1995) Epidemiological features of HTLV-I and adult T cell leukemia. Intervirology 38(3–4):238–246

Takatsuki K (2005) Discovery of adult T-cell leukemia. Retrovirology 2:16
Takino H, Li C, Hu S, Kuo TT, Geissinger E, Muller-Hermelink HK, Kim B, Swerdlow SH, Inagaki H (2008) Primary cutaneous marginal zone B-cell lymphoma: a molecular and clinicopathological study of cases from Asia, Germany, and the United States. Mod Pathol 21(12):1517–1526
Tee W, Mijch A (1998) Campylobacter jejuni bacteremia in human immunodeficiency virus (HIV)-infected and non-HIV-infected patients: comparison of clinical features and review. Clin Infect Dis 26(1):91–96
Tsujimoto Y, Cossman J, Jaffe E, Croce CM (1985) Involvement of the bcl-2 gene in human follicular lymphoma. Science 228(4706):1440–1443
Tsukiyama-Kohara K, Sekiguchi S, Kasama Y, Salem NE, Machida K, Kohara M (2011) Hepatitis C virus-related lymphomagenesis in a mouse model. ISRN Hematol 2011:167501, Epub 2011 Jul 26
Turner NC, Dusheiko G, Jones A (2003) Hepatitis C and B cell lymphoma. Ann Oncol 14(9):1341–1345
van Maldegem F, van Dijk R, Wormhoudt TA, Kluin PM, Willemze R, Cerroni L, van Noesel CJ, Bende RJ (2008) The majority of cutaneous marginal zone B-cell lymphomas expresses class-switched immunoglobulins and develops in a T-helper type 2 inflammatory environment. Blood 112(8):3355–3361
Vanrompay D (2007) Chlamydophila psittaci transmission from pet birds to humans. Emerg Infect Dis 13:1108–1110
Vilcek J (2006) Fifty years of interferon research: aiming at a moving target. Immunity 25(3):343–348
Viswanatha DS, Dogan A (2007) Hepatitis C virus and lymphoma. J Clin Pathol 60(12):1378–1383
Vrsalovic MM, Korac P, Dominis M, Ostojic S, Mannhalter C, Kusec R (2004) T- and B-cell clonality and frequency of human herpes viruses-6, -8 and Epstein Barr virus in angioimmunoblastic T-cell lymphoma. Hematol Oncol 22(4):169–177
Wotherspoon AC, Ortiz-Hidalgo C, Falzon MR, Isaacson PG (1991) Helicobacter pylori-associated gastritis and primary B-cell gastric lymphoma. Lancet 338:1175–1176
Wotherspoon AC, Doglioni C, Diss TC, Pan L, Moschini A, de Boni M, Isaacson PG (1993) Regression of primary low-grade B-cell gastric lymphoma of mucosa-associated lymphoid tissue type after eradication of Helicobacter pylori. Lancet 342:575–577
Wyatt JI, Rathbone BJ (1988) Immune response of the gastric mucosa to Campylobacter pylori. Scand J Gastroenterol 142(Suppl):44–49
Yoneyama M, Kikuchi M, Natsukawa T et al (2004) The RNA helicase RIG-I has an essential function in double-stranded RNA-induced innate antiviral responses. Nat Immunol 5(7):730–737
Zein NN (2000) Clinical significance of hepatitis C virus genotypes. Clin Microbiol Rev 13:223–235
Zhang H, Yang XY, Hong T, Feldman T, Bhattacharyya PK (2010) Kaposi sarcoma-associated herpesvirus (human herpesvirus type 8)-associated extracavitary lymphoma: report of a case in an HIV-positive patient with simultaneous kaposi sarcoma and a review of the literature. Acta Haematol 123(4):237–241, Epub 2010 May 19
Zhou Y, Attygalle AD, Chuang SS, Diss T, Ye H, Liu H, Hamoudi RA, Munson P, Bacon CM, Dogan A, Du MQ (2007) Angioimmunoblastic T-cell lymphoma: histological progression associates with EBV and HHV6B viral load. Br J Haematol 138(1):44–53
Zignego AL, Macchia D, Monti M et al (1992) Infection of peripheral mononuclear blood cells by hepatitis C virus. J Hepatol 15(3):382–386

Hodgkin Lymphoma: From Molecular Pathogenesis to Targeted Therapy

Ádám Jóna, Árpád Illés, and Anas Younes

Introduction

Hodgkin lymphoma (HL) is a rare human B-cell lymphoid malignancy representing 11.4% of all lymphomas in the United States (Jemal et al. 2009) and with an incidence rate of 2.2 per 100,000 in the more developed countries. (Jemal et al. 2011). The treatment of patients with relapsed and refractory HL, especially those who have disease relapse after autologous stem cell transplantation, remains challenging. However current therapy is expected to cure more than 80% of these patients.

Balancing the side effects of chemo- and radiotherapy against the need for definitive treatment is crucial. Recent efforts have been made focusing on improving success rates while decreasing chemotherapy cycles and radiation doses, and in some cases chemotherapy or radiotherapy are even eliminated, avoiding long-term toxicities. Despite the rather rare successful treatment of HL in medical oncology, the current treatment continues to lack specificity and induce unacceptable long-term toxicities that paradoxically shortens patient survival and decreases health-related quality of life (perception of physical, psychological, and social well-being and fatigue).

Within the first 10 years of diagnosing HL, the leading cause for mortality is the disease itself (van Rijswijk et al. 1987). After 10 years, HL survivors are affected by an increased risk of late medical complications, including secondary malignancies, cardiovascular diseases, pulmonary toxicity, hypothyroidism, infertility, osteoporosis,

Á. Jóna • Á. Illés
Department of Internal Medicine, University of Debrecen Medical and Health Science Center, Debrecen, Hungary

A. Younes, M.D. (✉)
Department of Lymphoma and Myeloma, The University of Texas MD Anderson Cancer Center, 1515 Holcombe Blvd., Houston, TX 77030, USA
e-mail: ayounes@mdanderson.org

J. Tao and E. Sotomayor (eds.), *Hematologic Cancers: From Molecular Pathobiology to Targeted Therapeutics*, Cancer Growth and Progression 14,
DOI 10.1007/978-94-007-5028-9_10,

renal toxicity, and dental caries (Abrahamsen et al. 1999; Ng et al. 2002a; Matasar et al. 2009; Brierley et al. 1998). The most common secondary malignancies are hematological: acute myelogenous leukemia (relative risk [RR]: 82.5), non-Hodgkin lymphoma (RR: 16.5), and multiple myeloma (RR:9.4). Solid malignancies occur mainly in bone and soft tissue (RR: 11.7), stomach (RR: 9.5), lung (RR:6.7), breast (RR:6.1), the oral cavity (RR:5.1), colon (RR:4.3), thyroid (RR:3.1), kidney (RR: 3.1), female genitals, (RR: 2.2) and skin (melanoma; RR: 1.6) (Ng et al. 2002b; Hodgson et al. 2007). Cardiovascular morbidity often includes coronary artery disease, myocardial infarction, and heart failure (Adams et al. 2004). Pulmonary toxicity can be radiation pneumonitis and fibrosis. Radiation and blemocycin are known to be a major cause of pulmonary toxicity (Martin et al. 2005).

Patients who are not cured with front-line or second-line therapy, including stem cell transplantation, have an estimated median survival of less than 3 years (Horning et al. 2008). As the median age of this patient population is mid-1930s, the impact of early mortality on the number of years lost from productive life is more significant than for many other cancers. However, because HL is a rare disease that is highly curable, the development of new drugs for the treatment of HL has been very slow; the United States Food and Drug Administration (FDA) has not approved any new drugs for treating HL in more than three decades. Thus, there is a clear unmet medical need for developing new drugs for HL. With recent advances in our understanding of HL pathology, biology, and immunology, several therapeutic targets have been identified (Küppers 2009). For example, new treatment strategies that are based on targeting oncogenic signaling pathways are currently being explored (Jona and Younes 2010) (Fig. 1). The aim of drug development for HL is to not only further improve the cure rate but also decrease the toxic effects of therapy. This chapter will focus on emerging new treatment modalities targeting the molecular targets of classical HL that are currently under investigation for patients with relapsed disease.

Current Standard of Care for Hodgkin Lymphoma

The HD10 study conducted by the German Hodgkin Study Group has shown recently that the intensity of treatment can be reduced in patients with early-stage HL with favorable prognoses. The study compared four versus two cycles of the adriamycin, bleomycin, vinblasine, and dacarbazine (ABVD) regimen followed by 20- or 30-Gy involved field radiation therapy (IFRT). The weakest combination was as effective as stronger combinations, but less toxic. The rates of freedom from treatment failure (FFTF) were 93% with four cycles of ABVD and 91% with the two-cycle regimen, regardless of whether 20- or 30-Gy IFRT was given. Therefore, two cycles of ABVD followed by 20-Gy IFRT is considered the standard of care for patients with early-stage HL with favorable prognoses (Engert et al. 2010a).

In patients with early-stage HL with unfavorable prognoses the European Organization for Research and Treatment of Cancer (EORTC)/GELA H9-U trial showed that there was no significant difference in outcomes when either four or six cycles of ABVD followed by IFRT was given. Event-free survival (EFS) was 87 and 91%,

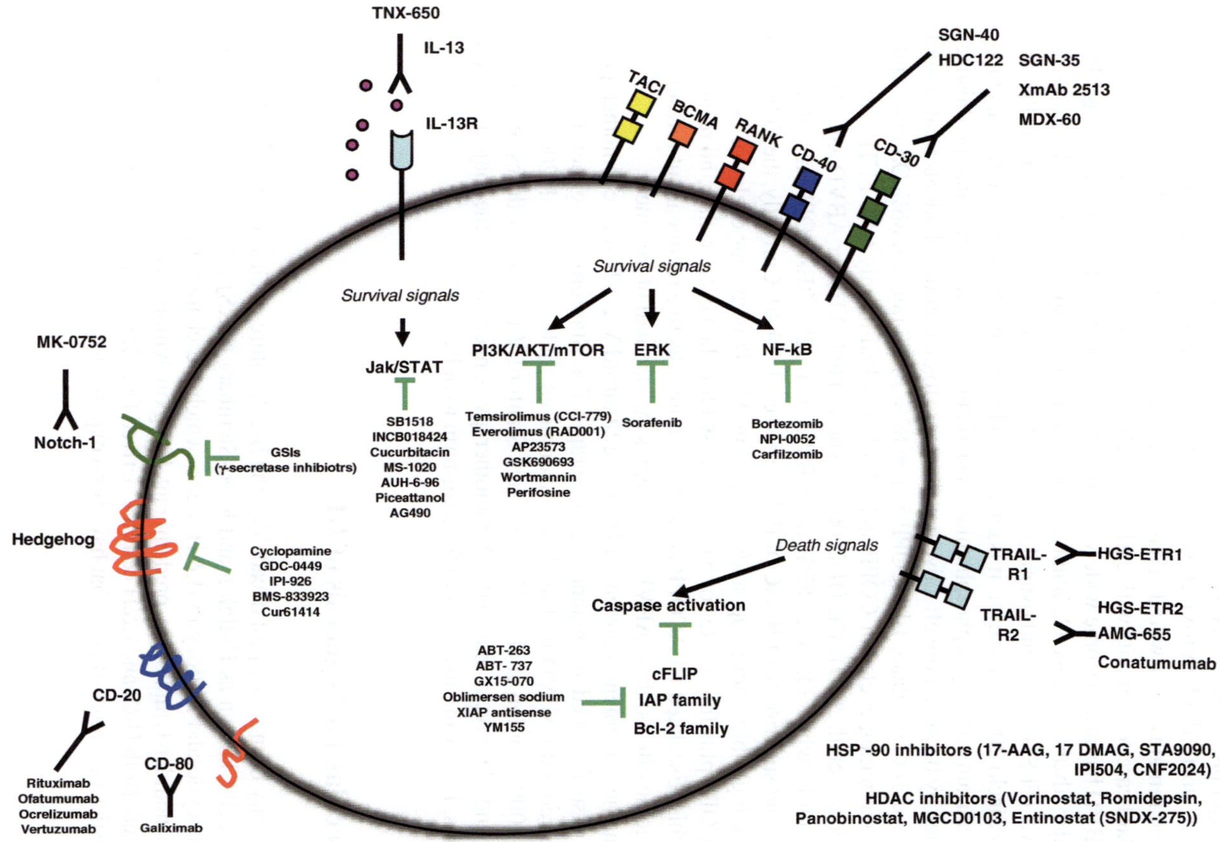

Fig. 1 Targeted therapy of HRS cells. HRS cells express a variety of receptors and antigens that can be targeted by monoclonal antibodies. Many of these receptors trigger well-defined signaling pathways that promote HRS cell survival. These signaling pathways can be targeted by a variety of small molecules

respectively, at 4 years (Thomas et al. 2007). Concerning the intensity of chemotherapy, two (Thomas et al. 2007; Borchmann et al. 2009) of three (Engert et al. 2010b) trials failed to show a benefit from dose escalation. Although the last of these trials showed 6% longer progression-free survival (PFS), one could argue whether this improvement is worth the late toxicities incurred. The question of whether IFRT can be omitted remains unanswered. Thus, four cycles of ABVD followed by IFRT is the standard care for patients with early-stage HL with unfavorable prognoses.

ABVD has also become widely accepted for the treatment of advanced-stage HL. A major advantage is its tolerability. A long-term follow-up study showed failure-free survival (FFS) of 47% and overall survival (OS) of 59% after 14 years (Canellos and Niedzwiecki 2002). However, it is challenged by the BEACOPP (bleomycin, etoposide, adriamycin, cyclophosphamide, vincristine, procarbazine, and prednisone) regimen by being a more effective but also more toxic regimen. Patients treated with eight cycles of escalated BEACOPP had an 82% FFTF rate and 86% OS rate, while eight alternating cycles of COPP and ABVD resulted in only 64% FFTF and 75% OS (Engert et al. 2009). The HD2000 trial showed 68% FFS for ABVD and 81% for BEACOPP, while OS was 83% for ABVD and 92% BEACOPP, revealing the significant superiority of BEACOPP (Federico et al. 2009). Currently, the EORTC is conducting a big intergroup study comparing ABVD with four cycles of escalated BEACOPP plus four cycles of baseline BEACOPP; however, to date, no results are available. Escalated BEACOPP is clearly a more effective regimen, but it results in more toxic effects; thus, ABVD still remains the standard of treatment.

Another treatment regimen for HL is the Stanford V, a 12-week chemotherapeutic course followed by planned radiotherapy. The Stanford V regimen has been compared with ABVD in three large randomized trials (Gobbi et al. 2005; Hoskin et al. 2009; Gordon et al. 2010). Although Stanford V uses lower doses of doxorubicin and bleomycin than ABVD, radiotherapy is necessary in the majority of the patients, increasing the concern for long-term toxicity.

Molecular Pathogenesis of HL

It has only been in the past decade that the B-cell nature of classical HL's pathognomical cells, known as Hodgkin and Reed-Sternberg (HRS) cells and their variants, lymphocytic and histiocytic (L&H) cells, which are characteristic of another subgroup nodular-lymphocyte predominant Hodgkin lymphoma (NLPHL), has been revealed. Both HRS and L&H cells are derived from B cells, that is, they show characteristics of mature B cells that have been exposed to antigens (Küppers 2009).

Clonality and Origin of HRS Cells

HRS cells are unique in the hematopoietic system, as they express markers of several lineages. HRS cells are most likely derived from pre-apoptotic germinal-centre

B cells, since they carry rearranged and somatically mutated *IgG V* genes, which are seen in germinal-center B cells, and normally induce immediate cell death (Kanzler et al. 1996; Kuppers and Rajewsky 1998). However, sequence analysis of HRS cells has only identified a subset of these unfavorable mutations, suggesting that these cells can avoid apoptosis. Further, there are several characteristics of HRS cells that are only seen in antigen-activated B cells, including complete Ig heavy and light chain rearrangements, somatic mutations occurring only in antigen-activated B cells, and Ig class switch recombination in HL cell lines and primary HRS cells. In addition, both HL and mature B-cell lymphomas carry somatically mutated *IgG V* genes, demonstrating their common precursor (Küppers 2009). Nevertheless, some HL cases show T-cell characteristics, and these cases are derived from T cells. L&H cells may also originate from the germinal center, since they express several B-cell markers and grow in a follicular pattern (Hansmann et al. 1999). L&H cells are likely derived from antigen-selected germinal-centre B cells, since they resemble an intermediate developmental stage between germinal-centre and memory B cells (Brune et al. 2008).

Signaling Pathways

The rescue of HRS cells from apoptosis may play a key role in the pathogenesis of HL. Several signaling pathways that activate or inhibit apoptotic pathways have been recently revealed to play a role in the pathogenesis of HL. The genetic lesions that lead to dysregulated signaling pathways are found most commonly in the janus kinase–signal transducer and activator of transcription (JAK-STAT) and nuclear factor-κB (NF-κB) pathways. Many cytokines transmit their signal through members of the JAK family, which phosphorylate STATs on activation. The phosphorylated STATs then dimerize and translocate to the nucleus where they function as transcription factors. There are frequent genomic gains of *JAK2* in HRS cells, and suppressor of cytokine signaling 1 (SOCS1), a negative regulator of JAK–STAT signaling, is often somatically mutated and inactivated in classical HL and NLPHL (Joos et al. 2000; Weniger et al. 2006; Mottok et al. 2007). The more important alterations of the NF-κB pathway are affected by the transcription factor REL in about 50% of all HL cases (Barth et al. 2003), the positive regulator Bcl-3 (Martin-Subero et al. 2006), the negative regulators NFKBIA (Jungnickel et al. 2000) and NFKBIE (Emmerich et al. 2003), and TNFAIP3, the encoder of A20 that ubiquitinates and deubiquitinates members of the NF-κB signaling pathway and negatively regulates NF-κB activity. Interestingly, all HL cases where mutations prevented TNFAIP3 function were found to be Epstein-Barr virus (EBV)-negative, indicating that TNFAIP mutations and EBV infection are alternative mechanisms of HL pathogenesis.

EBV infection plays an important role in about 40% of all classical HL cases. For EBV-infected HL patients, the EBV proteins LMP-1 and LMP-2 seem to promote the survival of the infected HRS cells. LMP-1 induces constitutive NF-κB signaling (Kilger et al. 1998), while LMP-2 functions as a surrogate of the B-cell receptor (Alber et al. 1993). These results indicate that these mechanisms are alternative methods of HL pathogenesis (Schmitz et al. 2009).

Besides the constitutive activation of the JAK-STAT and NF-κB pathways, autocrine and paracrine signaling events also contribute to the pathogenesis of HL (Scheeren et al. 2008). In HRS cells, STATs are activated by expression of IL-13 and IL-21 and their receptors (Kapp et al. 1999). HRS cells also express the tumor necrosis–factor receptor family proteins, including CD30, CD40, TACI, BCMA, and RANK, which are all involved in NF-κB signaling (Fiumara et al. 2001; Chiu et al. 2007). Other signaling pathways can be also dysregulated in HRS cells, such as the phosphatidylinositol 3-kinase (PI3K)–Akt and extracellular signal-regulated kinase (ERK) pathways (Georgakis et al. 2006a; Zheng et al. 2003). The PI3K pathway may be activated through CD30, CD40, and RANK. The ERK pathway, which regulates apoptosis and cell proliferation and differentiation, may also be activated through CD30, CD40, and RANK in HRS cells (Zheng et al. 2003).

Tumor Microenvironment: Cytokines, Chemokines, and the Dysregulated Immune Response

The tumor microenvironment seems to be crucial for the survival of HRS cells, which represent about 98% of the tumor tissue. HRS cells secrete cytokines and chemokines that actively attract B cells, T cells, eosinophils, plasma cells and mast cells, which make up the tumor microenvironment. CCL5 (RANTES), CCL17 (thymus and activation-regulated chemokine [TARC]), and CCL22 (MDC) attract T-helper 2 and T-regulatory cells (Poppema 2005). Moreover, HRS cells also secrete a number of cytokines, including IL-5, -6, -9, -10, and −13. In particular, IL-13 and its receptor IL-13R were found to promote the differentiation of T-helper cells into T-helper 2 cells (Skinnider et al. 2001). Evidence suggests that changes in immunosurveillance allow HRS cells to survive; there is a shift from an antitumor T-helper 1 response to a protumor T-helper 2 response (Tan and Coussens 2007). HRS cells also express a variety of immunosuppressive factors, such as IL-10, TGF-ß, galectin-1, and prostaglandin E_2 (Gandhi et al. 2007; Newcom and Gu 1995; Chemnitz et al. 2006). They also express the programmed cell death protein 1 ligand, which inhibits the cytotoxic function of T-cells (Yamamoto et al. 2008).

Epigenetic Therapy: Histone Deacetylase Inhibitors

Post-transcriptional histone modification plays an important role in regulating gene transcription and is mediated by several enzymes, including histone acetyltransferases (HATs) and histone deacetylases (HDACs) (Glozak and Seto 2007). These enzymes mediate acetylation and deacetylation of specific lysine amino acid

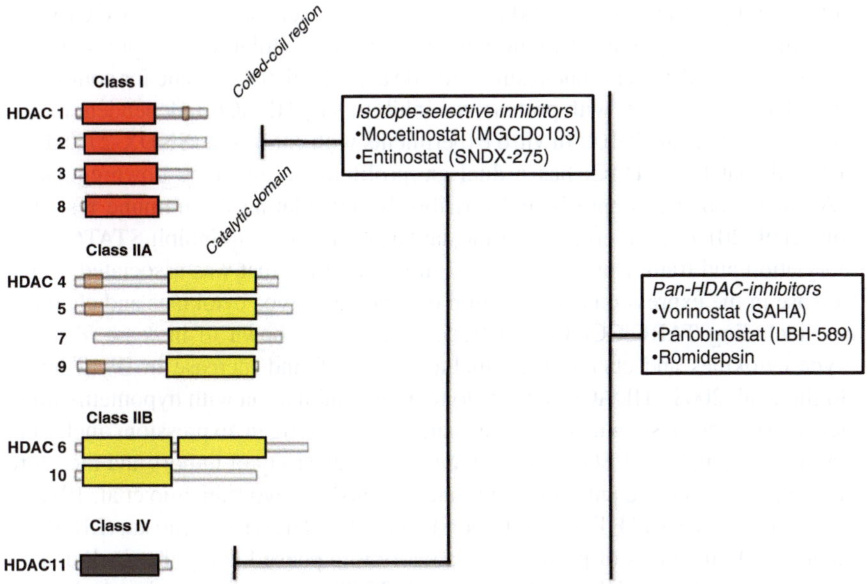

Fig. 2 Human zinc-dependent histone deacetylases (HDACs) and their inhibitors. Isotope-selective inhibitors inhibit certain classes of HDACs, while pan-HDAC-inhibitors inhibit all of them

residues on histone and non-histone proteins that regulate a variety of proteins involved in cell proliferation and survival, angiogenesis, and immunity (Heider et al. 2006; Wang et al. 2006; Brogdon et al. 2007). To date, 18 HDACs have been identified in humans and grouped into two major categories: zinc-dependent HDACs and nicotinamide adenine dinucleotide (NAD)-dependent HDACs (Bolden et al. 2006; Minucci and Pelicci 2006). HDACs are further classified into four major classes: Class I (HDACs 1, 2, 3, 8, and 11), Class II (HDACs 4, 5, 6, 7, 9, and 10), Class III (SIRTs 1–7), and Class IV (HDAC 11) (Fig. 2). Class III HDACs are NAD-dependent, whereas class I, II, and IV HDACs are zinc dependent. Several clinical-grade pharmacologic inhibitors of zinc-dependent HDACs are now being tested in clinical trials, but only two inhibitors, vorinostat and romidepsin, have been approved by the FDA for treating relapsed cutaneous T-cell lymphoma. Vorinostat (SAHA) and panobinostat (LBH589) inhibit class I and II HDACs (pan-HDAC inhibitors). MGCD0103 and entinostat (SNDX-275, formerly MS-275) preferentially inhibit class I HDACs (isotype-selective HDAC inhibitors).

There are several rationales for using HDAC inhibitors (HDACis) for the treatment of HL. For example, although HRS cells are of B-cell origin, they infrequently express B-cell antigens (Schwering et al. 2003). This loss of B-cell phenotype has been reported to be epigenetically regulated and may be therapeutically reversible (Ushmorov et al. 2004, 2006). Further, several HDAC inhibitors have antiproliferative activity in HL-derived cell lines in vitro. In a recent study, vorinostat was shown to induce cell-cycle arrest and apoptosis in HL cell lines and to synergize with chemotherapy (Buglio et al. 2008). Similarly, the isotype-selective MGCD0103 (mocetinostat) has potent antilymphoma activity; it modulates the expression of a variety of survival proteins and results provide a mechanistic rationale for combining class I HDAC inhibitors with proteasome inhibitors by HDAC6-independent mechanisms (Buglio et al. 2010). In vitro experiments with entinostat (SNDX-275) demonstrated that this HDACi has a dual antiproliferative effect by downregulating XIAP and inducing apoptosis and possibly by modulating the immune response (Jona et al. 2010). Furthermore, vorinostat has been shown to inhibit STAT6 phosphorylation and transcription in HL cell lines, an effect that was associated with a decrease in the expression and secretion of T-helper 2-type cytokines and chemokines, including TARC/CCL17 and IL-5. It was also shown to increase T-helper 1-type cytokines and chemokines, including a profound increase in IP-10 levels (Buglio et al. 2008). HDACis, either alone or in combination with hypomethylating agents, have been shown to induce cancer testis antigen expression, including MAGE, SSX, and NY-ESO-1 family members, in a variety of tumors and therefore may induce a favorable antitumor immune response in vivo (Shichijo et al. 1996).

The most successful HDACi compound for treating HL is panobinostat (LBH589) (Table 1). On the basis of promising results from a phase I study that included 13 patients with relapsed HL (Dickinson et al. 2009), a large international phase II study was initiated. In that pivotal study, oral panobinostat was administered at a dose of 40 mg three times per week, every week, in 21-day cycles. Dose delays and modifications for the management of adverse events was permitted, but the lowest dose allowed on study was 20 mg. Efficacy was evaluated every two cycles by imaging studies. Patients were enrolled in the study for less than 1 year. The median age was 32 years (range, 18–75 years), and the median number of prior chemotherapeutic regimens was four (range, 2–7 years). Importantly, the median time to relapse after the first autologous stem cell transplantation was only 8 months, which predicts a poor prognosis. Moreover, 37% of the patients had not responded to their last prior therapy. Twelve patients had also received a prior allogeneic transplant. In the final analysis, 129 patients were evaluable for a response or discontinued early. A total of 35 responses (5 complete responses [CRs] and 30 partial responses [PRs]) with an overall response rate (ORR) of 27% and a disease-control rate (CR+PR+stable disease [SD]) of 82% were observed. Preliminary PFS was more than 5.7 months. Common drug-related grade 1 and 2 adverse effects were diarrhea, nausea, fatigue, vomiting, and anorexia. Common drug-related grade 3 and 4 adverse events were thrombocytopenia, anemia, and neutropenia. The thrombocytopenia was manageable and reversible with dose hold and modification (Sureda et al. 2010).

MGCD0103 is a novel oral nonhydroxymate benzamide–based HDACi that selectively inhibits HDACs 1 and 2 (and to a lesser extent, HDACs 3 and 11) (Khan et al. 2008). Its IC_{50} for inhibiting recombinant HDAC 1 activity is 0.082 mM compared with >30 mM for HDAC6 (Beckers et al. 2007; Riester et al. 2007). The safety and efficacy of MGCD0103 given orally three times per week (85 mg to 110 mg starting doses) was recently evaluated in a phase II study in patients with relapsed and refractory HL. Patients were allowed to receive therapy for 1 year in the absence of disease progression or prohibitive toxicity. Of the 23 patients who were treated

Table 1 Summary of results of studies of selected novel agents in relapsed Hodgkin lymphoma (HL)

Agent	Target	Route	Phase	N	PR	CR	PR+CR	1st author
Panobinostat (Dickinson et al. 2009)	HDACs	Oral	II	13	7	0	7 (58%)	Dickinson
Panobinostat (Sureda et al. 2010)	HDACs	Oral	II	129	30	5	35 (27%)	Sureda
MGCD0103 (Younes et al. 2010a)	HDACs	Oral	II	51	12	2	14 (29%)	Younes
Vorinostat (Kirschbaum et al. 2007)	HDACs	Oral	II	25	1	0	1 (4%)	Kirshbaum
SNDX275 (Younes et al. 2010b)	HDACs	Oral	II	27	3	0	3 (11%)	Younes
Everolimus (Johnston et al. 2010)	mTOR	Oral	II	19	8	1	9 (47%)	Johnston
MDX060 (Ansell et al. 2007)	CD30	IV	II	47	2	2	4 (8%)	Ansell
SGN30 (Forero-Torres et al. 2009)	CD30	IV	II	38	0	0	0 (0%)	Forrero-Torres
SGN35 (Younes et al. 2010d)	CD30	IV	I	45	6	11	17 (37%)	Younes
SGN35 (Fanale et al. 2009)	CD30	IV	I	31	6	10	16 (46%)	Fanale
SGN35 (Chen et al. 2010)	CD30	IV	II	102				Chen
Lenalidomide (Fehniger et al. 2009)	?	Oral	II	35	5	1	6 (17%)	Fehniger
Lenalidomide (Kuruvilla et al. 2008)	?	Oral	II	15	2	0	2 (13%)	Kuruvilla
Lenalidomide (Boll et al. 2010)	?	Oral	II	24	11	1	12 (50%)	Bolls

HDACs indicates histone deacetylases, *mTOR* mammalian target of rapamycin with the 110-mg dose, eight (35%) achieved partial or complete remissions. However, this dose was poorly tolerated, resulting in dose interruptions and reductions and discontinuation of therapy after a median of 4.5 months. Subsequently, the study was revised to allow a lower starting dose of 85 mg at the same schedule. Six of the 28 (21%) patients enrolled on the reduced dose achieved partial remissions. Grade 3 and 4 toxicities were mainly thrombocytopenia and fatigue, which occurred in around 20% of the patients. Overall, 80% of the 30 evaluable patients had some decrease in their tumor measurements. Although none of the patients developed significant electrocardiogram abnormalities, two patients developed significant pericardial effusions requiring discontinuation of therapy. Collectively, MGCD0103 was active as a single agent in patients with heavily pretreated relapsed/refractory HL. The response rate reported in this study is among the best for a single-agent HDACi in HL, especially in the context of the minimal hematological toxicity observed (Younes et al. 2010a). Finally, the Southwest Oncology Group (SWOG) conducted a phase 2 trial of vorinostat in patients with relapsed HL (Kirschbaum et al. 2007). Twenty-five patients were treated with 200 mg vorinostat given orally twice per day for 14 days every 21 days. One patient (4%) achieved a partial remission.

The interim results of a phase II trial with entinostat (SNDX-275), another class 1–selective HDACi, have been reported recently. Entinostat was administered at a dose of either 10 or 15 mg given orally every 14 days on a 28-day cycle. Of 27 evaluable patients, three (11%) had a PR, and there were no CRs; however disease control (CR+PR+SD) was 63%. Common grade 3 and 4 adverse events were thrombocytopenia (59%), neutropenia (28%), and anemia (34%). Overall, entinostat was well tolerated and the preliminary efficacy data are encouraging. The trial is being continued and patients are being enrolled into an alternative dosing schedule (Younes et al. 2010b).

mTOR Inhibitor: Everolimus (RAD001)

The PI3K/Akt/mTOR signaling pathway (Fig. 3) is one of the most frequently dysregulated survival pathways in cancer, making it an important target for drug development (Ihle and Powis 2009; Franke 2008). This pathway is negatively regulated by the tumor suppressor protein PTEN. Unlike most cancers in which PI3K activation is frequently associated with PTEN deletion or mutation, other mechanisms have been reported to activate this pathway in HL, including activation of the CD30, CD40, and RANK receptors, mutations in the p85a subunit of PI3K, and inactivation of PTEN by phosphorylation (Georgakis et al. 2006b; Jucker et al. 2002; Morrison et al. 2004; Nagel et al. 2005; Renne et al. 2007; Dutton et al. 2005). In vitro experiments have demonstrated that inhibition of PI3K, Akt, or mTOR with various small molecules can induce cell-cycle arrest, autophagy, and apoptosis in HRS-derived cell lines (Georgakis et al. 2006a, c; Jundt et al. 2005). In addition to a direct antitumor effect, mTOR inhibitors may induce clinical responses by enhancing the immune response and inhibiting angiogenesis (Zheng et al. 2007; Del Bufalo et al. 2006).

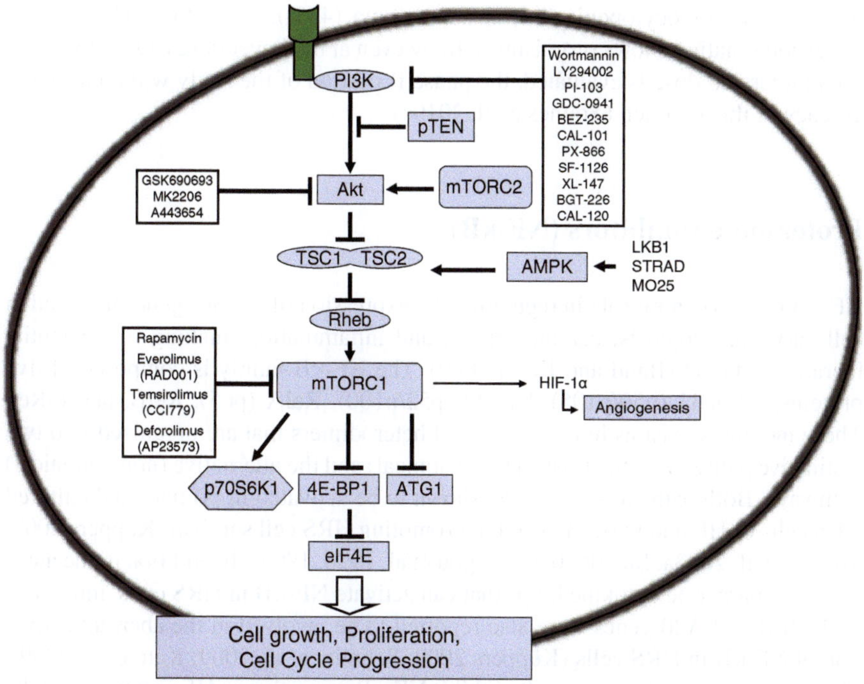

Fig. 3 The PI3 kinase/Akt/mTOR pathway is frequently activated in HRS cells. Pharmacologic inhibition of mTOR has produced clinical responses in patients with relapsed classical Hodgkin lymphoma (HL)

The therapeutic value of inhibiting the PI3K/Akt/mTOR axis has been recently studied using the oral mTOR inhibitor everolimus (Fig. 3) (Johnston et al. 2010). In one study, 19 evaluable patients with relapsed HL were treated with daily doses of 10 mg everolimus, and the ORR rate was 47%; eight patients achieved partial remission and one, complete remission (Table 1). Grade 3 adverse events included thrombocytopenia and anemia. If confirmed in a larger number of patients, everolimus may be considered one of the most active agents for treating relapsed HL. Because HRS cells frequently demonstrate aberrant and simultaneous activation of several survival pathways, including NF-κB, ERK, and PI3K/Akt (Fig. 3), rationally designed combination strategies will be required to improve the response rate and prolong the response duration of mTOR inhibitors. In vitro experiments have suggested that mTOR inhibitors may synergize with chemotherapy, PI3K inhibitors, and HDACis in a variety of tumor models, including HL (Georgakis et al. 2006c; Yazbeck et al. 2008; Lemoine et al. 2010). A phase I/II clinical trial combining the HDAC inhibitor panobinostat with the mTOR inhibitor everolimus is currently enrolling patients with NHL and HL. To date 12 patients have been enrolled, and 10 of 11 evaluable patients had their tumors reduced by 31 to 63%. Twenty milligrams panobinostat and 10 mg everolimus was used to determine the safety of these drugs. Adverse events were manageable, with the following being most commonly observed: thrombocytopenia (45%), neutropenia (45%), and fatigue (18%). This novel combination shows promising activity even at the lower doses. Once the maximum tolerated dose is evaluated, the phase II portion of the study will confirm the efficacy of this regimen (Younes et al. 2010c).

Proteasome Inhibitors (NF-κB)

NF-κB plays a central role in regulating the expression of various genes involved in cell survival, apoptosis, carcinogenesis, and inflammation, making it a potential therapeutic target (Baud and Karin 2009). The NF-κB family is composed of five proteins: NF-κB1 (p50/p105), NF-B2 (p52/p100), RelA (p65), RelB, and c-Rel. These members exist as homodimers and heterodimers that are organized into two distinctive pathways: the classical (or canonical) and the alternative (non-canonical) pathways. Both pathways have been shown to be activated in primary and cultured HRS cells of HL and to be involved in promoting HRS cell survival (Küppers 2009; Younes et al. 2003a; Staudt 2000; Bargou et al. 1996, 1997). In addition to the autocrine and paracrine cytokine loops that can activate NF-κB in HRS cells, mutations in the IκB and A20 genes were also reported to be involved in the aberrant activation of NF-κB in HRS cells (Küppers 2009; Schmitz et al. 2009; Kato et al. 2009). The first attempt to therapeutically inhibit NF-κB activation in HL patients used the proteasome inhibitor bortezomib. By inhibiting the degradation of cytoplasmic IκBα, bortezomib inhibits the activation of NF-κB. Furthermore, bortezomib has been reported to alter the levels of p21, p27, Bcl-2, Bax, XIAP, survivin, and p53, leading to cell-cycle arrest and apoptosis in several tumor types (Adams 2003). In preclinical studies, bortezomib inhibited the proliferation of HL cells in culture and induced cell-cycle arrest and apoptosis in a time- and dose-dependent manner. Further, bortezomib was effective even in HL cell lines that harbored mutations in the IκBα gene (Zheng et al. 2004). Despite these favorable preclinical results, bortezomib demonstrated no significant clinical activity when tested in patients with relapsed HL (Younes et al. 2006; Blum et al. 2007).

On the basis of preclinical experiments that demonstrated synergy between bortezomib and chemotherapy, two independent groups evaluated bortezomib-based combinations in patients with relapsed classical HL. In the first study, a phase I trial was conducted to evaluate the combination of bortezomib with the ifosomide, carboplatin, etoposide (ICE) regimen (Fanale et al. 2011). Escalating doses of bortezomib were given on days 1 and 4 of each ICE cycle. Thirteen patients were enrolled, and six achieved PRs and four achieved CRs, for an ORR of 77%. The treatment was well tolerated and associated with reversible grade 4 neutropenia and thrombocytopenia in 18 and 35% of the patients, respectively. On the basis of these encouraging data, a phase II randomized trial is currently enrolling patients to prospectively evaluate response rates to determine the PFS of patients treated with bortezomib-ICE versus those treated with ICE alone. In a second study, bortezomib was combined with gemcitabine for the treatment of patients with relapsed HL (Mendler et al. 2008). Bortezomib (1 mg/m^2) was given on days 1, 4, 8, and 11, and gemcitabine (800 mg/m^2) was given on days 1 and 8. Treatments were repeated every 21 days. The ORR in 18 patients was 22%. However, because of the relatively low response rate and treatment-related liver toxicity, the authors concluded that this regimen should not be further developed for the treatment of HL.

Brentuximab Vedotin (SGN-35)

The dense expression of CD30 by HRS cells coupled with its highly restricted expression to a relatively small population of activated B cells and T cells and a small portion of eosinophils, makes it an obvious target for a therapeutic monoclonal antibody (Younes and Carbone 1999; Younes and Aggarwall 2003). Results from two clinical studies using first-generation naked anti-CD30 monoclonal antibodies in patients with relapsed HL have been disappointing, perhaps reflecting their poor antigen binding and/or effector cell activation properties (Table 1) (Ansell et al. 2007; Forero-Torres et al. 2009). In an alternate strategy, the anti-CD30 antibody cAC10 was conjugated to a synthetic anti-microtubule agent, monomethyl auristatin E, resulting in the novel immunotoxin conjugate brentuximab vedotin (SGN-35) (Oflazoglu et al. 2008). Brentuximab vedotin was recently evaluated in two phase I clinical trials in patients with relapsed HL and anaplastic large-cell lymphoma. In the first phase I study, brentuximab vedotin was administered on a 3-week schedule. Forty-five patients with relapsed HL and anaplastic large-cell lymphoma were treated with escalating doses (0.1 to 3.6 mg/kg) by intravenous infusions every 3 weeks. The treatment was reasonably well tolerated, with neutropenia and hyperglycemia being the only dose-limiting toxicities. Neuropathy was also observed in some patients, especially after repeated dosing. Remarkably, 36 of 42 patients (86%), who could have been evaluated by computed tomography patients had reductions in their tumor size, and 17 of 45 patients (38%) achieved partial or complete remissions (Younes et al. 2010d).

In a second phase I study, 37 patients (31 with HL) were treated with brentuximab vedotin that was administered on a weekly schedule for 3 consecutive weeks in 4-week cycles. Dose-limiting toxicities included grade 3 gastrointestinal toxicity and grade 4 hyperglycemia. The ORR was 46% (29% CRs) (Fanale et al. 2009). These encouraging results led to a pivotal phase II trial, which recently completed enrollment of 102 HL patients who had undergone autologous stem cell transplantation. The patients were then treated with 1.8 mg/kg brentuximab vedotin given every 3 weeks for up to 16 cycles. Tumor shrinkage was observed in 95% of patients. The most common adverse events of any grade were peripheral sensory neuropathy (43%), fatigue (40%), nausea (35%), neutropenia (19%), diarrhea (18%), and pyrexia (16%). The only Grade 4 treatment-related events were neutropenia (4%) and thrombocytopenia, abdominal pain, and pulmonary embolism (1% each). In the 35 patients who had B symptoms (fever, night sweats and weight loss), at baseline, the symptoms resolved in 29 (83%) (Chen et al. 2010).

Rituximab

Although HRS cells are of B-cell origin, CD20 antigen is infrequently expressed in these cells (15–30% in classical HL). However, HRS cells express a variety of receptors, including CD30, CD40, and RANK. Hematopoietic cells in the microenvironment, including B and T cells, express the ligand for these receptors, which activates survival signals in different oncogenic pathways (NF-κB, Erk, and Akt). Rituximab is a monoclonal antibody against CD20, and the central hypothesis for using rituximab in HL patients, regardless of the levels of CD20 expression in HRS cells, is that eliminating the HL microenvironment may deprive the malignant cells of survival signals and make them more susceptible to chemotherapy. An additional rationale is that putative classical HL stem cells also express CD20. A recent study showed that a blood sample from HL patients consisted of a very small subpopulation of cells with a memory B-cell phenotype (Ig light chain, CD27, CD20, and the stem-cell marker aldehyde dehydrogenase). These cells were responsible for maintaining the predominant HRS cell population. Interestingly this cell subpopulation could not be found in blood samples from healthy donors (Jones et al. 2009). Finally, a study using tumor-bearing mice suggested that depleting normal B cells enhances the antitumor response while diminishing IL-10 production. The data suggested that rituximab may deplete reactive B cells, which would increase the host immune response against HL. However, to date, this hypothesis has not been validated in patients receiving rituximab (Inoue et al. 2006).

Targeting CD20 may be beneficial for HL patients, despite the fact that CD20 expression isn't often seen in HRS cells, but eliminating the HL microenvironment may deprive the malignant cells of survival signals. In a study of CD20 expression in HRS cells and the infiltrating small B cells, the investigators hypothesized that higher CD20 expression correlates with better treatment outcome. Findings from that study confirmed that there is a correlation between an increased number of tumor-infiltrating CD20-positive small B cells and longer PFS ($p=0.02$) and disease-free survival ($p=0.02$). Increased number of CD20-positive small B cells coupled with low CD68 expression in tumor-infiltrating macrophages may identify a very favorable patient population who would benefit from minimizing treatment. However, no correlation was found between the number of CD20-positive HRS cells and treatment outcome (Steidl et al. 2010).

In recent years, rituximab has been evaluated in patients with classical HL as single agent, in combination with gemcitabine, or in combination with ABVD chemotherapy (Oki and Younes 2010). In a pilot study, investigators from The University of Texas MD Anderson Cancer Center treated 22 patients with relapsed classical HL with 6 weekly doses of rituximab. Six (27%) of the 22 patients had HRS cells that expressed CD20 (Younes et al. 2003b). Five (23%) patients achieved PR or CR, and eight additional patients had SD. Clinical remissions were observed in patients regardless of CD20 expression in HRS cells and were limited in patients whose disease was confined to the lymph nodes.

A phase II study investigated the effect of the rituximab-gemcitabine combination in heavily pretreated HL patients. Thirty-three patients received rituximab at a dose of 375 mg/m^2 given intravenously every week for 6 weeks and gemcitabine at a dose of 1,250 mg/m^2 given intravenously on days 1 and 8, with a week of rest on day 15 of a 21-day cycle. Sixteen patients (48%) showed objective response after two cycles regardless of whether the HRS cells expressed CD20 (Oki et al. 2008). Further, an Italian group conducted a study investigating the combination of gemcitabine and rituximab with other agents (R-GIFOX [rituximab, gemcitabine, ifosfamide, and oxaliplatin]) in patients with relapsed HL. Twenty-one patients received three courses of R-GIFOX followed by stem-cell mobilization and high-dose therapy if eligible for autologous stem cell transplantation. Eighty-six percent of the patients responded, with two PRs and 16 CRs (Corazzelli et al. 2009).

In another study, investigators at MD Anderson Cancer Center combined rituximab with ABVD chemotherapy to treat patients with newly diagnosed classical HL (Wedgwood et al. 2007). Fifty-two patients with newly diagnosed classical HL were treated in a phase II study. With a median follow-up of 32 months, the estimated EFS was 82% and OS was 100%. Importantly, the EFS was improved for all risk categories: for patients with a international prognostic score (IPS) from 0 to 1 the EFS was 92%; for prognostic scores of 2, the EFS was 86%; and for scores from 3 to 5, the EFS was 73%. These data are currently being confirmed in a multicenter randomized study comparing ABVD alone with rituximab plus ABVD. In the final report of this study, 104 patients, with a median age of 35 years, were enrolled. At a 5-year median follow-up (range, 6–94 months), the projected EFS for R-ABVD was 87%, which is significantly better than institutional results with ABVD alone ($p=0.0036$). Improvement in EFS was observed with R-ABVD in patients with an IPS from 0 to 2 (EFS 89 vs. 71%, $p=0.0248$) and those with an IPS >2 (80 vs. 55%; $p=0.0532$). These data serve as the rationale for a multicenter randomized trial comparing ABVD with R-ABVD in newly diagnosed patients with stage III and IV classical HL with IPS >2. The study is currently enrolling patients (Copeland et al. 2009).

Lenalidomide

Two independent groups have evaluated the safety and efficacy of lenalidomide in patients with relapsed HL. In the first study, Fehniger et al. reported their experience with 25 mg/day of lenalidomide on days 1 to 21 of a 28-day cycle (Fehniger et al. 2009). Treatment continued until progressive disease or an unacceptable adverse event occurred. Despite the liberal dose reductions that were allowed for hematologic and non-hematologic toxicity, six of the 35 evaluable patients responded (one CR and five PRs). Grade 3 and 4 neutropenia was observed in 40% of patients, anemia in 24%, leukopenia in 21%, and thrombocytopenia in 16% of patients. In a second study, Kuruvilla and colleagues treated 15 patients who had relapsed HL using the same dose and schedule of lenalidomide as in the previous study (Kuruvilla et al. 2008). Two patients achieved PRs and seven achieved SD, with a median time to progression of 3.2 months. Six patients discontinued therapy because of disease progression and five discontinued because of toxicity. Four patients developed grade

3 or 4 neutropenia and thrombocytopenia and five developed grade 1 or 2 skin rash. Collectively, these data suggest that lenalidomide has promising single-agent activity in relapsed HL. A French group conducted a small study of eight patients with refractory or relapsed HL receiving the etoposide, solumedrol, ara-C, cisplatin (ESAP) salvage regimen in combination with lenalidomide. Response was evaluated by positron emission tomography or computed tomography. Seven CRs and one PR were seen. The toxicities seen were nearly all hematological (neutropenia and thrombocytopenia) (Tempescul et al. 2009).

The German Hodgkin Study Group investigated the effect of lenalidomide in heavily pretreated relapsed or refractory HL patients. Patients were treated with 25 mg lenalidomide given orally in 21- or 28-day cycles. Twenty-four patients were eligible for analysis. Eight of the analyzed patients had SD and 12 achieved clinical responses (11 PRs, 1 CR) as measured by computed tomography. No hematological or other toxicities above grade 2 were reported (Boll et al. 2010).

Summary and Future Directions

Several compounds have been identified as promising agents for the treatment of patients with relapsed classical HL. Brentuximab vedotin and panobinostat are currently being examined in clinical trials seeking potential approval by the FDA. If approved, these compounds will be incorporated with conventional chemotherapy regimens that will likely change the standard of care for this disease. As more novel drugs are identified, future investigations should focus on identifying predictive markers that will lead to a more personalized therapeutic approach.

References

Abrahamsen AF, Loge JH, Hannisdal E et al (1999) Late medical sequelae after therapy for supradiaphragmatic Hodgkin's disease. Acta Oncol 38(4):511–515

Adams J (2003) Potential for proteasome inhibition in the treatment of cancer. Drug Discov Today 8:307–315

Adams MJ, Lipsitz SR, Colan SD et al (2004) Cardiovascular status in long-term survivors of Hodgkin's disease treated with chest radiotherapy. J Clin Oncol 22(15):3139–3148

Alber G, Kim KM, Weiser P et al (1993) Molecular mimicry of the antigen receptor signalling motif by transmembrane proteins of the Epstein-Barr virus and the bovine leukaemia virus. Curr Biol 3(6):333–339

Ansell SM, Horwitz SM, Engert A (2007) Phase I/II study of an anti-CD30 monoclonal antibody (MDX-60) in Hodgkin's lymphoma and anaplastic large-cell lymphoma. J Clin Oncol 25:2764–2769

Bargou RC, Leng C, Krappmann D (1996) High-level nuclear NF-kappa B and Oct-2 is a common feature of cultured Hodgkin/Reed-Sternberg cells. Blood 87:4340–4347

Bargou RC, Emmerich F, Krappmann D (1997) Constitutive nuclear factor-kappaB-RelA activation is required for proliferation and survival of Hodgkin's disease tumor cells. J Clin Invest 100:2961–2969

Barth TFE, Martin-Subero JI, Joos S et al (2003) Gains of 2p involving the REL locus correlate with nuclear c-Rel protein accumulation in neoplastic cells of classical Hodgkin lymphoma. Blood 101(9):3681–3686

Baud V, Karin M (2009) Is NF-kappaB a good target for cancer therapy? Hopes and pitfalls. Nat Rev Drug Discov 8:33–40

Beckers T, Burkhardt C, Wieland H (2007) Distinct pharmacological properties of second generation HDAC inhibitors with the benzamide or hydroxamate head group. Int J Cancer 121:1138–1148

Blum KA, Johnson JL, Niedzwiecki D et al (2007) Single agent bortezomib in the treatment of relapsed and refractory Hodgkin lymphoma: cancer and leukemia Group B protocol 50206. Leuk Lymphoma 48:1313–1319

Bolden JE, Peart MJ, Johnstone RW (2006) Anticancer activities of histone deacetylase inhibitors. Nat Rev Drug Discov 5:769–784

Boll B, Fuchs M, Reiners KS, Engert A, Borchmann P (2010) Lenalidomide in patients with relapsed or refractory Hodgkin lymphoma. Blood (ASH Ann Meet Abstr) 116(21):2828

Borchmann P, Diehl V, Goergen H et al (2009) Combined modality treatment with intensified chemotherapy and dose-reduced involved field radiotherapy in patients with early unfavourable Hodgkin Lymphoma (HL): final analysis of the German Hodgkin Study Group (GHSG) HD11 trial. Blood (ASH Ann Meet Abstr) 114(22):717

Brierley JD, Rathmell AJ, Gospodarowicz MK et al (1998) Late effects of treatment for early-stage Hodgkin's disease. Br J Cancer 77(8):1300–1310

Brogdon JL, Xu Y, Szabo SJ (2007) Histone deacetylase activities are required for innate immune cell control of Th1 but not Th2 effector cell function. Blood 109:1123–1130

Brune V, Tiacci E, Pfeil I et al (2008) Origin and pathogenesis of nodular lymphocyte-predominant Hodgkin lymphoma as revealed by global gene expression analysis. J Exp Med 205(10):2251–2268

Buglio D, Georgiakis GV, Hanabuchi S (2008) Vorinostat inhibits STAT6-mediated TH2 cytokine and TARC production and induces cell death in Hodgkin lymphoma cell lines. Blood 112:1424–1433

Buglio D, Mamidipudi V, Khaskhely NM et al (2010) The class-I HDAC inhibitor MGCD0103 induces apoptosis in Hodgkin lymphoma cell lines and synergizes with proteasome inhibitors by an HDAC6-independent mechanism. Br J Haematol 151(4):387–396

Canellos GP, Niedzwiecki D (2002) Long-term follow-up of Hodgkin's disease trial. N Engl J Med 346(18):1417–1418

Chemnitz JM, Driesen J, Classen S et al (2006) Prostaglandin E2 impairs CD4+ T cell activation by inhibition of lck: implications in Hodgkin's lymphoma. Cancer Res 66(2):1114–1122

Chen R, Gopal AK, Smith SE et al (2010) Results of a pivotal phase 2 study of brentuximab vedotin (SGN-35) in patients with relapsed or refractory Hodgkin lymphoma. Blood (ASH Ann Meet Abstr) 116(21):283

Chiu A, Xu W, He B et al (2007) Hodgkin lymphoma cells express TACI and BCMA receptors and generate survival and proliferation signals in response to BAFF and APRIL. Blood 109(2):729–739

Copeland AR, Cao Y, Fanale M et al (2009) Final report of a phase-II study of rituximab plus ABVD for patients with newly diagnosed advanced stage classical Hodgkin lymphoma.: results of long follow up and comparison to institutional historical data. Blood (ASH Ann Meet Abstr) 114(22):1680

Corazzelli G, Frigeri F, Marcacci G et al (2009) Rituximab plus gemcitabine, ifosfamide, oxaliplatin (R-GIFOX) as salvage therapy for recurrent Hodgkin lymphoma. ASCO Meet Abstr 27(15 S):8579

Del Bufalo D, Ciuffreda L, Trisciuoglio D (2006) Antiangiogenic potential of the mammalian target of rapamycin inhibitor temsirolimus. Cancer Res 66:5549–5554

Dickinson M, Ritchie D, DeAngelo DJ et al (2009) Preliminary evidence of disease response to the pan deacetylase inhibitor panobinostat (LBH589) in refractory Hodgkin Lymphoma. Br J Haematol 147(1):97–101

Dutton A, Reynolds GM, Dawson CW, Young LS, Murray PG (2005) Constitutive activation of phosphatidyl-inositide 3 kinase contributes to the survival of Hodgkin's lymphoma cells through a mechanism involving Akt kinase and mTOR. J Pathol 205:498–506

Emmerich F, Theurich S, Hummel M et al (2003) Inactivating I kappa B epsilon mutations in Hodgkin/Reed-Sternberg cells. J Pathol 201(3):413–420

Engert A, Diehl V, Franklin J et al (2009) Escalated-dose BEACOPP in the treatment of patients with advanced-stage Hodgkin's lymphoma: 10 years of follow-up of the GHSG HD9 study. J Clin Oncol 27(27):4548–4554

Engert A, Plutschow A, Eich HT et al (2010a) Reduced treatment intensity in patients with early-stage Hodgkin's lymphoma. N Engl J Med 363(7):640–652

Engert A, Borchmann P, Pluetschow A et al (2010b) Dose-escalation with BEACOPP escalated is superior to ABVD in the combined-modality treatment of early unfavorable Hodgkin lymphoma: final analysis of the German Hodgkin Study Group (GHSG) HD14 trial. Blood (ASH Ann Meet Abstr) 116(21):765

Fanale M, Bartlett NL, Forero-Torres A et al (2009) The antibody-drug conjugate brentuximab vedotin (SGN-35) induced multiple objective responses in patients with relapsed or refractory CD30-positive lymphomas in a phase 1 weekly dosing study. Blood (ASH Ann Meet Abstr) 114:2731

Fanale M, Fayad L, Pro B et al (2011) Phase I study of bortezomib plus ICE (BICE) for the treatment of relapsed/refractory Hodgkin lymphoma. Br J Haematol 154:284–286

Federico M, Luminari S, Iannitto E et al (2009) ABVD compared with BEACOPP compared with CEC for the initial treatment of patients with advanced Hodgkin's lymphoma: results from the HD2000 Gruppo Italiano per lo Studio dei Linfomi Trial. J Clin Oncol 27(5):805–811

Fehniger TA, Larson S, Trinkaus K et al (2009) A phase II multicenter study of lenalidomide in relapsed or refractory classical Hodgkin lymphoma. Blood (ASH Ann Meet Abstr) 114(22):3693

Fiumara P, Snell V, Li Y (2001) Functional expression of receptor activator of nuclear factor kappaB in Hodgkin disease cell lines. Blood 98:2784–2790

Forero-Torres A, Leonard JP, Younes A et al (2009) A Phase II study of SGN-30 (anti-CD30 mAb) in Hodgkin lymphoma or systemic anaplastic large cell lymphoma. Br J Haematol 146(2):171–179

Franke TF (2008) PI3K/Akt: getting it right matters. Oncogene 27:6473–6488

Gandhi MK, Moll G, Smith C et al (2007) Galectin-1 mediated suppression of Epstein-Barr virus specific T-cell immunity in classic Hodgkin lymphoma. Blood 110(4):1326–1329

Georgakis GV, Li Y, Rassidakis GZ et al (2006a) Inhibition of the phosphatidylinositol-3 kinase/Akt promotes G1 cell cycle arrest and apoptosis in Hodgkin lymphoma. Br J Haematol 132:503–511

Georgakis GV, Yazbeck VY, Li Y, Younes A (2006b) Preclinical rationale for therapeutic targeting of mTOR by CC-I779 and rapamycin in Hodgkin lymphoma [abstract]. Proc ASCO 24:10070

Georgakis GV, Yazbeck VY, Li Y, Younes A (2006c) The mTOR inhibitor temsirolimus (CCI-779) induces cell cycle arrest and autophagy in Hodgkin lymphoma (HL) cell lines and enhances the effect of the PI3-kinase inhibitor LY294002 [abstract]. Blood 108:2259

Glozak MA, Seto E (2007) Histone deacetylases and cancer. Oncogene 26:5420–5432

Gobbi PG, Levis A, Chisesi T et al (2005) ABVD versus modified stanford V versus MOPPEBVCAD with optional and limited radiotherapy in intermediate- and advanced-stage Hodgkin's lymphoma: final results of a multicenter randomized trial by the Intergruppo Italiano Linfomi. J Clin Oncol 23(36):9198–9207

Gordon LI, Hong F, Fisher RI et al (2010) A randomized phase III trial of ABVD vs. Stanford V +/− radiation therapy in locally extensive and advanced stage Hodgkin's lymphoma: an intergroup study coordinated by the Eastern Cooperatve Oncology Group (E2496). Blood (ASH Ann Meet Abstr) 116(21):415

Hansmann ML, Weiss LM, Stein H, Harris NL, Jaffe ES (eds) (1999) Pathology of lymphocyte predominance Hodgkin's disease. Lippincott Williams Wilkins, Philadelphia, pp 168–180

Heider U, Kaiser M, Sterz J (2006) Histone deacetylase inhibitors reduce VEGF production and induce growth suppression and apoptosis in human mantle cell lymphoma. Eur J Haematol 76:42–50

Hodgson DC, Gilbert ES, Dores GM et al (2007) Long-term solid cancer risk among 5-year survivors of Hodgkin's lymphoma. J Clin Oncol 25(12):1489–1497

Horning S, Fanale M, deVos S (2008) Defining a population of Hodgkin lymphoma patients for novel therapeutics: an international effort [abstract]. Ann Oncol 20:118

Hoskin PJ, Lowry L, Horwich A et al (2009) Randomized comparison of the stanford V regimen and ABVD in the treatment of advanced Hodgkin's lymphoma: United Kingdom National Cancer Research Institute Lymphoma Group Study ISRCTN 64141244. J Clin Oncol 27(32):5390–5396

Ihle NT, Powis G (2009) Take your PIK: phosphatidylinositol 3-kinase inhibitors race through the clinic and toward cancer therapy. Mol Cancer Ther 8:1–9

Inoue S, Leitner WW, Golding B, Scott D (2006) Inhibitory effects of B cells on antitumor immunity. Cancer Res 66(15):7741–7747

Jemal A, Siegel R, Ward E et al (2009) Cancer statistics, 2009. CA Cancer J Clin 59(4):225–249

Jemal A, Bray F, Center MM et al (2011) Global cancer statistics. CA Cancer J Clin 61(2):69–90

Johnston PB, Inwards DJ, Colgan JP et al (2010) A phase II trial of the oral mTOR inhibitor everolimus in relapsed Hodgkin lymphoma. Am J Hematol 85(5):320–324

Jona A, Younes A (2010) Novel treatment strategies for patients with relapsed classical Hodgkin lymphoma. Blood Rev 24(6):233–238

Jona A, Khaskhely N, Buglio D et al (2010) The histone deacetylase inhibitor entinostat (SNDX-275) targets Hodgkin lymphoma through a dual mechanism of immune modulation and apoptosis induction. Blood (ASH Ann Meet Abstr) 116(21):2850

Jones RJ, Gocke CD, Kasamon YL (2009) Circulating clonotypic B cells in classic Hodgkin lymphoma. Blood 113:5920–5926

Joos S, Kupper M, Ohl S et al (2000) Genomic imbalances including amplification of the tyrosine kinase gene JAK2 in CD30+ Hodgkin cells. Cancer Res 60(3):549–552

Jucker M, Sudel K, Horn S (2002) Expression of a mutated form of the p85alpha regulatory subunit of phosphatidylinositol 3-kinase in a Hodgkin's lymphoma-derived cell line (CO). Leukemia 16:894–901

Jundt F, Raetzel N, Muller C (2005) A rapamycin derivative (everolimus) controls proliferation through down-regulation of truncated CCAAT enhancer binding protein beta and NF-{kappa} B activity in Hodgkin and anaplastic large cell lymphomas. Blood 106:1801–1807

Jungnickel B, Staratschek-Jox A, Brauninger A et al (2000) Clonal deleterious mutations in the I{kappa} b{alpha} gene in the malignant cells in Hodgkin's lymphoma. J Exp Med 191(2):395–402

Kanzler H, Kuppers R, Hansmann ML, Rajewsky K (1996) Hodgkin and Reed-Sternberg cells in Hodgkin's disease represent the outgrowth of a dominant tumor clone derived from (crippled) germinal center B cells. J Exp Med 184(4):1495–1505

Kapp U, Yeh WC, Patterson B (1999) Interleukin 13 is secreted by and stimulates the growth of Hodgkin and Reed-Sternberg cells. J Exp Med 189:1939–1946

Kato M, Sanada M, Kato I (2009) Frequent inactivation of A20 in B-cell lymphomas. Nature 459:712–716

Khan N, Jeffers M, Kumar S (2008) Determination of the class and isoform selectivity of small-molecule histone deacetylase inhibitors. Biochem J 409:581–589

Kilger E, Kieser A, Baumann M, Hammerschmidt W (1998) Epstein-Barr virus-mediated B-cell proliferation is dependent upon latent membrane protein 1, which simulates an activated CD40 receptor. EMBO J 17(6):1700–1709

Kirschbaum MH, Goldman BH, Zain JM (2007) Vorinostat (suberoylanilide hydroxamic acid) in relapsed or refractory Hodgkin lymphoma: SWOG 0517 [abstract]. Blood 110:2574

Küppers R (2009) The biology of Hodgkin's lymphoma. Nat Rev Cancer 9:15–27

Kuppers R, Rajewsky K (1998) The origin of Hodgkin and Reed/Sternberg cells in Hodgkin's disease. Annu Rev Immunol 16:471–493

Kuruvilla J, Taylor D, Wang L et al (2008) Phase II trial of lenalidomide in patients with relapsed or refractory Hodgkin lymphoma [abstract]. Blood 112:3052

Lemoine M, Buglio D, Jona A et al (2010) The pan-deacetylase inhibitor panobinostat downregulates HIF-1{alpha} and VEGF and, synergizes with everolimus in Hodgkin lymphoma cell lines. Blood (ASH Ann Meet Abstr) 116(21):2851

Martin WG, Ristow KM, Habermann TM et al (2005) Bleomycin pulmonary toxicity has a negative impact on the outcome of patients with Hodgkin's lymphoma. J Clin Oncol 23(30):7614–7620
Martin-Subero JI, Wlodarska I, Bastard C et al (2006) Chromosomal rearrangements involving the BCL3 locus are recurrent in classical Hodgkin and peripheral T-cell lymphoma. Blood 108(1):401–403
Matasar MJ, McCallen LN, Riedel ER et al (2009) Late mortality and morbidity of patients with Hodgkin lymphoma treated in adulthood. J Clin Oncol (ASCO Meet Abstr) 27(15S):8547
Mendler JH, Kelly J, Voci S (2008) Bortezomib and gemcitabine in relapsed or refractory Hodgkin's lymphoma. Ann Oncol 19:1759–1764
Minucci S, Pelicci PG (2006) Histone deacetylase inhibitors and the promise of epigenetic (and more) treatments for cancer. Nat Rev Cancer 6:38–51
Morrison JA, Gulley ML, Pathmanathan R, Raab-Traub N (2004) Differential signaling pathways are activated in the Epstein-Barr virus-associated malignancies nasopharyngeal carcinoma and Hodgkin lymphoma. Cancer Res 64:5251–5260
Mottok A, Renne C, Willenbrock K, Hansmann M-L, Brauninger A (2007) Somatic hypermutation of SOCS1 in lymphocyte-predominant Hodgkin lymphoma is accompanied by high JAK2 expression and activation of STAT6. Blood 110(9):3387–3390
Nagel S, Scherr M, Quentmeier H (2005) HLXB9 activates IL6 in Hodgkin lymphoma cell lines and is regulated by PI3K signalling involving E2F3. Leukemia 19:841–846
Newcom SR, Gu L (1995) Transforming growth factor beta 1 messenger RNA in Reed-Sternberg cells in nodular sclerosing Hodgkin's disease. J Clin Pathol 48(2):160–163
Ng AK, Bernardo MVP, Weller E et al (2002a) Second malignancy after Hodgkin disease treated with radiation therapy with or without chemotherapy: long-term risks and risk factors. Blood 100(6):1989–1996
Ng AK, Bernardo MP, Weller E et al (2002b) Long-term survival and competing causes of death in patients with early-stage Hodgkin's disease treated at age 50 or younger. J Clin Oncol 20(8):2101–2108
Oflazoglu E, Kissler KM, Sievers EL, Grewal IS, Gerber HP (2008) Combination of the anti-CD30-auristatin-E antibody-drug conjugate (SGN-35) with chemo-therapy improves antitumour activity in Hodgkin lymphoma. Br J Haematol 142:69–73
Oki Y, Younes A (2010) Does rituximab have a place in treating classic hodgkin lymphoma? Curr Hematol Malig Rep 5(3):135–139
Oki Y, Pro B, Fayad LE et al (2008) Phase 2 study of gemcitabine in combination with rituximab in patients with recurrent or refractory Hodgkin lymphoma. Cancer 112(4):831–836
Poppema S (2005) Immunobiology and pathophysiology of Hodgkin lymphomas. Hematology 1:231–238
Renne C, Willenbrock K, Martin-Subero JI (2007) High expression of several tyrosine kinases and activation of the PI3K/AKT pathway in mediastinal large B cell lymphoma reveals further similarities to Hodgkin lymphoma. Leukemia 21:780–787
Riester D, Hildmann C, Grunewald S, Beckers T, Schwienhorst A (2007) Factors affecting the substrate specificity of histone deacetylases. Biochem Biophys Res Commun 357:439–445
Scheeren FA, Diehl SA, Smit LA et al (2008) IL-21 is expressed in Hodgkin lymphoma and activates STAT5: evidence that activated STAT5 is required for Hodgkin lymphomagenesis. Blood 111(9):4706–4715
Schmitz R, Hansmann ML, Bohle V (2009) TNFAIP3 (A20) is a tumor suppressor gene in Hodgkin lym-phoma and primary mediastinal B cell lymphoma. J Exp Med 206:981–989
Schwering I, Brauninger A, Klein U (2003) Loss of the B-lineage-specific gene expression program in Hodgkin and Reed-Sternberg cells of Hodgkin lymphoma. Blood 101:1505–1512
Shichijo S, Yamada A, Sagawa K (1996) Induction of MAGE genes in lymphoid cells by the demethylating agent 5-aza-2′-deoxycytidine. Jpn J Cancer Res 87:751–756
Skinnider BF, Elia AJ, Gascoyne RD (2001) Interleukin 13 and interleukin 13 receptor are frequently expressed by Hodgkin and Reed-Sternberg cells of Hodgkin lymphoma. Blood 97:250–255
Staudt LM (2000) The molecular and cellular origins of Hodgkin's disease. J Exp Med 191:207–212

Steidl C, Lee T, Shah SP et al (2010) Tumor-associated macrophages and survival in classic Hodgkin's lymphoma. N Engl J Med 362(10):875–885

Sureda A, Younes A, Ben-Yehuda D et al (2010) Final analysis: phase II study of oral panobinostat in relapsed/refractory Hodgkin lymphoma patients following autologous hematopoietic stem cell transplant. Blood (ASH Ann Meet Abstr) 116(21):419

Tan TT, Coussens LM (2007) Humoral immunity, inflammation and cancer. Curr Opin Immunol 19(2):209–216

Tempescul A, Ianotto JC, Guillerm G, Berthou C (2009) ESAP-lenalidomide – a highly active regimen in refractory or relapsed Hodgkin's lymphoma. Blood (ASH Ann Meet Abstr) 114(22):4797

Thomas J, Ferme C, Noordijk E (2007) Results of the EORTC-GELA H9 randomized trials: the H9F trial (comparing 3 radiation dose levels) and H9-U trial (comparing 3 chemotherapeutic schemes) in patients with favorable or unfavorable early stage Hodgkin's lymphoma. Hematologica 92:27

Ushmorov A, Ritz O, Hummel M (2004) Epigenetic silencing of the immunoglobulin heavy-chain gene in classical Hodgkin lymphoma-derived cell lines contributes to the loss of immunoglobulin expression. Blood 104:3326–3334

Ushmorov A, Leithauser F, Sakk O (2006) Epigenetic processes play a major role in B-cell-specific gene silencing in classical Hodgkin lymphoma. Blood 107:2493–2500

van Rijswijk RE, Verbeek J, Haanen C et al (1987) Major complications and causes of death in patients treated for Hodgkin's disease. J Clin Oncol 5(10):1624–1633

Wang S, Yan-Neale Y, Cai R, Alimov I, Cohen D (2006) Activation of mitochondrial pathway is crucial for tumor selective induction of apoptosis by LAQ824. Cell Cycle 5:1662–1668

Wedgwood AR, Fanale MA, Fayad LE (2007) Rituximab+ABVD improves event-free survival (EFS) in patients with classical Hodgkin lymphoma in all International Prognostic Score (IPS) groups and in patients who have PET positive disease after 2–3 cycles of therapy [abstract]. Blood 110:215

Weniger MA, Melzner I, Menz CK et al (2006) Mutations of the tumor suppressor gene SOCS-1 in classical Hodgkin lymphoma are frequent and associated with nuclear phospho-STAT5 accumulation. Oncogene 25(18):2679–2684

Yamamoto R, Nishikori M, Kitawaki T et al (2008) PD-1-PD-1 ligand interaction contributes to immunosuppressive microenvironment of Hodgkin lymphoma. Blood 111(6):3220–3224

Yazbeck VY, Buglio D, Georgakis GV (2008) Temsirolimus downregulates p21 without altering cyclin D1 expression and induces autophagy and synergizes with vorinostat in mantle cell lymphoma. Exp Hematol 36:443–450

Younes A, Aggarwall BB (2003) Clinical implications of the tumor necrosis factor family in benign and malignant hematologic disorders. Cancer 98:458–467

Younes A, Carbone A (1999) CD30/CD30 ligand and CD40/ CD40 ligand in malignant lymphoid disorders. Int J Biol Markers 14:135–143

Younes A, Garg A, Agarwal BB (2003a) Nuclear transcription factor-kappa B in Hodgkin's disease. Leuk Lymphoma 44:929–935

Younes A, Romaguera J, Hagemeister F (2003b) A pilot study of rituximab in patients with recurrent, classic Hodgkin disease. Cancer 98:310–314

Younes A, Pro B, Fayad L (2006) Experience with bortezomib for the treatment of patients with relapsed classical Hodgkin lymphoma. Blood 107:1731–1732

Younes A, Bociek RG, Kuruvilla J et al (2010a) Mocetinostat (MGCD0103), an isotype-selective histone deacetylase (HDAC) inhibitor, produces clinical responses in relapsed/refractory Hodgkin lymphoma (HL): update from a phase II clinical study. Blood (ASH Ann Meet Abstr) 116(21):1763

Younes A, Hernandez F, Bociek RG et al (2010b) ENGAGE- 501: phase 2 study investigating the role of epigenetic therapy with entinostat (SNDX-275) in relapsed and refractory Hodgkin's lymphoma (HL), interim results. Blood (ASH Ann Meet Abstr) 116(21):3959

Younes A, Copeland A, Fanale MA et al (2010c) Phase I/II study of the novel combination of panobinostat (LBH589) and everolimus (RAD001) in relapsed/refractory Hodgkin and non-Hodgkin lymphoma. Blood (ASH Ann Meet Abstr) 116(21):3964

Younes A, Bartlett NL, Leonard JP et al (2010d) Brentuximab vedotin (SGN-35) for relapsed CD30-positive lymphomas. N Engl J Med 363(19):1812–1821
Zheng B, Fiumara P, Li YV (2003) MEK/ERK pathway is aberrantly active in Hodgkin disease: a signaling pathway shared by CD30CD40, and RANK that regulates cell proliferation and survival. Blood 102:1019–1027
Zheng B, Georgakis GV, Li Y (2004) Induction of cell cycle arrest and apoptosis by the proteasome inhibitor PS-341 in Hodgkin disease cell lines is independent of inhibitor of nuclear factor-kappaB mutations or activation of the CD30, CD40, and RANK receptors. Clin Cancer Res 10:3207–3215
Zheng Y, Collins SL, Lutz MA (2007) A role for mammalian target of rapamycin in regulating T cell activation versus anergy. J Immunol 178:2163–2170

Myelodysplastic Syndromes

Grant E. Nybakken and Adam Bagg

The myelodysplastic syndromes (MDS) are clonal hematopoietic stem cell disorders typically characterized by peripheral cytopenia(s) despite marrow hypercellularity, with associated morphologic dysplasia in one or more myeloid lineages. In their initial phases they display ineffective hematopoiesis; later, maturation arrest develops, resulting in the progression to acute myeloid leukemia in a sizable minority of patients. While pediatric MDS is increasingly recognized, they are prototypically diseases of the elderly with a majority (~85%) of the affected over the age of 60 years and a median age at diagnosis of 76 years. Officially, the incidence is relatively low at 3.5 per 100,000 overall (Rollison et al. 2008); however, this is likely an underestimate, particularly since the incidence is gradually increasing with the average age of the United States population. Indeed, by the age of 70 it now approaches 75 per 100,000 (Cogle et al. 2011). Men are affected at a somewhat higher rate than women. There is no apparent geographical bias to the distribution of MDS; on an age-adjusted basis, the incidence is similar throughout the world, although it should be noted that these studies were performed primarily in Western nations.

The World Health Organization (WHO) classifies MDS into subgroups with distinct characteristics and varying rates of progression to acute myeloid leukemia (Table 1). These subgroups include high risk (refractory anemia with excess blasts-2 [RAEB-2]), intermediate risk (refractory cytopenia with multilineage dysplasia [RCMD] with or without ringed sideroblasts, and refractory anemia with excess blasts-1 [RAEB-1]) and low risk (refractory cytopenia with unilineage dysplasia [RCUD], 5q- syndrome and refractory anemia with ringed sideroblasts [RARS]). As seen in other chapters of this book, many hematologic disorders have well-defined

G.E. Nybakken, M.D., Ph.D. • A. Bagg, M.D. (✉)
Department of Pathology and Laboratory Medicine, University of Pennsylvania, 7.103 Founders Pavilion 3400 Spruce Street, Philadelphia, PA 19104-4283, USA
e-mail: adambagg@mail.med.upenn.edu

J. Tao and E. Sotomayor (eds.), *Hematologic Cancers: From Molecular Pathobiology to Targeted Therapeutics*, Cancer Growth and Progression 14, DOI 10.1007/978-94-007-5028-9_11,

Table 1 WHO 2008 classification of MDS: Criteria and clinical associations

Disease	Blood findings	Bone marrow findings	Frequency (%)	Median survival	Rate of progression to AML
Refractory cytopenia with unilineage dysplasia (RCUD), Refractory anemia (RA), Refractory neutropenia (RN), refractory thrombocytopenia (RT)	Unicytopenia or bicytopenia No or rare blasts (<1%)	Unilineage dysplasia: ≥10% of the cells in one myeloid lineage <5% blasts <15% of erythroid precursors are ring sideroblasts	10–20	66 months	2% at 5 years
Refractory anemia with ring sideroblasts (RARS)	Anemia No blasts	≥15% of erythroid precursors are ring sideroblasts Erythroid dysplasia only <5% blasts	3–11	69–108 months	1–2%
Refractory cytopenia with multilineage dysplasia (RCMD)	Cytopenia(s) No or rare blasts (<1%) No Auer rods $<1 \times 10^9$/L Monocytes	Dysplasia in ≥10% of the cells in > two myeloid lineages <5% blasts in marrow No Auer rods ±15% ring sideroblasts	30	30 months	10% at 2 years

Refractory anemia with excess blasts-1 (RAEB-1)	Cytopenia(s) <5% blasts No Auer rods <1 × 10^9/L monocytes	Unilineage or multilineage dysplasia 5–9% blasts No Auer rods	40	16 months	~25%
Refractory anemia with excess blasts-2 (RAEB-2)	Cytopenia(s) 5–19% blasts Auer rods += <1 × 10^9/L monocytes	Unilineage or multilineage dysplasia 10–19% blasts Auer rods ±	Unknown	9 months	~33%
Myelodysplastic syndrome – unclassified (MDS-U)	Cytopenias ≤1% blasts	Unequivocal dysplasia in less than 10% of cells in one or more myeloid cell lines when accompanied by a cytogenetic abnormality considered as presumptive evidence for a diagnosis of MDS <5% blasts	Unknown	Unknown	Unknown
MDS associated with isolated del(5q)	Anemia Usually normal or increased platelet count No or rare blasts (<1%)	Normal to increased megakaryocytes with hypolobated nuclei <5% blasts Isolated del(5q) cytogenetic abnormality No Auer rods	5–10	145 months	<10%

Adapted from WHO 2008 (Swerdlow et al. 2008)

or extensively described genetic and molecular alterations that shape the disease phenotype, such as the *BCR-ABL1* fusion in chronic myelogenous leukemia and *JAK2* mutations in myeloproliferative neoplasms such as polycythemia vera. By contrast, although associated with a spectrum of cytogenetically detectable and typically numeric chromosomal abnormalities, the molecular basis of MDS remains largely unclear. The diagnosis rests upon clinical/hematological and morphological features, requiring identification of dysplasia within the erythroid, myeloid or megakaryocytic lineages, in addition to karyotypic abnormalities, albeit evident in only ~50% of de novo cases. Although the molecular underpinnings of MDS are beginning to be elucidated, unifying alterations remain elusive. Regardless, thc distinct behaviors of the noted subtypes are likely due to genetic and epigenetic alterations in hematopoietic precursors. Understanding the development of the disease at a mechanistic level will aid in diagnosis, determination of prognosis, and in the generation of novel, directed therapies for MDS. To this end, this chapter will review the current cellular and genetic factors believed to be implicated in the genesis of MDS, followed by a description of both current therapies and newer directed therapies in development and their mechanisms.

Cellular Features of MDS

MDS has been thought to represent, iteratively, a group of diseases characterized by varying degrees of ineffective hematopoiesis, impaired differentiation and abnormal proliferation. In addition to the quantitative defects (cytopenias) of MDS, there are marked qualitative defects in MDS, both in terms of characteristic microscopically detectable morphologic changes (as detailed in Table 2) and defective function (e.g. diminished platelet and neutrophil function). These ineffective end functions of mature blood components suggest impaired differentiation as the principal defect in MDS; however, the hypercellular marrow of MDS suggests that the main disease process is instead atypical and ineffective proliferation. Ultimately, a variety of factors are currently believed to contribute to MDS, and are discussed below.

Impaired Apoptosis

Initial studies demonstrated that the development of MDS is associated with upregulated apoptosis (Raza et al. 1995a, b). Early, low-risk MDS is typified by markedly increased apoptosis in bone marrow cells, including CD34+ stem cells (Bouscary et al. 1997). Mechanistically, this apoptosis may simply be a physiologic response to the abnormal differentiation of hematopoietic precursors. With progression to higher-risk types of MDS, mutations are thought to accumulate, overcoming apoptotic signals, shifting the balance toward increased proliferation (Bouscary et al. 1997; Parker et al. 1998, 2000).

Table 2 Characteristic dysplastic features

Lineage	Features
Erythroid	Nuclear features: Budding, internuclear bridging, karyorrhexis, multinuclearity and megalobastoid changes
	Cytoplasmic features: Ring sideroblasts, vacuolization and periodic acid-Schiff positivity, either diffuse or granular
Megakaryocytic	Micromegakaryocytes
	Nuclear hypolobation
	Multinucleation
Myeloid	Small or unusually large size
	Irregular nuclear hypolobation
	Irregular hypersegmentation
	Reduced granules or agranularity
	Pseudo Chediak-Higashi granules
	Auer rods

Adapted from WHO 2008 (Swerdlow et al. 2008)

Apoptosis can be induced through both intrinsic and extrinsic pathways. The extrinsic pathway begins at the cell surface with a signaling cascade often initiated by engagement of the Fas receptor (CD95) (Gersuk et al. 1996). Hematopoiesis is negatively regulated by signaling molecules in the extrinsic pathway, which are activated following binding of ligands such as tumor necrosis factor (TNF)-α, Fas and TNF-related apoptosis-inducing ligand (TRAIL) (Zang et al. 2001; Gersuk et al. 1998). These ligands are produced by CD68+ stromal macrophages in the marrow, implicating these cells in the pathogenesis of MDS (Kitagawa et al. 1997; Stifter et al. 2005). In contrast, the intrinsic pathway is generally initiated in mitochondria and is mediated through caspases and BCL2-related proteins. In early MDS, these pro-apoptotic pathways are upregulated and there is increased activity of the individual members of the various pathways (Parker et al. 1998; Davis and Greenberg 1998).

With the above observations, an examination of the extrinsic pathway demonstrated altered modulatory adaptor proteins in MDS. For example, FLIP (FLICE [FAS-associated death-domain-like IL-1β-converting enzyme]-inhibitory protein) is a protein with long and short splice variants that is thought to block caspase 8 activation. Increased expression of $FLIP_{short}$ has been associated with increased apoptosis; conversely, increased expression of $FLIP_{long}$ has been associated with decreased apoptosis. Interestingly, the levels of $FLIP_{long}$ correlate with disease progression, consistent with the observed decrease in apoptotic activity in late disease (Benesch et al. 2003; Seal et al. 2008). Furthermore, low-risk MDS stem cells appear very sensitive to Fas mediated activation of caspase 3 (Mundle et al. 1999). Unsurprisingly, the number of myeloblasts in MDS negatively correlates with the Fas expression on bone marrow stem cells (Bouscary et al. 1997). Although specific cause and effect associations are not well characterized, it appears that down regulation of Fas occurs as the disease progresses. This may hold true not only for Fas ligand but also for other pro-apoptotic signals.

TRAIL has an interesting differential effect profile. Although inducing apoptotic effects on bone marrow progenitors, it appears to do so in a selective manner. TRAIL has only limited effects on normal bone marrow cells, but induces a marked apoptosis in clonal MDS cells (Zang et al. 2001; Plasilova et al. 2002). The distinct responses may be mediated by differential surface expression of TRAIL receptors TRAIL-R1, TRAIL-R2, TRAIL-R3 and TRAIL-R4 (Zang et al. 2001).

The intrinsic signaling pathway is initiated with the release of cytochrome c from mitochondria into the cytoplasm, activating caspase 9, which then in turn cleaves (and thus activates) caspase 3. Spontaneous relocation of cytochrome c to the cytoplasm and consequent constitutive activation of caspase 3 and caspase 9 has been demonstrated in low-risk MDS (Boudard et al. 2000; Bouscary et al. 2000). Although the mechanism behind cytochrome c release is unclear, G-CSF administration inhibits cytochrome c release. This rebalancing of apoptotic signals may partially explain the increase in peripheral blood counts in low-risk MDS following G-CSF administration (Hellstrom-Lindberg et al. 1997a; Tehranchi et al. 2003).

NF-κB is an apoptosis-related transcription factor suggested to be involved in hematopoietic neoplasms. Increased expression of NF-κB is positively correlated with the expression of several anti-apoptotic genes in higher-risk MDS. Downstream targets of NF-κB, including members of the inhibitors of apoptosis family (including survivin, cIAP1, NAIP and XIAP), are markedly increased with progression to AML (Grosjean-Raillard et al. 2009; Sanz et al. 2002).

It is thus clear that apoptosis plays a critical role in the development and progression of MDS. The MDS stem cell's ability to limit apoptosis includes both the intrinsic and the extrinsic pathways. These altered apoptotic signals also may contribute to the altered differentiation seen in MDS.

Immune Dysfunction

Immune dysregulation has been proposed as an important step in the pathobiology of MDS. An autoimmune phenomenon has been suggested to result in the generation of polyclonal CD4+ T cell and oligoclonal CD8+ T cell responses both in the bone marrow and blood of some patients with MDS (Melenhorst et al. 2002). These clones are associated with development of hematopoietic suppression in MDS (Voulgarelis et al. 2004; Epperson et al. 2001; Matsutani et al. 2003). Such a hypothesis raises the possibility for immunosuppression as a potential treatment strategy in MDS patients. Effective treatment with immunosuppression requires careful patient selection. The patients in this cohort of low-risk MDS patients typically have bone marrow hypocellularity, HLA-DR15 phenotype, younger age, lower platelet counts, and a shorter duration of transfusion requirements (Sloand et al. 2008; Saunthararajah et al. 2002). Interestingly, a cytogenetic abnormality, trisomy 8, is also associated with ineffective hematopoiesis with T cell autoreactive clones (Sloand et al. 2005).

T regulatory (T_{reg}) cells with a CD4+ CD25+ FOXP3+ immunophenotype are physiologic immunosuppressive cells thought to be essential in limiting autoimmune dysfunction. As such, they have a dual role in the setting of cancer; they may induce self-tolerance while at the same time reducing anti-tumor response. These cells have recently been shown to be dysfunctional in early MDS patients where CXCR4 is down regulated, resulting in reduced homing of T_{reg} cells to the bone marrow. By contrast, T_{reg} cells are markedly increased in higher risk MDS in both bone marrow and blood, suggesting a role for T_{reg} cells in generating a microenvironment conducive to MDS progression (Kordasti et al. 2007). The expansion of T_{regs} is coincident with increased IL-17 and a Th17 phenotype, as seen in solid tumors (Bouchliou et al. 2011).

The Role of the Microenvironment

The microenvironment has been suggested to play a critical role in the development of MDS (Marcondes et al. 2009). The specific mechanisms by which the microenvironment promotes the development of MDS are uncertain. Some have proposed that the stromal cells themselves are abnormal and induce the creation of a hematopoietic clone. This theory is supported by a murine model that demonstrates the microenvironment alone is capable of generating a myeloproliferative phenotype (Walkley et al. 2007). The role of bone marrow stromal cells in MDS, however, is still controversial. Some studies have suggested that they may have inherent defects, such as cytogenetic abnormalities (Blau et al. 2007; Flores-Figueroa et al. 2002), whereas other studies support the idea that they are simply bystanders (Aizawa et al. 2000).

Another theory is that the malignant clone effects changes upon the stromal cells, developing a microenvironment that in turn supports the proliferation of malignancy. In the setting of MDS, proinflammatory cytokines such as IL-1β, IL-32 and TNF-α are upregulated and may play a role in limiting maturation and inducing apoptosis (Raza et al. 1996). These cytokines have been demonstrated to change the gene expression patterns of stromal cell lines derived from normal bone marrow (Stirewalt et al. 2008). These alterations in the stromal cells generate a pro-apoptotic environment, including expression of high levels of IL-32 and other cytokines, similar to that seen in early MDS. Contrastingly, the IL-32 level is low in myeloproliferative neoplasms, mirroring the differences in apoptosis (Marcondes et al. 2008). Therefore the microenvironment may, in part, explain the disease phenotype.

The vascular environment also plays a prominent role in the development of MDS. Compared with control patients, the bone marrow of patients with MDS has a markedly increased microvascular density. The microvascular density is also increased as the disease progresses to acute myeloid leukemia (Korkolopoulou et al. 2001). Interestingly, the microvascular density is increased beyond that of de novo acute myeloid leukemia. These changes in bone marrow vascularity are accompanied by increased expression of angiogenic factors, including vascular endothelial growth factor (VEGF), basic fibroblast growth factor (bFGF), tumor necrosis factor alpha (TNFα), hepatocyte growth factor (HGF), and angiopoietin-1 and -2 (Ang-1 and Ang-2)

(Stifter et al. 2005; Brunner et al. 2002; Alexandrakis et al. 2005). Corresponding receptors are also increased, including VEGF receptor 2 (VEGFR2) and the tyrosine kinase receptor, TIE2 (Hu et al. 2004; Keith et al. 2007). Both positive and negative changes in the expression of anti-angiogenic factor transforming growth factor-β (TGF-β) have been demonstrated. Recently activation of downstream effectors have been demonstrated to be important in the pathogenesis of MDS (Zhou et al. 2008). This suggests a role for it in inhibiting the vascular microenvironment of MDS.

The importance of these alterations in the vascular environment is highlighted by therapy driven changes in angiogenesis. A number of therapeutics, including lenalidomide, have anti-angiogenic effects that may contribute to their mechanism of action. Currently some therapeutic trials are specifically aimed at altering the microenvironment in the treatment of MDS, including those for SCIO-469.

The Genetic Basis of MDS

The genetic alterations identified in MDS run the gamut from gross karyotypic abnormalities and copy number changes to gene specific mutations targeting oncogenes and tumor suppressors, epigenetic changes such as methylation status, and acquired uniparental disomy. These may occur (presumably) de novo, while in other scenarios they are associated with known genotoxic agents. In addition, there are a number of inherited bone marrow failure syndromes that are associated with an increased risk for the development of MDS. The genetic factors and scenarios associated with the genesis of MDS as well as different testing modalities are discussed below.

DNA-Damaging Environmental Factors and Therapeutic Agents

MDS can be either de novo or secondary. Some of the de novo cases are associated with exposure to certain agents including benzene, cigarettes, solvents, pesticides, building materials, and hair dyes. Most of the agents are thought to be mutagens that allow hematopoietic precursors to accumulate genetic alterations that are permissive for the development of MDS.

Secondary MDS (therapy related-MDS, or t-MDS), on the other hand, typically follows better-defined pathways following exposure to specific types of chemotherapy (Leone et al. 2010). Although there is a correlation with ionizing radiation (Iwanaga et al. 2011), typical courses of limited field radiation therapy are associated with a low risk of t-MDS. Within the category of chemotherapy, alkylating agents and topoisomerase II inhibitors, in particular the former, have been associated with increased risk of developing MDS, in addition to AML (Pedersen-Bjergaard and Philip 1991). AML associated with alkylating agents typically evolves through a preceding MDS phase. By contrast, topoisomerase II inhibitors often lead directly to AML, although a subset of cases do have a preceding period of MDS. The rates of developing a therapy-related myeloid neoplasm with the use of these drugs is remarkably high with upwards of 10% of patients developing one within 10 years. These rates are lower for patients that are being treated for solid tumors than those treated for hematopoietic neoplasms (Pedersen-Bjergaard et al. 2006).

Alkylating agents, such as melphalan and cyclophosphamide, mechanistically damage DNA through cross-linking separate DNA strands and through methylation of DNA. A 5–7-year latency between treatment and the development of MDS is typical. Patients who develop t-MDS associated with these drugs have classical cytogenetic abnormalities (see below), including monosomy 5/5q- and/or monosomy 7/7q-. Historically, patients with t-MDS were grouped with respect to treatment (alkylating agents vs. topoisomerase II inhibitors). However, since many patients are exposed to both types of drugs simultaneously or sequentially, this separation is no longer emphasized (Pedersen-Bjergaard et al. 2006).

In addition, studies have subclassified these t-MDS cases into different molecular pathways (Pedersen-Bjergaard et al. 2006). Patients who develop –5/5q- in the absence of chromosome 7 abnormalities will often have concurrent monosomy 17/17p-, dicentric chromosomes, duplication or amplification of chromosome band 11q23 and a complex karyotype. These patients commonly have mutations within *TP53* (Christiansen et al. 2001). A second class of patients has -7/7q- abnormalities, with chromosome 5 unaffected. In contrast to the first pathway, these patients often have *CDKN4B* gene promoter methylation alterations and *RUNX1* mutations. The clinical importance of separating these two groups is uncertain, although they may suggest common pathways in the development of certain de novo cases of MDS as well.

Topoisomerase II is a gyrase that supercoils or relaxes the supercoiling of DNA through the generation of a strand breakage of double stranded DNA. Topoisomerase II inhibitors, such as doxorubicin, etoposide and teniposide, function through blocking the religation of the DNA strands, resulting in a DNA breakage. These strand breaks often result in cell death, but in rare instances the cell may be "saved" through the formation of a chromosomal translocation. These translocations may predispose to the development of MDS and/or AML. Unlike treatment with alkylating agents, the post-treatment latency for development of AML is typically only 1–3 years (Pedersen-Bjergaard et al. 2006; Christiansen et al. 2001).

The use of topoisomerase II inhibiters prototypically leads to the development of t-AML with 11q23 translocations, affecting the *MLL* gene (Bloomfield et al. 2002). By contrast, t-MDS develops along two alternative primary pathways. One pathway involves the t(3;21)(q26;q22) translocation linking the *EVI1* and *RUNX1* genes (Slovak et al. 2002). These cases typically also show monosomy 7 or 7q-. A second, rarer pathway involves balanced translocations of 11p15 involving the *NUP98* gene. *NUP98* encodes a nuclear pore complex protein; like *MLL*, it is a promiscuous translocation site that may involve numerous partners (Arai et al. 1997). Beyond these pathways, however, there are a number of patients that progress to MDS without any of these hallmark cytogenetic changes (or any cytogenetic abnormalities at all) (Pedersen-Bjergaard et al. 2006).

Inherited Bone Marrow Failure Syndromes

There are a number of constitutional genetic syndromes associated with the development of MDS, including Fanconi anemia (FA), Diamond-Blackfan syndrome (DBA), Dyskeratosis congenita (DC), and Shwachman-Diamond syndrome (SDS). Additionally, GATA1 mutations are also associated with MDS, with the mutations directly causing myelodysplasia.

Fanconi anemia (Cioc et al. 2010) is an autosomal recessive disorder (with the exception of FANCB, which is X-linked) characterized by, amongst other features, bone marrow failure, genomic instability and predisposition to hematologic and/or solid tumors. Although patients with FA are prone to the development of MDS, their diagnosis is complicated by the frequent presence of baseline dysplastic changes. The finding of dyserythropoiesis is sometimes complicated by the presence of treatment-related changes seen with frequent dosing with G-CSF and other bone marrow stimulants. MDS in FA most commonly involves only the red cell lineage. Thirteen different genes have been associated with FA, most commonly DNA repair genes. Eight of the proteins (FANCA, FANCB, FANCC, FANCE, FANCF, FANCG, FANCL and FANKM) form the FA core complex, an ubiquitin ligase. This complex ubiquitinates other FA proteins, FANCD2 and FANCI.

Diamond Blackfan anemia is an autosomal dominant disorder typified by a macrocytic anemia usually with normal leukocytes and platelets, congenital malformations and growth retardation. The vast majority (90%) display hematologic manifestations as infants, and have a greatly increased risk of MDS as well as AML and solid tumors. Nine proteins have been associated with DBA, all of which are ribosomal. However, only just over one-half (53%) of patients with DBA have an identifiable mutation. The most commonly identified mutations are within *RPS19* (25%) with *RPL5*, *RPS10*, *RPL11*, *RPL35A*, *RPS26*, *RPS24*, *RPS17* and *RPS7* being less frequently affected. The proteins encoded by these genes are all members of either the 40S or 60S ribosomal subunits, and haploinsufficiency of one of the subunits leads to improper ribosomal function and increased apoptosis in the bone marrow, findings that are similar to those seen in MDS (Choesmel et al. 2007).

Schwachman-Diamond syndrome (Hashmi et al. 2011) is an autosomal recessive syndrome affecting multiple organ systems. In addition to pancreatic and bone abnormalities, there are marked associated hematopoietic phenotypes, including cytopenia, MDS and AML. The genetic deficiency is a mutation in *SBDS* on 7q11.21. SBDS is a ribosomal maturation protein that couples GTP hydrolysis by the GTPase elongation factor-like 1 (EFL1) on the ribosome to eurkaryotic initiation factor 6 (EIF6) release. Mutations of *SBDS* lead to inhibition of this enzymatic activity, with the generation of a ribosomopathy, somewhat analogous to DBA (Finch et al. 2011).

Dyskeratosis congenita is a condition with a classic triad of dysplastic nails, oral leukoplakia and a lacy reticular hyper- or hypopigmentation of the upper chest/neck. These patients are prone to develop bone marrow failure, MDS (median age of onset of 35 years) or AML, and solid tumors. Patients with DC have a dramatically increased risk for the development of MDS (>2,000-fold). DC has been associated with mutations within six genes, *DKC1*, *TERC*, *TERT*, *TINF2*, *NHP2*, and *NOP10*, with mutations in *DKC1* and *TINF2* being both the most prevalent and severe. The inheritance of dyskeratosis congenita is dependent on the gene with *DKC1* being X-linked, TINF2 and *TERC* being autosomal dominant, *TERT* being autosomal dominant or autosomal recessive and *NHP2* and *NOP10* being autosomal recessive. These proteins are all telomere associated and patients with DC have shortened telomeres for their age. Interestingly, in light of the frequent ribosomal abnormalities noted above, DKC1, NHP2 and NOP10 are involved in processing rRNA. TERT and TERC are both involved in chromosome lengthening through the addition of TTAGGG nucleotide repeats to telomeres. *TINF2* encodes part of shelterin, the telomere protection complex. This disorder has unique aspects in regards to the relationship between telomere length and MDS. A few studies have recently begun to examine the length of telomeres in the development of de novo MDS and have found that telomere lengths often shorten preceding the development of MDS and then lengthen prior to the development of AML (Chakraborty et al. 2009).

GATA1 is a gene located on chromosome Xp11.23 encoding a GATA zinc finger transcription factor. It is highly expressed in a number of hematopoietic precursors, including mast cells, erythroid precursors, megakaryocytes and eosinophils. It plays a critical role in erythropoiesis and megakaryocytopoiesis by coordinating lineage-specific genes to induce differentiation and by inhibiting genes that promote retention of the undifferentiated state. Acquired *GATA1* mutations have been noted in Down syndrome patients who develop transient abnormal myelopoiesis and acute megakaryoblastic leukemia.

Certain inherited *GATA1* mutations (V205M, G208R, G208S, D218G and D218Y) result in severe macrothrombocytopenia and dyserythropoietic anemia. These mutations are found in a highly conserved region of *GATA1* and prevent the association with friend of GATA1 (FOG1), and with zinc fingers 1, 6 and 9, although interactions with palindromic DNA sites remain intact. V205M, G208R and D218Y mutations induce a more severe phenotype than the G208S and D218G mutations. Although all of the mutations inhibit the GATA1-FOG1 interaction, the milder mutations are thought to interfere less with the protein binding (Ciovacco et al. 2008).

The 5q- Syndrome

Within the general classification of MDS, only one is (cyto)genetically defined: the 5q- syndrome. WHO defines the 5q- syndrome as an isolated 5q- cytogenetic abnormality, associated with macrocytic anemia, a normal or elevated platelet count, unilobular megakaryocytes and a low propensity to progress to AML. Importantly, however, the chromosomal aberration 5q- is not per se limited to the 5q- syndrome MDS. In fact, it is the most common cytogenetic abnormality in MDS, found in around 15% of all cases. Interestingly, the commonly deleted regions (CDR) in "conventional" (i.e., not 5q- syndrome) MDS and 5q- syndrome MDS are subtly distinct, with 5q32-5q33.3 (CDR2) most commonly targeted in the specific entity of the 5q- syndrome while 5q31.1-5q31.3 (CDR1) is typically deleted in other forms of MDS (and AML). However, this distinction is not absolute as both regions may be deleted in these clinically and hematologically distinct scenarios.

Critical genes in the CDRs of both "conventional" MDS and 5q- syndrome have been identified. The 5q- region, however, more broadly encodes numerous proteins important for hematopoiesis, including IL-3, IL-4, IL-5, IL-9, IL-13, IL-17, GM-CSF, CSF-1R, PDGFRA and PDGFRB. Initial studies explored patients with 5q- syndrome by looking for mutations in the other allele within the deleted region. These studies proved largely unilluminating, due to the lack of mutations in the complementary allele. Additionally, uniparental disomy is not observed, suggesting the mechanism is not through biallelic inactivation of a tumor suppressor gene. Therefore, more recent studies have begun to look at haploinsufficiency as the driving mechanism for the 5q- syndrome and for other forms of MDS with 5q-. Thus far, these studies have found a number of candidate genes, including *RPS14* (5q33, CDR2), *SPARC* (5q33, CDR2), *EGR1* (5q31.2, CDR1), *CTNNA1* (5q31.2, CDR1), *CDC25C* (5q31, CDR1), *APC* (5q22, neither CDR) and *NPM1* (5q35.1, neither CDR).

RPS14, found at the distal aspect of the CDR2, encodes ribosomal protein S14, part of the 40S subunit of the ribosome. 5q- syndrome patient samples have been demonstrated to have haploinsufficiency of the RPS14 protein and reduced expression of *RPS14* through an RNAi screen leads to impaired erythropoiesis and relatively retained non-erythroid hematopoiesis (Ebert et al. 2008a; Pellagatti et al. 2008). Forced expression of RPS14 in patient derived bone marrow cells rescues erythropoiesis. Combined, these aspects suggest that *RPS14* is the leading candidate gene for the development of MDS in patients with a 5q- or monosomy 5. The ribosomopathy is generated by haploinsufficiency of RPS14 though inhibition of HDM2, resulting in erythroid specific accumulation of p53 (Barlow et al. 2010; Dutt et al. 2011). As noted above, ribosomal proteins have been linked to the generation of MDS in a number of congenital bone marrow disorders, like Diamond-Blackfan anemia and Shwachman-Diamond syndrome. Together, these findings suggest a central pathogenetic role for ribosomopathies in the development of MDS.

SPARC, another gene found at the distal aspect of the CDR2, encodes osteonectin. Osteonectin is a tumor suppressor gene that has been linked to ovarian cancer and leukemia, in addition to other cancers (Tai and Tang 2008). Osteonectin has diverse functions, including inhibiting VEGF, resulting in blocked angiogenesis (Kupprion et al. 1998). Cells from MDS patients with 5q- typically have *SPARC* transcript levels that are markedly decreased as compared with those of healthy controls (Lehmann et al. 2007; Boultwood et al. 2007). Notably, as osteonectin inhibits angiogenesis, proliferation and cellular adhesion, it has an effector profile similar to that of lenalidomide, an approved therapy for MDS (Kupprion et al. 1998). Additionally, patients with deletion of the CDR2 region (containing *SPARC*) respond well to lenalidomide. Lenalidomide treatment results in increased expression of *SPARC*, returning the transcript levels to those of normal controls (Pellagatti et al. 2007).

EGR1 is found in CDR1 on 5q and encodes a zinc-finger transcription factor. It is involved in stem cell quiescence and prevents stem cells from leaving the bone marrow compartment. Deletion or haploinsufficiency of EGR1 renders murine models more susceptible to myeloid neoplasms when treated with DNA-alkylating agents (Joslin et al. 2007).

CTNNA1, found at 5q31.2, has reduced expression in cells from patients with MDS and AML with 5q-. Additionally, overexpression of CTNNA1 in HL-60 cells reduces proliferation and increases apoptosis, perhaps generating a survival advantage (Liu et al. 2007). *CTNNA1* is methylated not only within the 5q- population but also within patients with high-risk MDS, suggesting that CTTNA1 suppression may support progression to AML (Ye et al. 2009).

CDC25C and *PP2A*, both found at 5q31, encode phosphatases that control entry into the cell cycle. As with the proteins above, they have reduced expression in cell lines with loss of 5q. Both phosphatases are inhibited by lenalidomide in the haploinsufficient state allowing cell cycle arrest and apoptosis (Wei et al. 2009).

APC is a tumor suppressor gene located at 5q23, proximal to both CDR1 and CDR2, that is commonly deleted or mutated in colorectal carcinomas. Loss of at least one copy of *APC* is found in over 95% of patients with 5q- (Pedersen 1996). Mice with heterozygous *APC* deletion have severe macrocytic anemia and dyserythropoiesis, similar to that of MDS patients with 5q- (Lane et al. 2010). In addition to its function as a tumor suppressor, reduced expression of APC can lead to bone marrow failure, suggesting a role in the hematopoietic stem cell (Wang et al. 2010).

NPM1 is found on 5q35 outside the CDRs noted above. *NPM1* mutations are frequently identified in AML and are associated with a normal karyotype and good prognosis. *NPM1* deletions are relatively uncommon in patients with 5q- (Pedersen 1996). Mice with *NPM1* haploinsufficiency develop a variety of hematopoietic malignancies, including some with a phenotype suggestive of MPD/AML (Sportoletti et al. 2008).

Karyotypic Abnormalities

Karyotypic abnormalities, that is, those detected by traditional metaphase chromosomal analysis (see Fig. 1), are found in ~50% of all cases of de novo MDS; patients with secondary MDS are more likely to harbor cytogenetic aberrations (~80%) (Olney and Le Beau 2007).

In contrast to many other hematologic neoplasms that are typified (and indeed sometimes defined) by characteristic translocations, such genetic phenomena are the exception in MDS; rather, numeric abnormalities, in particular deletions, are much more common. Interstitial deletion of chromosome 5 (5q-), the most common deletion, is found in around 15% of all cases (Haase et al. 2007). Other frequently encountered abnormalities are 7q-/monosomy 7 (~10%) and trisomy 8 (~10%) (Haase et al. 2007). Less common are loss of 18q-/monosomy 18, 20q-, monosomy 5, loss of chromosome Y, 17p-/monosomy 17/isochromosome 17q (all at about 5%) (Haase et al. 2007). Complex karyotypes (≥3 cytogenetic abnormalities) are found in 10–20% of patients with primary MDS at the time of diagnosis (Olney and

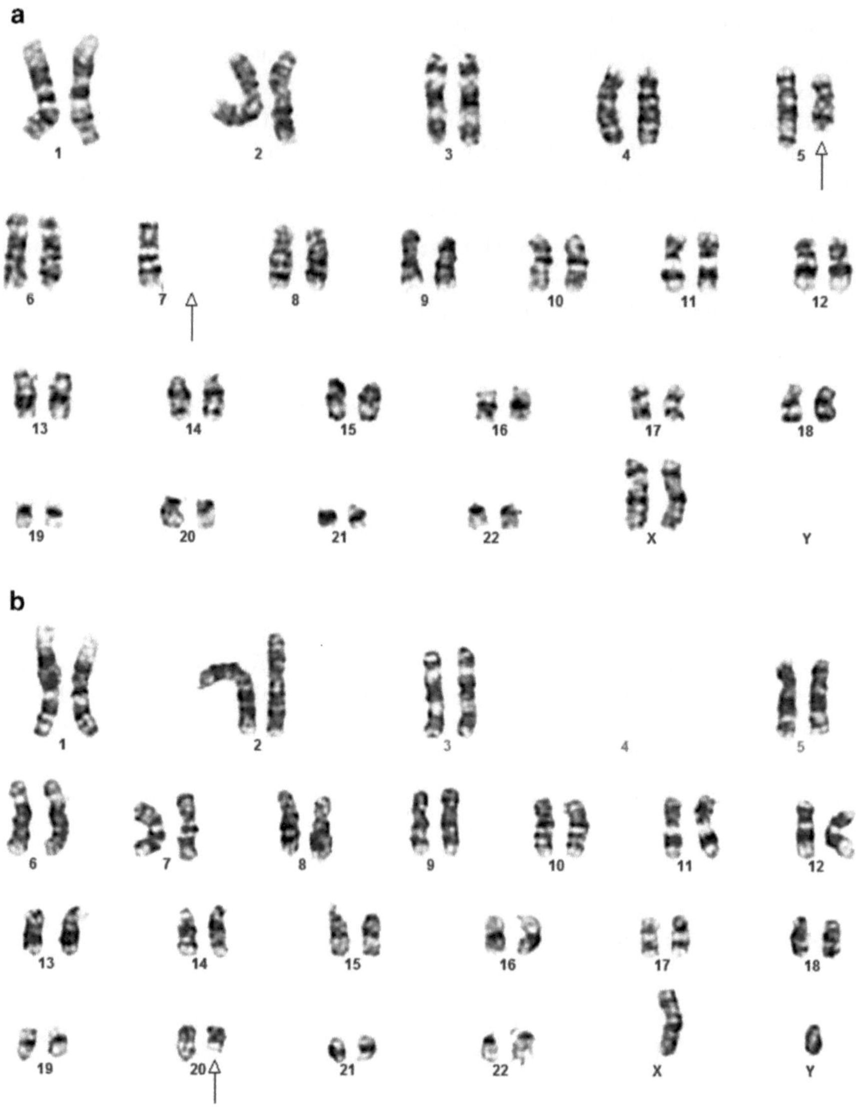

Fig. 1 Cytogenetics in MDS. A. Karyogram with the two most common aberrations in MDS, 5q- and monosomy 7, highlighted by *arrows*. B. Karyogram with a 20q-, highlighted by *arrow*, and complete loss of chromosome 4

Le Beau 2007). Although multiple translocations have been described in MDS, they are infrequent; their main utility to date is in providing presumptive evidence of MDS in patients with refractory cytopenia and an absence of dysplasia. Cytogenetic abnormalities in MDS are not static, rather they tend to accumulate with disease progression.

Table 3 Prognostically important karyotypic abnormalities in de novo MDS

Prognostic subgroup	Cytogenetic abnormality	Frequency (%)	Survival/Risk of progression to AML 25% (years)
Very good	-Y, del(11q)	4	5.4/NR
Good	Normal, del(5q), del(12q), del(20q), double including del(5q)	72	4.8/9.4
Intermediate	Del(7q), +8, +19, i(17q), any other single or double independent clones	13	2.7/2.5
Poor	-7, inv(3)/t(3q)/del(3q), double including -7/del(7q), complex: 3 abnormalities	4	1.5/1.7
Very poor	Complex: >3 abnormalities	7	0.7/0.7

MDS cytogenetic abnormalities are recognized to have an important role in disease progression with many having prognostic implications. Cytogenetics are considered in most prognostic systems, including the IPSS (Greenberg et al. 1997) and the WPSS. A retrospective analysis of a large cohort of patients suggested that the IPSS underweighted cytogenetics and that poor prognostic cytogenetics are of equivalent risk as a high blast count (Schanz et al. 2011). The recently published IPSS revision (IPSS-R) addresses these concerns, by stratifying cytogenetics into five different prognostic groupings. These prognostic groupings include very good, good, intermediate, poor and very poor (See Table 3) and have very distinct risks of progression to AML.

Other than for 5q- detailed above, there remains little mechanistic understanding of the role of these cytogenetic abnormalities in the development of MDS. It is not known whether they are primary or secondary events that are accumulated in the progression of the disease. Monosomy 7 and loss of 7q have been examined for critical genetic regions. Three commonly deleted regions have been found, but the corresponding genes remain unidentified. Trisomy 8, the only commonly seen amplification in MDS, is seen in about 10% of patients. Interestingly, however, patients with this karyotype often respond well and have sustained responses to immunomodulatory therapy. Once again, however, the molecular mechanism behind trisomy 8 remains unclear.

FISH

Fluorescence in situ hybridization (FISH) is an alternative method for determining the cytogenetics of a neoplasm. Initial smaller studies generally demonstrated a role for FISH in facilitating a diagnosis of MDS (Bernasconi et al. 2003; Romeo et al. 2002). More recently, larger studies have demonstrated that they detect around 70% of the abnormalities detected by cytogenetics (Coleman et al. 2011). Furthermore, FISH had only limited ability to detect abnormalities that are undetectable by

Table 4 Common mutations in MDS

Gene	Protein function	Frequency (%)
TET2	Alpha ketoglutarate-dependent dioxygenase	~20
RUNX1	Member of transcription factor protein complex	~15
TP53	Multiple: DNA damage repair, apoptosis	~10–15
ASXL1	Chromatin binding protein	~10–20
NRAS	GTPase signal transducer controlling cell growth	~10
CBL	Tyrosine kinase associated ubiquitin ligase	~1
EZH2	Histone methylating protein	~6

metaphase cytogenetics (2.7%) with >20 cells analyzed. The primary utility of FISH is in the setting of inadequate cell yield in cytogenetics (Coleman et al. 2011; Mallo et al. 2008; Pitchford et al. 2010; Yang et al. 2010).

Mutations

Molecular alterations occurring in specific genes in MDS are currently not well understood. Although mutations are found in about half of MDS patients, no specific mutation is present in more than about 20% of patients with MDS (Bejar et al. 2011). Patients with t-MDS have mutation rates that exceed those of de novo MDS. The genes most frequently affected by point mutations include *TET2*, *RUNX1*, *TP53*, *NRAS* and *ASXL1*, and less commonly *CBL* and *EZH2* (Table 4). Each of these genes is also commonly mutated in AML; despite this, their role in the pathogenesis of MDS is not well understood.

Mutations within *TET2* are among the most common mutations found in MDS (20%), although the mechanism by which *TET2* mutations lead to disease is unknown (Delhommeau et al. 2009; Langemeijer et al. 2009). TET2 functions as an alpha ketoglutarate-dependent dioxygenase and converts 5-methylcystosine to 5-hydroxymethylcystosine. This enzymatic activity alters the product of DNA methyl transferases (DNMTs). As is thematic below, they are not unique to MDS since mutations are also found in MPN, CMML and AML. The presence of *TET2* mutations in a spectrum of hematopoietic neoplasms with dysplasia and their association with low risk MDS suggests a role in the early development of myeloid neoplasms (Smith et al. 2010). Following progression to AML, *TET2* mutations are associated with a poor prognosis, but their prognostic role in non-transformed MDS is uncertain (Bejar et al. 2011; Smith et al. 2010; Kosmider et al. 2009). *TET2* mutations have been associated with response to azacitidine (Itzykson et al. 2011)

RUNX1 is one component of a core-binding factor protein complex that plays a central role in hematopoiesis. The gene is involved by translocation in both the t(8;21) in AML and t(12;21) in ALL. These translocations are not typically seen in MDS; rather *RUNX1* typically acquires point mutations altering the DNA- or protein-binding domains of the protein. These mutations are found in around 15% of MDS and are associated with increased risk of progression to AML (Chen et al. 2007; Dicker et al. 2010). They are found in not only MDS, but also in AML, CMML, and less commonly in MPN (Harada and Harada 2011). N-terminal mutations affecting DNA binding are found in MDS, MDS/AML, de novo AML and the familial RUNX1 disorder and are often found in therapy related MDS. C-terminal mutations affect the trans-activation domain, are typically frame shift mutations, and are specific for high risk MDS and MDS/AML (although not de novo AML or therapy related MDS). Mutations in *RUNX1* are associated with a significantly worse prognosis (Bejar et al. 2011; Harada et al. 2004).

Similar *RUNX1* point mutations are associated with an autosomal dominant disorder, familial platelet disorder with propensity to myeloid malignancy. These patients have hematopoietic abnormalities, including myeloid dysplasia, thrombocytopenia and dysfunctional platelets (Song et al. 1999). As leukemia is a late occurrence, *RUNX1* mutations are thought to represent an early event with additional mutations being required for progression (Osato 2004). Interestingly, different *RUNX1* point mutations in murine models demonstrate different hematologic phenotypes characterized by multilineage dysplasia. Those that have transcripts with early termination in the RUNT homology domain frequently have an MDS-like phenotype with more prominent myeloid dysplasia, leukocytosis and marked hepatosplenomegaly with increased rates of progression to AML with cooperative high expression of EVI1. Mutations found in the C-terminal portion of RUNX1, also associated with progression to AML, were noted to have more prominent erythroid dysplasia and pancytopenia (Watanabe-Okochi et al. 2008).

Mutations within the tumor suppressor *TP53* occur in roughly 10–15% of MDS patients and typically result in inhibition of the tumor suppressor function of the protein (Padua et al. 1998; Kita-Sasai et al. 2001; Misawa et al. 1998). These mutations are frequently found in patients with complex cytogenetics and seem to be associated with chromosomal instability (Kaneko et al. 1995). These patients are typically resistant to therapy and have a poor prognosis (Bejar et al. 2011). There are typically MDS subclones with *TP53* mutations and it is possible these subclones selectively persist post-therapy, possibly explaining instances of refractory disease (Jadersten et al. 2009).

ASXL1, additional sex-comb like-1, is mutated in 10–20% of cases of MDS, with a typical distal heterozygous frame shift mutation. Once again, these mutations are commonly found in MPN, AML and CMML (Gelsi-Boyer et al. 2009) and portend a poor prognosis (Bejar et al. 2011; Thol et al. 2011). ASXL1 is a chromatin binding protein and modulates gene expression at an epigenetic level. It interacts with both an acetyltransferase, SRC-1 (Cho et al. 2006), and a histone demethylase, LSD1 (Lee et al. 2010), to affect gene expression.

NRAS mutations are found in approximately 10% of MDS patients (Bacher et al. 2007). These mutations generally disrupt GTPase activity leading to constitutive activation of the serine/threonine kinase. These patients have increased rates of progression to AML and decreased survival (Bejar et al. 2011).

CBL, which encodes a tyrosine kinase associated ubiquitin ligase, is mutated in less than 1% of MDS but is mutated in around 10% of MPN/MDS diseases such as CMML (Makishima et al. 2009). The mutations inhibit the ubiquitin ligase activity of the protein, but are of uncertain functional and prognostic significance. Germline *CBL* mutations have recently been described in patients with juvenile myelomonocytic leukemia (Perez et al. 2010), suggesting that it plays a role in the development of this MDS/MPN.

EZH2 encodes a histone methylating protein that is mutated in approximately 6% of MDS patients. It is encoded on 7q36.1 and likely plays a role in the poor prognosis noted in MDS patients with monosomy 7 and 7q-. These mutations reduce the function of the histone 3 lysine 27 methyltransferase activity (Nikoloski et al. 2010). EZH2 and other polycomb group genes RING1 and BMI1 are often overexpressed in MDS and are also associated with a worsened prognosis (Xu et al. 2011). Why both inactivating mutations and overexpression of EZH2 is found in MDS is not currently understood.

JAK2 mutations, which have a key role in the development of MPN, are found mutated in 79% of cases of RARS-T, a provisional entity in the 2008 WHO classification scheme. These mutations, like those seen in MPNs, result in constitutive activation of tyrosine kinase functionality and are likely related to the thrombocytosis seen in this overlap MDS/MPN category, with the mutation and elevated platelet count much more MPN-like than MDS-like (Gurevich et al. 2011).

A large study looking at mutations in MDS has recently confirmed the above findings and suggested using a mutation within *TP53, EZH2, ETV6, RUNX1* or *ASXL1* as an independent prognostic factor. In addition to poor prognosis, these mutations were associated with distinct clinical phenotypes and may be, in part, responsible for the patient-to-patient variability seen in MDS. Two genes, *ETV6* and *GNAS*, cited in the study had not previously been identified as targets in MDS. ETV6 is a transcription factor that found in a number of leukemia associated translocations and GNAS is an oncogene encoding the stimulatory G-protein alpha subunit (Bejar et al. 2011).

Further sequencing studies have highlighted the numerous mutations in the RNA spliceosome in patients with MDS[22]. These studies have demonstrated high rates of mutations within U2AF35[23], ZRSR2[24], SRSF2 and SF3B1[25]. These mutations were found in 45 to 85% of myelodysplasia patients, the variation depending upon the particular subtype of disease. The role of RNA splicing is particularly interesting given the relationship of RNA machinery to MDS, with haploinsufficiency of RPS14 in isolated del(5q) syndrome and the mutations in MDS related syndromes including Shwachman-Bodian-Diamond and Diamond-Blackfan anemia. A number of other mutations have been described in genes that are also seen in AML. These mutations, however, are rarely found in MDS, and include *ATRX* (Gibbons et al. 2003), *NPM1* (Bains et al. 2011), *IDH1* (Thol et al. 2010), *IDH2* (Kosmider et al. 2010), *CEBPA* (Shih et al. 2005), *WT1* (Hosoya et al. 1998), *MLL* (Dicker et al. 2010), *PTPN11* (Loh et al. 2005), *FLT3* (Bains et al. 2011), and *KIT* (Lorenzo et al. 2006). These mutations are typically of uncertain significance, however, a number of them, including *IDH1*, *IDH2*, *FLT3*, and *KIT* are associated with more advanced disease and progression to AML. As such, these mutations may be more characteristic of the evolving AML than of the MDS itself.

High-throughput DNA sequencing may help to finally elucidate the underpinnings of MDS at a more comprehensive molecular level. These studies will be able to pick out important mutations that have been missed in the more directed searches that have been performed to date. Additionally, they will be better able to examine the non-exonic sequences that may be crucial in the development of the disease.

Mitochondrial Mutations

Mitochondria have been demonstrated to have aberrant phenotypes in the hematopoietic cells of MDS. Ultrastructurally, the mitochondria have abnormal iron deposition in addition to the abnormal leakage of apoptotic proteins (van de Loosdrecht et al. 2001). The iron deposition in ring sideroblasts is largely contained in mitochondria. These alterations may be secondary to changes in mitochondrial DNA (MtDNA). Interestingly, mutation scanning in MDS has demonstrated mutations that severely reduced expression of ABCB7, a protein responsible for transporting iron from the mitochondria to the cytoplasm and associated with inherited ringed sideroblast formation (Nikpour et al. 2010). Mutation scanning of the MtDNA of MDS patients has demonstrated mutation rates approaching 60%, with a higher incidence in older patients and those with advanced disease. The mutations do not, however, correlate with phenotype and were not enriched in particular regions of the MtDNA (Wulfert et al. 2008). Reduced expression of mitochondria encoded proteins has been demonstrated in MDS, although these changes are of uncertain significance.

Single Nucleotide Polymorphism Arrays

Single nucleotide polymorphism (SNP) arrays have extraordinary power to recognize loss of heterozygosity within tumors. SNP array analysis of MDS represents a technique to complement metaphase cytogenetics in identifying cytogenetic abnormalities. SNPs have the ability to recognize acquired uniparental disomy (aUPD) and as well as smaller regions of deletion that would be otherwise invisible by conventional cytogenetics. Roughly 75% of patients with MDS have chromosomal abnormalities detected by SNP array, increasing the typical 50% yield by conventional cytogenetics. Stated another way, upwards of 50% of patients with normal karyotypes have occult changes that are unmasked by SNP arrays. These patients also have been demonstrated to have a worse prognosis than patients with normal cytogenetic profiles. The true strength of SNP arrays will come in the analysis of the commonly deleted regions within MDS patients. Among patients with low risk MDS, a 500k SNP array demonstrated that UPD occurs in 46% of low-risk patients and 4q abnormalities are the most prominent, being seen in 25% of RARS cases, 12% of RCMD with normal cytogenetics cases, 17% of RAEB cases, and 6% of 5q− syndrome cases (Mohamedali et al. 2007). It should be noted, however that at least some of the prognostic utility of these studies is already encapsulated in conventional stratifying strategies like IPSS. Further studies will be needed to compare the results of SNP-defined losses of poor prognostic chromosomes like chromosome 7 with those identified by conventional cytogenetics. Further studies have begun to use SNPs to examine specific chromosomal regions for mutations and UPDs. One study revealed that 26% of patients had *TET2* mutations present in the vast majority of their hematopoietic compartments, including lymphoid cells. This suggests that it is a common mutation that may represent an early event in the development of some MDS (Langemeijer et al. 2009). SNP arrays are likely to become a complementary method to FISH and conventional cytogenetics in the analysis of diagnostic MDS specimens. A recent study has demonstrated the independent clinical utility of SNP analysis with its own prognostic import (Tiu et al. 2011). Additional studies have examined hypocellular MDS with respect to aplastic anemia and identified distinct cytogenetic abnormalities, suggesting a potential diagnostic use for SNP analysis in distinguishing these two disorders (Afable et al. 2011).

Comparative Genomic Hybridization

Comparative genomic hybridization has been used in a similar fashion to SNP arrays to examine copy number in MDS, albeit with a lower degree of resolution and inability to detect aUPD. Studies have demonstrated an ability to detect cryptic chromosomal duplications and losses that are of prognostic utility and thus they have been primarily been used to evaluate MDS patients with a normal karyotype. Cryptic deletions are frequently located at the sites of known common cytogenetic alterations, including 4q24 (*TET2*), 5q31.2, 7q22.1, 21q22.12. These cryptic alterations were present in nearly 40% of patients with normal cytogenetics and were associated with inferior survival (Thiel et al. 2011). These findings echo those of previous studies in patients with low-risk MDS (Starczynowski et al. 2008).

Epigenetic Changes

Epigenetic changes are modifications in chromatin altering the expression of genes without changing DNA sequence; DNA methylation and histone acetylation are two of the most prominent epigenetic modifications. These physiologic mechanisms play a role in the process of hematopoiesis; however, they are also implicated in the development of cancer, often by silencing of tumor suppressor genes.

Genes frequently have CG repeats in their promoter sequence (CpG islands) that can be methylated by DNA methyltransferases. Methylation at these sites results in reduced binding of transcription factors to the CpG islands, Reducing expression of the affected gene. Within cancer in general and MDS more specifically, there is a global reduction in methylation. Simultaneously, however, there is increased methylation at previously unmethylated regions, often corresponding to tumor suppressors and other regulators of proliferation. The increased methylation is mediated through DNMTs. Inhibition of these proteins is increasingly becoming a key treatment modality for MDS.

Alterations in DNA methylation have been demonstrated in MDS in particular. Reversion induced LIM (*RIL*) in the 5q31 region was identified as being hypermethylated in 36% of MDS patients in a search for tumor suppressors associated with MDS. In intermediate and high risk patients, approximately 50% of patients have altered methylation of the *RIL* gene with an apparent altered prognosis. Aberrant *RIL* methylation is associated with an inferior survival of 55 weeks compared to 119 weeks for a patient with typical methylation (Boumber et al. 2007). Additional genes with altered methylation in MDS include the cell-cycle regulators *CDKN2A* and *CDKN2B*. *CDKN2B* inhibits cyclin-depedent kinases 4 and 6, negatively regulating the cell cycle. Reduced expression of these genes is rarely due to deletion or mutation, but is instead due to increased methylation. Altered methylation of *CDKN2B* is associated with a worse prognosis and correlates positively with blast percentage, signaling disease progression. *CDH1*, *ER*, *HICl1* (Aggerholm et al. 2006), *SOCS1* (Wu et al. 2006; Brakensiek et al. 2005), *FHIT* (Lin et al. 2008) and *KLF11* (Potapova et al. 2010) are also noted to have altered methylation in patients with MDS and appear to correlate with disease progression and poor prognosis. Not all hypermethylated genes, however, are associated with adverse prognosis. The methylation of *DAPK1* (Qian et al. 2010), *DDIT3* (Lin et al. 2010) *MYOD1*, *MAFB*, and *KLF5* (Potapova et al. 2010) in MDS appears not to affect outcomes. Unsurprisingly, given the frequent association of individual gene methylation with poor prognosis, hypermethylation at ten different gene sites was also associated with poor prognosis (Shen et al. 2010).

More recent studies have examined the methylation status of MDS on a larger scale and found that abnormal promotor methylation is present in more than one-sixth (138 of 807) of the analyzed genes in high risk MDS but less common (60 of 807 genes) in low risk MDS, although this difference was not significant. A total of 51 CpG sites were hypermethylated in more than 50% of patients. *ALOX12*, *GSTM1*, *HIC1*, *FZD9* and *HS3ST2* were hypermethylated in over 70% of high risk MDS patients examined (Jiang et al. 2009). These genes are involved in DNA repair, cell-cycle control, and regulation of development, differentiation, and apoptosis. It has also been shown that the aberrant methylation in MDS segregates to particular chromosomal locations, including more frequently in Alu-poor genes. Additionally, this study demonstrated that treatment with DNMT inhibitors results in global reduction in promoter methylation; this effect is both profound and persistent, lasting at least a month following the cessation of treatment (Figueroa et al. 2009).

microRNA

MicroRNAs (miRNAs) are short single stranded RNA molecules that post-transcriptionally regulate protein expression levels. By binding complementary sequences in mRNA transcripts they induce an RNA-induced silencing complex (RISC) containing Dicer that destroys them. MiRNAs have been demonstrated to be involved in physiologic hematopoiesis. More recent studies have implicated them in playing an epigenetic role in the development of MDS. A recent study, using miRNA arrays, compared the expression levels between normal patients and those with MDS and those with AML that had progressed from MDS. A panel of 13 upregulated and 9 down regulated miRNAs were identified in MDS/AML patients.

There was further differentiation between patients that had MDS and those progressing to AML (Dostalova Merkerova et al. 2011; Sokol et al. 2011; Zuo et al. 2011).

Treatment of MDS

Most currently used treatments for MDS are supportive in nature, and in fact the first drugs released for the treatment of MDS received FDA approval only recently. The drugs approved for the treatment of MDS include lenalidomide, a derivative of thalidomide, and azacitidine and decitabine, which are hypomethylating agents. These medications have altered the natural course of the disease, improving survival and reducing supportive measures required for treatment. Additional therapies are currently being developed that, despite demonstrating promise, have yet to be proven effective in the treatment of MDS and thus are not currently FDA approved. To date, the only potentially curative procedure in the treatment of MDS is bone marrow transplant.

Supportive Care

Supportive care is often required for the chronic cytopenias experienced by MDS patients. Symptomatic chronic anemia is an indication for blood transfusion, and as a result MDS patients with erythroid dysplasia frequently are transfusion dependent. Transfusion dependency is associated with numerous potential difficulties, including the production of antibodies against alloantigens. The alloantibodies complicate matters not only by creating the potential for hemolysis with every blood transfusion, but also by limiting the pool of acceptable donors to those not expressing the given alloantigen. Furthermore, transfusion of as few as 20 units of packed red blood cells (pRBCs) can lead to the development of iron overload. Deposition in the liver, heart and endocrine organs can result in impaired function or even organ failure; unsurprisingly, iron overload is associated with a decreased overall survival in MDS patients.

To ameliorate the effects of iron overload, iron chelation remains a standard therapy. Traditionally iron chelation entailed intravenous or subcutaneous deferoxamine; more recently an oral agent has become available (deferasirox), greatly easing the treatment for patients. The most common side effects of deferasirox include gastrointestinal side effects and skin rash; however, deferasirox has also been noted to have more severe adverse effects including renal and liver toxicity, precluding its use in those with renal or liver failure (Gattermann et al. 2010). The side effects are pronounced with high rates of treatment withdrawal, as high has 50% in MDS patients (Gattermann et al. 2010). Unfortunately, deferasirox is contraindicated in patients with platelet counts below 50,000, limiting its use in some MDS populations. Although iron chelators have been demonstrated to reduce ferritin levels and iron deposits in liver biopsies in MDS, to date iron chelators have not shown a corresponding decrease in mortality. Iron chelation therapy should be adjusted with respect to the patient's ferritin levels and use limited to transfusion dependent patients with greater than 1-year life expectancy.

Hematopoietic Stimulants

Bone marrow stimulants can also be used to treat cytopenias. There are two red blood cell stimulants (erythropoietic stimulatory agents, ESA) currently on the market, recombinant human erythropoietin (rhEpo) and darbapoetin alfa, and they are considered functionally equivalent (Rizzo et al. 2010). Whereas rhEpo has a short half life and requires administration every few days, darbapoetin can be given every 1–3 weeks. A major erythroid response, defined as an increase in hemoglobin of 2 g/dl, is seen in around 40% of patients. Response rates are highly dependent on the risk status of the patient. Patients with low-risk MDS who have not been previously treated have response rates approaching 60%, whereas less well optimized patients that are intermediate risk with previous exposure to erythroid stimulating agents show responses in 20% (Kelaidi and Fenaux 2010; Moyo et al. 2008). Low serum EPO levels and transfusion independence are also associated with higher response rates (Park et al. 2008). The overall survival benefit of taking ESAs remains uncertain, some studies have demonstrated an effect in carefully selected patient populations (Park et al. 2008; Greenberg et al. 2009; Jadersten et al. 2008), while others have not (Musto et al. 2010).

Granulocyte-colony stimulating factor (G-CSF) can also be given in combination with ESAs. Interestingly, G-CSF may have a synergistic effect when administered in conjunction with erythropoietin, increasing both the erythroid response and the survival benefit, but this finding is not consistent in all studies (Park et al. 2008; Hellstrom-Lindberg et al. 1998). An online calculator has been developed that predicts response in low-, intermediate- and high-risk groups (Hellstrom-Lindberg et al. 1997b; http://www.qxmd.com/calculate-online/hematology/epo-gcsf-response-in-myelodysplastic-syndrome).

Romiplostim (AMG-531) is a platelet stimulating medication approved by the FDA (2008) for treatment of chronic idiopathic thrombocytopenic purpura. It consists of a thombopoietin analog fused to an Fc fragment. It stimulates the thrombopoietin receptor and is being used in the treatment of low-risk MDS. Delivered subcutaneously weekly, the medication has a favorable adverse event profile, with fatigue and headache being the most common. Over an 8 week trial period, 57 and 8% of patients had complete (platelet count increased to $>100 \times 10^9$/L) or partial/major (platelets count increases by $>30 \times 10^9$/L) responses, respectively (Sekeres et al. 2011). The patients that respond require fewer platelet transfusions and had fewer bleeding events (Sekeres et al. 2011; Kantarjian et al. 2010a). Another study, examining platelet transfusion requirements in the setting of treatment with the hypomethylating agent azacitidine (see below), demonstrated a trend toward reduced clinically significant thrombocytopenic events and platelet transfusions in patients administered romiplostim, however perhaps due to small sample size was not statistically significant (Kantarjian et al. 2010b).

Lenalidomide

Lenalidomide is a derivative of thalidomide that is effective in the treatment of plasma cell myeloma. Thalidomide, although effective against MDS, is limited by high toxicity (Moreno-Aspitia et al. 2006). Compared to thalidomide, lenalidomide has fewer side effects and greater potency. Although originally shown to be effective in 5q- syndrome (List et al. 2005, 2006), more recently lenalidomide has demonstrated effectiveness in lower risk MDS patients without 5q abnormalities (Raza et al. 2008). It has anti-angiogenic and immunomodulatory activity. Studies have demonstrated high response rates, with 67% (99/148) of patients becoming transfusion independent within 4.6 weeks and lasting over 2 years. Nearly half (45%) of patients achieved a complete cytogenetic remission (List et al. 2006). These responses were less likely in patients with thrombocytopenia or high transfusion requirements (List et al. 2006).

Unfortunately, response rates in patients lacking the del(5q) are lower. 43% of patients achieved complete or partial responses, with 26% achieving transfusion independence after 4.8 weeks. The duration of response was also considerably shorter at only 41 weeks (Raza et al. 2008). Therefore, in this patient population, it is most helpful as a secondary or supplemental medication. The side effects of lenalidomide, although less severe than those of thalidomide, remain significant. Patients with del(5q) are more likely to develop neutropenia (55%) or thrombocytopenia (44%) than other MDS patients (25 and 20%, respectively). These effects arise early in the course and often force drug reduction or interruption. As with thalidomide there is a potential risk of teratogenicity, so lenalidomide is administered under a restricted distribution program.

The mechanism of action of lenalidomide is probably multifactorial. In patients with del(5q), it is thought to specifically suppress the clone. The precise mechanism is uncertain, although it is likely through interactions with the specific genes associated with del(5q). Lenalidomide appears to block proliferation through G_1 arrest mediated indirectly through Cdc25A and PP2A (Wei et al. 2009). PP2A inhibition results in decreased MDM2 which the results in decreased p53, perhaps counteracting the p53 accumulation seen in 5q syndrome (Wei et al. 2009). In patients lacking the 5q abnormality, lenalidomide has a number of effects. In addition to its inhibition of angiogenesis (Lu et al. 2009), a prominent factor in the progression of MDS, it also seems to alleviate the ineffective erythropoiesis of MDS. Gene expression profiling of MDS patients treated with lenalidomide has demonstrated that drug responsive patients often have pro-erythroid targets of STAT5 and GATA1 down-regulated (Ebert et al. 2008b). In addition, lenalidomide promotes erythropoiesis in vitro.

Hypomethylating Agents

Targeted hypermethylation is quite common in MDS, leading to the suppression of normal regulatory genes, including those limiting entry into the cell cycle. This may be decreased or reversed by blocking methylation enzymes, allowing expression of tumor suppressors and proteins promoting differentiation.

Two medications approved by the FDA for use in MDS to inhibit methylation are azacitidine and decitabine (Kaminskas et al. 2005). These pyrimide nucleoside analogues permanently inhibit DNA methyltransferase (DNMT) following incorporation. Azacytidine was originally developed for the treatment of leukemia as a cytotoxic agent when used at high doses; at lower doses it is effective for the treatment of MDS. Patients treated with azacitidine have demonstrated complete response rates (IWG 2006 criteria: ≤5% myeloblasts with normal maturation of all cell lines in bone marrow; Hgb ≥11 g/dL, platelets $\geq 100 \times 10^9$/L, neutrophils $\geq 1.5 \times 10^9$/L, 0% blasts in blood) in around 20%, partial response rates (IWG 2006 criteria: as with complete response, but with blasts decreased by 50%) in around 10% and additional improvement in around 20% of patients (Silverman et al. 2006; Muller-Thomas et al. 2009; Lyons et al. 2009). These patients also have lower rates of AML transformation, higher survival rates, and improved quality of life (Silverman et al. 2002; Kantarjian et al. 2007a; Gurion et al. 2010). These studies have demonstrated survival rates of 24.5 months with treatment compared to 15 months for patients receiving best supportive care. Azacitidine was the first drug to demonstrate overall survival benefit in the treatment of MDS (Kantarjian et al. 2007a; Gurion et al. 2010).

Decitabine operates in a similar manner to that of azacitidine, however, unlike azacitidine, which contains a ribose sugar moiety, decitabine contains a deoxyribose moiety. As such, decitabine is more commonly integrated into DNA. The response rate in initial trials is similar to that of azacitidine (Garcia et al. 2010; Kantarjian et al. 2007b, 2008; Kumar et al. 2010). Unlike azacitidine, however, decitabine has not demonstrated a survival benefit or decreased transformation to AML in a randomized trial. Additionally, no prospective comparison between azacitidine and decitabine has been performed, hindering the comparison.

These medications are well tolerated, even in the elderly population. As with many of the MDS drugs, the most common adverse effect is myelosuppression. Interestingly, in patients who respond, the myelosuppression decreases as therapy is continued and concurrent treatment with growth factors can shorten the duration of myelosuppression.

Hypomethylating agents are typically useful in patients with high risk MDS. They have reduced effectiveness, however, in patients with complex cytogenetics. Notably, treatment with azacitidine or decitabine can lead to the loss of −7 with a return to a normal karyotype. (Jadersten et al. 2009; Kantarjian et al. 2008; Lubbert et al. 2001; Ades et al. 2009). Additionally, it has been demonstrated that the hypermethylation of genes associated with poor prognosis in MDS have decreased levels of methylation following treatment with DNMTs and this decreased methylation is associated with clinical response (Shen et al. 2010). Treatment with DNMTs results in global reduction in promoter methylation; this effect remains intact even a month after treatment is stopped (Figueroa et al. 2009).

A current goal is to develop prognostic factors that can predict response to demethylation agents. Some potential targets have recently been identified. *miR-29b* is a microRNA that targets DNMTs modulating their expression (Garzon et al. 2009). It has been demonstrated in preliminary studies that higher pretreatment levels of *miR-29b* are associated with a clinical response to decitabine (Blum et al. 2010). A more recent analysis has failed to replicate this result, however (Yang et al. 2011). The expression of phosphoinositide-phospholipase C beta1 (PLCB1) has also been suggested to be a prognostic factor. A reduction of PLCB1 methylation correlates with the response to azacitidine (Follo et al. 2009). These initial findings of prognostic elements in relation to treatment with DNMTs require further investigation, especially in the setting of MDS.

HDAC Inhibitors

The success of drugs altering the methylation status of genes has raised interest in histone deacetylase (HDAC) inhibitors. These drugs, with similar effects on protein expression to DNMTs, operate in a complementary manner. Histones are proteins critical to the quaternary structure of DNA. They are extensively post-translationally modified, allowing modulation of the accessibility of DNA to transcription at an epigenetic level. The major post-translational modification that occurs on histones is acetylation. Acetylation of residues K9 and K14 in histone 3, catalyzed by histone lysine acetyltransferases, can result in increased transcription of the associated DNA. These acetyl groups are removed by histone deacetylases, which may be inhibited by certain drugs.

Although the general mechanism of HDAC inhibitors is understood, the specific mechanism for altering MDS progression is not. Indeed, no specific genes are currently recognized as having reduced acetylation in MDS. HDAC inhibitors, although they most notably affect the acetylation status of histones, may also affect the acetylation status of a number of other proteins. Therefore, the hypothesis that HDAC inhibitors directly increase expression of tumor suppressors may not be the complete story. Rather, they may operate through additional cellular mechanisms.

Vorinostat and Romidepsin are two HDAC inhibitors that have been approved by the FDA in 2006 and 2009, respectively, for treatment of cutaneous T-cell lymphoma. In addition to these agents, there are numerous other drugs that are currently undergoing study for treatment of a variety of solid and hematologic malignancies. These medications have a diverse range of specificities, including only class I HDACs (entinostat) or broadly inhibitory (panobinostat). The significance of class specificity as it relates to both treatment effectiveness and side effect profile is currently under study. By tailoring the specificity to MDS, it may be possible to limit the common side effects, such as gastrointestinal symptoms, severe fatigue and thrombocytopenia.

Interestingly, a number of medications initially developed for other purposes have been found to have effects on the HDAC axis, including valproic acid and sodium phenylbutyrate.

Initial studies have demonstrated HDAC inhibitors to have bone marrow suppression as a primary side effect, occurring in about 40% of patients, in addition to constitutional and gastrointestinal side effects. As single-agent drugs, many of the HDAC inhibitors that have been developed appear to have limited activity, with a few studies demonstrating a maximal response rate of about 10% (response defined as follows: complete response: absolute neutrophil count of 1×10^9/L or more, platelet count of 100×10^9/L or more, no blasts in the peripheral blood, bone marrow cellularity of 20% or more with normal trilineage maturation, bone marrow blasts of 5% or less, and absence of extramedullary involvement. Partial response: normalization of peripheral blood counts as for CR, and the complete disappearance of peripheral blasts was observed with more than 5%, but less than 25% blasts in the marrow) (Garcia-Manero et al. 2008). Valproic acid, a mood-stabilizer and anticonvulsant, also has activity as an HDAC inhibitor. Two recent studies have demonstrated hematologic responses (IWG 2000 criteria: Hgb increase of >1.5 g/dL (pretreatment of <11 g/dL); >4 unit decrease in RBC transfusions/8 weeks versus pretreatment; platelets increased by $>30 \times 10^9$/L (pretreatment of $20–100 \times 10^9$/L) or platelets increased by >100% to above 20×10^9/L (pretreatment of $<20 \times 10^9$/L); neutrophils increased by >100% to $>0.5 \times 10^9$/L (pretreatment $<1 \times 10^9$/L)) in 25–30% of patients with MDS (Kuendgen et al. 2005; Pilatrino et al. 2005). In addition to the single-agent setting, multi-drug therapy in tandem with azacitidine or decitabine shows promise as an effective treatment for MDS. A number of *in vitro* studies have demonstrated that treatment with DNMTs with subsequent treatment with HDACs is necessary for optimal re-expression of silenced genes (Cameron et al. 1999). A number of small studies have begun to explore combination therapy and have demonstrated partial or complete response (by IWG 2006 criteria) in about 30% patients (Gore et al. 2006; Voso et al. 2009).

Immunosuppressive Therapy

In a subset of MDS patients, the cytopenias are thought to be secondary to immunosuppression by clonal T cells. These patients typically are younger with low or intermediate-1 disease, with hypocellular bone marrows, shorter duration of RBC transfusion requirements, express HLA-DR15 or harbor a paroxysmal nocturnal hemoglobinuria clone (Sloand et al. 2008; Saunthararajah et al. 2002). This distinct purported mechanism of disease is particularly interesting in light of treatment with immunosuppressive therapy with antithymocyte globulin (ATG) with or without the addition of cyclosporine and/or corticosteroids. Around 30% of unselected patients respond, as defined by transfusion independence, with variable lengths of remission (Molldrem et al. 2002; Passweg et al. 2011). The criteria for selecting appropriate patients for this treatment are relatively strict due to the high incidence of side-effects and low tolerance of ATG in many patients.

Halogenated Purine Nucleoside Analogs

Halogenated purine nucleoside analogs, including clofarabine, cladribine and fludarabine, are chemotherapeutic agents with two primary mechanisms of action. In addition to the commonly recognized apoptosis secondary to incorporation into DNA, they also inhibit DNA synthesis and ribonucleotide reductase. Cladribine and fludarabine have both been approved by the FDA for treatment of lymphoid neoplasms, like hairy cell leukemia and chronic lymphocytic leukemia. Although more prominently studied in lymphoid neoplasms, they also have shown some efficacy against myeloid neoplasms, including MDS. Clofarabine has been examined in high-risk MDS in a small study of 32 patients that demonstrated an overall response rate of 43% (CR: neutrophils $\geq 1 \times 10^9$/L, platelets $\geq 100 \times 10^9$/L with $\leq 5\%$ marrow blasts. Hematologic Improvement: all criteria for CR except for platelet $\geq 100 \times 10^9$/L. Clinical benefit: 50% increase of platelets to more than 30×10^9/L; and/or 100% neutrophil increase and to more than 1×10^9/L; and/or hemoglobin increase by 2 g/dL; and/or transfusion independence.). The side effects were typically gastrointestinal or liver and not severe (Faderl et al. 2010; Steensma 2010).

Glutathione Analogs

Ezatiostat (TLK199) is a tripeptide glutathione analog that affects the mitogenic Jun-N-terminal kinase (JNK) pathway. TLK199, a precursor drug, is metabolized to TLK117 and binds Glutathione S-transferase P1-1 (GST P1-1), preventing its inhibition of Jun-N-terminal kinase (JNK) and favoring cell growth and activation of the JNK pathway. Its role in the treatment of MDS is, therefore, as a novel bone marrow stimulant; studies comparing it to the above bone marrow stimulants are necessary.

Preliminary studies have demonstrated some effectiveness. A small study using intravenous liposomal ezatiostat showed efficacy in MDS with a response, according to IWG 2000 criteria (detailed above), in 25–50% of patients, depending on the cell lineage affected (best in platelet counts, least in hemoglobin) (Raza et al. 2009a). The side effects were relatively infrequent and mild.

An oral preparation has been developed for lower-risk MDS patients with hematologic improvement noted (as defined by IWG, 2000) in about one-third of patients with only mild, primarily gastrointestinal, side effects (Raza et al. 2009b). Ezatiostat has also demonstrated effectiveness in a patient refractory to treatment with lenalidomide in a recent case report (Quddus et al. 2010). Its role as a pro-proliferative agent raises concerns for potential AML development in patients with MDS, but, to date, with the caveat of small sample size, no studies have found evidence to support this. Therefore ezatiostat is a promising new therapeutic in the treatment of MDS and further studies, including ones with combination therapy, are being considered.

Farnesyltransferase Inhibitors

Tipifarnib (R115777, Zarnestra) and lonafarnib are farnesyltransferase inhibitors that inhibit multiple proteins that require farnesylation, a form of posttranslational modification. The prototypical protein affected, and the most likely target is Ras, a small GTPase involved in proliferation signal transduction. The drug inhibits cell growth in multiple cell lines, including those without Ras overexpression, which is not surprising since they block farnesylation of proteins other than Ras. Despite being denied approval for the treatment of AML in 2005, studies are now examining its use in MDS.

Recent studies have demonstrated that in patients with intermediate and high risk MDS, oral tipifarnib results in a complete response rate of about 20% (bone marrow blasts <5%, no identifiable karyotypic abnormalities and peripheral blood counts reaching and holding at: hemoglobin >11 g/dL, neutrophils $>1\times10^9$/L and platelets $>100\times10^9$/L without transfusion or bone marrow stimulants) (Fenaux et al. 2007). Lonafarnib appears to have similar efficacy and side effects (Feldman et al. 2008), although a smaller study failed to show benefit (Ravoet et al. 2008). The most common adverse effects include myelosuppression and gastrointestinal distress, but these agents are much better tolerated at lower doses (Kurzrock et al. 2004). A recent study suggests that due to differential efficacy between MDS progenitors and normal progenitors at low concentration of drug, lower doses may remain effective while improving the side effect profile (Kotsianidis et al. 2008).

SCIO-469

SCIO-469 is a selective small molecule inhibitor of the serine/threonine kinase, p38 MAPK. p38 MAPK is a mitogenic protein involved in multiple pathways for cytokine signaling. SCIO-469 is thought to primarily act through inhibition of CD34+ stem cell apoptosis. A recent study, however, has demonstrated that it also limits production of proinflammatory cytokines *in vitro* and may affect the bone marrow microenvironment (Navas et al. 2008). The drug is currently in phase I/II trials.

References

Ades L et al (2009) Efficacy and safety of lenalidomide in intermediate-2 or high-risk myelodysplastic syndromes with 5q deletion: results of a phase 2 study. Blood 113:3947–3952

Afable MG et al (2011) SNP array-based karyotyping: differences and similarities between aplastic anemia and hypocellular myelodysplastic syndromes. Blood 117:6876–6884

Aggerholm A, Holm MS, Guldberg P, Olesen LH, Hokland P (2006) Promoter hypermethylation of p15INK4B, HIC1, CDH1, and ER is frequent in myelodysplastic syndrome and predicts poor prognosis in early-stage patients. Eur J Haematol 76:23–32

Aizawa S et al (2000) Establishment of stromal cell line from an MDS RA patient which induced an apoptotic change in hematopoietic and leukemic cells in vitro. Exp Hematol 28:148–155

Alexandrakis MG et al (2005) Serum evaluation of angiogenic cytokine basic fibroblast growth factor, hepatocyte growth factor and TNF-alpha in patients with myelodysplastic syndromes: correlation with bone marrow microvascular density. Int J Immunopathol Pharmacol 18:287–295

Arai Y et al (1997) The inv(11)(p15q22) chromosome translocation of de novo and therapy-related myeloid malignancies results in fusion of the nucleoporin gene, NUP98, with the putative RNA helicase gene, DDX10. Blood 89:3936–3944

Bacher U, Haferlach T, Kern W, Haferlach C, Schnittger S (2007) A comparative study of molecular mutations in 381 patients with myelodysplastic syndrome and in 4130 patients with acute myeloid leukemia. Haematologica 92:744–752

Bains A, Luthra R, Medeiros LJ, Zuo Z (2011) FLT3 and NPM1 mutations in myelodysplastic syndromes: frequency and potential value for predicting progression to acute myeloid leukemia. Am J Clin Pathol 135:62–69

Barlow JL et al (2010) A p53-dependent mechanism underlies macrocytic anemia in a mouse model of human 5q- syndrome. Nat Med 16:59–66

Bejar R et al (2011) Clinical effect of point mutations in myelodysplastic syndromes. N Engl J Med 364:2496–2506

Benesch M, Platzbecker U, Ward J, Deeg HJ, Leisenring W (2003) Expression of FLIP(Long) and FLIP(Short) in bone marrow mononuclear and CD34+ cells in patients with myelodysplastic syndrome: correlation with apoptosis. Leukemia 17:2460–2466

Bernasconi P et al (2003) Is FISH a relevant prognostic tool in myelodysplastic syndromes with a normal chromosome pattern on conventional cytogenetics? A study on 57 patients. Leukemia 17:2107–2112

Blau O et al (2007) Chromosomal aberrations in bone marrow mesenchymal stroma cells from patients with myelodysplastic syndrome and acute myeloblastic leukemia. Exp Hematol 35:221–229

Bloomfield CD et al (2002) 11q23 balanced chromosome aberrations in treatment-related myelodysplastic syndromes and acute leukemia: report from an international workshop. Genes Chromosomes Cancer 33:362–378

Blum W et al (2010) Clinical response and miR-29b predictive significance in older AML patients treated with a 10-day schedule of decitabine. Proc Natl Acad Sci U S A 107:7473–7478

Bouchliou I et al (2011) Th17 and Foxp3(+) T regulatory cell dynamics and distribution in myelodysplastic syndromes. Clin Immunol 139:350–359

Boudard D et al (2000) Expression and activity of caspases 1 and 3 in myelodysplastic syndromes. Leukemia 14:2045–2051

Boultwood J et al (2007) Gene expression profiling of CD34+ cells in patients with the 5q- syndrome. Br J Haematol 139:578–589

Boumber YA et al (2007) RIL, a LIM gene on 5q31, is silenced by methylation in cancer and sensitizes cancer cells to apoptosis. Cancer Res 67:1997–2005

Bouscary D et al (1997) Fas/Apo-1 (CD95) expression and apoptosis in patients with myelodysplastic syndromes. Leukemia 11:839–845

Bouscary D et al (2000) Activity of the caspase-3/CPP32 enzyme is increased in "early stage" myelodysplastic syndromes with excessive apoptosis, but caspase inhibition does not enhance colony formation in vitro. Exp Hematol 28:784–791

Brakensiek K, Langer F, Schlegelberger B, Kreipe H, Lehmann U (2005) Hypermethylation of the suppressor of cytokine signalling-1 (SOCS-1) in myelodysplastic syndrome. Br J Haematol 130:209–217

Brunner B et al (2002) Blood levels of angiogenin and vascular endothelial growth factor are elevated in myelodysplastic syndromes and in acute myeloid leukemia. J Hematother Stem Cell Res 11:119–125

Cameron EE, Bachman KE, Myohanen S, Herman JG, Baylin SB (1999) Synergy of demethylation and histone deacetylase inhibition in the re-expression of genes silenced in cancer. Nat Genet 21:103–107

Chakraborty S et al (2009) Accelerated telomere shortening precedes development of therapy-related myelodysplasia or acute myelogenous leukemia after autologous transplantation for lymphoma. J Clin Oncol 27:791–798

Chen CY et al (2007) RUNX1 gene mutation in primary myelodysplastic syndrome–the mutation can be detected early at diagnosis or acquired during disease progression and is associated with poor outcome. Br J Haematol 139:405–414

Cho YS, Kim EJ, Park UH, Sin HS, Um SJ (2006) Additional sex comb-like 1 (ASXL1), in cooperation with SRC-1, acts as a ligand-dependent coactivator for retinoic acid receptor. J Biol Chem 281:17588–17598

Choesmel V et al (2007) Impaired ribosome biogenesis in Diamond-Blackfan anemia. Blood 109:1275–1283

Christiansen DH, Andersen MK, Pedersen-Bjergaard J (2001) Mutations with loss of heterozygosity of p53 are common in therapy-related myelodysplasia and acute myeloid leukemia after exposure to alkylating agents and significantly associated with deletion or loss of 5q, a complex karyotype, and a poor prognosis. J Clin Oncol 19:1405–1413

Cioc AM, Wagner JE, MacMillan ML, DeFor T, Hirsch B (2010) Diagnosis of myelodysplastic syndrome among a cohort of 119 patients with fanconi anemia: morphologic and cytogenetic characteristics. Am J Clin Pathol 133:92–100

Ciovacco WA, Raskind WH, Kacena MA (2008) Human phenotypes associated with GATA-1 mutations. Gene 427:1–6

Cogle CR, Craig BM, Rollison DE, List AF (2011) Incidence of the myelodysplastic syndromes using a novel claims-based algorithm: high number of uncaptured cases by cancer registries. Blood 117:7121–7125

Coleman JF, Theil KS, Tubbs RR, Cook JR (2011) Diagnostic yield of bone marrow and peripheral blood FISH panel testing in clinically suspected myelodysplastic syndromes and/or acute myeloid leukemia: a prospective analysis of 433 cases. Am J Clin Pathol 135:915–920

Davis RE, Greenberg PL (1998) Bcl-2 expression by myeloid precursors in myelodysplastic syndromes: relation to disease progression. Leuk Res 22:767–777

Delhommeau F et al (2009) Mutation in TET2 in myeloid cancers. N Engl J Med 360:2289–2301

Dicker F et al (2010) Mutation analysis for RUNX1, MLL-PTD, FLT3-ITD, NPM1 and NRAS in 269 patients with MDS or secondary AML. Leukemia 24:1528–1532

Dostalova Merkerova M et al (2011) Distinctive microRNA expression profiles in CD34+ bone marrow cells from patients with myelodysplastic syndrome. Eur J Hum Genet 19:313–319

Dutt S et al (2011) Haploinsufficiency for ribosomal protein genes causes selective activation of p53 in human erythroid progenitor cells. Blood 117:2567–2576

Ebert BL et al (2008a) Identification of RPS14 as a 5q- syndrome gene by RNA interference screen. Nature 451:335–339

Ebert BL et al (2008b) An erythroid differentiation signature predicts response to lenalidomide in myelodysplastic syndrome. PLoS Med 5:e35

Epperson DE, Nakamura R, Saunthararajah Y, Melenhorst J, Barrett AJ (2001) Oligoclonal T cell expansion in myelodysplastic syndrome: evidence for an autoimmune process. Leuk Res 25:1075–1083

Faderl S et al (2010) Oral clofarabine in the treatment of patients with higher-risk myelodysplastic syndrome. J Clin Oncol 28:2755–2760

Feldman EJ et al (2008) On the use of lonafarnib in myelodysplastic syndrome and chronic myelomonocytic leukemia. Leukemia 22:1707–1711

Fenaux P et al (2007) A multicenter phase 2 study of the farnesyltransferase inhibitor tipifarnib in intermediate- to high-risk myelodysplastic syndrome. Blood 109:4158–4163

Figueroa ME et al (2009) MDS and secondary AML display unique patterns and abundance of aberrant DNA methylation. Blood 114:3448–3458

Finch AJ et al (2011) Uncoupling of GTP hydrolysis from eIF6 release on the ribosome causes Shwachman-Diamond syndrome. Genes Dev 25:917–929

Flores-Figueroa E, Gutierrez-Espindola G, Montesinos JJ, Arana-Trejo RM, Mayani H (2002) In vitro characterization of hematopoietic microenvironment cells from patients with myelodysplastic syndrome. Leuk Res 26:677–686

Follo MY et al (2009) Reduction of phosphoinositide-phospholipase C beta1 methylation predicts the responsiveness to azacitidine in high-risk MDS. Proc Natl Acad Sci U S A 106:16811–16816
Garcia JS, Jain N, Godley LA (2010) An update on the safety and efficacy of decitabine in the treatment of myelodysplastic syndromes. Onco Targets Ther 3:1–13
Garcia-Manero G et al (2008) Phase 1 study of the histone deacetylase inhibitor vorinostat (suberoylanilide hydroxamic acid [SAHA]) in patients with advanced leukemias and myelodysplastic syndromes. Blood 111:1060–1066
Garzon R et al (2009) MicroRNA-29b induces global DNA hypomethylation and tumor suppressor gene reexpression in acute myeloid leukemia by targeting directly DNMT3A and 3B and indirectly DNMT1. Blood 113:6411–6418
Gattermann N et al (2010) Deferasirox in iron-overloaded patients with transfusion-dependent myelodysplastic syndromes: results from the large 1-year EPIC study. Leuk Res 34:1143–1150
Gelsi-Boyer V et al (2009) Mutations of polycomb-associated gene ASXL1 in myelodysplastic syndromes and chronic myelomonocytic leukaemia. Br J Haematol 145:788–800
Gersuk GM, Lee JW, Beckham CA, Anderson J, Deeg HJ (1996) Fas (CD95) receptor and Fas-ligand expression in bone marrow cells from patients with myelodysplastic syndrome. Blood 88:1122–1123
Gersuk GM et al (1998) A role for tumour necrosis factor-alpha, Fas and Fas-Ligand in marrow failure associated with myelodysplastic syndrome. Br J Haematol 103:176–188
Gibbons RJ et al (2003) Identification of acquired somatic mutations in the gene encoding chromatin-remodeling factor ATRX in the alpha-thalassemia myelodysplasia syndrome (ATMDS). Nat Genet 34:446–449
Gore SD et al (2006) Combined DNA methyltransferase and histone deacetylase inhibition in the treatment of myeloid neoplasms. Cancer Res 66:6361–6369
Graubert TA et al (2011) Recurrent mutations in the U2AF1 splicing factor in myelodysplastic syndromes. Nat Genet 44:53–57
Greenberg PL (2011) Myelodysplastic syndromes: dissecting the heterogeneity. J Clin Oncol 29:1937–1938
Greenberg P et al (1997) International scoring system for evaluating prognosis in myelodysplastic syndromes. Blood 89:2079–2088
Greenberg PL et al (2009) Treatment of myelodysplastic syndrome patients with erythropoietin with or without granulocyte colony-stimulating factor: results of a prospective randomized phase 3 trial by the Eastern Cooperative Oncology Group (E1996). Blood 114:2393–2400
Greenberg PL et al (2012) Revised International Prognostic Scoring System (IPSS-R) for myelodysplastic syndromes. Blood. doi: 10.1182/blood-2012-03-420489
Grosjean-Raillard J et al (2009) ATM mediates constitutive NF-kappaB activation in high-risk myelodysplastic syndrome and acute myeloid leukemia. Oncogene 28:1099–1109
Gurevich I et al (2011) Refractory anemia with ring sideroblasts associated with marked thrombocytosis: a mixed group exhibiting a spectrum of morphologic findings. Am J Clin Pathol 135:398–403
Gurion R et al (2010) 5-azacitidine prolongs overall survival in patients with myelodysplastic syndrome–a systematic review and meta-analysis. Haematologica 95:303–310
Haase D et al (2007) New insights into the prognostic impact of the karyotype in MDS and correlation with subtypes: evidence from a core dataset of 2124 patients. Blood 110:4385–4395
Harada Y, Harada H (2011) Molecular mechanisms that produce secondary MDS/AML by RUNX1/AML1 point mutations. J Cell Biochem 112:425–432
Harada H et al (2004) High incidence of somatic mutations in the AML1/RUNX1 gene in myelodysplastic syndrome and low blast percentage myeloid leukemia with myelodysplasia. Blood 103:2316–2324
Hashmi S et al (2011) Comparative analysis of Shwachman-Diamond syndrome to other inherited bone marrow failure syndromes and genotype-phenotype correlation. Clin Genet 79:448–458
Hellstrom-Lindberg E, Kanter-Lewensohn L, Ost A (1997a) Morphological changes and apoptosis in bone marrow from patients with myelodysplastic syndromes treated with granulocyte-CSF and erythropoietin. Leuk Res 21:415–425

Hellstrom-Lindberg E et al (1997b) Erythroid response to treatment with G-CSF plus erythropoietin for the anaemia of patients with myelodysplastic syndromes: proposal for a predictive model. Br J Haematol 99:344–351

Hellstrom-Lindberg E et al (1998) Treatment of anemia in myelodysplastic syndromes with granulocyte colony-stimulating factor plus erythropoietin: results from a randomized phase II study and long-term follow-up of 71 patients. Blood 92:68–75

Hosoya N, Miyagawa K, Mitani K, Yazaki Y, Hirai H (1998) Mutation analysis of the WT1 gene in myelodysplastic syndromes. Jpn J Cancer Res 89:821–824

http://www.qxmd.com/calculate-online/hematology/epo-gcsf-response-in-myelodysplastic-syndrome

Hu Q et al (2004) Soluble vascular endothelial growth factor receptor 1, and not receptor 2, is an independent prognostic factor in acute myeloid leukemia and myelodysplastic syndromes. Cancer 100:1884–1891

Itzykson R et al (2011) Impact of TET2 mutations on response rate to azacitidine in myelodysplastic syndromes and low blast count acute myeloid leukemias. Leukemia 25:1147–1152

Iwanaga M et al (2011) Risk of myelodysplastic syndromes in people exposed to ionizing radiation: a retrospective cohort study of Nagasaki atomic bomb survivors. J Clin Oncol 29:428–434

Jadersten M et al (2008) Erythropoietin and granulocyte-colony stimulating factor treatment associated with improved survival in myelodysplastic syndrome. J Clin Oncol 26:3607–3613

Jadersten M et al (2009) Clonal heterogeneity in the 5q- syndrome: p53 expressing progenitors prevail during lenalidomide treatment and expand at disease progression. Haematologica 94:1762–1766

Jiang Y et al (2009) Aberrant DNA methylation is a dominant mechanism in MDS progression to AML. Blood 113:1315–1325

Joslin JM et al (2007) Haploinsufficiency of EGR1, a candidate gene in the del(5q), leads to the development of myeloid disorders. Blood 110:719–726

Kaminskas E et al (2005) Approval summary: azacitidine for treatment of myelodysplastic syndrome subtypes. Clin Cancer Res 11:3604–3608

Kaneko H, Misawa S, Horiike S, Nakai H, Kashima K (1995) TP53 mutations emerge at early phase of myelodysplastic syndrome and are associated with complex chromosomal abnormalities. Blood 85:2189–2193

Kantarjian HM et al (2007a) Survival advantage with decitabine versus intensive chemotherapy in patients with higher risk myelodysplastic syndrome: comparison with historical experience. Cancer 109:1133–1137

Kantarjian H et al (2007b) Results of a randomized study of 3 schedules of low-dose decitabine in higher-risk myelodysplastic syndrome and chronic myelomonocytic leukemia. Blood 109:52–57

Kantarjian H et al (2008) Decitabine improves patient outcomes in myelodysplastic syndromes: results of a phase III randomized study. Cancer 106:1794–1803

Kantarjian H et al (2010a) Safety and efficacy of romiplostim in patients with lower-risk myelodysplastic syndrome and thrombocytopenia. J Clin Oncol 28:437–444

Kantarjian HM et al (2010b) Phase 2 study of romiplostim in patients with low- or intermediate-risk myelodysplastic syndrome receiving azacitidine therapy. Blood 116:3163–3170

Keith T et al (2007) Regulation of angiogenesis in the bone marrow of myelodysplastic syndromes transforming to overt leukaemia. Br J Haematol 137:206–215

Kelaidi C, Fenaux P (2010) Darbepoetin alfa in anemia of myelodysplastic syndromes: present and beyond. Expert Opin Biol Ther 10:605–614

Kita-Sasai Y et al (2001) International prognostic scoring system and TP53 mutations are independent prognostic indicators for patients with myelodysplastic syndrome. Br J Haematol 115:309–312

Kitagawa M et al (1997) Overexpression of tumor necrosis factor (TNF)-alpha and interferon (IFN)-gamma by bone marrow cells from patients with myelodysplastic syndromes. Leukemia 11:2049–2054

Kordasti SY et al (2007) CD4+CD25high Foxp3+ regulatory T cells in myelodysplastic syndrome (MDS). Blood 110:847–850

Korkolopoulou P et al (2001) Prognostic evaluation of the microvascular network in myelodysplastic syndromes. Leukemia 15:1369–1376

Kosmider O et al (2009) TET2 mutation is an independent favorable prognostic factor in myelodysplastic syndromes (MDSs). Blood 114:3285–3291

Kosmider O et al (2010) Mutations of IDH1 and IDH2 genes in early and accelerated phases of myelodysplastic syndromes and MDS/myeloproliferative neoplasms. Leukemia 24:1094–1096

Kotsianidis I et al (2008) In vitro effects of the farnesyltransferase inhibitor tipifarnib on myelodysplastic syndrome progenitors. Acta Haematol 120:51–56

Kuendgen A et al (2005) Results of a phase 2 study of valproic acid alone or in combination with all-trans retinoic acid in 75 patients with myelodysplastic syndrome and relapsed or refractory acute myeloid leukemia. Ann Hematol 84(Suppl 1):61–66

Kumar A, List AF, Hozo I, Komrokji R, Djulbegovic B (2010) Decitabine versus 5-azacitidine for the treatment of myelodysplastic syndrome: adjusted indirect meta-analysis. Haematologica 95:340–342, author reply 343–4

Kupprion C, Motamed K, Sage EH (1998) SPARC (BM-40, osteonectin) inhibits the mitogenic effect of vascular endothelial growth factor on microvascular endothelial cells. J Biol Chem 273:29635–29640

Kurzrock R et al (2004) Phase II study of R115777, a farnesyl transferase inhibitor, in myelodysplastic syndrome. J Clin Oncol 22:1287–1292

Lane SW et al (2010) The Apc(min) mouse has altered hematopoietic stem cell function and provides a model for MPD/MDS. Blood 115:3489–3497

Langemeijer SM et al (2009) Acquired mutations in TET2 are common in myelodysplastic syndromes. Nat Genet 41:838–842

Lee SW et al (2010) ASXL1 represses retinoic acid receptor-mediated transcription through associating with HP1 and LSD1. J Biol Chem 285:18–29

Lehmann S et al (2007) Common deleted genes in the 5q- syndrome: thrombocytopenia and reduced erythroid colony formation in SPARC null mice. Leukemia 21:1931–1936

Leone G, Fianchi L, Pagano L, Voso MT (2010) Incidence and susceptibility to therapy-related myeloid neoplasms. Chem Biol Interact 184:39–45

Lin J et al (2008) Methylation status of fragile histidine triad (FHIT) gene and its clinical impact on prognosis of patients with myelodysplastic syndrome. Leuk Res 32:1541–1545

Lin J et al (2010) Aberrant methylation of DNA-damage-inducible transcript 3 promoter is a common event in patients with myelodysplastic syndrome. Leuk Res 34:991–994

List A et al (2005) Efficacy of lenalidomide in myelodysplastic syndromes. N Engl J Med 352:549–557

List A et al (2006) Lenalidomide in the myelodysplastic syndrome with chromosome 5q deletion. N Engl J Med 355:1456–1465

Liu TX et al (2007) Chromosome 5q deletion and epigenetic suppression of the gene encoding alpha-catenin (CTNNA1) in myeloid cell transformation. Nat Med 13:78–83

Loh ML et al (2005) Acquired PTPN11 mutations occur rarely in adult patients with myelodysplastic syndromes and chronic myelomonocytic leukemia. Leuk Res 29:459–462

Lorenzo F et al (2006) Mutational analysis of the KIT gene in myelodysplastic syndrome (MDS) and MDS-derived leukemia. Leuk Res 30:1235–1239

Lu L et al (2009) The anti-cancer drug lenalidomide inhibits angiogenesis and metastasis via multiple inhibitory effects on endothelial cell function in normoxic and hypoxic conditions. Microvasc Res 77:78–86

Lubbert M et al (2001) Cytogenetic responses in high-risk myelodysplastic syndrome following low-dose treatment with the DNA methylation inhibitor 5-aza-2′-deoxycytidine. Br J Haematol 114:349–357

Lyons RM et al (2009) Hematologic response to three alternative dosing schedules of azacitidine in patients with myelodysplastic syndromes. J Clin Oncol 27:1850–1856

Makishima H et al (2009) Mutations of e3 ubiquitin ligase cbl family members constitute a novel common pathogenic lesion in myeloid malignancies. J Clin Oncol 27:6109–6116

Mallo M et al (2008) Fluorescence in situ hybridization improves the detection of 5q31 deletion in myelodysplastic syndromes without cytogenetic evidence of 5q. Haematologica 93:1001–1008

Marcondes AM et al (2008) Dysregulation of IL-32 in myelodysplastic syndrome and chronic myelomonocytic leukemia modulates apoptosis and impairs NK function. Proc Natl Acad Sci U S A 105:2865–2870

Marcondes AM, Ramakrishnan A, Deeg HJ (2009) Myeloid malignancies and the marrow microenvironment: some recent studies in patients with MDS. Curr Cancer Ther Rev 5:310–314

Matsutani T et al (2003) Determination of T-cell receptors of clonal CD8-positive T-cells in myelodysplastic syndrome with erythroid hypoplasia. Leuk Res 27:305–312

Melenhorst JJ et al (2002) Molecular and flow cytometric characterization of the CD4 and CD8 T-cell repertoire in patients with myelodysplastic syndrome. Br J Haematol 119:97–105

Misawa S et al (1998) Significance of chromosomal alterations and mutations of the N-RAS and TP53 genes in relation to leukemogenesis of acute myeloid leukemia. Leuk Res 22:631–637

Mohamedali A et al (2007) Prevalence and prognostic significance of allelic imbalance by single-nucleotide polymorphism analysis in low-risk myelodysplastic syndromes. Blood 110:3365–3373

Molldrem JJ et al (2002) Antithymocyte globulin for treatment of the bone marrow failure associated with myelodysplastic syndromes. Ann Intern Med 137:156–163

Moreno-Aspitia A et al (2006) Thalidomide therapy in adult patients with myelodysplastic syndrome. A North Central Cancer Treatment Group phase II trial. Cancer 107:767–772

Moyo V, Lefebvre P, Duh MS, Yektashenas B, Mundle S (2008) Erythropoiesis-stimulating agents in the treatment of anemia in myelodysplastic syndromes: a meta-analysis. Ann Hematol 87:527–536

Muller-Thomas C, Schuster T, Peschel C, Gotze KS (2009) A limited number of 5-azacitidine cycles can be effective treatment in MDS. Ann Hematol 88:213–219

Mundle SD et al (1999) Correlation of tumor necrosis factor alpha (TNF alpha) with high Caspase 3-like activity in myelodysplastic syndromes. Cancer Lett 140:201–207

Musto P et al (2010) Response to recombinant erythropoietin alpha, without the adjunct of granulocyte-colony stimulating factor, is associated with a longer survival in patients with transfusion-dependent myelodysplastic syndromes. Leuk Res 34:981–985

Navas T et al (2008) Inhibition of p38alpha MAPK disrupts the pathological loop of proinflammatory factor production in the myelodysplastic syndrome bone marrow microenvironment. Leuk Lymphoma 49:1963–1975

Nikoloski G et al (2010) Somatic mutations of the histone methyltransferase gene EZH2 in myelodysplastic syndromes. Nat Genet 42:665–667

Nikpour M et al (2010) Gene expression profiling of erythroblasts from refractory anaemia with ring sideroblasts (RARS) and effects of G-CSF. Br J Haematol 149:844–854

Olney HJ, Le Beau MM (2007) Evaluation of recurring cytogenetic abnormalities in the treatment of myelodysplastic syndromes. Leuk Res 31:427–434

Osato M (2004) Point mutations in the RUNX1/AML1 gene: another actor in RUNX leukemia. Oncogene 23:4284–4296

Padua RA et al (1998) RAS, FMS and p53 mutations and poor clinical outcome in myelodysplasias: a 10-year follow-up. Leukemia 12:887–892

Papaemmanuil E et al (2011) Somatic SF3B1 mutation in myelodysplasia with ring sideroblasts. N Engl J Med 365:1384–1395

Park S et al (2008) Predictive factors of response and survival in myelodysplastic syndrome treated with erythropoietin and G-CSF: the GFM experience. Blood 111:574–582

Parker JE et al (1998) 'Low-risk' myelodysplastic syndrome is associated with excessive apoptosis and an increased ratio of pro- versus anti-apoptotic bcl-2-related proteins. Br J Haematol 103:1075–1082

Parker JE et al (2000) The role of apoptosis, proliferation, and the Bcl-2-related proteins in the myelodysplastic syndromes and acute myeloid leukemia secondary to MDS. Blood 96:3932–3938

Passweg JR et al (2011) Immunosuppressive therapy for patients with myelodysplastic syndrome: a prospective randomized multicenter phase III trial comparing antithymocyte globulin plus cyclosporine with best supportive care--SAKK 33/9. J Clin Oncol 29:303–309

Pedersen B (1996) Anatomy of the 5q- deletion: different sex ratios and deleted 5q bands in MDS and AML. Leukemia 10:1883–1890

Pedersen-Bjergaard J, Philip P (1991) Two different classes of therapy-related and de-novo acute myeloid leukemia? Cancer Genet Cytogenet 55:119–124

Pedersen-Bjergaard J, Christiansen DH, Desta F, Andersen MK (2006) Alternative genetic pathways and cooperating genetic abnormalities in the pathogenesis of therapy-related myelodysplasia and acute myeloid leukemia. Leukemia 20:1943–1949

Pellagatti A et al (2007) Lenalidomide inhibits the malignant clone and up-regulates the SPARC gene mapping to the commonly deleted region in 5q- syndrome patients. Proc Natl Acad Sci U S A 104:11406–11411

Pellagatti A et al (2008) Haploinsufficiency of RPS14 in 5q- syndrome is associated with deregulation of ribosomal- and translation-related genes. Br J Haematol 142:57–64

Perez B et al (2010) Germline mutations of the CBL gene define a new genetic syndrome with predisposition to juvenile myelomonocytic leukaemia. J Med Genet 47:686–691

Pilatrino C et al (2005) Increase in platelet count in older, poor-risk patients with acute myeloid leukemia or myelodysplastic syndrome treated with valproic acid and all-trans retinoic acid. Cancer 104:101–109

Pitchford CW, Hettinga AC, Reichard KK (2010) Fluorescence in situ hybridization testing for –5/5q, -7/7q, +8, and del(20q) in primary myelodysplastic syndrome correlates with conventional cytogenetics in the setting of an adequate study. Am J Clin Pathol 133:260–264

Plasilova M et al (2002) TRAIL (Apo2L) suppresses growth of primary human leukemia and myelodysplasia progenitors. Leukemia 16:67–73

Potapova A et al (2010) Epigenetic inactivation of tumour suppressor gene KLF11 in myelodysplastic syndromes*. Eur J Haematol 84:298–303

Qian J et al (2010) Methylation of DAPK1 promoter: frequent but not an adverse prognostic factor in myelodysplastic syndrome. Int J Lab Hematol 32:74–81

Quddus F et al (2010) Oral Ezatiostat HCl (TLK199) and Myelodysplastic syndrome: a case report of sustained hematologic response following an abbreviated exposure. J Hematol Oncol 3:16

Ravoet C et al (2008) Farnesyl transferase inhibitor (lonafarnib) in patients with myelodysplastic syndrome or secondary acute myeloid leukaemia: a phase II study. Ann Hematol 87:881–885

Raza A et al (1995a) Apoptosis in bone marrow biopsy samples involving stromal and hematopoietic cells in 50 patients with myelodysplastic syndromes. Blood 86:268–276

Raza A et al (1995b) Simultaneous assessment of cell kinetics and programmed cell death in bone marrow biopsies of myelodysplastics reveals extensive apoptosis as the probable basis for ineffective hematopoiesis. Am J Hematol 48:143–154

Raza A et al (1996) Novel insights into the biology of myelodysplastic syndromes: excessive apoptosis and the role of cytokines. Int J Hematol 63:265–278

Raza A et al (2008) Phase 2 study of lenalidomide in transfusion-dependent, low-risk, and intermediate-1 risk myelodysplastic syndromes with karyotypes other than deletion 5q. Blood 111:86–93

Raza A et al (2009a) Phase 1-2a multicenter dose-escalation study of ezatiostat hydrochloride liposomes for injection (Telintra, TLK199), a novel glutathione analog prodrug in patients with myelodysplastic syndrome. J Hematol Oncol 2:20

Raza A et al (2009b) Phase 1 multicenter dose-escalation study of ezatiostat hydrochloride (TLK199 tablets), a novel glutathione analog prodrug, in patients with myelodysplastic syndrome. Blood 113:6533–6540

Rizzo JD et al (2010) American Society of Clinical Oncology/American Society of Hematology clinical practice guideline update on the use of epoetin and darbepoetin in adult patients with cancer. J Clin Oncol 28:4996–5010

Rollison DE et al (2008) Epidemiology of myelodysplastic syndromes and chronic myeloproliferative disorders in the United States, 2001–2004, using data from the NAACCR and SEER programs. Blood 112:45–52

Romeo M, Chauffaille Mde L, Silva MR, Bahia DM, Kerbauy J (2002) Comparison of cytogenetics with FISH in 40 myelodysplastic syndrome patients. Leuk Res 26:993–996

Sanz C, Richard C, Prosper F, Fernandez-Luna JL (2002) Nuclear factor k B is activated in myelodysplastic bone marrow cells. Haematologica 87:1005–1006

Sauntharajah Y et al (2002) HLA-DR15 (DR2) is overrepresented in myelodysplastic syndrome and aplastic anemia and predicts a response to immunosuppression in myelodysplastic syndrome. Blood 100:1570–1574
Schanz J et al (2011) Coalesced multicentric analysis of 2,351 patients with myelodysplastic syndromes indicates an underestimation of poor-risk cytogenetics of myelodysplastic syndromes in the international prognostic scoring system. J Clin Oncol 29:1963–1970
Seal S et al (2008) Differential responses of FLIPLong and FLIPShort-overexpressing human myeloid leukemia cells to TNF-alpha and TRAIL-initiated apoptotic signals. Exp Hematol 36:1660–1672
Sekeres MA et al (2011) Subcutaneous or intravenous administration of romiplostim in thrombocytopenic patients with lower risk myelodysplastic syndromes. Cancer 117:992–1000
Shen L et al (2010) DNA methylation predicts survival and response to therapy in patients with myelodysplastic syndromes. J Clin Oncol 28:605–613
Shih LY et al (2005) Heterogeneous patterns of CEBPalpha mutation status in the progression of myelodysplastic syndrome and chronic myelomonocytic leukemia to acute myelogenous leukemia. Clin Cancer Res 11:1821–1826
Silverman LR et al (2002) Randomized controlled trial of azacitidine in patients with the myelodysplastic syndrome: a study of the cancer and leukemia group B. J Clin Oncol 20:2429–2440
Silverman LR et al (2006) Further analysis of trials with azacitidine in patients with myelodysplastic syndrome: studies 8421, 8921, and 9221 by the Cancer and Leukemia Group B. J Clin Oncol 24:3895–3903
Sloand EM et al (2005) Preferential suppression of trisomy 8 compared with normal hematopoietic cell growth by autologous lymphocytes in patients with trisomy 8 myelodysplastic syndrome. Blood 106:841–851
Sloand EM, Wu CO, Greenberg P, Young N, Barrett J (2008) Factors affecting response and survival in patients with myelodysplasia treated with immunosuppressive therapy. J Clin Oncol 26:2505–2511
Slovak ML et al (2002) 21q22 balanced chromosome aberrations in therapy-related hematopoietic disorders: report from an international workshop. Genes Chromosomes Cancer 33:379–394
Smith AE et al (2010) Next-generation sequencing of the TET2 gene in 355 MDS and CMML patients reveals low-abundance mutant clones with early origins, but indicates no definite prognostic value. Blood 116:3923–3932
Sokol L et al (2011) Identification of a risk dependent microRNA expression signature in myelodysplastic syndromes. Br J Haematol 153:24–32
Song WJ et al (1999) Haploinsufficiency of CBFA2 causes familial thrombocytopenia with propensity to develop acute myelogenous leukaemia. Nat Genet 23:166–175
Sportoletti P et al (2008) Npm1 is a haploinsufficient suppressor of myeloid and lymphoid malignancies in the mouse. Blood 111:3859–3862
Starczynowski DT et al (2008) High-resolution whole genome tiling path array CGH analysis of CD34+ cells from patients with low-risk myelodysplastic syndromes reveals cryptic copy number alterations and predicts overall and leukemia-free survival. Blood 112:3412–3424
Steensma DP (2010) Novel therapies for myelodysplastic syndromes. Hematol Oncol Clin North Am 24:423–441
Stifter G, Heiss S, Gastl G, Tzankov A, Stauder R (2005) Over-expression of tumor necrosis factor-alpha in bone marrow biopsies from patients with myelodysplastic syndromes: relationship to anemia and prognosis. Eur J Haematol 75:485–491
Stirewalt DL et al (2008) Tumour necrosis factor-induced gene expression in human marrow stroma: clues to the pathophysiology of MDS? Br J Haematol 140:444–453
Swerdlow SH, International Agency for Research on Cancer, World Health Organization, Louis A. Duhring Fund (2008) WHO classification of tumours of haematopoietic and lymphoid tissues. International Agency for Research on Cancer, Lyon
Tai IT, Tang MJ (2008) SPARC in cancer biology: its role in cancer progression and potential for therapy. Drug Resist Updat 11:231–246

Tehranchi R et al (2003) Granulocyte colony-stimulating factor inhibits spontaneous cytochrome c release and mitochondria-dependent apoptosis of myelodysplastic syndrome hematopoietic progenitors. Blood 101:1080–1086
Thiel A et al (2011) Comprehensive array CGH of normal karyotype myelodysplastic syndromes reveals hidden recurrent and individual genomic copy number alterations with prognostic relevance. Leukemia 25:387–399
Thol F et al (2010) IDH1 mutations in patients with myelodysplastic syndromes are associated with an unfavorable prognosis. Haematologica 95:1668–1674
Thol F et al (2011) Prognostic significance of ASXL1 mutations in patients with myelodysplastic syndromes. J Clin Oncol 29:2499–2506
Thol F et al (2012) Frequency and prognostic impact of mutations in SRSF2, U2AF1, and ZRSR2 in patients with myelodysplastic syndromes. Blood 119:3578–3584
Tiu RV et al (2011) Prognostic impact of SNP array karyotyping in myelodysplastic syndromes and related myeloid malignancies. Blood 117:4552–4560
van de Loosdrecht AA et al (2001) Mitochondrial disruption and limited apoptosis of erythroblasts are associated with high risk myelodysplasia. An ultrastructural analysis. Leuk Res 25:385–393
Voso MT et al (2009) Valproic acid at therapeutic plasma levels may increase 5-azacytidine efficacy in higher risk myelodysplastic syndromes. Clin Cancer Res 15:5002–5007
Voulgarelis M, Giannouli S, Ritis K, Tzioufas AG (2004) Myelodysplasia-associated autoimmunity: clinical and pathophysiologic concepts. Eur J Clin Invest 34:690–700
Walkley CR et al (2007) A microenvironment-induced myeloproliferative syndrome caused by retinoic acid receptor gamma deficiency. Cell 129:1097–1110
Wang J, Fernald AA, Anastasi J, Le Beau MM, Qian Z (2010) Haploinsufficiency of Apc leads to ineffective hematopoiesis. Blood 115:3481–3488
Watanabe-Okochi N et al (2008) AML1 mutations induced MDS and MDS/AML in a mouse BMT model. Blood 111:4297–4308
Wei S et al (2009) A critical role for phosphatase haplodeficiency in the selective suppression of deletion 5q MDS by lenalidomide. Proc Natl Acad Sci U S A 106:12974–12979
Wu SJ et al (2006) Clinical implications of SOCS1 methylation in myelodysplastic syndrome. Br J Haematol 135:317–323
Wulfert M et al (2008) Analysis of mitochondrial DNA in 104 patients with myelodysplastic syndromes. Exp Hematol 36:577–586
Xu F et al (2011) Overexpression of the EZH2, RING1 and BMI1 genes is common in myelodysplastic syndromes: relation to adverse epigenetic alteration and poor prognostic scoring. Ann Hematol 90:643–653
Yang W et al (2010) FISH analysis in addition to G-band karyotyping: utility in evaluation of myelodysplastic syndromes? Leuk Res 34:420–425
Yang H et al (2011) Levels of miR-29b do not predict for response in patients with acute myelogenous leukemia treated with the combination of 5-azacytidine, valproic acid, and ATRA. Am J Hematol 86:237–238
Ye Y et al (2009) Progressive chromatin repression and promoter methylation of CTNNA1 associated with advanced myeloid malignancies. Cancer Res 69:8482–8490
Yoshida K et al (2011) Frequent pathway mutations of splicing machinery in myelodysplasia. Nature 478:64–69
Zang DY, Goodwin RG, Loken MR, Bryant E, Deeg HJ (2001) Expression of tumor necrosis factor-related apoptosis-inducing ligand, Apo2L, and its receptors in myelodysplastic syndrome: effects on in vitro hemopoiesis. Blood 98:3058–3065
Zhou L et al (2008) Inhibition of the TGF-beta receptor I kinase promotes hematopoiesis in MDS. Blood 112:3434–3443
Zuo Z et al (2011) Circulating microRNAs let-7a and miR-16 predict progression-free survival and overall survival in patients with myelodysplastic syndrome. Blood 118:413–415

Chronic Myeloproliferative Disorders: From Molecular Pathogenesis to Targeted Therapy

Richard A. Walgren and Josef Prchal

Introduction

The elucidation of the molecular pathogenesis of a variety of serious or life-threatening conditions has led to the expectation that the development of safe and effective therapies will soon follow. This principle has previously been demonstrated in malignant hematology by the effectiveness of *ABL* kinase inhibitors in chronic myelogenous leukemia (CML). A new opportunity to deliver on this premise is underway in the closely related non-CML myeloproliferative neoplasms (MPNs).

The MPNs are a group of hematologic malignancies characterized by the clonal or oligoclonal proliferation of one or more myeloid lineages that arise from a polyclonal stem cell pool. The WHO classification (Swerdlow et al. 2008) of MPNs includes chronic myelogenous leukemia (CML), polycythemia vera (PV), essential thrombocythemia (ET), primary myelofibrosis (PMF), chronic neutrophilic leukemia, chronic eosinophilic leukemia not otherwise specified, mastocytosis, and MPN unclassifiable. This chapter will focus on the insights, research developments, and recent clinical trial results related to PV, ET, and PMF.

The modern classification of PV, ET, and PMF can be traced back to the description in 1892 by Louis Henri Vasquez of a patient with marked erythrocytosis and hepatosplenomegaly (Vasquez 1892). Vasquez postulated that this polycythemia vera resulted from a hematopoietic cell proliferation. Gustav Hueck added to the field the first description of the presence of bone marrow fibrosis and extramedullary hematopoiesis (Hueck 1879). In 1934, Emil Epstein and Alfred Goedel

R.A. Walgren, M.D., Ph.D. (✉)
Lilly Oncology, Eli Lilly and Company, Lilly Corporate Center, Indianapolis, IN, USA
e-mail: walgren_richard_a@lilly.com

J. Prchal, M.D.
Huntsman Cancer Hospital, Myeloproliferative Disorders Clinic,
1950 Circle of Hope, Salt Lake City, UT, USA

J. Tao and E. Sotomayor (eds.), *Hematologic Cancers: From Molecular Pathobiology to Targeted Therapeutics*, Cancer Growth and Progression 14,
DOI 10.1007/978-94-007-5028-9_12,

(Epstein and Goedel 1934) described a distinct clinical syndrome characterized by thrombocytosis in the absence of marked erythrocytosis. The association of these three diseases as a set of phenotypically related "myeloproliferative disorders" was contributed by William Dameshek (1951), who also noted that PV patients can later develop bone marrow fibrosis, leuko-erythroblastic changes, and increasing splenomegaly. Dameshek drew the conclusion that these changes were consistent with progression to "terminal myelofibrosis" and that it may be useful to consider the myeloproliferative disorders as "closely interrelated" disorders of bone marrow proliferation related to an "undiscovered stimulus."

The discovery of the Philadelphia chromosome and its later refinement as a balanced 9;22 translocation and the BCR/ABL transcript provided direction for understanding the molecular pathogenesis of CML (Nowell and Hungerford 1960; Rowley 1973). Subsequently, the stem cell origin of the malignant clone in CML could be clearly demonstrated by the detection of either the characteristic t(9;22) cytogenetic abnormality and its BCR/ABL transcript in all hematopoietic lineages by fluorescent in situ hybridization (FISH) or reverse transcription-polymerase chain reaction (RT-PCR) techniques. Targeting of the BCR/ABL transcript served as the rationale for the development of imatinib. Imatinib therapy has subsequently been shown to result in a reversal of the clonal dominance supported by the BCR/ABL fusion protein and results in normalization of cell morphology, peripheral blood counts, and molecular cytogenetics (including RT-PCR). A dramatic improvement in survival has been demonstrated with most therapy-compliant patients having sustained clinical remission, many with chromosomal and molecular remission. The increased understanding of the etiology of CML and the eventual therapeutic developments for CML drew a clear diagnostic and management separation between CML and the other classically described MPDs.

The magnitude of success that stemmed from targeting the BCR/ABL fusion protein offers promise that that this approach can be successful if applied to other clonal diseases associated with a unique somatic genetic lesion. To be true, three key requirements should be met. The first is the identification of a drugable target that is responsible for initiating and maintaining the disease phenotype. To achieve maximal benefit, interference with the function of that target must result in both cessation and reversal of the disease process (i.e. tumor cells can no longer survive or replicate and relevant disease associated consequences need to be reversible). Second, the intended target must be a common event present in most of the individuals affected by that disease. The frequency of the disease needs to be sufficient enough to support the clinical, scientific, and financial challenges that are inherent in drug development and clinical practice. A third key feature is the ability to selectively interfere with the unique processes and metabolism of the tumor and thus avoid unintended consequences (Walgren et al. 2005). The ease of achieving this requirement is target dependent and tissue dependent. Some targets may play significant roles in both normal homeostasis and in the neoplastic process. Other targets may have little that differentiates them from non-mutated or closely homologous forms. An inability to selectively target the form involved in the disease process may lead to dose limiting toxicities and thereby reduce the ability to maximize clinical effectiveness. This is a potential challenge that may have two different effects when one considers genetic lesions. On one hand, the ability to selectively target only the aberrant cellular protein or process limits toxicity, improves efficacy rates, and potentially enhances the ability to modulate (for example silence) the target for improved efficacy in a selected population. On the other hand, this high degree of selectivity potentially restricts therapeutic benefit to only those patients with that particular target. The sections that follow will first briefly review characteristics of PV, ET, and PMF and then discuss the ongoing research efforts while attempting to illustrate how well these therapeutic challenges are being addressed for the non-CML MPN patients.

Non-CML MPN Disease Characteristics

Polycythemia Vera

Polycythemia vera is a chronic neoplasm of the blood that is characterized by increased red blood cell production that occurs independently of normal regulatory mechanisms. PV has been described to have three phases:

1. a prodromal phase during which a mild erythrocytosis may be detected (often retrospectively). However, some patients with early PV have normal blood counts yet may present with typical PV complications such as Budd-Chiari syndrome and the full PV phenotype eventually develops.
2. an overt polycythemic phase with a noticeably increased red blood cell mass and common symptoms and signs such as rubor and pruritus.
3. a post-PV myelofibrotic phase (post-PV MF, a form of secondary MF) in which patients exhibit disease transformation characterized by progressive cytopenias, bone marrow fibrosis, and extramedullary hematopoiesis.

Prominent clinical features of PV include hypertension, vascular abnormalities such as rubor due to high RBC mass, venous or arterial thromboses such as MI, stroke, DVT, PE, portal or splenic vein thrombosis, headache, visual disturbances, protracted dizziness, pruritus, gout, and erythromelalgia. The clinical phenotype of PV is dominated by the vascular complications. A minority of PV patients may also experience an evolution of their disease into post-PV MF (the third phase of the disease) or acute leukemia (Passamonti et al. 2004).

Original reports of disease survival suggested a median survival of less than 2 years in non-phlebotomized patients (Tefferi 2003). Changes in medical practice now facilitate much earlier diagnosis and with modern therapies and improved supportive care the median expected survival is reported in the range of 10 years or more. However, even with these improvements in care, patients with PV have been reported to have a 1.6-fold higher risk of death than the general population and a 3.3-fold higher risk of death has been reported for patients who are younger than

50 years of age at the time of disease diagnosis (Passamonti et al. 2004; Gruppo Italiano Studio 1995; Anía et al. 1994). Moreover, all currently available therapies have significant side effects. For example, alkylating myelosuppressive agents are associated with a >3-fold increase in cancer-related mortality (see Current Therapy Section for further discussion) (Gruppo Italiano Studio 1995). While transformation to acute leukemia is a rare event in PV patients, occurring with an estimated incidence of about 5 per 1,000 person-years, the outcome is disappointingly poor when leukemic transformation does occur, with a median survival of 2.9 months, independent of treatment strategy chosen (best supportive care or intensive chemotherapy) (Passamonti et al. 2005). The short median survival of these patients impedes their ability to successfully identify a suitable donor and receive hematopoietic stem cell transplant.

Essential Thrombocythemia

Similar to PV, essential thrombocythemia is a chronic neoplasm of the blood. However, ET is associated with marked elevation of peripheral platelet counts in the absence of an increased red blood cell mass. Beginning with the onset of the disease, there is an increased frequency of both major hemorrhagic events and major thrombotic events. In the absence of opportunities for routine complete blood cell count screening, many ET patients will first present with a potentially life threatening vascular complication. Major thrombotic events have been found to be prevalent in 7.6–29.4% of ET patients upon diagnosis and in 5.3–30.7% of patients during follow-up, with the occurrence of arterial thrombosis accounting for more than two-thirds of thrombotic events. Ischemic stroke still remains the most common thrombotic event among patients with ET (Papadakis et al. 2010). A retrospective study by GIMEMA (Gruppo Italiano Malattie Ematologiche dell'Adulto) found the recurrence of thrombosis in PV or ET patients with a history of major thrombosis (n = 494) to be high, with a probability of 17.7% at 2 years after the first thrombotic event. The risk of recurrence increased further to 30.8% at 5 years, and 49.9% at 10 years. Despite medical diagnosis and management of the ET patients, the nature of recurrent thrombosis continues to be life threatening. In their study, De Stefano et al. report that the most frequent first recurrent thrombosis was cerebrovascular disease (i.e. ischemic stroke or transient ischemic attacks) in 191 cases, followed by venous thromboembolism (160/494), acute coronary syndrome (106/494), and peripheral arterial thrombosis (44/494) (De Stefano et al. 2008).

The survival of patients with ET is primarily defined by disease-related complications, the primary ones being vascular disease (thrombosis and hemorrhage) and transformation to either myelofibrosis or leukemia (Cervantes et al. 2008). In a study conducted in Olmsted County, Minnesota, 5 and 10-year survival were a respectable 74.4 and 61.3%, correspondingly, but both were significantly lower than expected for age-matched controls. The cause of death was related to complications of ET, such as leukemia transformation, thrombotic complications, and gastrointestinal hemorrhage (Mesa et al. 1999). Similar to the disease history of PV, as the time from diagnosis extends, the cumulative risk of transformation to leukemia and myelofibrosis is increased for patients with ET. In one study (Tefferi 2003), the cumulative probability for AML transformation was reported to be 1.4% at 10 years and 8.1% at 20 years, with a median time to AML transformation of 13.8 years. A higher risk was seen for transformation to myelofibrosis (i.e. post-ET MF) of 3.8% at 10 years and 19.9% at 20 years with a median time to transformation of 12.4 years.

Primary Myelofibrosis

Primary myelofibrosis (PMF) is a clonal neoplasm of the blood which is characterized by a progressive evolution or worsening of bone marrow fibrosis. The bone marrow of PMF patients may also have proliferation of megakaryocytes and granulocytes. In the fibrotic stage of the disease, the bone marrow has marked reticulin or collagen fibrosis often with osteosclerosis. Concurrently, the fibrotic stage may be characterized by leukoerythroblastosis and extramedullary hematopoiesis associated with splenomegaly caused by clonal circulating progenitor cells as demonstrated by the concordance between bone marrow and splenic cytogenetic clones (Mesa et al. 2001; Wolf and Neiman 1987). Some clinical investigators recognize an early stage of the disease, termed the "prefibrotic phase", that may present with a hypercellular bone marrow, thrombocytosis, and atypical megakaryocytes with minimal or no appreciable reticulin fibrosis.

A diagnosis of PMF is associated with a poorer prognosis than that for PV or ET. PMF is associated with a substantial reduction in life expectancy, on the order of 31% compared to gender- and age-adjusted controls (Cervantes et al. 2008; Rozman et al. 1991), with reports of median overall survival ranging from 3.5 to 7 years (Cervantes et al. 1997, 2001, 2008, 2009; Cervantes and Barosi 2005; Tefferi and Elliott 2007). The largest prognostic study reported for PMF included over 1,000 patients from 7 institutions with a reported median overall survival of 5.7 years. At the time the study was reported, 517 deaths had occurred. Among the 276 cases where the cause of death was reported, the most frequent cause of death was transformation to acute leukemia (86 cases, 31%), followed by progression of myelofibrosis without leukemic transformation (50 cases, 18%), thrombosis and cardiovascular complications (37 cases, 13%), infection (n=29, 11%), bleeding outside of the setting of acute transformation (n=14, 5%), portal hypertension (n=12, 4%), and other causes (n=48, including 12 cases of second neoplasias) (Cervantes et al. 2009).

A study of 337 patients with primary or secondary myelofibrosis in chronic phase found the risk of progressing into blast phase (persistent elevation of blasts ≥20% in blood or bone marrow) to be 4, 11, and 22% at 1, 3, and 5 years, respectively (Tam et al. 2009). The results are more disappointing when one considers that progression into blast phase is uniformly fatal (Vaidya et al. 2009). Analysis of the PMF population from the Olmsted County, MN study reveals a median time to progression (defined as a decrease in hemoglobin of 2 g/dL, progressive splenomegaly, or onset of hypercatabolic symptoms) of 7 months, and 3-year survival of 52.4%, which was significantly lower than expected for age-matched controls (Mesa et al. 1999). Also concerning is the observation that the risk of developing a second disease of the marrow is high in PMF patients. In a Swedish study of 1,368 patients with PMF spanning the interval from 1958 to 2004, the standardized incidence ratio (SIR; calculated as the ratio of observed to expected number of cases) for any second leukemia was 26.6 with a median time of 2 years. Of the reported secondary leukemias, acute myeloid leukemia was the most frequent and was diagnosed a median of 3 years after PMF, with a SIR of 73.0 (Hemminki et al. 2009).

Allogeneic or syngeneic hematopoietic stem cell transplantation (HSCT) is the only identified curative treatment for patients with PMF (Hoffman et al. 2007). There are a number of associated challenges that limit the widespread use of HSCT including availability of a suitable matched donor, selection of appropriate conditioning or graft versus host disease (GVHD) prophylaxis regimens, and selection of recipients. In addition, the average age at diagnosis for PMF is 67, and even nonmyeloablative allogeneic HSCT is rarely done after age 65. Unfortunately, no randomized prospective clinical trials are available to facilitate these decisions and the optimal use of HSCT is not clear. Retrospective analyses of published data indicate that HSCT risks are significant (reviewed in Mesa 2010) and depend on patient age, donor compatibility, conditioning regimen, and other factors. The frequency of treatment related mortality ranges between 10 and 30%. Rates of acute GVHD and chronic GVHD are higher at 10–60% and up to 85% respectively, and overall survival ranges from 30 to 60%. Interestingly, the use of HSCT has been associated with resolution of two key chronic features of MF – splenic enlargement and bone marrow fibrosis (Ciurea et al. 2008). In a small case series, there was a progressive resolution of bone marrow fibrosis noted at 3 and 12 months post-HSCT (reducing from a median pre-treatment of grade 2–3 to a 12 month median < grade 1). Similarly, spleen sizes were noted to have “progressively decreased in every patient evaluable at each time point.” While HSCT has not yet become an optimized or even a broadly applicable treatment modality, these observations suggest that with effective elimination of the PMF clone these two key chronic and debilitating features of PMF may be reversible.

MPN Research

In 1974, the observation was made that bone marrow cells from PV patients, but not normal volunteers, were able to give rise to erythroid colonies in the absence of exogenous cytokines. This capacity is defined as endogenous erythroid colony (EEC) formation (Prchal and Axelrad 1974). EEC formation was also observed in subsets of ET and PMF patients. These results provided new insights into the pathophysiology of MPNs and served to confirm the clinical conjectures from the proceeding century. The wider availability of sequencing and array techniques in the past decade provided new opportunities to gain insight into the pathophysiology of MPN. Studies examining gene expression profiles led to the identification of several genes that were dysregulated in MPN such as *MPL* (Moliterno et al. 1998), *BCL-XL* (Silva et al. 1998), and *PRV-1* (Temerinac et al. 2000). In 2005, results emerged from several labs that pointed toward one frequent genetic lesion as a common etiology across PV, ET, and PMF.

Using a liquid culture system to study CD34+ PV cells, Ugo et al (2004) observed that whereas erythroid differentiation was an erythropoietin (EPO)-independent phenomenon, it was still mediated by signaling pathways identical to those in EPO-induced differentiation. This signaling pathway was sensitive to inhibitors of EPO-induced differentiation such as the PI3K inhibitor LY294002, the Src kinase family inhibitor PP2, and the JAK2[1] inhibitor AG490. As the highest upstream kinase identified, the role of JAK2 gained further attention. Use of short interfering RNA (siRNA) decreased JAK2 protein levels, and in cells from PV patients, JAK2 siRNA impaired spontaneous erythroid differentiation while markedly inhibiting EEC formation thus confirming the results seen with the small molecule inhibitors (James et al. 2005). These results prompted the same group, led by William Vainchenker, to look for activating mutations in the *JAK2* gene and directly lead to the identification of a G-to-T mutation at nucleotide 1849. This coding mutation leads to a substitution of phenylalanine for valine at position 617 (V617F). Confirmation of this result demonstrated that the same mutation was present in samples from 40 to 45 PV patients. However, the mutation was not observed in 15 controls, nor was it found in samples from 35 patients with secondary erythrocytosis. Further examination showed that this mutation was an acquired phenomenon.

Independently, Anthony Green and his colleagues (James et al. 2005) sequenced the coding exons of *JAK2* in 73 PV patients, 51 ET patients, 16 PMF patients, and 90 controls. The V617F mutation was not observed in any controls, but it was identified in 97, 57, and 50% of the PV, ET, and PMF patients, respectively. Their results also agreed that the gain of this mutation is an acquired event that arose in a multipotent progenitor that is capable of giving rise to erythroid and myeloid cells. They were able to demonstrate that the V617F mutation was present in all EPO-independent erythroid colonies.

Based on the speculation that loss of heterozygosity (LOH) could be a molecular basis of PV, Kralovics et al. (2002) found that LOH on the short arm of chromosome 9 (9pLOH) was a recurrent event in MPN suggesting that 9p could harbor a mutation responsible for the clonal expansion of hematopoietic cells in MPN. To test their hypothesis, Kralovics et al. (2005) performed microsatellite mapping of the 9pLOH region and DNA sequencing in 244 patients with myeloproliferative disorders

[1] Using HUGO Gene Nomenclature, human gene symbols are italicized, with all letters in uppercase (*JAK2*). Protein designations are the same as the gene symbol but are not italicized (JAK2). Mouse gene symbols are italicized, with only the first letter in uppercase (*Jak2*). Murine proteins are designation in the same fashion but are not italicized (Jak2).

Table 1 Incidence rate and frequency of mutation of MPD

Disease	US age-adjusted incidence rate/100,000[a]	V617F mutation frequency
CML	1.5	
Non-CML MPN (PV+ET+PMF)	1.57	
PV	0.79	~95%[b]
ET	0.53	60%[b]
PMF	0.25	63.5%[c]

Data source: [a]SEER and NAACCR (Rollison et al. 2008; Ries et al. 2008)
[b]GIMEMA-MPD WP (Vannucchi et al. 2007)
[c]GIMEMA-Italian Registry of MF (Barosi et al. 2007)

(128 PV, 93 ET, and 23 PMF). Their mapping identified a 9pLOH region that included *JAK2*. When the 9pLOH was present, *JAK2* had a homozygous G to T transversion, resulting in the V617F substitution. All 51 patients with 9pLOH had the V617F mutation. Their functional studies indicated that the V617F mutation provided hematopoietic precursors with proliferative and survival advantages. Furthermore, the patients they studied with the V617F mutation had a significantly longer duration of disease, a higher rate of complications (fibrosis, hemorrhage, and thrombosis), and more frequently required treatment with cytoreductive therapy than patients with wild-type *JAK2*.

Using a high-throughput DNA resequencing technique to evaluate the kinome of MPN patients, a group lead by Gilliland also reported this recurrent non-synonymous variant of the *JAK2* gene (Levine et al. 2005). Their results demonstrated that the *JAK2* V617F mutation was present in 74% of their granulocyte DNA samples from 164 PV patients and approximately one-third of these individuals were homozygous for the mutation. In their sample set, the mutation was also present in granulocyte DNA samples from 37 of 115 ET and 16 of 46 PMF patients, but was not observed in 269 normal individuals.

Regardless of the approach used, each group arrived at the conclusion that an acquired point mutation in *JAK2* resulting in a valine to phenylalanine substitution was a common event in PV, ET, and PMF. The exact frequency of this mutation in each disease is debatable as evidenced by the reported frequencies in these initial reports. In part, this may be due to factors such as sample sizes, referral biases, and methods of detection. Subsequent studies suggest that frequencies may be higher (Table 1). In terms of feasibility for new therapy development, it was suggestive that the frequency of this mutation was sufficient to warrant additional research investment from academic and industry groups. However, did this observation reduce the attractiveness of this indication as a potential target or did this potentially drugable target have a large enough population to support trial enrollments and recovery of development costs? This translates to the basic question of how many patients have this disease? Data on the incidence and prevalence of MPNs are limited. Availability of this information in the European Union is variable across member nations. US population-based registries (SEER and the NAACR) began monitoring MPDs in 2001 when these disorders were officially reclassified as neoplasms. As a result, these sources are able to provide incidence rates but are not yet able to provide prevalence rates. The age-adjusted incidence of PV, ET, and PMF are similar and the combine incidence approximates that of CML (Table 1) (Rollison et al. 2008). In contrast to the prognosis of CML before imatinib therapy, the median survival of MPNs is longer, especially for PV and ET, and as a result the prevalence of MPNs is likely larger based on the multiple of incidence and median survival time. Hence, the results reported in 2005 demonstrated the existence of a new potential therapeutic target for a majority of MPN patients and new questions began to feed research interests.

Wild-Type and Mutant JAK2 Function

The JAK protein family is a group of cytosolic non-receptor tyrosine kinases that facilitate signal transduction from activated cytokine receptors. Upon cytokine stimulation, JAK protein interaction with the ligand bound receptor results in autophosphorylation of the JAK protein at its activation loop, in the kinase domain, resulting in an activated JAK. Activated JAK proteins subsequently phosphorylate the cytoplasmic domains of cognate receptors and additional signalling proteins such as signal transducer and activator of transcription family (STAT), the MAPK pathway, and PI3K-Akt (Yamaoka et al. 2004; Vainchenker et al. 2008). The mammalian JAK proteins, comprising JAK1, JAK2, JAK3, and TYK2, are structurally unique among protein tyrosine kinases (PTK) and are characterized by sharing a complex multidomain structure which consists of seven distinct domains termed the JAK homology (JH1–JH7) domains. Two of these domains at the C-terminal end (JH1 and JH2) are similar but not identical domains. The JH1 domain is a highly conserved PTK domain that is responsible for ATP binding and is critically important for the protein's physiological function. The cis JH2 domain, also called the pseudokinase domain or kinase-like domain, lacks catalytic activity but plays a crucial role in the regulation of the JH1 PTK domain (Boudeau et al. 2006; Saharinen and Silvennoinen 2002; Saharinen et al. 2000, 2003). When expressed *in vitro*, the JAK2 V167F protein has constitutive kinase activity in the absence of cytokines and is constitutively phosphorylated unlike wild-type JAK2 protein (Levine et al. 2005). The wild-type JAK2's phosphotransferase activity of the kinase domain is inhibited by the JH2 domain and deletion of this region in JAK2 or JAK3 significantly increases kinase activity. This JH2-mediated inhibition appears to be an inherent structurally dependent feature of the JAK protein family that does not require other regulatory proteins. The aminoterminal FERM domain plays an additional role in regulating JAK2 function both through interactions with the EPO-receptor (Funakoshi-Tago et al. 2008). The JH2 domain is capable of inhibiting the kinase activity noncompetitively, decreasing the enzyme's maximum velocity (V_{max}) without changing the substrate affinity (K_m) (Saharinen et al. 2000). The non-synonymous V617F mutation in this autoinhibitory JH2 pseudokinase domain abrogates this cis negative regulation by lowering its K_m value for substrates (Zhao et al. 2010) (Fig. 1).

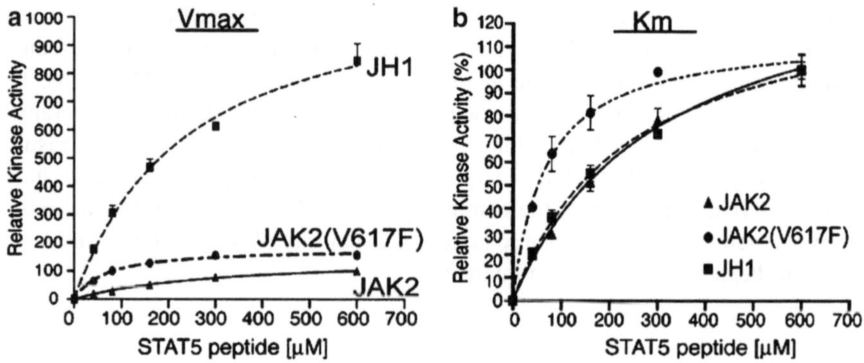

Fig. 1 JAK2(V617F) has a lower K_m towards a STAT5-derived peptide than JAK2 or JAK2 kinase domain (From Zhao et al. (2010)) (**a**) Comparison of V_{max} values for HA–JAK2, HA–JAK2(V617F) and HA–JH1. Data were normalized to maximal wild-type JAK2 activity. (**b**) Comparison of K_m values for HA–JAK2, HA–JAK2(V617F) and HA–JH1. The activity of the kinases was normalized to the maximal activity of each construct

Using biochemical methods to characterize enzymatic activity of JAK2 wild type and V617F mutant enzymes, Zhao et al (2010) demonstrated that the K_m of the V617F mutant was much lower than that of either the intact wild-type protein or the JH1 fragment across a number of substrates. In the case of STAT5 the K_m was 67 µM for JAK2 V617F, 306 µM for JAK2, and 239 µM for the JH1 domain. A similar trend was seen for a JAK2-derived peptide (a surrogate for measuring autophosphorylation). Interestingly, while the wild-type construct lacking the FERM domain was more active than the intact wild-type JAK2 (suggesting the FERM domain has an autoinhibitory role in JAK2), the activity of the JAK2 V617F mutant construct lacking the FERM domain was dramatically lower than the intact JAK2 V617F. Collectively, these results help to explain the constitutive activity seen in the V617F mutant through the demonstration that under normal intracellular conditions where concentrations of substrates are typically below saturation, the JAK2 V167F enzyme exhibits hyperactivity compared to the wild-type JAK2 enzyme. Moreover, the demonstration of selectivity in affinity between the wild-type and mutant proteins raises the possibility for selective inhibition of the mutant at the catalytic site.

How the JH1 and JH2 domains interact is not known precisely. Only a portion of the Jak2 (murine) kinase domain has been crystalized (Lucet et al. 2006). Computational models have been generated to aid in understanding this knowledge gap, and they suggest a possible JH1-JH2 interface between residues D994-E1024 in the JH1 domain and V617-E621 in the JH2 autoinhibitory domain (Lindauer et al. 2001; Giordanetto and Kroemer 2002).

An additional reported change in the regulation of JAK2 V617F is the ability to escape normal negative feedback regulation. For example, the suppressor of cytokine signalling proteins, SOCS1 and SOCS3, are normally capable of binding to wild-type JAK2 and inhibiting its kinase activity (Nicholson et al. 1999; Sasaki et al. 2000). In stark contrast, expression of SOCS3 appears to paradoxically increase

Table 2 Cytokine receptors and Janus kinase binding preferences

Receptor sub-type	Cytokine receptor	Janus kinase
Type I		
Homodimeric cytokine receptors	EPOR, TPOR, GHR, PRLR, G-CSFR	JAK2
Cytokine receptors sharing βc subunit	IL-3R, IL-5R, GM-CSFR	JAK2
Cytokine receptors sharing gp130 subunit	IL-6R, IL-11R, OSMR, LIFR	JAK1, JAK2, TYK2
Cytokine receptors sharing γc subunit	IL-2R, IL-4R, IL-7R, IL-9R, IL-15R, IL-21R	JAK1, JAK3
Type II		
Interferon & others	IFNα, IFNβ, IFNγ, IL-10R, IL-19R, IL-20R, IL-22R, IL-24R, IL-28R, IL-29R	JAK1, JAK2, TYK2

JAK2 V617F protein stability resulting in increased SOCS3 phosphorylation and increased JAK2 V617F phosphorylation (Hookham et al. 2007).

Three Diseases from One Genetic Lesion?

Based on the observation that the *JAK2* V617F mutation is not present in all of the PV, ET, and PMF patients one could question if this is a causal lesion. However, in aggregate the data suggest that there are multiple pathways for activation of JAK2 that appear to give rise to a functionally similar etiology that is defined by dysregulation of JAK2. The source for three different disease manifestations of this acquired genetic lesion appears to rest on two contributors – genetic background and gene dosage.

One consideration that has implications on this question is the diversity of roles that JAK2 plays in normal hematopoiesis. JAK2 is a binding partner for numerous Type I and Type II cytokine receptors (Table 2). Given the range of cytokine receptors and associated ligands that rely on JAK2, either alone or in combination with other Janus kinases, it is easy to see the crucial role that JAK2 can play in hematopoiesis starting from the level of long-term and short-term hematopoietic stem cell differentiation to the more lineage committed differentiation (Fig. 2). It is interesting to note that the role JAK2 plays is more limited along the avenues of differentiation that give rise to lymphocytes, both T-cells and B-cells.

A number of studies have demonstrated that retroviral expression of Jak2 V617F in murine bone marrow transplant assays results in significant PV-like phenotype in the transplant recipient mice (James et al. 2005; Lacout et al. 2006; Wernig et al. 2006; Zaleskas et al. 2006; Bumm et al. 2006). Recipient animals demonstrate a peripheral erythrocytosis which can progress to myelofibrosis, but thrombocytosis was not demonstrated in these studies. Leukocytosis was observed in the Balb/C

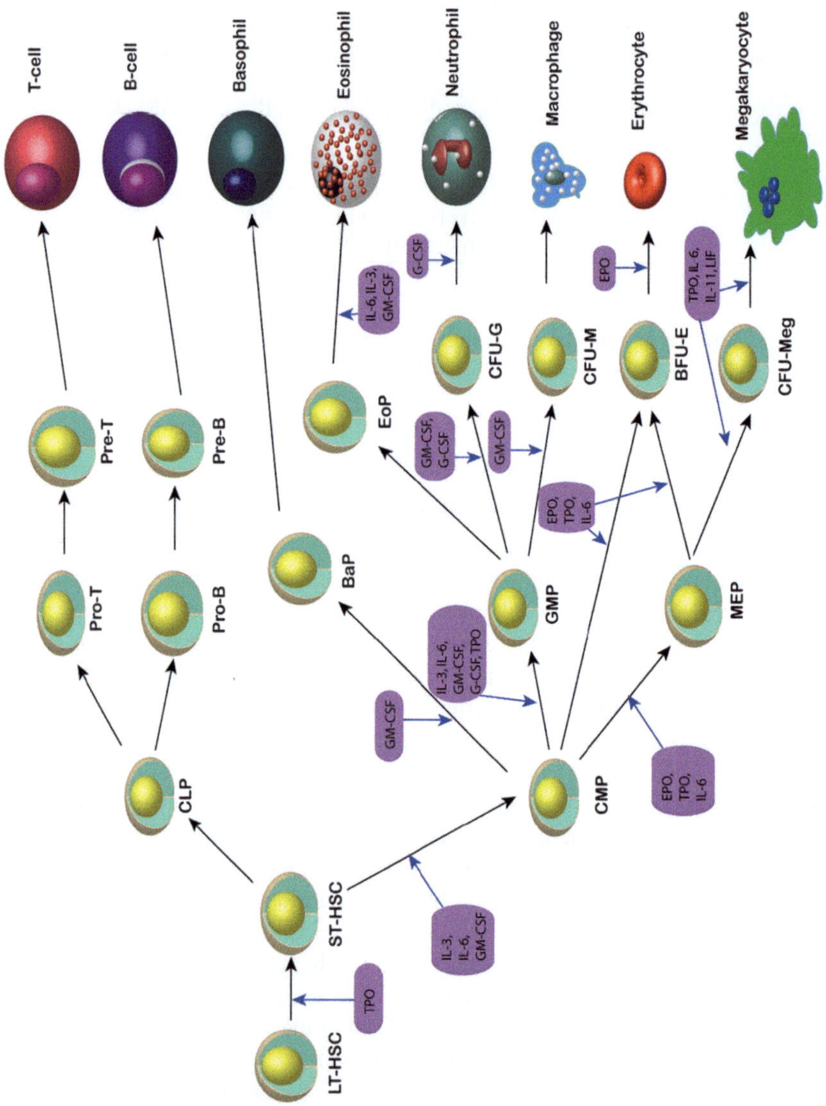

Fig. 2 Role of JAK2 dependent cytokine signaling in hematopoiesis

strain but was not present in studies with the C57Bl/6 strain. It has been suggested that the failure to observe an ET-like phenotype in these studies together with the observed difference in susceptibility to leukocytosis between genetic strains could stem from the need for additional genetic events to support the development of the different MPN phenotypes. Potential criticisms of these studies include their use of a murine *Jak2* cDNA in which the G to T mutation at codon 617 was introduced and the failure to evaluate the role of the natural promotor for *JAK2*. To address these issues, Tiedt et al. (2008) employed an inducible transgenic technique to demonstrate that human *JAK2* V617F is also capable of inducing an MPN phenotype in mice. Through the use of a combination of inducible promotor systems, they were able to develop transgenic strains with varying human JAK2 V617F expression levels. In mice with expression levels of JAK2 V617F that were lower than the endogenous wild-type Jak2, the resulting phenotype bore resemblance to ET with elevated platelet counts and moderate neutrophilia. In a transgenic strain that expressed JAK2 V617F at a level approximating that of the wild-type Jak2, a PV-like phenotype was observed with increased hemoglobin, thrombocytosis, and neutrophilia.

Employing a transgenic approach combined with the use of a tissue specific promoter, Xing et al. (2008) generated murine lines that model ET, PV, and PMF. *JAK2* V617F copy number varied in these three derived transgenic lines and was determined to be 13 ± 1.4, 1.9 ± 0.3, and 24 ± 2.7, respectively, for their A, B, and C mice. Transgenic line A mice, which had a high level of JAK2 V617F expression, demonstrated a marked increase in platelet number as well as increases in white blood cell counts, hemoglobin, and hematocrit, thus having a phenotype resembling that of ET or PV. The transgenic line B mice exhibited more moderate elevations in cell counts that did not reach the levels more characteristic of ET or PV. Spleen weights in the line A mice were 2 to 8-times larger than control animals and 10 of 26 line A mice developed fibrosis in the bone marrow and spleen after 30 weeks of age. These findings were not noted in examination of animals less than 30 weeks of age. Development of fibrosis was reported to be a much less frequent event in the lower copy number line B mice. The line C founder mouse, which had the highest gene copy number, demonstrated a more aggressive phenotype, and this animal died at 4 weeks. At that time the line C mouse was noted to have a spleen 10 times the size of comparably aged control mice.

Examination of heterozygous *Jak2* V617F Gene knock-in mice demonstrated constitutive Jak2 activation and autonomous erythroid progenitor cell growth. These mice manifested a severe PV-like disorder that progressed to PMF with decreased bone marrow cellularity and splenomegaly associated with marked increases in erythroid (88-fold) and myeloid (82-fold) precursors (Marty et al. 2010). Interestingly, the embryonic development of this disease was not fatal; however, the authors noted that the severity of the disease could contribute to embryo-lethality of a similar inherited disorder in humans. Akada et al. (2010) independently employed a Gene knock-in mouse model yielding heterozygous and homozygous Jak2 V167F expression patterns. Heterozygous expression resulted in a phenotype characterized by polycythemia due to excessive production of erythrocytes despite low serum EPO levels, increased hematocrit and hemoglobin, leukocytosis with neutrophilia, thrombocytosis, and extramedullary hematopoiesis with splenomegaly. These results demonstrate that heterozygous Jak2 V617F expression is sufficient for induction of PV. Homozygous expression was associated with greater increases in reticulocytes and peripheral cell counts as well as a more pronounced fibrosis of the bone marrow and splenomegaly. Basal activation of the Stat5, Akt, and Erk1/2 pathways was greater in the homozygous state suggesting that wild-type Jak2 may be able to compete with the mutant Jak2 V617F when coexpressed.

Greater understanding of the influence of genetic complexity of MPN is also emerging. Standard karyotyping techniques have shown abnormal cytogenetics are present in approximately 30% of MPN patients, occurring more frequently in PMF but less frequently in PV and ET (Gangat et al. 2008, 2009; Hussein et al. 2009; Panani 2007; Reilly 2008). As previously described in this chapter, microsatellite studies identified uniparental disomy of the short arms of both chromosomes 9 and 1 (9pUPD and 1pUPD), contributing to the identification of *JAK2* V617F and mutations in *MPL* respectively (Kralovics et al. 2002). The sensitivity of cytogenetic evaluation that is now possible has increased with the availability of high-resolution DNA microarrays. This has facilitated identification of deletions and UPDs of 4q associated with *TET2* (Delhommeau et al. 2009; Tefferi et al. 2009), as well as abnormalities in 11q associated with *CBL* (Dunbar et al. 2008; Sanada et al. 2009). In contrast with the 9p and 1p UPDs, which are known to be associated with enhance JAK2 signaling, the influence of mutations in *TET2* and *CBL* is presently not clear. Additional mutations in *ASXL1* (Carbuccia et al. 2009) and *IDH1/2* (Green and Beer 2010) and deletions in *IKZF1* (Jager et al. 2010) have been identified in association with leukemic transformation (post-MPN AML).

When higher resolution techniques have been employed the frequency of genomic aberrations appears to be more common. In a sample of 408 MPN patients (162 PV, 80 ET, 79 PMF, 29 post-MPN AML, and 58 patients with secondary MF or accelerated phase) analyzed by high-resolution SNP microarrays only 37.5% of patients had a wild-type karyotype and the remainder harbored at least one chromosomal aberration (Klampfl et al. 2011). Of these patients, 297 (72.8%) exhibited a *JAK2* mutation and all patients with 9pUPD (n = 169) were positive for the *JAK2* V167F mutation. Six of the seven patients with 1pUPD were positive for *MPL* W515L. An additional 25 aberrations were recurrent in 3 or more patients.

When Klampfl et al. (2011) further examined the frequency of chromosomal aberration and other disease factors, they noted no association with disease duration. However, patient age at the time of sample collection was positively associated with the number of defects. Interestingly, patients with *JAK2* mutations were not found to carry more chromosomal aberrations then their *JAK2 V617F* negative counterparts. No aberration was specifically associated with *JAK2* V617F negative MPN. *JAK2* V617F homozygosity (9pUPD) was associated with secondary myelofibrosis or accelerated phase (sMF/AP) suggesting that gene dosage may predispose for a higher risk of secondary myelofibrosis in both PV and ET (post-PV MF or post-ET MF). Chromosome 1q amplifications, all of which amplified *MDM4*, were also associated with sMF/AP as well as post-MPN AML. The *MDM4* gene product is an inhibitor of p53 and its amplification may set the stage for MPN disease progression as loss of p53 function is associated with a number of malignancies and has been shown to participate in leukemic transformation in MPN (Beer et al. 2010; Laurie et al. 2006; Riemenschneider et al. 2003).

Current Therapy

Polycythemia Vera (PV)

Current treatment practices do not cure PV but, rather, attempt to control the disease related symptoms and decrease the likelihood of complications. Historically, agents such as radioactive phosphorus and alkylating agents (melphalan and busulfan) were used to reduce red blood cell mass. Use of these agents is associated with an increased risk of leukemic transformation (Gruppo Italiano Studio 1995), and as a result, these agents are typically reserved for use in patients >70 yo. Currently, phlebotomy is the frontline modality employed by many practitioners with the goal of reducing the hematocrit to ≤45%. However, a prospective study of 1,638 PV patients found no difference in risk of death, thrombotic events, or hematologic progression when hematocrits were in a range between 35 and 55. Thus, the role of hematocrit in PV thrombosis and benefits of phlebotomies are unclear (Di Nisio et al. 2007). Findings from this same population did suggest that high platelet counts might be associated with a decreased risk of hematologic transformation and myelofibrosis. When compared to patients with a platelet count of $\leq 300 \times 10^9$/L, patients whose platelet counts were in the range of either $301–500 \times 10^9$/L or $>500 \times 10^9$/L, the risk of myelofibrosis was 54% lower (HR 0.46; 95% CI: 0.21–1.02, $P=0.550$) and 66% lower (HR 0.34; 95% CI: 0.12–0.97, $P=0.431$), respectively. This later finding is of particular interest in light of the previously discussed murine MPN model studies that demonstrated loss of thrombocytosis and more rapid progression of myelofibrotic features in models with higher gene dosage. However, when platelet count exceeds 100×10^{10}/L, there is an increased risk of bleeding attributable, at least in part, to an acquired von Willebrand disease (Michiels 1999).

After the initial reduction in red blood cells is achieved, patients are typically treated with maintenance phlebotomy with the goal to maintain the same target hematocrit. During maintenance therapy, the frequency of phlebotomy requirements will typically decline as patients develop an intentional iatrogenic iron deficiency state. The development of iron deficiency may assist in subsequent control of red blood cell counts. Phlebotomy may increase the number of platelets, and while it may attenuate some disease related symptoms, it typically does not reduce the size of an enlarged liver or spleen. Thus, even patients who respond to phlebotomy may also require pharmacologic intervention.

Non-specific cytoreductive therapy can be achieved with hydroxyurea (hydroxycarbamide). Myelosuppression with hydroxyurea offers a greater reduction in symptoms than phlebotomy. Although hydroxyurea can help to decrease splenomegaly, it rarely results in complete resolution. Treatment with hydroxyurea usually is well-tolerated but does require routine monitoring of cell counts and often requires dose modification or holidays to avoid significant myelosuppression. In some studies hydroxyurea therapy has been associated with poor tolerance due to the development of leg ulcers, buccal aphthous ulcers, gastric pain, or diarrhea (Najean and Rain 1997a, b). In addition, when used for many years there is concern that hydroxyurea therapy may increase the risk of transformation to leukemia. Despite numerous attempts to define this risk it remains the subject of debate (Finazzi et al. 2005). It is unclear whether this leukemogenesis is secondary to hydroxyurea or results from progression of the underlying MPN disease.

Alternative drugs for lowering the number of platelets, such as interferon-alpha and anagrelide, are sometimes used in younger people who may need treatment for long periods. The use of interferon and its pegylated forms for the management of MPNs is an area of active research. In exploratory studies, pegylated interferon-alpha has been reported to result in hematologic improvements in up to 80% of PV and ET patients. Pegylated interferon-alpha may reduce the allele burden of *JAK2* V617F and may restore polyclonal hematopoiesis with extended therapy (Liu et al. 2003; Kiladjian et al. 2006, 2008; Quintas-Cardama et al. 2009).

Low-dose aspirin is frequently used in PV as an antiplatelet therapy intended to reduce the risk of thrombotic events. However, use of aspirin has been controversial. At one time aspirin therapy was avoided due to the observation by the Polycythemia Vera Study Group that a high incidence of gastrointestinal bleeding occurred in patients who received 900 mg of aspirin daily (Tartaglia et al. 1986). A pilot study (Gruppo Italiano Studio Policitemia (GISP) 1997) suggested that the use of low-dose aspirin was not associated with bleeding complications in PV patients, and these results enabled further study of low-dose aspirin by the European Collaboration on Low-Dose Aspirin in Polycythemia Vera (ECLAP) (Finazzi 2004; Landolfi et al. 2004). The ECLAP projected included 1,638 patients, enrolling 1,120 into a prospective, observational cohort study. The remaining 518 patients were enrolled in a parallel randomized, double-blinded, placebo-controlled trial to assess the efficacy and safety of enteric coated low-dose aspirin (100 mg per day). Patients were excluded from the randomized trial and directed to the observational study if they had a recognized need for anti-thrombotic therapy, a contraindication to aspirin, or were unwilling to participate in the randomized study. In the observational study (Finazzi 2004), 40% of deaths were due to a cardiovascular event, and the incidence of fatal, major and minor thrombosis was 5.5 events per 100 patients per year. Factors associated with an increased risk of cardiovascular events included age greater than 65 years, history of thrombosis, smoking, hypertension, and congestive heart failure. Antiplatelet therapy was the only variable associated with a lower risk of thrombosis. In ECLAP's randomized study (Landolfi et al. 2004), treatment with aspirin was associated with a reduced, but not statistically significant, risk of nonfatal myocardial infarction, nonfatal stroke, or death from cardiovascular causes (RR 0.41, 95% CI 0.15–1.15; $P=0.09$). However, when secondary analysis also included pulmonary embolism and major venous thrombosis in the combined endpoint, the relative risk was significantly decreased (RR 0.40, 95% CI 0.18–0.91; $P=0.03$). This study did confirm the pilot study's finding that low-dose aspirin was not associated with a significant risk of major bleeding episodes (RR 1.62, 95% CI 0.27–9.71). In a systematic review of the literature that include the ECLAP study, Squizzato et al. (2008), concluded that the use of aspirin in PV was associated with a lower, but not statistically significant, risk of fatal thrombotic events in PV patients (OR 0.20, 95% CI 0.03–1.14).

Stemming from increased turn-over of nucleic acids and an associated increased production of uric acid, PV and other MPN patients may develop hyperuricemia, gout, or renal injury that can be managed with allopurinol therapy. Other drugs may be considered to assist in management of symptom burden such as treatment of itching with antihistamines, and use of NSAIDs for headaches, bone pain, and burning sensations in the hands and feet.

Essential Thrombocythemia (ET)

Therapy for ET is primarily focused on management of peripheral cell counts with cytotoxic agents and reduction of arterial and venous thrombotic events. Drugs frequently used include hydroxyurea, anagrelide, and interferon-alpha. Treatment with one of these drugs and aspirin is typically started when clotting complications develop. Therapy may be initiated prophylactically when indicated by a risk assessment. Independent predictive risk factors include: age, platelet count $>1{,}000\times10^9$/L, leukocyte count $>12\times10^9$/L, prior history of thrombosis, smoking, and diabetes (De Stefano et al. 2008; Tam et al. 2009; Landolfi et al. 2006; Elliott and Tefferi 2005). While not yet independently validated, De Stefano et al (2008) have also reported the presence of the *JAK2* V617F mutation to be associated with a 3.8-fold enhanced risk for thrombosis in patients less than 60 years of age. If drug treatment does not sufficiently slow platelet production in emergent situations drug therapy may be replaced by platelet pheresis.

The use of either hydroxyurea and aspirin or anagrelide and aspirin has been compared in a large randomized study. In high-risk patients with ET, hydroxyurea with aspirin was demonstrated to have an overall superiority at the composite endpoint of risk of arterial thrombosis, venous thrombosis, serious hemorrhage, or death from thrombotic or hemorrhagic causes (Harrison et al. 2005). While both treatment assignments were equivalent in long-term control of platelet counts, the anagrelide plus aspirin treatment was associated with significantly higher rates of arterial thrombosis (OR 2.16, 95% CI 1.27–3.69, $P=0.004$), serious hemorrhage (OR 2.61, 95% CI 1.27–5.33, $P=0.008$), and transformation to myelofibrosis (OR 2.92, 85% CI 1.24–6.86, $P=0.01$). Hydroxyurea plus aspirin was noted to be better tolerated with a lower treatment discontinuation rate. In contrast, anagrelide was noted to have better activity preventing venous thrombosis (OR 0.27, 95% CI 0.11–0.71, $P=0.006$). Looking at the risk of hematologic transformation, 16 of 405 ET patients receiving anagrelide plus aspirin transformed to post-ET MF compared to

5 of 404 on the hydroxyurea plus aspirin arm. The estimated actuarial risk of secondary myelofibrosis 5 years after trial entry was 2% for the hydroxyurea group (95% CI 0–5) and 7% for the anagrelide group (95% CI 3–10).

Myelofibrosis (MF)

Although classified differently by the WHO, primary myelofibrosis (PMF) and secondary myelofibrosis are clinically managed similarly. Generally speaking, outcomes in MF are determined by the degree of retained bone marrow function, the risk of arterial and venous cardiovascular events, and by the risk of leukemic transformation. MF may progress slowly as some individuals may live for 10 years or longer. In others with higher risk or lower levels of bone marrow function, the disorder can worsen rapidly. A number of risk stratifications techniques are available to help individualize a patient's risk assessment and treatment strategy. The Lille scoring system has been a widely used method of risk stratification based on two adverse prognostic factors (Dupriez et al. 1996). Scoring based on the presence of a hemoglobin <10 g/dL and a WBC of either <4 or $>30 \times 10^9$/L, allows separation into one of three groups with *low* (0 factors), *intermediate* (1 factor), and *high risk* (2 factors), associated with respective median survivals of 93, 26, and 13 months. Other more recent risk stratification tools for MF include the International Prognostic Scoring System (IPSS) (Cervantes et al. 2009), the Dynamic International Prognostic Scoring System (DIPSS) (Passamonti et al. 2010) and the Dynamic International Prognostic Scoring System Plus (DIPSS Plus) (Gangat et al. 2011). The IPSS uses five adverse risk factors determined at the time of diagnosis. Age >65 years, the presence of constitutional symptoms, hemoglobin <10 g/dL, leukocyte count $>25 \times 10^9$/L, and circulating blasts ≥1%. Each risk factor is assign 1 adverse risk point. Detection of 0, 1, 2, and ≥3 risk factors defines *low*, *intermediate-1*, *intermediate-2*, or *high-risk* and the corresponding association with median survivals of 11.3, 7.9, 4, and 2.3 years. The limitation of this tool is its restriction for use only at the time of initial diagnosis.

The introduction of the DIPSS, which was also developed by the International Working Group for Myeloproliferative Neoplasms Research and Treatment, allows application of the tool throughout the course of disease (Passamonti et al. 2010). DIPSS scoring assesses the same risks used in the IPSS, but DIPSS assigns a risk score of 2 points for hemoglobin <10 g/dL and adjusts the assignment of risk category to the following score totals: *low* (0 adverse points), *intermediate-1* (1 or 2 points), *intermediate-2* (3 or 4 points), and *high risk* (5 or 6 points). With the DIPSS, the corresponding median survivals were not reached for the low risk group, and were 14.2, 4, and 1.5 years respectively for intermediate-1, intermediate-2, and high risk scores. Whereas the DIPSS was validated for use during clinical course to assist in treatment decisions, the DIPSS does not allow modification of risk scoring for cytogenetic abnormalities, platelet counts, or transfusion status. These deficiencies have been addressed in the DIPSS Plus with the assignment of 1 point each for unfavorable karyotype, platelets lower than 100×10^9/L, and need for red cell transfusion (Gangat et al. 2011). Unfavorable karyotypes were defined as complex karyotype or single or two abnormalities including +8, −7/7q−, i(17q), −5/5q−, 12p−, inv(3), or 11q23 rearrangement. In this tool, categories were assigned as *low risk* with 0 points, *intermediate-1* with 1 point, *intermediate-2* with 2–3 points and *high risk* with 4 or more points. The median survival times associated with each DIPSS Plus risk group are 180, 80, 35, and 16 months respectively. Each of these models has merits for application with current treatment options, and yet each has omitted potentially predictive factors in order to reduce complexity and maintain usability. The future will undoubtably present new refinements to these models and new models are likely, especially as new treatment options and refinements in disease understanding become available.

Once the risk is assessed, treatment options for MF are limited and not satisfactory. Currently, only stem cell transplantation can cure the disorder. However, HSCT is not a treatment option for most MF patients and is often performed under research protocols. Alternate treatments used in MF are intended to relieve symptoms or prevent complications and little is available to delay the progression of the disorder.

A prominent disease feature requiring treatment is anemia. Anemia in MF is often a challenge to treat due to its complex origins. In a disease characterized by excessive JAK2-STAT signalling it is not surprising that EPO stimulating agents are often ineffective (Huang and Tefferi 2009). Use of single agent or combination use of an androgen and/or prednisone temporarily lessens the severity of the anemia in about one third of people with myelofibrosis (Cervantes et al. 2007). The immunomodulatory IMiDs (i.e. thalidomide, lenalidomide, and pomalidomide) which inhibit TNF-α and have been tested for their ability to improve anemia in myelodysplastic syndrome are being explored for use in MF (Begna et al. 2011; Lacy and Tefferi 2011; Mesa et al. 2010). Use of the DNA hypomethylating agent 5-azacitidine has also been examined in phase II studies in MF patients. Despite demonstrating the induction of global hypomethylation, 5-azacitidine therapy was associated with limited clinical activity in MF patients (Mesa et al. 2008; Quintas-Cardama et al. 2008). Although potential therapies for anemia are available, the efficacy of these therapies in MF is limited, and many patients will require red blood cell transfusions. Platelet transfusions may also be required to treat thrombocytopenia with bleeding. The degree of transfusion requirements may also be a confounding issue for individuals who are transplant candidates due to the development of transfusion-associated allosensitization. As with other neoplasms, MF related myelosuppression can predispose patients to bacterial infections and the resultant need for antibiotic therapy.

Symptomatic splenomegaly is another characteristic problem in MF that is thought to develop from a combination of factors including sequestration of immature circulating myeloid progenitors and proliferation of extramedullary hematopoiesis (Mesa et al. 2001; Zhang and Lewis 1989). Available treatment options include hydroxyurea, surgery, and splenic irradiation. Use of hydroxyurea has been discussed in the sections on PV and ET above. Many of the same tolerability and safety issues apply in the management of MF and may even be exacerbated by the need to use higher doses of hydroxyurea (i.e. 2–3 g/day) (Mesa 2009). The resulting cytopenias may in fact be problematic limiting the ability to continue high doses. Even with the use of higher doses, it is not common to achieve a sustained 50% reduction in spleen size which is a bench mark criteria for clinical improvement as defined by the International Working Group for Myelofibrosis Research and Treatment (Tefferi et al. 2006). Therapy with alkylating agents can be considered for control of splenic enlargement in some MF patients. This approach should be undertaken with caution due to the potential risk of therapy associated leukemia. With oral melphalan therapy, two-thirds of patients may have reductions in spleen size, but in one study, 26% of treated patients developed acute leukemia (Petti et al. 2002).

Therapeutic splenectomy is another potential option. Removal of the spleen may increase the number of red blood cells and reduce the need for transfusions in some patients. Splenectomy has no clear positive impact on survival or disease course and may, in fact, have a net negative effect due to the associated rates of complication (27.7%) and fatality (6.7%) (Mesa et al. 2006).

A third option for management of splenomegaly is radiotherapy. Extramedullary hematopoiesis is as sensitive to radiation as is medullary hematopoiesis. Thus, involved sites such as the spleen, lungs, or paraspinal masses are sites suitable for external beam radiotherapy. Unfortunately, the associated myelosuppression from this procedure can be severe, the long term benefit is often limited, and adjacent tissue injury can create additional problems (e.g. abdominal visceral adhesions) (Mesa 2009).

Evolution in Therapeutic Understanding

Whether JAK2 dysregulation is the initial disease defining event or a consequence remains unanswered. However, the evolving *in vivo* pre-clinical data does clearly illustrate that constitutive activity of JAK2 V617F is sufficient to recapitulate the MPN phenotypes. As such, there is interest in determining whether inhibition of JAK2 or selective inhibition of JAK2 V167F could result in clinical benefits. In 2007, the first JAK inhibitor was tested in MF patients. This is a relatively quick response time compared to that for CML where decades passed between the identification of the genetic lesion and the introduction of a targeted inhibitor. The ability to rapidly move an experimental therapeutic into an MPN focused clinical trial was facilitated by the availability of multitargeted kinase inhibitors that had activity against the Janus kinase family. At the time that JAK inhibitors started clinical testing it was unclear whether responses in MPN patients would parallel that observed in CML with imatinib. To date this has not been the case. However, lack of selectivity may limit the ability to adequately inhibit the mutant clones due to toxicities stemming from non-selective inhibition. Recognizing this possibility, other drug development groups have taken an approach of identifying new molecules that possess higher degrees of selectivity for JAK2 (TG101348) or for the JAK2 V617F mutant (LY2784544) (Hood et al. 2007; Ma et al. 2010).

Non-clinical testing of the JAK2 inhibitors has been based on a combination of *in vitro* enzymatic biochemical assays, *in vitro* cell based assays, *in vitro* model testing, and *ex vivo* testing of patient samples such as the endogenous erythroid progenitor (EEC) assay. Testing of potential candidate molecules typically begins with biochemical enzyme screening. Molecules of interest are then selected for further testing. Most cell based proliferation studies have verified *in vitro* enzyme assay results, demonstrating effective inhibition of JAK2 V617F expressing cells. The observed potency in enzymatic assays is often higher (i.e. lower IC_{50}) than what is observed in cell based assays (Table 3). Interrogations of signalling pathways, also, typically confirm reductions in JAK2 dependent signalling pathways such as STAT3 or STAT5 phosphorylation. Several agents (including TG101348 (Lasho et al. 2008), CEP-701 (Hexner et al. 2008), SGI-1252 (Ahmed et al. 2011), LY274544 (Florensa et al. 2010), CYT387 (Pardanani et al. 2009)) have demonstrated the ability to inhibit growth-factor independent progenitor cell growth in patient derived samples.

Much of the available non-clinical study results for the JAK2 targeting molecules currently in clinical development come from *in vitro* work and less information is available regarding testing in the *in vivo* murine models. In part, this is a reflection of the fact that development of many of these molecules occurred before the development of the murine models. As a result, few of these agents have been tested in the context of different gene dose levels or across multiple genetic backgrounds.

Using a transgenic mouse model, ruxolitinib (INCB018424), an inhibitor of JAK1 and JAK2 (Table 3), was shown to decrease splenomegaly and eliminate neoplastic cells from the bone marrow, spleen and liver (Fridman et al. 2007). XL019 and SGI-1252 have been studied in a tumor xenograft model with human erythroid leukemia cells (HEL), a *JAK2* V617F homozygous cell line, injected subcutaneously in athymic nu/nu mice. In this model, a reduction in STAT5 phosphorylation but not total STAT5 levels was observed when SGI-1252 was dosed at 400 mg/kg orally 3 times a week (Ahmed et al. 2011). Dosing with XL019 twice daily led to an increase in apoptosis and decreased tumor vasculature (Verstovsek et al. 2007). TG101348 was studied in mice who had received bone marrow cells from donor mice after the cells had been retrovirally transfected with either *Jak2* wild-type or *Jak2* V617F. Transplanted mice who received TG101348 at 120 mg/kg oral gavage twice a day demonstrated a decrease in hematocrit, splenomegaly, and reticulin fibrosis in the bone marrow and had a prolonged survival (Wernig et al. 2008).

Another MPN model employs the use of SCID mice injected with GFP-positive BaF/3 cells expressing JAK2 V617F. With this model, normal murine peripheral blood cell counts (i.e. endogenous wild-type bone marrow derived cells) are suppressed. In addition, BaF/3 *JAK2* V617F cells preferentially localize to the spleen, resulting in splenomegaly. Oral treatment with TG101209 at 100 mg/kg twice daily for 10 days prolonged survival and reduced STAT5 phosphorylation (Lasho et al. 2008). Oral administration of LY2784544 for 7 or 14 days in the BaF/3 *JAK2* V617F xenograft model demonstrated dose-dependent reductions in STAT5 phosphorylation, reductions in splenic enlargement, and a decreased BaF/3 *JAK2* V617F-GFP tumor burden with a TED50 of 13.7 mg/kg (Ma et al. 2010). LY2784544, demonstrated an apparent dose dependent selectivity for the mutant JAK2 V617F kinase (Table 3) and in the BaF/3 *JAK2* V617F xenograft model, treatment with LY2784544 showed no effect on normal murine erythroid progenitors (CD71/Ter119 positive

Table 3 Kinase selectivity of JAK2 inhibitors in clinical development

Inhibitor	Assay type	Literature reported IC_{50} (μM)					Other reported targets	References
		JAK1	JAK2	JAK2 V617F	JAK3	TYK2		
XL019	Enzyme	0.130	0.002		0.250	0.340		Paquette et al. (2008) and Verstovsek et al. (2007)
ruxolitinib (INCB018424)	Enzyme	0.0033	0.0028		0.428	0.019		Quintas-Cardama et al. (2010)
	Cell prolif.			0.127				
lestaurtinib (CEP701)	Enzyme		0.001				FLT3, PDGFR, Trk-A, RET	Dobrzanski et al. (2006) and Verstovsek (2010)
TG101348	Enzyme	0.105	0.003		0.996	0.405		Hood et al. (2007) and Wernig et al. (2008)
	Cell prolif.			0.270				
SB1518	Enzyme		0.023	0.019			FLT3	Verstovsek et al. (2010a)
LY2784544	Cell prolif.		1.36	0.003				Ma et al. (2010)
CYT387	Enzyme	0.011	0.018		0.155	0.017		Pardanani et al. (2009) and Tyner et al. (2010)
	Cell prolif.		1.424	1.5				

cells) or neutrophils/monocytes (CD11b/Gr-1 positive cells) after treatment for 7 days. Treatment with LY2784544 at 80 mg/kg restored the murine platelet counts to the levels seen in untreated non-tumor bearing animals.

In a murine model of MPN in which lethally irradiated Balb/c mice were transplanted with bone marrow donor cells that had been retrovirally transduced with *JAK2* V617F, twice daily oral gavage treatment with CYT387 was associated with a dose-dependent reduction or normalization of white cell counts, hematocrit, and spleen size (Tyner et al. 2010). Treatment at the highest dose (50 mg/kg twice daily) was associated with a statistically significant reduction in mutant positive cells in the spleen but failed to eliminate the *JAK2* V617F/GFP-positive cell population.

Emerging Clinical Trial Results

Clinical trial results for a number of the JAK2 targeting agents are emerging in abstract form, and peer-reviewed results have been published describing the exploratory experience in primary and secondary MF with three of these agents (lestaurtinib, ruxolitinib, and TG101348) (Verstovsek et al. 2010b; Pardanani et al. 2007; Santos et al. 2009).

Clinical development of lestaurtinib (CEP-701) in MF began in phase II prior to development as a FLT-3 inhibitor for use in AML. Studies in AML had previously defined a recommended oral dose regimen of 80 mg daily with the most frequently observed adverse events being nausea, emesis, anorexia, and diarrhea (Smith et al. 2004). In the first MPN phase II study of lestaurtinib (NCT00494585) 22 *JAK2* V617F positive primary or secondary (post-PV or post-ET) MF patients were studied (Santos et al. 2009). All but two had received prior therapy. Six previously treated subjects had a clinical improvement response by the International Working Group Myelofibrosis Research and Treatment criteria (IWG-MRT). The IWG-MRT is a composite set of MF disease response criteria that define complete remission (CR), partial remission (PR), clinical improvement (CI), progressive disease (PD), stable disease (SD), and relapse based on physical examination, laboratory, and histologic evaluation elements (Tefferi et al. 2006). Of the patients who received CEP-701 and achieved a response, three had a decrease in spleen size greeter than 50%, two became transfusion independent, and another had a greater than 50% decrease in spleen size with an improvement in platelet and neutrophil counts. No patients had improvements in bone marrow fibrosis. Three of the patients who achieved a response had an abnormal karyotype, but none demonstrated a cytogenetic response to therapy. The lestaurtinib therapy study also failed to demonstrate an effect on *JAK2* V617F allele burden. GI toxicities were the most commonly reported study related adverse event with 72% having diarrhea of any grade, 50% nausea (grade 1–2), 27% vomiting (grade 1–2) and 23% with flatulence (grade 1–2). Of those treated, eight experienced grade 3 or 4 toxicities of thrombocytopenia

(23%), anemia (14%), and diarrhea (9%) with both thrombocytopenia and diarrhea resulting in dose reductions in some patients.

Ruxolitinib (INCB018424) was examined in a phase I-II trial (NCT00509899) in primary or secondary MF patients regardless of *JAK2* V617F status (Verstovsek et al. 2010b). Study participation required being refractory to or relapsing after prior therapy, or having intermediate or high risk according to the Lille risk score (Dupriez et al. 1996), or having splenomegaly or hepatomegaly if post-splenectomy. Dose exploration began with a standard 3 + 3 dose escalation schema but transitioned to a dose titration paradigm after establishing maximum tolerated doses of 25 mg twice daily or 100 mg once daily. Thrombocytopenia was identified as the dose limiting toxicity and at the 100 mg daily dose level, 2 of 6 patients developed clinically significant thrombocytopenia during the second month of therapy. Further exploration established that a 15 mg twice daily oral starting dose followed by individualized dose titration was the most tolerated regimen. Of the studies patients treated with this regimen, 52% achieved a ≥50% reduction in splenomegaly (occurring during the first 3 months of therapy) with <10% experiencing grade 3 or 4 toxicities. Splenic responses were durable beyond 1 year in the subset of patients with data for 1.5–2 years of follow-up. Although the best splenic response was seen in the group who received 25 mg twice daily, 60% of these patients required a dose reduction due to thrombocytopenia. Subgroup analysis revealed no differences in response between patients with or without the *JAK2* V617F mutation. No difference was seen among patients with primary or secondary MF. In addition to the objective improvements observed in spleen size, ruxolitinib therapy was associated with significant improvements in total and individual symptom scores as collected by the Myelofibrosis Symptom Assessment Form (MFSAF). These symptom improvements were durable through 6 months of therapy.

Within the group of patients receiving the highest ruxolitinib dose intensity regimens (either 15 mg twice daily or 25 mg twice daily), 28 patients were transfusion dependent at baseline. Four of these patients achieved transfusion independence after a median duration of 12 weeks. Study eligibility criteria excluded MF patients with a platelet count of $\leq 100 \times 10^9$/L and those with an absolute neutrophil count of $\leq 1.5 \times 10^9$/L. As a result the study could not evaluate the ability of ruxolitinib therapy to improve these two disease associate features. Across all assigned doses, 23% of the patients who were transfusion-independent at baseline developed anemia. A dose dependent association was suggested by the observation that the frequency of study related anemia ranged from 8% among the patients treated at 15 mg twice daily to 27% among the patients who received 25 mg twice daily.

Results of the TG101348 phase I dose escalation study have been reported (NCT00724334) (Pardanani et al. 2007). This trial also employed a 3 + 3 cohort design followed by a cohort expansion for dose-confirmation to study once daily oral administration of TG101348 in high- or intermediate-risk primary or secondary MF patients across the range of doses from 30 to 800 mg. Intrapatient dose escalation was permitted after completion of three treatment cycles at the assigned starting dose. Risk assessment was determined by the Mayo PSS (Elliott et al. 2007) which is similar to the previously described risk assessment tools. The Mayo PSS defines high-risk as having two criteria and intermediate-risk as having one criteria where criteria include: hemoglobin <10 g/dL, WBC count <4 or >30 × 10^9/L, platelet count <100 × 10^9/L, absolute monocyte count ≥1 × 10^9/L. The study enrolled 59 patients (75% PMF, 20% post-PV MF, and 5% post-ET MF), and the majority (86%) were *JAK2* V617F positive.

The study's declared maximum tolerated dose was 680 mg daily based on the development of reversible grade 3 or 4 hyperamylasemia at 800 mg. Across all patients, the most common adverse events were related to the gastrointestinal system. In the MTD cohort (680 mg/d), frequent GI events included nausea (77.5% grade 1–2, 5% grade 3–4), diarrhea (62.5% grade 1–2, 12.5% grade 3–4), and vomiting (67.5% grade 1–2, 2.5% grade 3–4). Hyperlipasemia was also frequent at this dose level (37.5% all grades, 15% grade 3–4). Treatment related grade 3 and 4 hematologic adverse events were observed. Of the patients treated with 680 mg per day who were not transfusion dependent at baseline, 54.2% developed grade 3 or 4 anemia and 27.5% developed grade 3 or 4 thrombocytopenia.

Sixty-one percent of the cohort who received 680 mg per day experienced a minimum 25% decrease in palpable splenomegaly within the first two cycles and 45% of patients receiving this dose had spleen size reduction qualifying as a criteria for clinical improvement by IWG-MRT (i.e. at least a 50% reduction lasting at least 8 weeks). This response was durable at cycle 12 with 50% having a clinical improvement (CI) defined by the spleen response. In addition, 72% of the patients with a baseline leukocytosis who received 680 mg/d for 6 cycles (13/18) achieved a normal WBC.

The TG101348 phase I trial is the first to report a statistically significant reduction in *JAK2* V617F allele burden. In patients harboring the *JAK2* V617F mutation, the allele burden at base line ranged from 3 to 100% with a median of 20%. Across all baseline values, without respect to study drug dose, there was a statistically significant reduction of allele burden to 17% (P=0.04) and 19% (P=0.01) after 6 and 12 cycles respectively. Looking only at those individuals with a baseline allele burden above 20%, the median was 60% prior to therapy and decreased to 31% after 6 cycles (P=0.002) and 32% at 12 cycles (P=0.002). In this group 45% (9 patients) had a ≥50% reduction in *JAK2* V617F allele burden, but 20% exhibited an increase.

Conclusions and Path Forward

Each of the three early clinical development studies for patients with MF discussed illustrate that the first group of experimental JAK2 inhibitors may offer clinical benefits through reductions in disease related symptoms and control of splenomegaly. To date, the studies have failed to demonstrate evidence of disease modification. These three studies have not yet reported a clinically significant ability to improve histiologic changes in bone marrow fibrosis, nor have they demonstrated a significant positive impact on cytopenias. Apparent compound-related adverse events are being observed during use of these agents.

The MPNs remain diseases where the approved therapy is primarily palliative, minimally effective, and complicated by safety concerns. In this setting the advances seen with the first wave of JAK2 inhibitors in development are suggestive of significant therapeutic progress and may offer new hope to patients. The results have led to interests in the study of combination therapies as a next step in the advancement of MPN therapy. The iterative effort of combining non-selective JAK2 inhibitors with other agents currently being studied in MPN such as the IMiDs or hypomethylation agents is a promising approach. However, our growing awareness of the molecular pathophysiology of this disease suggests that these efforts may not cure the MPNs, and these combinations are likely to be accompanied by at least additive side effect profiles.

The results of our current clinical data set may be looked at in a different light. HSCT has shown that effective eradication of the malignant clone can improve disease related symptoms, splenomegaly, and reverse the accumulated bone marrow fibrotic injuries. JAK2-directed therapy has demonstrated that disease symptoms are improved with inhibition of the JAK2/STAT pathway and the resultant modulation of cytokine signaling. In addition, JAK2 therapies have demonstrated that there is a relationship between dose, target inhibition, improvements in splenomegaly, and modulation of peripheral cell counts. Thus, it appears that maximally effective disease modification will require two things: (1) selective inhibition or eradication of the tumor cells and (2) retention or restoration of normal marrow reconstitution. Achieving both may lead to reversal of the disease manifestations similar to what has been demonstrated in HSCT. In the published clinical trials reviewed here, the ability to test this hypothesis was limited by target dependent inhibition of normal marrow. The observed dose limiting toxicities suggest that effective disease targeting will require exploration of the role of selective *JAK2* V617F inhibitors.

A potential additional confounder for defining effective therapy is the underlying heterogeneity of the MPN phenotypes. A single therapy used only at a single dose or regimen may not achieve the same degree of clinical effect in all subjects. This observation is supported by the observed genetic variability of the disease as well by the variability of responses seen with the available JAK2 therapies. This heterogeneity suggests that tailoring of therapy will be required for optimal care. Different patients may require potentially different therapies for first- or second-line therapy. With introduction of effective therapy it is predictable that some patterns of resistance will be identified requiring combination therapy or changes in drug selection. Genetic heterogeneity has implications for trial designs, regulatory approval, and companion diagnostic needs. There is debate about the appropriateness of current diagnostic classifications systems and whether they should be reconsidered in light of the new understandings of molecular pathogenesis (Harrison 2010).

There are additional challenges related to patient selection and evaluation of clonal burden reductions. Newer risk models such as the DIPSS and DIPSS Plus offer promise for risk assessments and benchmarks for evaluating the effect of therapy on overall survival. This is an important consideration given the duration of time and number of patients needed to demonstrate an improvement in overall survival or progression-free survival in MPN. An unavoidable challenge in the evaluation of any potentially disease modifying therapy is the need to evaluate the long term effects of therapy. This is especially true for the MPNs where there is diversity in overall survival times and potential heterogeneity for disease responsiveness. It is clear that the most advanced MF patients have a significant reduction in overall survival. It is not clear, if the disease process is equally amenable to therapy for both a newly diagnosed MPN patient or for an advanced MPN patient. In the case of CML, patients in chronic phase are more responsive to imatinib therapy then those in blast phase. Tools analogous to the DIPSS are needed to appropriately define and identify those ET and PV patients who are at greatest risk from their disease. Hopefully, the availability of patient selection discoveries will help to mitigate some of these issues, but this is not likely to resolve the conflict between ongoing and immediate need for therapy vs need for long-term outcome evaluations.

Non-clinical studies have suggested that gene dosage plays a role in defining disease phenotype and clinical complication risks. This will almost certainly be of interest in evaluating and selecting potential therapies. However, the ability to effectively validate whether gene dosage plays a role in the human disease process is hampered by the lack of a standardized reference test to measure *JAK2* V617F allelic burden in patient samples. This problem is more confounding to drug efficacy evaluations. The published JAK2 clinical trials have used different tests and each has its own unique characteristics such as differences in cell population analyzed (granulocytes vs mononuclear cells), source of control (cell lines vs plasmids), range of control alleles used in standardization, primers, etc. As a result, each *JAK2* assay has a different performance profile including variations in limits of detection, sensitivity, and specificity. While different clinical laboratories often pragmatically use such diverse tests, regulatory approvals that depend on laboratory test results for efficacy evaluation or patient stratification may require more rigorous validation.

MPN patients have a clear need for better therapy today. The basic and clinical research progress in the MPN field has truly accelerated in the past decade. More importantly, this progress looks poised to deliver new treatment options for MPN patients. It appears that soon, the first wave of JAK2 inhibitors will be reviewed for regulatory approval. If approved, patients and physicians could have access to a new tool for disease and symptom burden reduction. Based on the available data, this should not be a stopping point but should rather be viewed as the front runner of even more promising future therapies which are capable of meeting the needs of MPN patients and their families.

Author's Note

Following completion of this chapter, the US FDA granted approval in November 2011 for ruxolitinib phosphate for the treatment of patients with intermediate or high-risk myelofibrosis, including primary myelofibrosis, post-polycythemia vera myelofibrosis and post-essential thrombocythemia myelofibrosis.

References

Ahmed R, Begam K, Warner SL et al (2011) In vitro and in vivo characterization of SGI-1252, a small molecule inhibitor of JAK2. Exp Hematol 39:14–25. http://linkinghub.elsevier.com/retrieve/pii/S0301472X10004947?showall=true

Akada H, Yan D, Zou H et al (2010) Conditional expression of heterozygous or homozygous Jak2V617F from its endogenous promoter induces a polycythemia vera-like disease. Blood 115:3589–3597. http://bloodjournal.hematologylibrary.org/cgi/content/abstract/bloodjournal;115/17/3589

Anía BJ, Suman VJ, Sobell JL et al (1994) Trends in the incidence of polycythemia vera among Olmsted county, Minnesota residents, 1935–1989. Am J Hematol 47:89–93. http://dx.doi.org/10.1002/ajh.2830470205

Barosi G, Bergamaschi G, Marchetti M et al (2007) JAK2 V617F mutational status predicts progression to large splenomegaly and leukemic transformation in primary myelofibrosis. Blood 110:4030–4036. http://www.ncbi.nlm.nih.gov/entrez/query.fcgi?cmd=Retrieve&db=PubMed&dopt=Citation&list_uids=17712047

Beer PA, Delhommeau F, LeCouedic JP et al (2010) Two routes to leukemic transformation after a JAK2 mutation-positive myeloproliferative neoplasm. Blood 115:2891–2900. http://www.ncbi.nlm.nih.gov/entrez/query.fcgi?cmd=Retrieve&db=PubMed&dopt=Citation&list_uids=20008300

Begna KH, Mesa RA, Pardanani A et al (2011) A phase-2 trial of low-dose pomalidomide in myelofibrosis. Leukemia 25:301–304. http://www.ncbi.nlm.nih.gov/entrez/query.fcgi?cmd=Retrieve&db=PubMed&dopt=Citation&list_uids=21052089

Boudeau J, Miranda-Saavedra D, Barton GJ et al (2006) Emerging roles of pseudokinases. Trends Cell Biol 16:443–452. http://www.ncbi.nlm.nih.gov/entrez/query.fcgi?cmd=Retrieve&db=PubMed&dopt=Citation&list_uids=16879967

Bumm TG, Elsea C, Corbin AS et al (2006) Characterization of murine JAK2V617F-positive myeloproliferative disease. Cancer Res 66:11156–11165. http://www.ncbi.nlm.nih.gov/entrez/query.fcgi?cmd=Retrieve&db=PubMed&dopt=Citation&list_uids=17145859

Carbuccia N, Murati A, Trouplin V et al (2009) Mutations of ASXL1 gene in myeloproliferative neoplasms. Leukemia 23:2183–2186. http://www.ncbi.nlm.nih.gov/entrez/query.fcgi?cmd=Retrieve&db=PubMed&dopt=Citation&list_uids=19609284

Cervantes F, Barosi G (2005) Myelofibrosis with myeloid metaplasia: diagnosis, prognostic factors, and staging. Semin oncol 32:395–402. http://linkinghub.elsevier.com/retrieve/pii/S0093775405001557

Cervantes F, Pereira A, Esteve J et al (1997) Identification of 'short-lived' and 'long-lived' patients at presentation of idiopathic myelofibrosis. Br J Haematol 97:635–640. http://www.ncbi.nlm.nih.gov/entrez/query.fcgi?cmd=Retrieve&db=PubMed&dopt=Citation&list_uids=9207412

Cervantes F, Barosi G, Hernandez-Boluda JC et al (2001) Myelofibrosis with myeloid metaplasia in adult individuals 30 years old or younger: presenting features, evolution and survival. Eur J Haematol 66:324–327. http://www.ncbi.nlm.nih.gov/entrez/query.fcgi?cmd=Retrieve&db=PubMed&dopt=Citation&list_uids=11422412

Cervantes F, Mesa R, Barosi G (2007) New and old treatment modalities in primary myelofibrosis. Cancer J 13:377–383. http://www.ncbi.nlm.nih.gov/entrez/query.fcgi?cmd=Retrieve&db=PubMed&dopt=Citation&list_uids=18032975

Cervantes F, Passamonti F, Barosi G (2008) Life expectancy and prognostic factors in the classic BCR/ABL-negative myeloproliferative disorders. Leukemia 22:905–914. http://www.ncbi.nlm.nih.gov/entrez/query.fcgi?cmd=Retrieve&db=PubMed&dopt=Citation&list_uids=18385755

Cervantes F, Dupriez B, Pereira A et al (2009) New prognostic scoring system for primary myelofibrosis based on a study of the International Working Group for Myelofibrosis Research and Treatment. Blood 113:2895–2901. http://www.ncbi.nlm.nih.gov/entrez/query.fcgi?cmd=Retrieve&db=PubMed&dopt=Citation&list_uids=18988864

Ciurea SO, Sadegi B, Wilbur A et al (2008) Effects of extensive splenomegaly in patients with myelofibrosis undergoing a reduced intensity allogeneic stem cell transplantation. Br J Haematol 141:80–83. http://dx.doi.org/10.1111/j.1365-2141.2008.07010.x

Dameshek W (1951) Some speculations on the myeloproliferative syndromes. Blood 6:372–375. http://www.ncbi.nlm.nih.gov/entrez/query.fcgi?cmd=Retrieve&db=PubMed&dopt=Citation&list_uids=14820991

De Stefano V, Za T, Rossi E et al (2008) Recurrent thrombosis in patients with polycythemia vera and essential thrombocythemia: incidence, risk factors, and effect of treatments. Haematologica 93:372–380. http://www.ncbi.nlm.nih.gov/entrez/query.fcgi?cmd=Retrieve&db=PubMed&dopt=Citation&list_uids=18268279

Delhommeau F, Dupont S, Della Valle V et al (2009) Mutation in TET2 in myeloid cancers. N Engl J Med 360:2289–2301. http://www.ncbi.nlm.nih.gov/entrez/query.fcgi?cmd=Retrieve&db=PubMed&dopt=Citation&list_uids=19474426

Di Nisio M, Barbui T, Di Gennaro L et al (2007) The haematocrit and platelet target in polycythemia vera. Br J Haematol 136:249–259. http://dx.doi.org/10.1111/j.1365-2141.2006.06430.x

Dobrzanski P, Hexner E, Serdikoff C et al (2006) CEP-701 is a JAK2 inhibitor which attenuates JAK2/STAT5 signaling pathway and the proliferation of primary cells from patients with myeloproliferative disorders. ASH Annu Meet Abstr 108:3594. http://abstracts.hematologylibrary.org/cgi/content/abstract/ashmtg;108/11/3594

Dunbar AJ, Gondek LP, O'Keefe CL et al (2008) 250 K single nucleotide polymorphism array karyotyping identifies acquired uniparental disomy and homozygous mutations, including novel missense substitutions of c-Cbl, in myeloid malignancies. Cancer Res 68:10349–10357. http://www.ncbi.nlm.nih.gov/entrez/query.fcgi?cmd=Retrieve&db=PubMed&dopt=Citation&list_uids=19074904

Dupriez B, Morel P, Demory JL et al (1996) Prognostic factors in agnogenic myeloid metaplasia: a report on 195 cases with a new scoring system. Blood 88:1013–1018. http://www.ncbi.nlm.nih.gov/entrez/query.fcgi?cmd=Retrieve&db=PubMed&dopt=Citation&list_uids=8704209

Elliott MA, Tefferi A (2005) Thrombosis and haemorrhage in polycythaemia vera and essential thrombocythaemia. Br J Haematol 128:275–290. http://www.ncbi.nlm.nih.gov/entrez/query.fcgi?cmd=Retrieve&db=PubMed&dopt=Citation&list_uids=15667529

Elliott MA, Verstovsek S, Dingli D et al (2007) Monocytosis is an adverse prognostic factor for survival in younger patients with primary myelofibrosis. Leukemia Res 31:1503–1509. http://linkinghub.elsevier.com/retrieve/pii/S0145212607000756?showall=true

Epstein E, Goedel A (1934) Hemorrhagic thrombocythemia with a vascular, sclerotic spleen. Virchows Arch 293:233–248

Finazzi G (2004) A prospective analysis of thrombotic events in the European collaboration study on low-dose aspirin in polycythemia (ECLAP). Pathol Biol (Paris) 52:285–288. http://www.ncbi.nlm.nih.gov/entrez/query.fcgi?cmd=Retrieve&db=PubMed&dopt=Citation&list_uids=15217715

Finazzi G, Caruso V, Marchioli R et al (2005) Acute leukemia in polycythemia vera: an analysis of 1638 patients enrolled in a prospective observational study. Blood 105:2664–2670. http://www.ncbi.nlm.nih.gov/entrez/query.fcgi?cmd=Retrieve&db=PubMed&dopt=Citation&list_uids=15585653

Florensa L, Bellosillo B, Arenillas L et al (2010) LY2784544, a novel JAK2 inhibitor, decreases in vitro growth of hematopoietic human progenitors from JAK2 V617F positive polycythemia vera patients. ASH Annu Meet Abstr 116:5054. http://abstracts.hematologylibrary.org/cgi/content/abstract/ashmtg;116/21/5054

Fridman J, Nussenzveig R, Liu P et al (2007) Discovery and preclinical characterization of INCB018424, a selective JAK2 inhibitor for the treatment of myeloproliferative disorders. ASH Annu Meet Abstr 110:3538. http://abstracts.hematologylibrary.org/cgi/content/abstract/ashmtg;110/11/3538

Funakoshi-Tago M, Pelletier S, Moritake H et al (2008) Jak2 FERM domain interaction with the erythropoietin receptor regulates Jak2 kinase activity. Mol Cell Biol 28:1792–1801. http://mcb.asm.org/cgi/content/abstract/28/5/1792

Gangat N, Strand J, Lasho TL et al (2008) Cytogenetic studies at diagnosis in polycythemia vera: clinical and JAK2V617F allele burden correlates. Eur J Haematol 80:197–200. http://www.ncbi.nlm.nih.gov/entrez/query.fcgi?cmd=Retrieve&db=PubMed&dopt=Citation&list_uids=18081705

Gangat N, Tefferi A, Thanarajasingam G et al (2009) Cytogenetic abnormalities in essential thrombocythemia: prevalence and prognostic significance. Eur J Haematol 83:17–21. http://www.ncbi.nlm.nih.gov/entrez/query.fcgi?cmd=Retrieve&db=PubMed&dopt=Citation&list_uids=19236446

Gangat N, Caramazza D, Vaidya R et al (2011) DIPSS plus: a refined dynamic international prognostic scoring system for primary myelofibrosis that incorporates prognostic information from karyotype, platelet count, and transfusion status. J Clin Oncol 29:392–397. http://www.ncbi.nlm.nih.gov/entrez/query.fcgi?cmd=Retrieve&db=PubMed&dopt=Citation&list_uids=21149668

Giordanetto F, Kroemer RT (2002) Prediction of the structure of human Janus kinase 2 (JAK2) comprising JAK homology domains 1 through 7. Protein Eng 15:727–737. http://www.ncbi.nlm.nih.gov/entrez/query.fcgi?cmd=Retrieve&db=PubMed&dopt=Citation&list_uids=12456871

Green A, Beer P (2010) Somatic mutations of IDH1 and IDH2 in the leukemic transformation of myeloproliferative neoplasms. N Engl J Med 362:369–370. http://www.ncbi.nlm.nih.gov/entrez/query.fcgi?cmd=Retrieve&db=PubMed&dopt=Citation&list_uids=20107228

Gruppo Italiano Studio Policitemia (1995) Polycythemia vera: the natural history of 1213 patients followed for 20 years. Ann Intern Med 123:656–664. http://www.ncbi.nlm.nih.gov/entrez/query.fcgi?cmd=Retrieve&db=PubMed&dopt=Citation&list_uids=7574220

Gruppo Italiano Studio Policitemia (GISP) (1997) Low-dose aspirin in polycythaemia vera: a pilot study. Br J Haematol 97:453–456. http://www.ncbi.nlm.nih.gov/entrez/query.fcgi?cmd=Retrieve&db=PubMed&dopt=Citation&list_uids=9163613

Harrison C (2010) Rethinking disease definitions and therapeutic strategies in essential thrombocythemia and polycythemia vera. Hematology 129–134. http://asheducationbook.hematologylibrary.org/cgi/content/abstract/bloodbook;2010/1/129

Harrison CN, Campbell PJ, Buck G et al (2005) Hydroxyurea compared with anagrelide in high-risk essential thrombocythemia. N Engl J Med 353:33–45. http://www.ncbi.nlm.nih.gov/entrez/query.fcgi?cmd=Retrieve&db=PubMed&dopt=Citation&list_uids=16000354

Hemminki K, Zhang H, Sundquist J et al (2009) Myeloproliferative disorders in Sweden: incidence trends and multiple tumors. Leuk Res 33:e14–e16. http://www.ncbi.nlm.nih.gov/entrez/query.fcgi?cmd=Retrieve&db=PubMed&dopt=Citation&list_uids=18501963

Hexner EO, Serdikoff C, Jan M et al (2008) Lestaurtinib (CEP701) is a JAK2 inhibitor that suppresses JAK2/STAT5 signaling and the proliferation of primary erythroid cells from patients with myeloproliferative disorders. Blood 111:5663–5671. http://www.ncbi.nlm.nih.gov/entrez/query.fcgi?cmd=Retrieve&db=PubMed&dopt=Citation&list_uids=17984313

Hoffman R, Prchal JT, Samuelson S et al (2007) Philadelphia chromosome-negative myeloproliferative disorders: biology and treatment. Biol Blood Marrow Transplant 13:64–72. http://www.ncbi.nlm.nih.gov/entrez/query.fcgi?cmd=Retrieve&db=PubMed&dopt=Citation&list_uids=17222772

Hood J, Doukas J, Martin M et al (2007) TG101348, a potent, highly selective JAK2 inhibitor, inhibits colony formation in stem cells from polycythemia vera patients and prevents JAK2V617F-mediated splenomegaly and death in a mouse model. J Clin Oncol ASCO Annu Meet Proc 25:7031

Hookham MB, Elliott J, Suessmuth Y et al (2007) The myeloproliferative disorder-associated JAK2 V617F mutant escapes negative regulation by suppressor of cytokine signaling 3. Blood 109:4924–4929. http://www.ncbi.nlm.nih.gov/entrez/query.fcgi?cmd=Retrieve&db=PubMed&dopt=Citation&list_uids=17317861

Huang J, Tefferi A (2009) Erythropoiesis stimulating agents have limited therapeutic activity in transfusion-dependent patients with primary myelofibrosis regardless of serum erythropoietin level. Eur J Haematol 83:154–155. http://www.ncbi.nlm.nih.gov/entrez/query.fcgi?cmd=Retrieve&db=PubMed&dopt=Citation&list_uids=19366369

Hueck G (1879) Two cases of leukemia with peculiar blood and bone marrow findings, respectively. Arch Pathol Anat 78:475–496

Hussein K, Van Dyke DL, Tefferi A (2009) Conventional cytogenetics in myelofibrosis: literature review and discussion. Eur J Haematol 82:329–338. http://www.ncbi.nlm.nih.gov/entrez/query.fcgi?cmd=Retrieve&db=PubMed&dopt=Citation&list_uids=19141119

Jager R, Gisslinger H, Passamonti F et al (2010) Deletions of the transcription factor Ikaros in myeloproliferative neoplasms. Leukemia 24:1290–1298. http://www.ncbi.nlm.nih.gov/entrez/query.fcgi?cmd=Retrieve&db=PubMed&dopt=Citation&list_uids=20508609

James C, Ugo V, Le Couedic J-P et al (2005) A unique clonal JAK2 mutation leading to constitutive signalling causes polycythaemia vera. Nature 434:1144–1148. http://dx.doi.org/10.1038/nature03546. http://www.nature.com/nature/journal/v434/n7037/suppinfo/nature03546_S1.html

Kiladjian JJ, Cassinat B, Turlure P et al (2006) High molecular response rate of polycythemia vera patients treated with pegylated interferon alpha-2a. Blood 108:2037–2040. http://www.ncbi.nlm.nih.gov/entrez/query.fcgi?cmd=Retrieve&db=PubMed&dopt=Citation&list_uids=16709929

Kiladjian JJ, Cassinat B, Chevret S et al (2008) Pegylated interferon-alfa-2a induces complete hematologic and molecular responses with low toxicity in polycythemia vera. Blood 112:3065–3072. http://www.ncbi.nlm.nih.gov/entrez/query.fcgi?cmd=Retrieve&db=PubMed&dopt=Citation&list_uids=18650451

Klampfl T, Harutyunyan A, Berg T et al (2011) Genome integrity of myeloproliferative neoplasms in chronic phase and during disease progression. Blood 118:167–176. http://bloodjournal.hematologylibrary.org/content/118/1/167.abstract

Kralovics R, Guan Y, Prchal JT (2002) Acquired uniparental disomy of chromosome 9p is a frequent stem cell defect in polycythemia vera. Exp Hematol 30:229–236. http://www.ncbi.nlm.nih.gov/entrez/query.fcgi?cmd=Retrieve&db=PubMed&dopt=Citation&list_uids=11882360

Kralovics R, Passamonti F, Buser AS et al (2005) A gain-of-function mutation of JAK2 in myeloproliferative disorders. N Engl J Med 352:1779–1790. http://www.ncbi.nlm.nih.gov/entrez/query.fcgi?cmd=Retrieve&db=PubMed&dopt=Citation&list_uids=15858187

Lacout C, Pisani DF, Tulliez M et al (2006) JAK2V617F expression in murine hematopoietic cells leads to MPD mimicking human PV with secondary myelofibrosis. Blood 108:1652–1660. http://bloodjournal.hematologylibrary.org/cgi/content/abstract/bloodjournal;108/5/1652

Lacy MQ, Tefferi A (2011) Pomalidomide therapy for multiple myeloma and myelofibrosis: an update. Leuk Lymphoma 52:560–566. http://informahealthcare.com/doi/abs/10.3109/10428194.2011.552139

Landolfi R, Marchioli R, Kutti J et al (2004) Efficacy and safety of low-dose aspirin in polycythemia vera. N Engl J Med 350:114–124. http://www.ncbi.nlm.nih.gov/entrez/query.fcgi?cmd=Retrieve&db=PubMed&dopt=Citation&list_uids=14711910

Landolfi R, Cipriani MC, Novarese L (2006) Thrombosis and bleeding in polycythemia vera and essential thrombocythemia: pathogenetic mechanisms and prevention. Best Pract Res Clin Haematol 19:617–633. http://www.ncbi.nlm.nih.gov/entrez/query.fcgi?cmd=Retrieve&db=PubMed&dopt=Citation&list_uids=16781491

Lasho TL, Tefferi A, Hood JD et al (2008) TG101348, a JAK2-selective antagonist, inhibits primary hematopoietic cells derived from myeloproliferative disorder patients with JAK2V617F, MPLW515K or JAK2 exon 12 mutations as well as mutation negative patients. Leukemia 22:1790–1792. http://www.ncbi.nlm.nih.gov/entrez/query.fcgi?cmd=Retrieve&db=PubMed&dopt=Citation&list_uids=18354492

Laurie NA, Donovan SL, Shih CS et al (2006) Inactivation of the p53 pathway in retinoblastoma. Nature 444:61–66. http://www.ncbi.nlm.nih.gov/entrez/query.fcgi?cmd=Retrieve&db=PubMed&dopt=Citation&list_uids=17080083

Levine RL, Wadleigh M, Cools J et al (2005) Activating mutation in the tyrosine kinase JAK2 in polycythemia vera, essential thrombocythemia, and myeloid metaplasia with myelofibrosis. Cancer Cell 7:387–397. http://www.ncbi.nlm.nih.gov/entrez/query.fcgi?cmd=Retrieve&db=PubMed&dopt=Citation&list_uids=15837627

Lindauer K, Loerting T, Liedl KR et al (2001) Prediction of the structure of human Janus kinase 2 (JAK2) comprising the two carboxy-terminal domains reveals a mechanism for autoregulation. Protein Eng 14:27–37. http://www.ncbi.nlm.nih.gov/entrez/query.fcgi?cmd=Retrieve&db=PubMed&dopt=Citation&list_uids=11287676

Liu E, Jelinek J, Pastore YD et al (2003) Discrimination of polycythemias and thrombocytoses by novel, simple, accurate clonality assays and comparison with PRV-1 expression and BFU-E response to erythropoietin. Blood 101:3294–3301. http://www.ncbi.nlm.nih.gov/entrez/query.fcgi?cmd=Retrieve&db=PubMed&dopt=Citation&list_uids=12515724

Lucet IS, Fantino E, Styles M et al (2006) The structural basis of Janus kinase 2 inhibition by a potent and specific pan-Janus kinase inhibitor. Blood 107:176–183. http://www.ncbi.nlm.nih.gov/entrez/query.fcgi?cmd=Retrieve&db=PubMed&dopt=Citation&list_uids=16174768

Ma L, Zhao B, Walgren R et al (2010) Efficacy of LY2784544, a small molecule inhibitor selective for mutant JAK2 kinase, in JAK2 V617F-induced hematologic malignancy models. ASH

Annu Meet Abstr 116:4087. http://abstracts.hematologylibrary.org/cgi/content/abstract/ashmtg;116/21/4087

Marty C, Lacout C, Martin A et al (2010) Myeloproliferative neoplasm induced by constitutive expression of JAK2V617F in knock-in mice. Blood 116:783–787. http://bloodjournal.hematologylibrary.org/cgi/content/abstract/bloodjournal;116/5/783

Mesa RA (2009) How I treat symptomatic splenomegaly in patients with myelofibrosis. Blood 113:5394–5400. http://bloodjournal.hematologylibrary.org/cgi/content/abstract/bloodjournal;113/22/5394

Mesa RA (2010) Assessing new therapies and their overall impact in myelofibrosis. Hematology 115–121. http://asheducationbook.hematologylibrary.org/cgi/content/abstract/bloodbook;2010/1/115

Mesa RA, Silverstein MN, Jacobsen SJ et al (1999) Population-based incidence and survival figures in essential thrombocythemia and agnogenic myeloid metaplasia: an Olmsted county study, 1976–1995. Am J Hematol 61:10–15. http://www3.interscience.wiley.com/cgi-bin/fulltext/61001800/PDFSTART

Mesa RA, Li C-Y, Schroeder G et al (2001) Clinical correlates of splenic histopathology and splenic karyotype in myelofibrosis with myeloid metaplasia. Blood 97:3665–3667. http://bloodjournal.hematologylibrary.org/cgi/content/abstract/bloodjournal;97/11/3665

Mesa RA, Nagorney DS, Schwager S et al (2006) Palliative goals, patient selection, and perioperative platelet management: outcomes and lessons from 3 decades of splenectomy for myelofibrosis with myeloid metaplasia at the Mayo Clinic. Cancer 107:361–370. http://www.ncbi.nlm.nih.gov/entrez/query.fcgi?cmd=Retrieve&db=PubMed&dopt=Citation&list_uids=16770787

Mesa RA, Verstovsek S, Rivera C et al (2008) 5-Azacitidine has limited therapeutic activity in myelofibrosis. Leukemia 23:180–182. http://dx.doi.org/10.1038/leu.2008.136

Mesa RA, Pardanani AD, Hussein K et al (2010) Phase1/-2 study of Pomalidomide in myelofibrosis. Am J Hematol 85:129–130. http://www.ncbi.nlm.nih.gov/entrez/query.fcgi?cmd=Retrieve&db=PubMed&dopt=Citation&list_uids=20052748

Michiels JJ (1999) Acquired von Willebrand disease due to increasing platelet count can readily explain the paradox of thrombosis and bleeding in thrombocythemia. Clin Appl Thromb Hemost 5:147–151. http://cat.sagepub.com/content/5/3/147.abstract

Moliterno AR, Hankins WD, Spivak JL (1998) Impaired expression of the thrombopoietin receptor by platelets from patients with polycythemia vera. N Engl J Med 338:572–580. http://www.ncbi.nlm.nih.gov/entrez/query.fcgi?cmd=Retrieve&db=PubMed&dopt=Citation&list_uids=9475764

Najean Y, Rain JD (1997a) Treatment of polycythemia vera: the use of hydroxyurea and pipobroman in 292 patients under the age of 65 years. Blood 90:3370–3377. http://www.ncbi.nlm.nih.gov/entrez/query.fcgi?cmd=Retrieve&db=PubMed&dopt=Citation&list_uids=9345019

Najean Y, Rain JD (1997b) Treatment of polycythemia vera: use of 32P alone or in combination with maintenance therapy using hydroxyurea in 461 patients greater than 65 years of age. The French Polycythemia Study Group. Blood 89:2319–2327. http://www.ncbi.nlm.nih.gov/entrez/query.fcgi?cmd=Retrieve&db=PubMed&dopt=Citation&list_uids=9116275

Nicholson SE, Willson TA, Farley A et al (1999) Mutational analyses of the SOCS proteins suggest a dual domain requirement but distinct mechanisms for inhibition of LIF and IL-6 signal transduction. Embo J 18:375–385. http://www.ncbi.nlm.nih.gov/entrez/query.fcgi?cmd=Retrieve&db=PubMed&dopt=Citation&list_uids=9889194

Nowell P, Hungerford D (1960) A minute chromosome in human chronic granulocytic leukemia. Science 132:1497–1498

Panani AD (2007) Cytogenetic and molecular aspects of Philadelphia negative chronic myeloproliferative disorders: clinical implications. Cancer Lett 255:12–25. http://linkinghub.elsevier.com/retrieve/pii/S0304383507000663?showall=true

Papadakis E, Hoffman R, Brenner B (2010) Thrombohemorrhagic complications of myeloproliferative disorders. Blood Rev 24:227–232. http://www.ncbi.nlm.nih.gov/entrez/query.fcgi?cmd=Retrieve&db=PubMed&dopt=Citation&list_uids=20817333

Paquette R, Sokol L, Shah NP et al (2008) A Phase I study of XL019, a selective JAK2 inhibitor, in patients with polycythemia vera. ASH Annu Meet Abstr 112:2810. http://abstracts.hematologylibrary.org/cgi/content/abstract/ashmtg;112/11/2810

Pardanani A, Hood J, Lasho T et al (2007) TG101209, a small molecule JAK2-selective kinase inhibitor potently inhibits myeloproliferative disorder-associated JAK2V617F and

MPLW515L/K mutations. Leukemia 21:1658–1668. http://www.ncbi.nlm.nih.gov/entrez/query.fcgi?cmd=Retrieve&db=PubMed&dopt=Citation&list_uids=17541402

Pardanani A, Lasho T, Smith G et al (2009) CYT387, a selective JAK1/JAK2 inhibitor: in vitro assessment of kinase selectivity and preclinical studies using cell lines and primary cells from polycythemia vera patients. Leukemia 23:1441–1445. http://dx.doi.org/10.1038/leu.2009.50

Passamonti F, Rumi E, Pungolino E et al (2004) Life expectancy and prognostic factors for survival in patients with polycythemia vera and essential thrombocythemia. Am J Med 117:755–761. http://www.ncbi.nlm.nih.gov/entrez/query.fcgi?cmd=Retrieve&db=PubMed&dopt=Citation&list_uids=15541325

Passamonti F, Rumi E, Arcaini L et al (2005) Leukemic transformation of polycythemia vera: a single center study of 23 patients. Cancer 104:1032–1036. http://www.ncbi.nlm.nih.gov/entrez/query.fcgi?cmd=Retrieve&db=PubMed&dopt=Citation&list_uids=16047334

Passamonti F, Cervantes F, Vannucchi AM et al (2010) A dynamic prognostic model to predict survival in primary myelofibrosis: a study by the IWG-MRT (International Working Group for Myeloproliferative Neoplasms Research and Treatment). Blood 115:1703–1708. http://bloodjournal.hematologylibrary.org/cgi/content/abstract/bloodjournal;115/9/1703

Petti MC, Latagliata R, Spadea T et al (2002) Melphalan treatment in patients with myelofibrosis with myeloid metaplasia. Br J Haematol 116:576–581. http://www.ncbi.nlm.nih.gov/entrez/query.fcgi?cmd=Retrieve&db=PubMed&dopt=Citation&list_uids=11849213

Prchal JF, Axelrad AA (1974) Letter: bone-marrow responses in polycythemia vera. N Engl J Med 290:1382. http://www.ncbi.nlm.nih.gov/entrez/query.fcgi?cmd=Retrieve&db=PubMed&dopt=Citation&list_uids=4827655

Quintas-Cardama A, Tong W, Kantarjian H et al (2008) A phase II study of 5-azacitidine for patients with primary and post-essential thrombocythemia//polycythemia vera myelofibrosis. Leukemia 22:965–970. http://dx.doi.org/10.1038/leu.2008.91

Quintas-Cardama A, Kantarjian H, Manshouri T et al (2009) Pegylated interferon alfa-2a yields high rates of hematologic and molecular response in patients with advanced essential thrombocythemia and polycythemia vera. J Clin Oncol 27:5418–5424. http://www.ncbi.nlm.nih.gov/entrez/query.fcgi?cmd=Retrieve&db=PubMed&dopt=Citation&list_uids=19826111

Quintas-Cardama A, Vaddi K, Liu P et al (2010) Preclinical characterization of the selective JAK1/2 inhibitor INCB018424: therapeutic implications for the treatment of myeloproliferative neoplasms. Blood 115:3109–3117. http://bloodjournal.hematologylibrary.org/cgi/content/abstract/bloodjournal;115/15/3109

Reilly JT (2008) Pathogenetic insight and prognostic information from standard and molecular cytogenetic studies in the BCR-ABL-negative myeloproliferative neoplasms (MPNs). Leukemia 22:1818–1827. http://www.ncbi.nlm.nih.gov/entrez/query.fcgi?cmd=Retrieve&db=PubMed&dopt=Citation&list_uids=18754027

Riemenschneider MJ, Knobbe CB, Reifenberger G (2003) Refined mapping of 1q32 amplicons in malignant gliomas confirms MDM4 as the main amplification target. Int J Cancer 104:752–757. http://www.ncbi.nlm.nih.gov/entrez/query.fcgi?cmd=Retrieve&db=PubMed&dopt=Citation&list_uids=12640683

Ries L, Melbert D, Krapcho M et al (2008) SEER cancer statistics review, 1975–2005. National Cancer Institute, Bethesda

Rollison DE, Howlader N, Smith MT et al (2008) Epidemiology of myelodysplastic syndromes and chronic myeloproliferative disorders in the United States, 2001–2004, using data from the NAACCR and SEER programs. Blood 112:45–52. http://www.ncbi.nlm.nih.gov/entrez/query.fcgi?cmd=Retrieve&db=PubMed&dopt=Citation&list_uids=18443215

Rowley JD (1973) Letter a new consistent chromosomal abnormality in chronic myelogenous leukaemia identified by quinacrine fluorescence and Giemsa staining. Nature 243:290–293. http://www.ncbi.nlm.nih.gov/entrez/query.fcgi?cmd=Retrieve&db=PubMed&dopt=Citation&list_uids=4126434

Rozman C, Giralt M, Feliu E et al (1991) Life expectancy of patients with chronic nonleukemic myeloproliferative disorders. Cancer 67:2658–2663. http://www.ncbi.nlm.nih.gov/entrez/query.fcgi?cmd=Retrieve&db=PubMed&dopt=Citation&list_uids=2015567

Saharinen P, Silvennoinen O (2002) The pseudokinase domain is required for suppression of basal activity of Jak2 and Jak3 tyrosine kinases and for cytokine-inducible activation of signal transduction. J Biol Chem 277:47954–47963. http://www.ncbi.nlm.nih.gov/entrez/query.fcgi?cmd=Retrieve&db=PubMed&dopt=Citation&list_uids=12351625

Saharinen P, Takaluoma K, Silvennoinen O (2000) Regulation of the Jak2 tyrosine kinase by its pseudokinase domain. Mol Cell Biol 20:3387–3395. http://www.ncbi.nlm.nih.gov/entrez/query.fcgi?cmd=Retrieve&db=PubMed&dopt=Citation&list_uids=10779328

Saharinen P, Vihinen M, Silvennoinen O (2003) Autoinhibition of Jak2 tyrosine kinase is dependent on specific regions in its pseudokinase domain. Mol Biol Cell 14:1448–1459. http://www.ncbi.nlm.nih.gov/entrez/query.fcgi?cmd=Retrieve&db=PubMed&dopt=Citation&list_uids=12686600

Sanada M, Suzuki T, Shih LY et al (2009) Gain-of-function of mutated C-CBL tumour suppressor in myeloid neoplasms. Nature 460:904–908. http://www.ncbi.nlm.nih.gov/entrez/query.fcgi?cmd=Retrieve&db=PubMed&dopt=Citation&list_uids=19620960

Santos FPS, Kantarjian HM, Jain N et al (2009) Phase II study of CEP-701, an orally available JAK2 inhibitor, in patients with primary or post polycythemia vera/essential thrombocythemia myelofibrosis. Blood:blood-2009-10-246363. http://bloodjournal.hematologylibrary.org/cgi/content/abstract/bloodjournal;blood-2009-10-246363v1

Sasaki A, Yasukawa H, Shouda T et al (2000) CIS3/SOCS-3 suppresses erythropoietin (EPO) signaling by binding the EPO receptor and JAK2. J Biol Chem 275:29338–29347. http://www.ncbi.nlm.nih.gov/entrez/query.fcgi?cmd=Retrieve&db=PubMed&dopt=Citation&list_uids=10882725

Silva M, Richard C, Benito A et al (1998) Expression of Bcl-x in erythroid precursors from patients with polycythemia vera. N Engl J Med 338:564–571. http://www.ncbi.nlm.nih.gov/entrez/query.fcgi?cmd=Retrieve&db=PubMed&dopt=Citation&list_uids=9475763

Smith BD, Levis M, Beran M et al (2004) Single-agent CEP-701, a novel FLT3 inhibitor, shows biologic and clinical activity in patients with relapsed or refractory acute myeloid leukemia. Blood 103:3669–3676. http://www.ncbi.nlm.nih.gov/entrez/query.fcgi?cmd=Retrieve&db=PubMed&dopt=Citation&list_uids=14726387

Squizzato A, Romualdi E, Middeldorp S (2008) Antiplatelet drugs for polycythaemia vera and essential thrombocythaemia. Cochrane Database Syst Rev CD006503. http://www.ncbi.nlm.nih.gov/entrez/query.fcgi?cmd=Retrieve&db=PubMed&dopt=Citation&list_uids=18425953

Swerdlow SH, Campo E, Harris NL, Jaffe ES, Pileri SA, Stein H, Thiele J, Vardiman JW (2008) International Agency for Research on Cancer, World Health Organization. WHO classification of tumours of haematopoietic and lymphoid tissues, 4th edn. International Agency for Research on Cancer (IARC), Lyon

Tam CS, Kantarjian H, Cortes J et al (2009) Dynamic model for predicting death within 12 months in patients with primary or post-polycythemia vera/essential thrombocythemia myelofibrosis. J Clin Oncol 27:5587–5593. http://www.ncbi.nlm.nih.gov/entrez/query.fcgi?cmd=Retrieve&db=PubMed&dopt=Citation&list_uids=19786661

Tartaglia AP, Goldberg JD, Berk PD et al (1986) Adverse effects of antiaggregating platelet therapy in the treatment of polycythemia vera. Semin Hematol 23:172–176. http://www.ncbi.nlm.nih.gov/entrez/query.fcgi?cmd=Retrieve&db=PubMed&dopt=Citation&list_uids=3749927

Tefferi A (2003) Polycythemia vera: a comprehensive review and clinical recommendations. Mayo Clin Proc 78:174–194. http://www.ncbi.nlm.nih.gov/entrez/query.fcgi?cmd=Retrieve&db=PubMed&dopt=Citation&list_uids=12583529

Tefferi A, Elliott M (2007) Thrombosis in myeloproliferative disorders: prevalence, prognostic factors, and the role of leukocytes and JAK2V617F. Semin Thromb Hemost 33:313–320. http://www.ncbi.nlm.nih.gov/entrez/query.fcgi?cmd=Retrieve&db=PubMed&dopt=Citation&list_uids=17525888

Tefferi A, Barosi G, Mesa RA et al (2006) International Working Group (IWG) consensus criteria for treatment response in myelofibrosis with myeloid metaplasia, for the IWG for Myelofibrosis Research and Treatment (IWG-MRT). Blood 108:1497–1503. http://www.ncbi.nlm.nih.gov/entrez/query.fcgi?cmd=Retrieve&db=PubMed&dopt=Citation&list_uids=16675707

Tefferi A, Pardanani A, Lim KH et al (2009) TET2 mutations and their clinical correlates in polycythemia vera, essential thrombocythemia and myelofibrosis. Leukemia 23:905–911. http://www.ncbi.nlm.nih.gov/entrez/query.fcgi?cmd=Retrieve&db=PubMed&dopt=Citation&list_uids=19262601

Temerinac S, Klippel S, Strunck E et al (2000) Cloning of PRV-1, a novel member of the uPAR receptor superfamily, which is overexpressed in polycythemia rubra vera. Blood 95:2569–2576. http://www.ncbi.nlm.nih.gov/entrez/query.fcgi?cmd=Retrieve&db=PubMed&dopt=Citation&list_uids=10753836

Tiedt R, Hao-Shen H, Sobas MA et al (2008) Ratio of mutant JAK2-V617F to wild-type Jak2 determines the MPD phenotypes in transgenic mice. Blood 111:3931–3940. http://bloodjournal.hematologylibrary.org/cgi/content/abstract/bloodjournal;111/8/3931

Tyner JW, Bumm TG, Deininger J et al (2010) CYT387, a novel JAK2 inhibitor, induces hematologic responses and normalizes inflammatory cytokines in murine myeloproliferative neoplasms. Blood 115:5232–5240. http://bloodjournal.hematologylibrary.org/content/115/25/5232.abstract

Ugo V, Marzac C, Teyssandier I et al (2004) Multiple signaling pathways are involved in erythropoietin-independent differentiation of erythroid progenitors in polycythemia vera. Exp Hematol 32:179–187. http://www.ncbi.nlm.nih.gov/entrez/query.fcgi?cmd=Retrieve&db=PubMed&dopt=Citation&list_uids=15102479

Vaidya R, Siragusa S, Huang J et al (2009) Mature survival data for 176 patients younger than 60 years with primary myelofibrosis diagnosed between 1976 and 2005: evidence for survival gains in recent years. Mayo Clin Proc 84:1114–1119. http://www.ncbi.nlm.nih.gov/entrez/query.fcgi?cmd=Retrieve&db=PubMed&dopt=Citation&list_uids=19955247

Vainchenker W, Dusa A, Constantinescu SN (2008) JAKs in pathology: role of Janus kinases in hematopoietic malignancies and immunodeficiencies. Semin Cell Dev Biol 19:385–393. http://www.ncbi.nlm.nih.gov/entrez/query.fcgi?cmd=Retrieve&db=PubMed&dopt=Citation&list_uids=18682296

Vannucchi AM, Antonioli E, Guglielmelli P et al (2007) Clinical profile of homozygous JAK2 617 V>F mutation in patients with polycythemia vera or essential thrombocythemia. Blood 110:840–846. http://www.ncbi.nlm.nih.gov/entrez/query.fcgi?cmd=Retrieve&db=PubMed&dopt=Citation&list_uids=17379742

Vasquez L (1892) On a special form of cyanosis accompanied by excessive persistent erythrocytosis. Comp Rend Soc Biol 12:384–388

Verstovsek S (2010) Therapeutic potential of Janus-activated kinase-2 inhibitors for the management of myelofibrosis. Clin Cancer Res 16:1988–1996. http://clincancerres.aacrjournals.org/content/16/7/1988.abstract

Verstovsek S, Pardanani AD, Shah NP et al (2007) A Phase I study of XL019, a selective JAK2 inhibitor, in patients with primary myelofibrosis and post-polycythemia vera/essential thrombocythemia myelofibrosis. ASH Annu Meet Abstr 110:553. http://abstracts.hematologylibrary.org/cgi/content/abstract/ashmtg;110/11/553

Verstovsek S, Deeg HJ, Odenike O et al (2010a) Phase 1/2 study of SB1518, a novel JAK2/FLT3 inhibitor, in the treatment of primary myelofibrosis. ASH Annu Meet Abstr 116:3082. http://abstracts.hematologylibrary.org/cgi/content/abstract/ashmtg;116/21/3082

Verstovsek S, Kantarjian H, Mesa RA et al (2010b) Safety and efficacy of INCB018424, a JAK1 and JAK2 inhibitor, in myelofibrosis. N Engl J Med 363:1117–1127. http://www.ncbi.nlm.nih.gov/entrez/query.fcgi?cmd=Retrieve&db=PubMed&dopt=Citation&list_uids=20843246

Walgren RA, Meucci MA, McLeod HL (2005) Pharmacogenomic discovery approaches: will the real genes please stand up? J Clin Oncol 23:7342–7349. http://www.ncbi.nlm.nih.gov/entrez/query.fcgi?cmd=Retrieve&db=PubMed&dopt=Citation&list_uids=16145062

Wernig G, Mercher T, Okabe R et al (2006) Expression of Jak2V617F causes a polycythemia vera-like disease with associated myelofibrosis in a murine bone marrow transplant model. Blood 107:4274–4281. http://bloodjournal.hematologylibrary.org/cgi/content/abstract/bloodjournal;107/11/4274

Wernig G, Kharas MG, Okabe R et al (2008) Efficacy of TG101348, a selective JAK2 inhibitor, in treatment of a murine model of JAK2V617F-induced polycythemia vera. Cancer Cell 13:311–320. http://www.sciencedirect.com/science/article/B6WWK-4S7B7Y8-6/2/de233c7814bd4b668c9de73d9dc8da76

Wolf BC, Neiman RS (1987) Hypothesis: splenic filtration and the pathogenesis of extramedullary hematopoiesis in agnogenic myeloid metaplasia. Hematol Pathol 1:77–80. http://www.ncbi.nlm.nih.gov/entrez/query.fcgi?cmd=Retrieve&db=PubMed&dopt=Citation&list_uids=3332874

Xing S, Wanting TH, Zhao W et al (2008) Transgenic expression of JAK2V617F causes myeloproliferative disorders in mice. Blood 111:5109–5117. http://bloodjournal.hematologylibrary.org/cgi/content/abstract/bloodjournal;111/10/5109
Yamaoka K, Saharinen P, Pesu M et al (2004) The Janus kinases (Jaks). Genome Biol 5:253. http://www.ncbi.nlm.nih.gov/entrez/query.fcgi?cmd=Retrieve&db=PubMed&dopt=Citation&list_uids=15575979
Zaleskas VM, Krause DS, Lazarides K et al (2006) Molecular pathogenesis and therapy of polycythemia induced in mice by JAK2 V617F. PLoS One 1:e18. http://dx.plos.org/10.1371/journal.pone.0000018
Zhang B, Lewis SM (1989) The splenomegaly of myeloproliferative and lymphoproliferative disorders: splenic cellularity and vascularity. Eur J Haematol 43:63–66. http://www.ncbi.nlm.nih.gov/entrez/query.fcgi?cmd=Retrieve&db=PubMed&dopt=Citation&list_uids=2767243
Zhao L, Ma Y, Seemann J et al (2010) A regulating role of the JAK2 FERM domain in hyperactivation of JAK2(V617F). Biochem J 426:91–98. http://www.ncbi.nlm.nih.gov/entrez/query.fcgi?cmd=Retrieve&db=PubMed&dopt=Citation&list_uids=19929856

Chronic Myeloid Leukemia

Kapil Bhalla, Celalettin Ustun, and Warren Fiskus

Epidemiology

It is estimated that 4,870 new cases of chronic myeloid leukemia (CML) will be diagnosed in 2010, representing approximately 11% of all leukemias (Jemal et al. 2010). The median age of diagnosis reported in 2003–2007 was 65 years of age (Altekruse et al. 2010). The etiology of CML remains unclear; evidence is lacking demonstrating hereditary influence. Individuals who have been exposed to ionizing radiation or benzene are at an increase risk of developing CML (Corso et al. 1995). Other risk factors which have been evaluated include smoking, benzene, and magnetic field exposure (Bjork et al. 2001; Hayes et al. 1997; Kheifets et al. 1997). To date, these are all inconclusive in correlating exposure to acquisition of CML. A recent M.D. Anderson study suggests that adulthood obesity and occupational exposure to agricultural chemicals increase CML risk (Strom et al. 2009).

Introduction

CML is a hematopoietic malignancy resulting from clonal expansion of primitive hematopoietic progenitor cells. CML constitutes the most common form of myeloproliferative neoplasms (MPN). CML is characterized by a specific chromosomal aberration called the Philadelphia chromosome (Ph) (Nowell and Hungerfold 1960). The Ph chromosome produced from a reciprocal translocation between chromosomes 9 and 22. The result of the translocation is the oncogenic BCR-ABL fusion

K. Bhalla (✉) • C. Ustun • W. Fiskus
The University of Kansas Medical Center, Kansas City, KS, USA
e-mail: kbhalla@kumc.edu

J. Tao and E. Sotomayor (eds.), *Hematologic Cancers: From Molecular Pathobiology to Targeted Therapeutics*, Cancer Growth and Progression 14,
DOI 10.1007/978-94-007-5028-9_13,

gene (Rowley 1973; deKlein et al. 1982). CML has become a highly treatable, chronic disease for most patients over the last decade since the discovery of imatinib, a tyrosine kinase inhibitor (TKI), and its clinical availability. This discovery has also become a model for "targeted" therapies that can "cure" hematologic malignancies.

Pathophysiology

Philadelphia Chromosome

The Ph chromosome is a shortened chromosome 22 resulting from a reciprocal translocation, t(9;22)-(q34;q11), between the long arms of chromosomes 9 and 22 and is found in 95% of chronic myeloid leukemia patients (Nowell and Hungerfold 1960). The Ph translocation adds a 3' segment of the *ABL* gene from chromosome 9q34 to the 5' part of the *BCR* gene on chromosome 22q11, creating a hybrid *BCR–ABL* gene that is transcribed into a chimeric *BCR–ABL* messenger RNA (Kurzrock et al. 1988). *c-ABL* is a cellular homolog of v-Abl, a transforming gene identified in the retrovirus Abelson murine leukemia virus located on the chromosome 9 at 9q34 (Goff et al. 1980). The breakpoints on the Abl gene at 9q34, occur towards the 5' end and have breakpoints in introns 1 or 2 of the ABL gene (Melo 1996). The breakpoints on chromosome 22 are clustered in a relatively small region named the *b*reakpoint *c*luster *r*egion, or *BCR* (Groffen et al. 1984). These breakpoints on BCR localize to one out three areas in patients with CML. In most patients with CML, the break occurs within a 5.8-kb region and is defined as the major breakpoint cluster region (M-*bcr*). Because of alternative splicing, fusion transcripts with either b2a2 or b3a2 junctions can be formed and from this mRNA translates into a 210-kDa chimeric protein (P210*BCR-ABL*). In the remaining patients with acute lymphocytic leukemia (ALL) and rarely in patients with CML, the breakpoints are further upstream in the 54.4-kb region between the alternative *BCR* exons e2' and e2, termed the minor breakpoint cluster region (m-*bcr*). The resultant e1a2 mRNA is translated into a 190-kDa protein (P190*BCR-ABL*) (Melo et al. 1994). A third breakpoint cluster region (m-*bcr*) was identified downstream of exon 19, giving rise to a 230-kDa fusion protein (P230*BCR-ABL*) associated with the rare Ph+chronic neutrophilic leukemia (Pane et al. 1996).

BCR-ABL Oncogene

Tyrosine kinases (TKs) are a group of approximately 100 proteins which transfer phosphate from ATP to tyrosine residues on specific cellular proteins. These are involved in controlling important cellular functions, including proliferation and differentiation processes (Robinson et al. 2000).

The *c-ABL* gene belongs to the family of the non-receptor TKs and is expressed ubiquitously in cells of various tissues, and the c-Abl protein may be found at the cytoplasmic and nuclear level (Pendergast 2002). C-Abl kinase has been associated with both negative and positive regulators of cell growth, depending on the cellular compartments (cytoplasmic or nuclear), depending on its phosphorylation status and level of expression. In quiescent cells, nuclear c-Abl is kept in an inactive state by binding to the retinoblastoma protein (pRB) (Welch and Wang 1993). Phosphorylation of pRB by cyclin D1 disrupts this complex and results in activation of c-Abl TK in S phase, during which c-Abl is able to stimulate the transcriptional activity of factors such as CREB and E2F-1 and promote the activity of RNA polymerase II (Baskaran et al. 1997). Therefore the ABL protein appears to be part of the general mechanism of regulation of gene transcription during cell cycle.

At a cytoplasmic level, there is a TK signaling cascade involving receptor (PDGFR) and non-receptor TKs (c-Src/c-Abl) important for mitogenesis and growth factor induced c-Myc expression (Furstoss et al. 2002). Along this pathway c-Abl is situated downstream of Src and contributes to transmitting the mitogenic signal activating c-Myc directly or through the Ras/Erk pathway. In contrast, it has been shown that the induction of cell cycle arrest in G1 phase of cell cycle can occur by over expression of wild type c-abl (Sawyers et al. 1994).

BCR is a 160-kDa cytoplasmic protein with several functional domains (Laurent et al. 2001). The N-terminal 426 amino acids of BCR, encoded by the first exon, are retained in all Bcr-Abl fusion protein isoforms. This region contains a serine-threonine kinase domain and two serine/threonine–rich regions that bind Src homology (SH)2 domains of other proteins, including ABL, p190BCR-ABL, and p210BCR-ABL (Maru and Witte 1991; Pendergast et al. 1993). The proximal SH2-binding domain is essential for transformation of rat fibroblasts by BCR-ABL (Pendergast et al. 1991). The two key regions of the first BCR exon are tyrosine 177 and the coiled-coil domain contained in amino acids 1–63. Phosphorylated tyrosine 177 forms a binding site for GRB2 (an adapter molecule that links Bcr to the Ras pathway) and is required for the induction of myeloid leukemia (He et al. 2002). The coiled-coil domain is crucial for the dimerization of BCR-ABL, which in turn is required for activation of ABL kinase activity and oncogenicity of BCR-ABL (McWhirter et al. 1993; Melo and Deininger 2004). BCR-ABL tyrosine kinase activity activates multiple signal pathways including Ras/Raf/MEK/ERK, JAK/STAT, and PI3K/Akt resulting in enhanced cell proliferation and survival.

PI3K Pathway

The PI3K pathway is an essential step in lymphoid transformation and also contributes to the transformation of cells by the BCR-ABL oncoprotein (Kharas et al. 2004). PI3K is a lipid kinase that phosphorylates phosphatidylinositol or its derivatives on the 3' hydroxyl position of the inositol head group. The PI3K signaling pathway plays an integral role in many normal cellular processes in multiple cell types, including survival, proliferation, differentiation, metabolism, and motility. PI3K enzymes are classified according to structure, substrate selectivity, and mechanism of regulation. Class 1A PI3K enzymes are activated downstream of tyrosine kinase activation and they are heterodimers consisting of catalytic and regulatory subunits (Kharas and Fruman 2005). Activated PI3K leads to formation of phosphatidylinositol-3,4,5-trisphosphate (PIP3) which has been known to accumulate in cells transformed by ABL oncogenes (Varticovski et al. 1991). Within the ABL protein sequence is a YXXM motif that when phosphorylated creates the optimal binding motif for the SH2 domains of class IA regulatory subunits; however, this motif is not absolutely essential because a mutation in the motif does not prevent a PI3K activation (Jain et al. 1996). A critical pathway for PI3K activation in BCR ABL-expressing cells is mediated by Y177 in the BCR portion and the adapter proteins, GRB2 and Gab2. Y177 is an autophosphorylation site for BCR-ABL and functions as a docking site for GRB2. GRB2 binds through its SH2 domain to Y177 on BCR-ABL, recruiting SOS which in turn activates Ras and the scaffold adapter GRB2-associated binding protein 2 (GAB2) (Sattler et al. 2002). A mutation of Y177 to phenylalanine (Y177F) exhibited either normal or decreased transformation capacity which was shown with mouse models demonstrating an inability to transform into a CML like disease (He et al. 2002). Similarly, Gab2–/– progenitor cells are resistant to transformation by Bcr-Abl.

RAS Pathway

Ras is a small GTP-binding protein, which is the common upstream molecule of several signaling pathways including Raf/MEK/ERK, PI3K/Akt and RalGEF-Ral.

GRB2 in association with SOS converts the inactive GDP bound Ras to its active GTP bound state. Ras is also activated by other proteins like Shc and CrkL that are substates of Bcr-Abl. Activated Ras binds to the serine/threonine kinase Raf-1, recruiting it to the plasma membrane where it is activated by tyrosine phosphorylation and initiates a signaling cascade by way of the mitogen-activated protein kinase (MAPK) pathway (Melo and Deininger 2004).

Stem Cells

Human hematopoietic stem cell (CD34+CD38-CD90+Lin-) acquires the Ph positivity and initiates CML (Fialkow et al. 1977). In mouse models, it was shown that chronic phase CML is characterized by the presence of both normal and leukemic stem cells whereas blast phase has CD34+ leukemic stem cell predominance (Sirard et al. 1996). In mice, the β-catenin pathway activation is necessary for self-renewal capacity. In CML- BP, the LSC, which resemble granulocyte-macrophage progenitor cells (GMP) have an aberrant activation of β-catenin. There appear to be two distinct cell populations responsible for the disease in CML, (1) the GMPs with their aberrant expression of β-catenin and (2) the quiescent leukemic stem cell population. The quiescent stem cells are responsible for the persistence of the disease even with treatment and can lead to disease progression (Quintas-Cardama and Cortes 2009).

Notch proteins regulate development and its signaling also plays a role in self renewal. Hes1 is a key notch target gene and combined expression of Bcr-Abl and hes1 in murine myeloid progenitors lead to a condition similar to blast crisis phase of CML. Increased expression of hes1 was also demonstrated in small group of patients in CML blast crisis (Nakahara et al. 2010). Another self-renewal pathway is the hedgehog pathway. Murine models of CML have demonstrated that high expression of Smoothened (Smo), a downstream protein in the hedgehog pathway, leads to a more accelerated disease (Zhao et al. 2009).

Genomic Instability

Accumulation of chromosomal abnormalities is believed to play an important role in transformation from chronic phase to blast phase. There is a modest increase in the genetic abnormalities, and chromosomal abnormalities are consistently increased in BCR-ABL expressing cells. BCR-ABL has been shown to induce the production of reactive oxygen species, which cause oxidative damage and mutations (Sattler et al. 2000). DNA damage may occur as a single nucleotide alterations, single strand breaks or double strand breaks.

DNA mutations can occur when there is an abnormality in DNA repair pathways or when there are mutations in the proteins necessary to recognize and repair DNA double strand breaks, or when there is failure of cell cycle checkpoints that allow subsequent replication of damaged DNA and BCR-ABL effects on all these pathways have been studied. Normally, single nucleotide alterations are repaired by mismatch repair or by nucleotide excision repair (NER), whereas strand breaks are repaired by high-fidelity homologous recombination (HR) or by non homologous end joining (NHEJ). BCR-ABL expression leads to decreased mismatched repair activity in CML cells. Tyrosine kinase inhibitor treatment leads to normalization of the mismatch repair activity which suggests that the BCR-ABL kinase function abrogates mismatch repair activity to cause a mutator phenotype and prevent apoptotic cell death (Stoklosa et al. 2008). The effects of BCR-ABL on nucleotide excision repair and the significance remains unclear. The effects of BCR-ABL on NER in different cell lines was variable. With Ba/F3 BCR-ABL cells, NER activity was decreased making them more susceptible to UV damage but in other cell lines the NER activity was increased. These effects were reversed with the use of imatinib suggesting that these are related to the Bcr-Abl kinase activity (Canitrot et al. 2003).

Telomeres and Telomerases in CML

Telomeres are specialized repeat structures located at the end of human chromosomes. They provide a hexameric repetitive sequence which forms a buffer of expendable

DNA and along with telomere associated core proteins help in maintaining chromosomal integrity and stability. As cells divide, or in cells without compensatory mechanisms, there is loss of telomere repeats that occurs with the division. This progressive loss results in critical shortening ultimately causing terminal cell cycle arrest or apoptotic cell death. Dysfunctional telomeres can also result in end to end fusions and repeated fusion and breakage cycles results in loss of heterozygosity and genomic instability which may result in activation of oncogenes or suppression of tumor-suppressor genes (Keller et al. 2009). Myeloid cells from CML patients demonstrate accelerated telomere shortening. Ph+ peripheral blood leucocytes have more rapid telomere shortening, as compared to age-matched controls, and this shortening is more accelerated in patients with high risk disease. Telomeres in accelerated phase and blast crisis were also found to be smaller than in chronic phase and high telomerase activity was detected in patients with blast crisis. This was also associated with the acquisition of new cytogenetic abnormalities indicating the presence of genomic instability and microsatellite changes (Drummond et al. 2004).

Natural History

CML has three clinical phases, chronic phase (CP) (early stage), blastic phase (advanced stage), and accelerated phase (AP) between them (clinically should be considered as advanced stage). Not all patients will have AP while CML progresses from CP to BP. The definition of CML phases is given in Table 1 (Sokal et al. 1988; Savage et al. 1997; Kantarjian et al. 1993). Untreated CML will progress to advanced stage and lead to death of the patient. Overall survival (OS) was approximately 3 years with no therapy. Therefore, allogeneic hematopoietic stem cell transplantation (alloHSCT) was the choice of treatment for CML patients who were transplantation candidates before TKI. Currently, the general consensus is that TKI therapy is the primary treatment option for CML patients (see management for detailed discussion).

Table 1 Staging of CML

CML phase	Definition
Chronic phase	Fewer than 10% blasts in the blood and bone marrow
Accelerated phase	Blasts 10–19% of WBCs in peripheral and/or bone marrow cells; persistent thrombocytopenia ($< 100 \times 10^9$/L) persistent thrombocytosis ($> 1000 \times 10^9$/L) which is unresponsive to therapy; splenomegaly unresponsive to therapy; cytogenetic evidence of clonal evolution; Triad of WBC $>50 \times 10^3/\mu$L, hematocrit 125% and persistent unexplained fever or bone pain
Blast crisis	≥ 20% blasts present in peripheral blood WBCs or nucleated bone marrow cells; extramedullary disease with immature blast proliferation; and large foci or clusters of blasts on bone marrow biopsy

Monitoring

Monitoring a patient with CML is essential to assess the efficacy of ongoing therapy. Definitions used in monitoring patients (hematologic, cytogenetic, and molecular responses) are found in Table 2. Responses are evaluated with recommended timelines (see Table 3); if responses are not achieved according to these criteria, this is a warning of suboptimal response or failure. In these incidences, current recommendations suggest changing TKI based therapy (Baccarani et al. 2009).

Table 2 Definitions of response in CML

Type of response	Definition of response	
Hematologic	Normalization of peripheral blood (PB) counts: Platelets < 450 × 109/L, WBCs < 10 × 109/L, No circulating immature myeloid cells, No palpable splenomegaly No symptoms of disease	
Cytogenetic	Complete	No Ph + cells
	Partial	1–35% Ph + cells
	Minor	36–65% Ph + cells
	Minimal	66–95% Ph + cells
	None	>95% Ph + cells
Molecular	Complete	BCR-ABL transcript undetectable
	Major	≥ 3 log reduction in BCR-Abl transcript

Table 3 Response monitoring timeline in CML

Months on therapy	Optimal response	Suboptimal response	Treatment failure
3	Complete hematologic response (CHR) and at least minor cytogenetic response (CyR)	No cytogenetic response	Less than CHR
6	At least partial cytogenetic response (PCyR)	< PCyR	No cytogenetic response
12	At least complete cytogenetic response (CCyR)	< CCyR	< PCyR
18	At least major molecular response (MMR)	< MMR	< CCyR
At any time during treatment	Stable or improving MMR	Additional cytogenetic abnormalities in Ph + cells, loss of MMR, mutation with a low level of insensitivity to imatinib	Loss of CHR, loss of CCyR, mutation with a high level of insensitivity to imatinib

Table 4 Monitoring of response in CML

Response	Criteria of response	Frequency of monitoring
Complete Hematologic Response (CHR)	Normalization of peripheral blood (PB) counts: Platelets < 450×10^9/L, WBCs < 10×10^9/L, No circulating immature myeloid cells, No palpable splenomegaly No symptoms of disease	Bi-weekly until CHR achieved and confirmed, then quarterly
Partial Hematologic Response (PHR)	Platelets <50 % of pretreatment counts but > 450×10^9/L, WBCs < 10×10^9/L, Splenomegaly <50 % of pretreatment size Presence of immature cells in PB	Bi-weekly until CHR achieved and confirmed, then quarterly
Minimal Cytogenetic Response (Minimal CyR)	66–95 % Ph+cells by conventional cytogenetic analysis of BMA	Every 6 months until CCyR achieved and confirmed, then annually
Minor Cytogenetic Response (Minor CyR)	36–65 % Ph+cells by conventional cytogenetic analysis of BMA	Every 6 months until CCyR achieved and confirmed, then annually
Partial Cytogenetic Response (PCyR)	1–35 % Ph+cells by conventional cytogenetic analysis of BMA	Every 6 months until CCyR achieved and confirmed, then annually
Complete Cytogenetic Response (CCyR)	No Ph+cells by conventional cytogenetic analysis of BMA	Every 6 months until CCyR achieved and confirmed, then annually
Major Molecular Response (MMR)	≥ 3-log reduction of BCR-ABL transcript	Peripheral blood tested every 3 months
Complete Molecular Response (CMR)	BCR-ABL transcript undetectable by RT- PCR	Peripheral blood tested every 3 months

Conventional cytogenetic studies are critical to follow patients on TKI not only for monitoring the leukemic cell mass and response to TKI, but also to identify novel chromosomal abnormalities which may evolve during treatment. Trisomy 8 is the most common abnormality and might be transient especially if it is not associated with myelodysplasia or leukemia. If patients are not losing response obtained on the treatment this minor abnormality or other new chromosomal abnormalities do not necessitate changing treatment (Hughes and Branford 2009). On the other hand, patients who have achieved CCyR or MMR and are not losing CHR do not benefit significantly from conventional cytogenetic analyses (Ross et al. 2006). Studies also revealed that results from peripheral blood (PB) and bone marrow (BM) are comparable in terms of leukemia level except for advanced stage disease where BM results are superior (Hughes and Branford 2009). In summary, cytogenetic studies from BM until CCyR is achieved are useful as shown in Table 4, then repeating the studies annually. Molecular studies can be performed from patients with CCyR once a year from BM or more often from PB.

In the clinical setting, an increase in the level of BCR/ABL is frequently associated with disease progression (Branford et al. 2004; Press et al. 2007; Wang et al. 2006). Fluorescence in situ hybridization (FISH) and real-time quantitative polymerase chain reaction (RQ-PCR) are useful to evaluate leukemic cell mass, the variant of BCR-ABL being expressed and the transcript levels of BCR-ABL. Moreover, molecular quantification is well correlated with cytogenetic status (Ross et al. 2006). While FISH can only provide information on changes of tumor dynamics up to two log, RQ-PCR can detect up to a 5 log change in BCR-ABL transcript levels (Hughes and Branford 2009). An initial National Cancer Institute (NCI) workshop addressed criteria for molecular, cytogenetic, and hematologic relapse. The importance of Quantitative PCR analysis posttransplant was emphasized in terms of an association of positive BCR/ABL and relapse posttransplant. In addition, it is recommended that Ph+ chromosome presence must evaluated by FISH in PB or BM or by conventional cytogenetics in BM in patients with molecular relapse (Kroger et al. 2010).

Tyrosine Kinase Inhibition

The realization that activated tyrosine kinases could play a role in initiation of cancer led to the development of compounds that would inhibit these enzymes. One of the earliest steps was the synthesis of specific inhibitors of serine kinases (Hidaka et al. 1984). This led to initial attempts at designing molecules that were specific inhibitors of tyrosine kinases called the tryphostins (tyrosine phosphorylation inhibitors). The first series of these tryphostins developed by Alex Levitzki were modeled after a naturally occurring tyrosine kinase inhibitor and these included selective inhibitors of BCR-ABL (Yaish et al. 1988). Using the known structure of the ATP binding site of protein kinases, a series of proteins that were 2-Phenylaminopyrimidine derivatives were synthesized and screened for their potency towards inhibition of protein kinases. CGP 53716 and CGP 57148B both had in-vitro activity against PDGF receptor. The CGP 57148B also had potent activity against BCR-ABL and this became known as STI571 (signal transduction inhibitor 571) or imatinib (Gleevec®) (Druker et al. 1996).

Imatinib

Pre-clinical

Druker et al. demonstrated that CGP 57148B was a potent inhibitor of BCR-ABL with an IC_{50} of 0.025 μM in *in vitro* cell based assays. However, there were no significant inhibitions of other protein kinases except the platelet derived growth factor receptor (PDGF) tyrosine kinase. A dose dependent inhibition of tumor formation in mice was demonstrated by showing that CGP 57148B prevents tumor formation after inoculation with 32Dp210 or 32Dv-Src cells which normally lead to development of tumors within 4 weeks. At a dose of 1 μM, CGP 57148B prevented colony formation when there was BCR-ABL expression by the cells obtained from peripheral blood or bone marrow from t(9:22) expressing CML patients. Interestingly, CGP 57148B was shown to have no effect on normal hemaotpoiesis (Druker et al. 1996).

Phase I Studies

Following pre-clinical investigations, a Phase I trial was conducted with STI 571 in CML patients who failed treatment with interferon alpha. The primary endpoint was safety and tolerability in this dose escalation study. Dosing of STI 571 ranged from 25 to 1,000 mg and 83 patients were enrolled in the study. Pharmacokinetic analysis showed that the mean maximal concentration of 2.3 μg per milliliter (4.6 μM) was reached at steady state by once-daily administration of 400 mg of STI571. The half-life of the drug in the circulation ranged from 13 to 16 h, and the levels of the drug increased by a factor of 2–3 at steady state with once-daily dosing. The mean plasma trough concentration was 0.72 μg per milliliter (1.46 μM) 24 h after the administration of 400 mg of STI571 at steady state. This dose exceeded the concentration required for the inhibition of cellular phosphorylation by BCR-ABL (IC_{50}, 0.025 μM), and this concentration caused the death of cell lines positive for BCR-ABL *in vitro*.

The rate of CHR increased as the daily dose increased from 85 to 250 mg and reached 98% in patients treated with ≥300 mg of STI571. CHR typically occurred within 4 weeks after the initiation of therapy. The most frequent treatment related adverse effects were nausea, edema, myalgias, and diarrhea; overall, most were mild. In seven patients, there were elevations of liver-enzyme levels of grade 2 or higher. The maximum tolerated dose was not identified but based on the pharmacokinetic profile, the 400 mg dose led to adequate serum concentrations that correlated well with *in vitro* studies showing these concentrations to be effective in killing CML cells (Druker et al. 2001a).

In CML-BC patients, STI571 was initiated at 300 mg with dose escalation (range of 300–1,000 mg/day). Fifty-eight patients were enrolled in this study and included 38 myeloid blast crisis patients and 20 patients with lymphoid blast crisis patients (lymphoid blast crisis patients and Ph+ALL patients). The most frequent adverse effects noted were nausea (55%), vomiting (41%), and edema (41%); most of which were grade 1–2 as noted in the previous studies. Grade 4 neutropenia was observed in 40% and thrombocytopenia in 33% of STI571-treated patients. Forty-six of the 58 patients had a 50% reduction in the peripheral blood blasts. The overall response rate was 55% in the myeloid blast crisis patients and 70% in the lymphoid blast crisis patients. Of the 38 patients with myeloid blast crisis, four patients achieved a CHR and 17 reported a decrease in blasts in the marrow to ≤15% with eight of these having a decrease to ≤5%. In patients with a positive response to the STI571, decreases in peripheral blood blasts were seen within a week after initiation (Druker et al. 2001b).

Phase II Studies

Three phase II trials were initiated in patients with CP, AP and BC of CML. In a multicenter phase II study with CP patients, 532 patients were enrolled with 454 patients having confirmed disease. Patients received imatinib at 400 mg/day orally. The dose was increased to 400 mg BID in patients in whom a CHR had not been achieved following 3 months of treatment, those whose disease relapsed within 3 months after the achievement of a CHR, and those in whom a MCyR had not been achieved after 12 months of therapy. CHR was reported for 430 of the 454 patients (95%) studied. The median time to a CHR was 0.7 months and 86% of patients who had a response did so within 3 months of initiation.

Of the 454 patients, 272 (60%) had a MCyR and 343 (76%) had a major, minor, or minimal cytogenetic response. Of the 272 patients with a MCyR, 188 (41%) had a CCyR. The time to onset of MCyR ranged from 2.4 to 19 months. The estimated rate of PFS at 18 months was 89% (95% CI, 86–92%). The achievement of a cytogenetic response at 3 months was associated with a higher rate of PFS (Kantarjian et al. 2002).

The phase II study enrolling AP patients was an open label non-randomized trial evaluating the clinical efficacy and safety of the drug. Two-hundred thirty-five patients were enrolled in the study; however, 181 patients had confirmed AP and the results were based on these patients. Imatinib induced HR lasting at least 4 weeks in 69% of patients, including CHR in 34% of patients. Treatment with imatinib also induced MCyR in 24% of patients, 12-month PFS in 59% of patients, and 12-month overall survival in 74% of patients. Imatinib was well tolerated and the non-hematological side effect profile was similar to those in other studies (Talpaz et al. 2002).

Sawyers et al. (2002) evaluated imatinib in BC patients in an open-label, nonrandomized, multicenter, phase II trial. This investigation sought to determine the rate of HR lasting at least 4 weeks, and the safety of treatment in this population. Initially, enrolled patients received treatment with imatinib 400 mg/day, but later were escalated to 600 mg/day. The primary end point in this study was sustained HR lasting at least 4 weeks and secondary efficacy end points were the induction of CyR, duration of HR, and overall survival. Two-hundred sixty patients were enrolled in the study and 37 (14%) started therapy with imatinib at 400 mg/day and the remaining 223 patients (86%) started treatment at 600 mg/day based on the phase I data. Efficacy results were reported on 229 patients that had confirmed CML-BC. Of the 229 patients, 119 (52%) had reductions in blast values in peripheral blood and bone marrow features corresponding to a HR on at least one occasion. Thirty-five patients (15%) had a CHR, 55 (24%) had a CHR or marrow response, and 64 (28%) met the criteria for a return to chronic phase. MCyR were reported for 37 patients (16%) and 7% of those responses were complete. A major, minor, or minimal CyR was reported in 71 patients (31%). Rates of sustained HR and MCyR were markedly higher in patients treated with an initial imatinib 600 mg/day than in those given 400 mg/day. The estimated median survival time was 7.5 months for previously untreated patients. This compared favorably with the median OS time of 3–5 months observed with other therapies available at the time of this study in patients with newly diagnosed myeloid blast crisis. The non hematological side effects observed were similar to those in other studies and were rarely severe. A fluid-retention syndrome involving disorders of pleural effusion, pulmonary edema, acute respiratory distress syndrome, ascites, congestive heart failure, or edema was identified as a possible adverse drug reaction. The episodes of severe cytopenias were frequent and were likey due to the pharmacological activity of the medication in this group of patients with very low bone marrow reserve.

Phase III Studies

The IRIS (International Randomized Study of Interferon and STI571) trial was a prospective, multicenter, open-label, phase III, randomized study comparing imatinib with the standard treatment of recombinant interferon with low dose cytarabine. The primary end point was progression; secondary end points were the rate of CHR, the rate of MCyR, safety, and tolerability. Patients were allowed to cross over if they had no response, had a loss of response, had an increase in WBC count, or could not tolerate treatment. Eleven-hundred six patients were enrolled with 553 patients in each group, the median dose of imatinib was 400 mg/day (range 114–700 mg/day). The study was reported with a median follow-up of 19 months. A total of 79 patients (14.3%) in the imatinib group and 493 patients (89.2%) in the combination-therapy group either discontinued treatment or crossed over to the alternative treatment group. Higher overall rates of CHR in the imatinib group were reported when compared to combination arm (95.3% vs. 55.5%, $p<0.001$). Furthermore, the responses were quicker in the imatinib arm (1 month vs 2.5 months). The rate of a MCyR was 85.2% in the imatinib group, as compared to 22.1% in the combination arm ($p<0.001$). The rates of CHR, MCyR and CCyR, freedom from progression to AP or BC CML, and tolerance of therapy were significantly improved in the imatinib arm (O'Brien et al. 2003).

Follow-up Analysis

An 8 year follow up from the original IRIS study revealed, 304 (55%) pts remained on imatinib treatment, and 45% had discontinued treatment due to adverse events (AEs)/safety (6%), unsatisfactory therapeutic outcome (16%), stem cell transplant (3%), death (3%) or other reasons (17% for withdrawal or lack of renewal of consent and miscellaneous). No new long-term safety issues or new adverse events were identified. Estimated EFS at 8 year was 81% and freedom from progression to AP/BC was 92%. Estimated OS was 85% at 8 year, and 93% when only CML-related deaths and those prior to stem cell transplant were considered. The annual rates of progression to accelerated AP or BC in 4–8 years following initiation of therapy were 0.9, 0.5, 0, 0, and 0.4%, respectively.

Most AP/BC events occurred early, with minimal risk after year 3 and no evidence of an increase over time. Minor CyR at 3, partial CyR at 6 and 12, and complete CyR at 18 month intervals were associated with stable complete CyR over the observation period. This data suggests that patients responding to imatinib are likely to maintain their responses over time with minimal long term drug induced adverse events (Deininger et al. 2009).

Imatinib Resistance

Pharmacokinetic Factors

Individual patient compliance should be investigated when patients do not respond appropriately to the TKIs. Patients not achieving a CCyR have less adherence (26% non-compliance vs 9%) to imatinib compared to those that achieve complete response at 12 month duration (Noens et al. 2009). Bioavailability of imatinib also varies based on gastrointestinal absorption and drug metabolism in the liver. Imatinib is metabolized by the CYP450 isoenzyme 4A. The activity of this enzyme differs between individuals and also has the potential for drug interactions. The mean trough concentrations were higher in patients with good response compared to patients with less favorable outcomes (Larson et al. 2008). Pharmacokinetic surveillance with serum trough level monitoring has been suggested as a way of ensuring patient compliance, although this practice is not routinely performed.

Imatinib binds to protein, predominantly albumin but also to alpha1-acid glycoprotein and increased binding to these proteins also may lead to impaired therapeutic effect (Breccia and Alimena 2009).

Resistance Mechanisms

Primary resistance to imatinib is defined as failure to achieve a target level of response at any given time points after starting treatment as defined by criteria published by the European LeukemiaNet expert panel. Secondary resistance is defined as a confirmed increase in leukemia load at any point during therapy after an initial response.

Kinase Domain Mutations

The most widely studied and clinically relevant mechanisms of imatinib resistance involves acquired point mutations in the kinase domain (KD) of BCR-ABL. Protein kinase inhibitors typically bind at the highly conserved nucleotide-binding pocket of the catalytic domain. The activation loop in protein kinases controls catalytic activity by switching between different states in a phosphorylation-dependent manner. In fully active kinases, the loop is stabilized in an open conformation by phosphorylation on serine, threonine or tyrosine residues within the loop. Imatinib binds to and stabilizes an inactive conformation of ABL in which the activation loop is in a "closed" substrate-mimicking position and there is extensive distortion of the ATP binding loop. Imatinib also interacts with the kinase through hydrogen bonds, some of which confer specificity (Schindler et al. 2000). Based on the position, mutations can be divided into four categories: mutations that directly impair imatinib binding, mutations in the p-loop, mutations in the a-loop and mutations in the catalytic domain (Breccia and Alimena 2009). Multiple mutations have been described but 15 amino acid substitutions account for more than 85% of the mutations, and the mutations responsible for 66% of reported cases occur at only seven sites. T315I was one of the earliest mutations to be described and this mutation leads to impaired binding of imatinib. This mutation is a result of single amino acid substitution at position 315. Threonine present at this position forms a crucial hydrogen bond with imatinib and substitution with isoleucine prevents this bond formation (Gorre et al. 2001). This mutation, in preclinical studies was also shown to induce a more aggressive leukemic phenotype (Liu et al. 2008).

Drug Transporters

Several drug efflux and influx mediators have been implicated in mechanisms of resistance to imatinib. One of the main drug transporters implicated is MDR1 gene product, Pgp protein, which was also implicated in development of resistance to chemotherapy in other cancers as well. Pgp over-expressing cells have significantly lower imatinib concentrations (Mahon et al. 2003). Other drug transporters such as breast cancer resistance protein (ABCG2), human organic cation transporter 1(OCT-1) have been also been described as a possible mechanism in imatinib resistance (Doyle and Ross 2003; White et al. 2007).

Increased Expression of BCR-ABL

Increased expression of BCR-ABL was shown clinically in 3 of 11 patients who had resistance to imatinib (Gorre et al. 2001). In BCR-ABL cell lines it was shown that expression of the Bcr-Abl gene oncoprotein is associated with marked enhancement of phosphorylation of its key substrates such as Stat5 and CrkL (Barnes et al. 2005). Increased expression of BCR-ABL led to increased resistance in those cell lines and also required less time to develop more resistant clones (Barnes et al. 2005). Mahon et al. (2000) showed that increased expression was BCR-ABL was the most frequent cause of resistance to imatinib to cell lines. The low clinical prevalence of this increased expression as a cause of resistance is probably due to the inhibitory effect of high BCR-ABL on the growth of cells.

Clonal Evolution

The decreased response to imatinib in blast crisis is well known. Clonal evolution is more common than kinase domain mutations in blast crisis CML patients (Lahaye et al. 2005). In a series of 171 patients with inadequate responses to imatinib, 24% of the patients exhibited clonal evolution. Patients with clonal evolution had a higher frequency of KD mutations as well (Jabbour et al. 2006a).

Dasatinib

The U.S. Food and Drug Administration (FDA) granted to dasatinib (Sprycel™, made by Bristol-Myers Squibb) approval for use in the treatment of adults with CML-CP, CML-AP, CML-BP with resistance or intolerance to prior therapy in June 2006. Dasatinib was later approved for the treatment of adults with Ph+ALL with resistance or intolerance to prior therapy.

Phase I Studies

In a phase I dose-escalating study (Talpaz et al. 2006), dasatinib (15–240 mg/day, once or twice daily, PO as 4-week cycles) was tested in 84 imatinib refractory (86%) or intolerant patients (14%). Most patients on the study had CML (n = 74) whereas 10 patients had CML-lymphoid blast crisis or Ph+ALL. OR rates were higher in CML-CP 92.7% (37 of 40 patients), compared to 82% (9 of 11 patients) in the CML-AP, 78% (18 of 23 patients) in the CML-BP, and 80% (8 out of 10) in the Ph+ALL or lymphoid blast. CHR rates were higher again in the CML-CP (92.7%) patients compared to CML-AP (45%), CML-BP (35%), or lymphoid crisis/Ph+ALL (70%) patients. Overall cytogenetic responses were observed in 62, 36, 52, and 90% of patients in CML-CP, CML-AP, CML-BP, and Ph+ALL/lymphoid crisis, respectively. Overall CCyR occurred in 30% of all patients. The duration of response was short (the median of 4 months) in CML-BP patients and Ph+ALL/lymphoid crisis patients. Relapse was seen in 90% of these advanced patients. Hematologic grade 3/4 adverse reactions were common, especially in patients with advanced disease. Pleural effusion was observed in 18% of the patients. Eight percent of the patients experienced elevated liver function tests, resolving spontaneously in the later courses. Patients with T315I mutation did not response to dasatinib.

The imatinib or nilotinib refractory 23 patients received a median of 4 months of dasatinib in a range of 70 mg PO twice day to 120 mg PO twice a day (Quintas-Cardama et al. 2007a). Four, ten, and nine patients were in CML-CP, CML-AP, CML-BP, respectively. Ten patients overall response (OR) was 57%, including CHR of 43%, cytogenetic response was observed in 7 (30%) patients, including 2 CCyR, 4 PCyR, and 1 miCyR.

Phase II Studies

In a phase II randomized trial, dasatinib (140 mg/day, n = 101) was compared with higher doses of imatinib (800 mg/day, n = 49) (Kantarjian et al. 2007a). The dasatinib arm enjoyed improved responses, including CHR (93% vs. 82%, p = 0.034), CCyR (40% vs. 16%, p = 0.004), and MMR (17% vs. 0%). DFS was also superior with dasatinib (>90% vs. >70% in the imatinib arm at 1 year, p < 0.001). This study allowed crossover between these two arms, post-crossover, 17 of the 38 patients (45%) who received dasatinib subsequently achieved a MCyR after an initial treatment failure with imatinib. In contrast, only 15% (two of the 13 patients) of the patients who received imatinib after dasatinib failure attained a MCyR. Pleural effusion was observed in ~17% of patients on the dasatinib arm but none of the patients on the imatinib arm. Myelosuppression was more common with dasatinib, but peripheral edema was more frequent with imatinib.

Dasatinib in Early Stage CML (CML-CP)

Hochhaus et al. (2009) focused on 1,067 CML-CP patients who received dasatinib on phase II/III trials because of imatinib-resistance/suboptimal response (n = 829) or -intolerance (n = 239). MMR was observed at a rate of 12, 22, 35, and 40% after 3, 6, 12, and 24 months, respectively. As expected, the response rate at 2-years was higher in intolerant patients (34% vs. 63%) with resistance or suboptimal response to imatinib (n = 829) and 63% in imatinib-intolerant patients (n = 238). Among patients who had achieved a complete cytogenetic response (CCyR), 72% also achieved MMR. Responses with dasatinib 100 mg once daily were similar to other doses. In landmark analyses, 24-month progression-free survival was higher in patients who had achieved MMR or CCyR at 12 months than in those without MMR or CCyR at 12-month.

In a phase II randomized START-R trial, dasatinib (70 mg twice a day, n = 101) or high-dose imatinib (800 mg/day, n = 49) was administered to imatinib-resistant (i.e., not responsive to imatinib 400–600 mg/day) CML-CP patients (Kantarjian et al. 2009a). Baseline BCR/ABL mutations, including T315I were more common in the dasatinib arm. At a minimum follow-up of 2 years, dasatinib resulted in higher rates of CHR (93% vs. 82%, p = 0.034), MCyR (53% vs. 33%, p = 0.017), CCyR (44% vs. 18%, p = 0.0025), and MMR (29% vs. 12%, p = 0.028). Dasatinib also provided improved PFS (86% vs. 65%, p = 0.0012). Crossover was allowed. Among patients who crossed over to the other treatment arm, a MCyR occurred more frequently with dasatinib (49%) than with high-dose imatinib (15%). The most common adverse events (>20%) were fluid retention, GIT events (diarrhea, nausea), fatigue, rash in both arms. Pleural effusion (25% vs. 0%), dyspnea (23% vs.4%), headache (26% vs. 10%), and musculoskeletal pain (21% vs. 12%) occurred more in the dasatinib arm. Grade 3/4 myelosuppression appear to be more common with dasatinib treatment as well. In contrast, vomiting was more common in the imatinib arm (24% vs. 10%). However, patients received substantially longer dasatinib treatment (median of 23 months) than imatinib treatment (median of 3 months), which provides difficulty comparing adverse events between the two arms.

In a phase II trial in patients with 186 CML-CP patients, dasatinib induced 90% CHR and 52% MCyR at 8-months follow-up (Hochhaus et al. 2007). Responses to dasatinib were durable. Only 4.7% of patients who had achieved a CHR had disease progression. PFS was 92.4%. The most common non-hematologic toxicities were headache, GI (diarrhea, nausea), fatigue/asthenia, and dyspnea. Grade 3/4 pleural effusion occurred in 3% of the patients. The durability of responses were shown in the updated results on more patients (n=387) in 2008. With median follow-up of 15.2 months, a CHR was similar to prior report (91%). A McyR was 59%, which was, as expected, higher in imatinib intolerant patients than imatinib resistant patients (80% vs. 52%). CCyR occurred in 49% of patients. PFS and OS were 90% and 96%, respectively. Grade 3/4 thrombocytopenia and neutropenia were reported in 48 and 49% of patients, respectively. The most common all grade non-hematologic toxicities were the same reported as in the Phase I trials.

Dasatinib in Advanced Stage CML (CML-AP/BP)

Dasatinib has been demonstrated to be efficacious in advanced CML patients. In a phase II study, imatinib-resistant (n=99) or -intolerant (n=8) patients with CML-AP received dasatinib (70 mg PO twice a day) (Guilhot et al. 2007). Overall hematologic response (OHR) was 81%, including MHR of 64% and CHR of 39%. CCyR was observed nearly 1/4th of the patients. A median time to reach MHR and MCyR was similar, 2 months. Progression-free survival (PFS) was 76% at 10 months. Cytopenias were frequently observed (60–100%). Non-hematologic toxicities described >20% of patients were headache, fatigue, and gastrointestinal tract (GIT) adverse event such as diarrhea and nausea. Pleural effusion was seen in 23% of the patients. BCR/ABL mutations occurred in 60% of these patients. "Gatekeeper" mutations like T315I inhibited dasatinib efficacy. Another trial exclusively evaluated CML patients in BP (Cortes et al. 2007a). Seventy-four patients were in myeloid blast crisis (MBC) 42 patients were in lymphoid blast crisis (LBC). Dasatinib induced similar MHR in MBC and LBC (approximately 30–35% MHR), but seemed to induce higher MCyR in the patients in LBC (50% vs. 31% with MBC). The patients with LBC appeared to have shorter time to the onset of response (approximately 1 month vs. 2 months). However, in accordance with other studies (Talpaz et al. 2006), disease progression occurred more frequently in the LBC patients (52% vs. 34%). All grades of pleural effusion occurred at a rate of 32.7%, including high grades in 10% of the patients. GIT toxicities, asthenia, fatigue other common toxicities. Most patients (57%) did not have BCR/ABL mutations, therefore dasatinib activity in these patients might be resulted from its activity on SRC family kinases (SFKs).

Dose of Dasatinib

In a phase 3 study, 670 imatinib-resistant, -intolerant, or -sub-optimally responsive CML-CP patients were randomized to dasatinib 100 mg once-daily, 50 mg twice-daily, 140 mg once-daily, or 70 mg twice-daily (Shah et al. 2010a). A 2-year minimum follow-up showed that dasatinib 100 mg once a day schedule resulted in comparable outcomes and less side effects compared to other dosing schedules. Dasatinib provided similar CHR (87–92%), MCyR (61–63%), and CCyR (50–54%) among the four dose schedule arms. MCyR was durable at 2 year (68–87% of MCyR lasted). Most of CCyR occurred within first 6 months of dasatinib treatment. No CCyR was seen in 15 patients with BCRABL/T315I mutation, and only 1 of 6 patients with the F317L mutation. MMR was obtained approximately 60% in all arms. PFS rates was 80% in the 100 mg once a day arm compared to others that were around 75% at 2 years. OS rate was 91% in the 100 mg once a day arm compared to others that were ranging from 88 to 94% at 2 years. Compared with the other three regimens, the 100 mg once daily regimen led the lowest rate of pleural effusions (14% vs. 23–25%, $p=0.049$), neutropenia ($p=0.034$), leukopenia ($p=0.017$), and grade 3/4 thrombocytopenia ($p=0.003$). The most common treatment-related non-hematologic adverse events were fluid retention, headache, diarrhea, nausea, fatigue, rash, and dyspnea. This study was updated after 4-year follow-up in the 2010 annual ASCO meeting. At 36-month, PFS and OS 73% and 87% with 100 mg daily dosing whereas 67, 60, and 72% in the 70 mg twice a day, 140 mg daily, and 50 mg twice a day dosing, respectively (Shah et al. 2010b). OS remained similar among all four arms. Exposure-response analysis has shown similar long-term outcomes achieved in CML-CP patients across dasatinib exposure (Roy et al. 2010). The incidence of Grade 3/4 non-hematologic adverse events were not related to differences in exposure.

A large study evaluating 1,067 CML-CP patients as a second line treatment concurred that 100 mg once a day results are comparable with other dosing schedules (Hochhaus et al. 2009).

In another phase III trial (Kantarjian et al. 2009b), 317 CML-AP patients were randomized to receive dasatinib 140 mg once daily or 70 mg twice a day. With a median follow-up of 15 months, the once day arm induced similar responses (CHR of 47% vs. 52% in the twice a day arm). MCyR of 39% vs. 43% in the twice a day arm (CCyR 32% vs. 33% in the twice a day arm). PFS and OS at 2-years were similar 51% vs. 55% and 63% vs. 72%, respectively. Pleural effusion was observed less with once daily dosing (20% vs. 39%, $p<0.001$). Recently, dasatinib was approved for 100 mg once-daily dosing for patients with CML-CP based on results following this optimization study (CA180-034, *NCT00123474*I).

Dasatinib in Patients Who Failed Both Imatinib and Nilotinib

Thirty-four patients with CML received dasatinib as a third line agent after imatinib and nilotinib failure. Sixteen patients were in CP, 18 patients in advanced stage (e.g., AP and BP). CHR was observed in 81 and 69% of the patients in CP and AP/BP, respectively. CCyR was attained in 31 and 22% of the patients in CP and AP/BP, respectively. MMR was attained in 13 and 11% of the patients in CP and AP/BP, respectively. Considering all patients (n=48), with a median follow-up of 13 months since the start of the third TKI, 25 patients (31%) are still alive, including 15 (44%) treated with dasatinib and 10 (71%) with nilotinib. The median OS and EFS were 20 months and 13 months, respectively. The duration of response was shorter compared to the patients in CP (i.e., 20 months vs. 5 months in AP and 3 months in BP). In summary, patients failing a second line TKI may respond to third TKI, however duration will be short in patients with advanced disease. An allogeneic stem cell transplant should be considered for these patients.

Dasatinib in Newly Diagnosed CML-CP Patients

Patients were randomized to receive dasatinib 100 mg once daily or 50 mg twice daily as initial therapy (Cortes et al. 2010a). Both arms were similar in terms of response rates and adverse events. CCyR and MMR were observed in 98 and 82% of 50 patients who were in the trial for at least 3 months, respectively. Responses occurred rapidly, with 94% of patients achieving CCyR by 6 months. The projected event-free survival rate and OS were 88 and 100% at 24 months. High grade neutropenia and thrombocytopenia occurred in 21 and 10% of patients, respectively. Nonhematologic toxicity was relatively low in severity.

Dasatinib vs. Imatinib in Newly Diagnosed CML-CP Patients

The result of DASISION trial were reported at 2010 annual ASCO meeting: 519 patients were randomized to receive dasatinib 100 mg daily (n=259) or to receive imatinib 400 mg daily (n=260) (Kantarjian et al. 2010a). The primary endpoint was 12-month rate of confirmed CCyR. Patients in both arms tolerated treatments well. Dasatinib induced higher CCyR (83% vs. 72%) and MMR (46% vs. 28%). Time to achieve MMR (HR 2.0, $p<0.0001$) and CCyR(HR 1.5, $p<0.0001$) was shorter with dasatinib. Rates of progression to AP/BC were 1.9% for dasatinib and 3.5% for IM. Although anemia and neutropenia were similar in both arms, thrombocytopenia was more common with dasatinib (19% vs. 10%). Non-hematologic adverse drug reactions (all grades) in ≥10% of pts (dasatinib vs. IM) were fluid retention (19% vs 42%; including pleural effusion, all grade 1 or 2, 10% vs. 0%), nausea (8% vs. 20%), vomiting (5% vs. 10%), myalgia (6% vs.12%), muscle inflammation (4% vs. 17%) and rash (11% vs. 17%).

Adverse Effects

Pleural Effusion

MD Anderson investigators reported severe pleural effusion occurred in 17% of patients while all grade frequency was 35% (Quintas-Cardama et al. 2007b). Advanced stage was an important risk factor by univariate but not in multivariate analysis (29% in CP, 50% in AP, and 33% in BP). A multivariate analysis showed history of cardiac disease, hypertension, and use of a twice-daily schedule were identified as risk factors for development of pleural effusions. Effusions were exudative in 78% of the assessable cases. In most patients, effusions were managed by holding dasatinib. The use of diuretics, steroids, and thoracentesis were utilized in some patients. The exact mechanism through which dasatinib induces pleural effusion is unknown but may be related to inhibition of platelet-derived growth factor receptor beta (PDGFR-β), Src-related kinases that mediate vascular permeability.

Another study from France exclusively sought pulmonary complications of dasatinib (70 mg twice a day) in 40 patients (Bergeron et al. 2007). Twenty-two percent developed respiratory symptoms, including dyspnea, cough, and chest pain. Fifteen percent of the patients had pleural effusion (all were exudates). In addition to the pleural effusion, there were paranchymal changes in the lung (17.5%) with ground-glass or alveolar opacities and septal thickening. Bronchoalveolar lavage (BAL) revealed lymphocyte predominance. Dasatinib interruption led to resolution of the lung complications in all patients. The authors suggested that dasatinib associated pulmonary complications might be secondary to immunologic reactions rather than simple fluid retention. In line with this study, de Lavallade et al. emphasized that immune mechanisms may be the cause of dasatinib-induced pleural effusion (de Lavallade et al. 2008). The cumulative incidence of pleural effusion at 1 year was 29.5%. Other parancymal changes, including lobe consolidations, nodular opacities, and septal thickness were noted in 4% of the patients. Higher incidence of pleural effusions was seen in patients with hypertension, a history of cardiac disease and a dasatinib dosage >100 mg/day. A prior history of auto-immune disease was strongly associated with the risk of developing a pleural effusion. It was also supported by the finding that the occurrence of a rash secondary to dasatinib was significantly associated with the risk of pleural effusion (RR = 7.07). In a multivariate analysis, a prior skin rash on dasatinib, a previous history of autoimmune disease, and hypercholesterolemia were found to be a risk factor to develop pleural effusion on dasatinib. Lymphatic network disorder was thought to be responsible for pleural effusions (Goldblatt et al. 2009). Pulmonary hypertension or right ventricular failure have also been reported secondary to dasatinib (Mattei et al. 2009; Rasheed et al. 2009).

Miscellaneous Issues: Dasatinib Use Before or After Allogeneic Stem Cell (AlloHSCT)

The use of dasatinib prior allogeneic stem cell transplantation (alloHSCT) has not demonstrated an increase in transplant-related morbidity or mortality (Jabbour et al. 2007; Breccia et al. 2010; Shimoni et al. 2009). Dasatinib did not impair graft-versus leukemia (GVL) effect (Garland et al. 2010). Dasatinib, like other TKIs, has demonstrated safety and efficacy in the post-transplant setting. However, algorithms for appropriate dosage, intervals from SCT, duration of application, and the combination with donor lymphocytes remain to be addressed (Klyuchnikov et al. 2010).

Dasatinib in Central Nervous System Leukemia

Dasatinib was shown to cross the blood–brain barrier and to be clinically active in central nervous system (CNS) Ph+ leukemias (Porkka et al. 2008; Abdelhalim et al. 2007). Seven patients (64%) responded to treatment, and four received dasatinib monotherapy (Porkka et al. 2008). CNS responses to dasatinib were generally durable and were maintained for at least 6 months in five patients, and for at least 12 months in two patients. On the other hand, systemic dasatinib did not prevent isolated CNS relapse in CML-BP posttransplantation (Frigeri et al. 2009).

Pregnancy and Breastfeeding

Pregnancy risk factor is labeled as D in the drug package insert. Pregnancy was detected in the first trimester in an 18-year old female while receiving dasatinib. Dasatinib was discontinued but progression occurred. The patient was treated with interferon and had a healthy infant (Conchon et al. 2010). A man was also reported to have conceived a healthy baby while receiving dasatinib (Oweini et al. 2010). In mice, dasatinib was extensively distributed in maternal tissues and secreted into milk (He et al. 2008).

Nilotinib

In October 2007, nilotinib (Tasigna®) received accelerated approval by the U.S. FDA for use in the treatment of CP and AP Ph+CML in adult patients resistant or intolerant to prior therapy that included imatinib. Early clinical studies demonstrated efficacy in obtaining notable hematologic, cytogenetic, and molecular response rates. As a result, nilotinib received accelerated approval in the treatment of newly diagnosed adult CML-CP patients in June 2010.

Phase I

Kantarjian et al. (2006), evaluated the efficacy and tolerability of nilotinib in 119 imatinib-resistant CML (106 patients; CP, 17; AP, 56; BP, 33) or Ph+ALL (13) patients. Nilotinib doses ranging from 50 to 1,200 mg daily or 400–600 mg twice daily were evaluated. Median duration of nilotinib exposure was 4.9 months (range, 1.4–9.3 months) and 5.1 months (range, 0.3–12.6 months) for CP and AP patients respectively. All five CP patients with active disease that received 400 mg twice daily dosing obtained a CHR, whereas, 38% (three patients) of all CP patients in this dosing cohort obtained a CCR. In Ph+ALL patients, limited clinical responses were reported (one partial HR and 1CMR in 13 patients). Response rates for 400 and 600 mg twice daily dosing were similar and were associated with reductions in phosphorylation of AKT ($P=0.036$), STAT-1 ($P=0.003$), and STAT-5 ($P=0.024$). No dose limiting toxicities were reported up to 600 mg daily. In dosing cohorts >600 mg/day, 18 patients (15%) experienced toxicities. These toxicities included elevations in aminotransferases, bilirubin, lipase and amylase, as well as, rash, pruritis, neutropenia, anemia, and thrombocytopenia. Analysis of >2,200 electrocardiograms revealed that nilotinib was associated with an increase of 5–15 msec of the corrected QT interval by Fredericia's formula (QTcF). One patient had a pleural effusion and atrial fibrillation without cardiac enzyme elevation. The dose of 400 mg twice daily was taken to Phase II trials (Kantarjian et al. 2007b; Le Coutre et al. 2008a) with a possible dose escalation to 600 mg BID in patients with insufficient response.

Phase II-(CML-CP)

Interim analysis was published for 280 CML-CP patients who were imatinib resistant or intolerant, with a follow-up period of at least 6 months (Kantarjian et al. 2007b). Patients with imatinib resistance had to have been treated with at least 600 mg/day of imatinib for 3 months prior to study enrollment. Intolerance was defined as any non-hematologic toxicity ≥ grade 3, a grade 2 toxicity lasting more than a month, or recurrence despite dose reduction and supportive care. Most patients had received previous treatment with hydroxyurea (83%), interferon-α (66%), or cytarabine (25%). Nilotinib was dosed at 400 mg BID with dose escalation of 600 mg BID in patients not achieving desired hematologic or cytogenetic responses at defined intervals.

Based on safety concerns from the Phase I trial, patients receiving medications which prolong the QT interval or inhibit CYP3A4 were excluded unless they could be switched to an alternate medication. BCR-ABL mutation status was assessed at study entry in 182 patients. A total of 28 different BCR-ABL mutations involving 23 amino acids were noted. The primary endpoint of the trial was to determine the rate of major cytogenetic response (MCyR). Secondary efficacy variables were time to MCyR, duration of MCyR, CHR, time to duration of CHR, and OS (Kantarjian et al. 2007b).

A MCyR was seen in 134/280 patients (48%; 95% CI, 41.9–53.9%). The rates of MCyR were similar in the imatinib resistant and imatinib intolerant groups (47% vs. 48%). The median time to obtain MCyR was 2.8 months. Eighty-eight patients (31%; 95% CI, 26–37.2%) achieved a CCyR whereas 137/185 patients (74%; 95% CI, 67.1–80.2%) achieved a CHR who were not in CHR at baseline. The median time to obtain a CHR was 1 month. The estimated 12 month survival was projected to be 95%. Responses were affected in patients harboring a BCR-ABL mutation and were also dependent on the type of mutation. Following 6 months of therapy, MCyR and CCyR were achieved in 42 and 23% patients with a BCR-ABL mutation at baseline, compared to 51 and 35% of patients who lacked a mutation at baseline. Rates of cytogenetic or hematologic response correlated with nilotinib's preclinical IC_{50} profile against BCR-ABL mutations. As anticipated, four patients (2.2%) who harbored the T315I mutation had no response to nilotinib (Kantarjian et al. 2007b). A 24 month follow-up analysis of this study reported that the overall MMR was 28% and was higher in patients with CHR at study entry. The median time to obtain a MMR was 5.6 months. Overall, 77 and 84% of responding patients maintained a MCyR and CCyR at 2 years, respectively (Kantarjian et al. 2009c).

Phase II-(CML-AP)

A multicenter, open-label study that assessed nilotinib 400 mg twice daily in the treatment of imatinib-resistant or intolerant AP patients was reported by Le Coutre et al. (2008a) Imatinib resistance was defined as one of the following despite receiving treatment of 600 mg/day of imatinib: (1) disease progression from CP to AP (2) 50% increase in peripheral WBCs, blasts, basophils, or platelets (3) not achieving a HR in the bone marrow following 4 weeks of treatment. In addition, patients receiving less than 600 mg/day of imatinib with one of the following BCR-ABL mutants (L248, G250, Q52, Y253, E255, T315, F317, and H396) were also eligible. One hundred and nineteen patients (81% imatinib resistant, 19% imatinib intolerant) were enrolled in this study. The primary endpoint was the rate of hematologic response on two consecutive visits, 4 weeks apart. Secondary endpoints were time to HR, duration of HR, MCyR, time and duration to MCyR and OS. Dose escalation to nilotinib 600 mg BID was permitted in the absence of toxicity at the investigator's discretion. Hematologic response was observed in 56 patients (47%; 95% CI, 38–56%); this included 31 patients (26%) with CHR, 11 patients (9%) with marrow responses or no evidence of leukemia, and 14 patients (12%) which returned to CP. A MCyR occurred in 35 patients (29.4%; 95% CI, 21–39%), 19 (16%) had a CCyR, and 16 (13.4%) obtained a PCyR. The projected overall survival rate at 12 months was 79% (95% CI, 70–87%). A 2 year follow-up from this study indicated that cytogenetic responses were durable, with 69% of patients remaining in MCyR at 18 months (Le Coutre et al. 2008b).

Phase II- (CML-BP)

Interim data from a phase II trial was presented at the European Hematology Association (EHA) by Giles et al. (2008) evaluating nilotinib in the treatment of imatinib resistant or intolerant CML-BP. Safety and efficacy were reported for 136 BC patients (105 myeloid, 31 lymphoid). For the patients on the study, prior therapies included interferon (34%), cytarabine (37%), and bone marrow transplant (13%). At study entry, authors noted that 53% of patients had additional chromosomal abnormalities. Nilotinib doses were initiated at 400 mg BID but could be escalated to 600 mg BID if patients did not adequately respond. Primary endpoints were consistent with previously mentioned Phase II trials. Hematologic response was observed in 25 (24%) of BC patients, CHR were reported in 11% of all patients (myeloid BC+lymphoid BC). The median treatment duration was 84 days (range, 3–666 mg/day) and the median dose intensity was 800 mg/day (range, 3.7–1113 mg/day). OS at 12 months was 42% in this advanced disease. Due to advanced disease, grade 3/4 cytopenias (neutropenia 67%, thrombocytopenia 62%, and anemia 42%) were increased in this study population. There was considerable discontinuation (54%) due to disease progression. Only 13 (10%) of patients remain on study. At interim analysis, death occurred in 6 (4%) patients.

Newly Diagnosed CML-CP Patients

Phase II

Cortes et al. (2008) evaluated nilotinib in previously untreated CML-CP patients. The primary objective of this study was to estimate the percentage of patients to obtain a MMR at 12 months. Forty-seven patients received nilotinib 400 mg BID for a median of 6.5 months. At 3 months, 46 (97%) patients had obtained a CCyR. This was impressive compared to historical control rates of 37% with imatinib 400 mg/day and 62% of imatinib 800 mg/day (O'Brien et al. 2003; Cortes et al. 2008). The 12 month MMR was 45% in patients who received nilotinib. This also compared favorably to imatinib data (400 mg/day – 24%; 800 mg/day – 47%) (O'Brien et al. 2003; Cortes et al. 2006). Updated results presented from these investigators suggest that the notable initial responses to nilotinib have continued. At 30 months, 92% CCyR and 75% MMR were projected (Cortes et al. 2010b). These findings suggested that nilotinib would be an effective first line agent in the treatment of CML.

Phase III

In a Phase III multicenter, open-label study called ENESTnd (Evaluating Nilotinib Efficacy and Safety in Clinical Trials- Newly Diagnosed Patients), 846 CML-CP patients were randomized to receive nilotinib (300 mg or 400 mg twice daily) or imatinib (400 mg once daily) (Saglio et al. 2010). Patients were randomized in a 1:1:1 ratio. Patients that received nilotinib produced significantly higher MMR rates (44% – 300 mg BID; 43% – 400 mg BID) compared to imatinib (22%), at 12 months ($p<0.001$). Similarly, nilotinib induced higher CCyR regardless of its dose (80% – 300 mg BID; 78% – 400 mg BID) compared to imatinib (65%), ($p<0.001$). PFS was significantly better with nilotinib (300 mg cohort; $p<0.0095$, 400 mg cohort; $p<0.0037$). Fourteen patients progressed to AP/BP on study, of which 11 (78%) had received imatinib. No patients who achieved a MCyR progressed to advanced disease. Patients that achieved a CCyR while receiving nilotinib did not progress, however, three imatinib patients that achieved a CCyR later progressed to AP. All study cohorts demonstrated tolerable adverse event profiles. Gastrointestinal and fluid-retention events were more frequent among patients receiving imatinib whereas dermatologic events and headache were more frequent in those receiving nilotinib. No patient in any of the study groups had a QT interval corrected for heart rate of more than 500 msec. No decrease from baseline in the mean left ventricular ejection fraction was observed at any time during the study.

Refractory Treatment (Post-Imatinib and Dasatinib Failure)

Jabbour et al. (2006b) were the first to report on the use of nilotinib in this clinical setting. In this small investigation ($n=7$), nilotinib 400 mg BID or 800 mg Q day were given to 1 CP, 1 AP, and 5 BP patients. Mutations F317L and E355G were found in two of four patients prior to nilotinib therapy. Nilotinib produced modest results in this study population, only 1 patient (AP) achieved a CHR. This patient had no mutations at baseline.

In an another analysis, 14 patients with CML received nilotinib as third line agent after imatinib and dasatinib failure (Garg et al. 2009). Nine patients were in CP, five patients in AP/BP. Following treatment with nilotinib, results were as followed – (CP: CHR 67%, CCyR 11%, MMR, 33%; AP/BP: CHR 80%, CCyR 20%, MMR 0%). Giles et al. (2007) presented a Phase II open label, multicenter study evaluating nilotinib in patients who were either intolerant or failed imatinib and dasatinib. A total of 67 patients were reported –CP (27), AP (15), and BC (25) ranging in age from 19 to 78 years. Specific baseline BCR-ABL mutations were not determined in this study. At a 4 month follow up visit, of the 13 patients in the AP arm, 6 of 13 (46%) achieved a HR and 1 of 13 (8%) achieved a MCyR. No patients in the AP arm had achieved a CHR at the time of follow-up, whereas four patients (20%) with BC achieved a CHR. Complete cytogenetic responses or MCyR were reported in 7/41 (17%) of AP and BC patients without a baseline CHR. The median duration of nilotinib exposure was 139 days (range, 22–542) in CP, 70 days (range, 2–264) in AP, and 41 days (range 4–198) in BC. A total of 7/27 (26%) CP patients, 4/47 (27%) AP, and 16/27 (64%) BC discontinued treatment due to progression of disease.

Adverse Events

Hematological

Neutropenia and thrombocytopenia are the most common hematological abnormalities reported following nilotinib treatment. Nilotinib administered at 300 mg BID is associated with 12% neutropenia and 10% thrombocytopenia (Grade 3/4 toxicity) in newly diagnosed CML-CP patients (Novartis Pharmaceuticals Corporation 2010). In advanced disease, hematological adverse events are more frequent. Grade 3/4 neutropenia and thrombocytopenia occurred in 21 and 35% of CML-AP patients, respectively (Le Coutre et al. 2008a). Advanced stage patients (AP/BP) often present with clonal evolution, which leads to a proliferation of the malignant phenotype, thereby compromising normal hematopoiesis. Hematological toxicities are manageable with dose interruption or reduction; however, patients may require hematopoietic growth factors or platelet transfusions.

Non-Hematological

All grade non-hematological adverse events associated with 300 mg BID dosing in newly diagnosed CP patients are rash (36%), headache (28%), nausea (19%), pruritus (19%), and fatigue (19%), however, toxicities are generally mild to moderate in severity. Grade ≥3 toxicities have been reported in 7% or less of patients (Novartis Pharmaceuticals Corporation 2010). Serum abnormalities rarely lead to the discontinuation of nilotinib. Hyperbilirubinemia has been reported, although elevations are usually transient and are resolved with continued treatment (Kantarjian et al. 2006, 2007b). The manufacturer recommends for ≥Grade 3 toxicities of serum lipase or amylase, bilirubin, or hepatic transaminases that nilotinib administration should be withheld and may be resumed at 400 mg/day upon toxicity resolution (Novartis Pharmaceuticals Corporation 2010).

Prolonged QT interval and sudden deaths have been reported. Kantarjian et al. (2006) reported nilotinib was associated with a mean increase of 5–15 msec in QT interval (QTcF) by Fridericia's formula. In a Phase II study, increases in QTcF of more than 60 ms compared to baseline were observed in 4% of patients, however no episodes of torsades de pointes were observed (Le Coutre et al. 2008b). The manufacturer recommends an ECG at baseline, a week following initiation, and periodically thereafter (Novartis Pharmaceuticals Corporation 2010). In March 2010, the US FDA approved a REMS (Risk Evaluation and Mitigation Strategy) to assist physicians and patients in reducing the risk of QT prolongation. To reduce this risk, it is recommended to avoid concurrent strong CYP3A4 inhibitors, take medication on an empty stomach, as well as correct hypokalemia and hypomagnesemia prior to administration.

Toxicities such as fluid retention, edema, cramps, and weight gain that are commonly associated with dasatinib and imatinib, are less commonly observed with nilotinib (Quintas-Cardama et al. 2007b; Novartis Pharmaceuticals Corporation 2010; Druker et al. 2006; Cohen et al. 2002). All grade peripheral edema was reported in 8% of newly diagnosed CP patients treated with nilotinib 300 mg BID, whereas, dasatinib-associated effusions occurred in 24% of patients receiving <100 mg/day (Novartis Pharmaceuticals Corporation 2010; Le Coutre et al. 2008c). As stated previously in this chapter, the increased rate of fluid retention associated with imatinib and dasatinib is hypothesized to be due to increased inhibition of PDGFR-β (Quintas-Cardama et al. 2007c).

Miscellaneous Hematopoietic Stem Cell Transplantation (HSCT)

With the emergence of data demonstrating second generation TKI improvement in newly diagnosed patients as well as imatinib resistance, more patients are receiving nilotinib prior to HSCT (Jabbour et al. 2007; Shimoni et al. 2009; Menzel et al. 2007). Early reports suggest that positive outcomes are possible when nilotinib is given prior to a HSCT (Jabbour et al. 2007; Shimoni et al. 2009; Menzel et al. 2007). Factors such as time to engraftment or donor chimerism have not been affected by the prior use of a TKI. Nilotinib has also been reported to restore full donor chimerism in a Ph+ALL patient following a sibling donor alloHSCT (Merante et al. 2009). For patients who have failed imatinib prior to transplant, other TKIs such as dasatinib or nilotinib are more appropriate in the post-transplant setting (National Comprehensive Cancer Network clincal practice guidelines in oncology 2010).

Pregnancy and Breastfeeding

Rats that were treated with nilotinib at oral doses of 10–100 mg/kg during mating and pre-mating periods had decreased viable fetuses at all doses tested. At doses of 100 mg/kg (5.7 times the AUC of 400 mg BID dosing), produced maternal toxicity that included decreased gestation weight, gravid uterine weight, and net weight gain. In rabbits, doses of 300 mg/kg/day (approximately 1/3 of human exposure based upon AUC) were associated with maternal toxicity. As a result, nilotinib has been classified as pregnancy category D. Females are advised to use contraception during treatment. It is unknown if nilotinib is excreted into breast milk (Novartis Pharmaceuticals Corporation 2010). A case report was published that discussed a successful pregnancy following taking nilotinib 200 mg BID within first trimester. Following positive pregnancy confirmation, the patient elected to withdrawal nilotinib therapy. A healthy baby boy weighing 3.2 kg with an Apgar score of 9 at 10 min at gestational week 33 was reported (Conchon et al. 2009).

Can TKIs Cure CML?

Despite the fact obtaining a CMR is common with the use of an oral TKI, patients with negative PCR might still have residual malignant cells. The presence of BCR-ABL+ stem and progenitor cells even in patients achieving CCyR with imatinib has been shown (Bhatia et al. 2003). In mice, leukemia cells could not be eradicated by a TKI (Hu et al. 2006). Clinically, data have shown that discontinuing TKI therapy is associated with increases in tumor burden and relapse in most patients despite achieving a CMR (Hughes and Branford 2009; Rousselot et al. 2007; Cortes et al. 2004; Merante et al. 2005). Some patients continued to have CMR more than 2 years after discontinuing TKI therapy. It is unknown whether prior interferon therapy had a role in this persistent CMR after imatinib discontinuation (Rousselot et al. 2007). The exact mechanisms for incomplete elimination of CML cells are unknown but may result from maintenance of quiescent, primitive CML cells that are not sensitive to imatinib-induced apoptosis (Graham et al. 2002; Holtz et al. 2005). As a result, the discontinuation of TKIs are not recommended even in patients remaining in persistent CMR.

Novel Agents

Farnesyltransferase Inhibitors

An MD Anderson study evaluated the efficacy of lonafarnib (SCH66336) in patients with imatinib-resistant or –intolerant CML-CP (n = 6) or CML-AP (n = 7) (Borthakur et al. 2006). Lonafarnib was administered at a dose of 200 mg, orally, twice daily for a median of 8 weeks. Two patients responded; one AP patient returned to CP for 3 months and one CP patient had normalization of leukocyte counts for 5 months without hydroxyurea. The most common adverse event was diarrhea. Lonafarnib is also being evaluated in combination with imatinib in imatinib-refractory patients. Tipifarnib (Zarnestra™) was also evaluated in combination with imatinib. The most common adverse effects were cytopenias and GIT adverse events. Therapy was administered for a median of 26 weeks. Overall HR occurred in 68% of the patients (17/25). Overall CyR was observed in nine patients (36%). One responsive patient presented with T315I mutation (Cortes et al. 2007b).

Histone Deacetylase Inhibitors (HDAC Inhibitors)

Several HDAC inhibitors such as panobinostat, vorinostat, and LAQ824 are being evaluated in preclinical or phase I/II trials in advanced leukemias including CML (Fiskus et al. 2006a, b; Spencer et al. 2007; Kadia et al. 2010).

The Role of AlloHSCT in Modern CML Treatment

For decades, AlloHSCT has been the only curative treatment for CML. Following the introduction of TKIs into clinical practice, the number of AlloHSCT in CML patients has significantly reduced. The Center of International Bone Marrow Registry data reported that AlloHSCT provided approximately 70% OS in CML patients undergoing transplant less than 1 year after diagnosis (www.cibmtr.org). Although the number of alloHSCT has decreased, OS rates have improved (Boehm et al. 2011). A recent study from a single institution reported that patients who underwent an alloHSCT between 2001 and 2007 had 76% OS rate compared to 50% OS in patients who received allograft between 1983 and 1994. AlloHSCT results were excellent in patients who achieved a CMR (86% at 3 years posttransplant) (Radich et al. 2003). In contrast, OS rates decrease significantly in advanced stage patients (43% for CML-AP, 11% for CML-BP) (Boehm et al. 2011; Clift et al. 1994; Biggs et al. 1992). In addition to disease stage, other factors affecting transplant-related mortality (TRM) are summarized by European Bone Marrow Transplantation (EBMT) (Table 5) (Baccarani et al. 2009). Low-risk (i.e., risk score of 0–2) patients report lower TRM (17%) in most recent years (in all years it was 31%) whereas TRM was high for intermediate score (i.e., risk score of 3–4) patients (50%) and high risk score (i.e., risk score of 5–6) patients (70%). Although favorable outcomes can be obtained with AlloHSCT alone, considering the efficacy and limited TRM associated with TKI therapy, most patients should be treated with a TKI prior to AlloHSCT. The subgroups in which alloHSCT should be considered are (1) pediatric patients, (2) young patients, (3) patients in accelerated phase, (4) patients in blast crisis, (5) following imatinib treatment, (6) patients who have fail to achieve optimal response according to European Leukemia Net (ELN) criteria, (7) patients with T315I Abl kinase domain mutation, (8) patients progressing to advance stage, (9) patients failing to second generation TKIs, (10) all patients/countries where cost effectiveness favors HSCT (Aguayo and Couban 2009). However, recommendations are subject to change because many critical questions remain: whether TKIs are curative and whether patients that develop resistant mutations to TKIs will have impaired long-term transplantation survival? The outcome of AlloHSCT in patients with T315I mutation also warrants further investigation

Table 5 Risk factors which can affect Transplant Related Mortality (TRM) in CML Risk Factor

Patient age	< 20 y/o = 0	20–40 y/o = 1	> 40 y/o = 2
Disease phase	Chronic = 0	Accelerated = 1	Blastic = 2
Donor type	HLA-matched sibling = 0	Other = 1	
Time from diagnosis	≤ 1 year = 0	> 1 year = 1	
Sex match	Male recipient/female donor = 1	Other = 0	
	Risk score	Level of risk	5-year survival
	0–2	Low	~60–70%
	3–4	Intermediate	~40–50%
	5–6	High	~20%

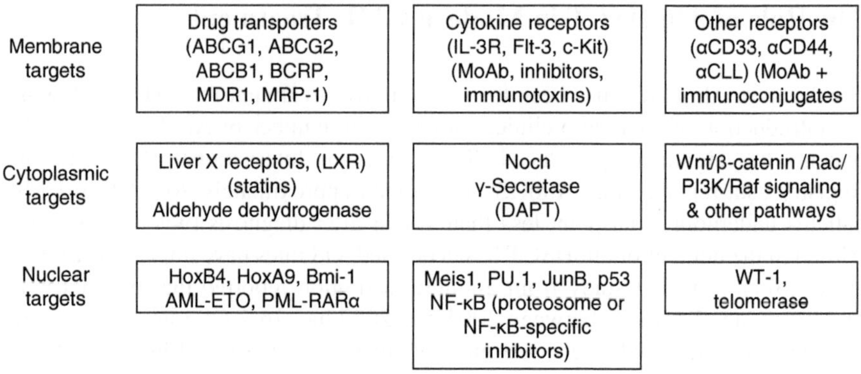

Fig. 1 Potential therapeutic targets on leukemic stem cells. Adapted with permission from Misaghian et al. (2009)

because follow-up, outcome information available for these patients is limited. A report from 8 patients with T315I mutations (2 patients in CP, 3 patients in AP, and 3 patients in BP) demonstrated improved results if patients underwent an alloSCT in early phases of CML. Both of CP patients remained in CMR. Although responses occurred in other patients as well, responses were short in duration. For all patients, failure-free survival was 32% at 12 months. Most patients received non-myeloablative conditioning regimens (Velev et al. 2010). The effect of disease stage appears more prominent on transplantation outcome rather than the direct negative effect of this mutation.

Management Guidelines

Treatment guidelines are useful to standardize treatment in CML patients and guide hematologists/oncologists how to manage these patients (Baccarani et al. 2009; National Comprehensive Cancer Network clincal practice guidelines in oncology 2010). An algorithm is presented in Fig. 1 based upon these recommendation statements. In CML-CP patients, imatinib 400 mg daily is the current standard therapy; however, additional treatment strategies for CML are rapidly evolving. The most recent studies with nilotinib or dasatinib as front line treatment show "superiority" (higher responses in shorter time) without significant toxicity compared to imatinib (Saglio et al. 2010; Kantarjian et al. 2010b). However, with approximately 10 years of impressive follow-up data with imatinib, physicians should be cautious in switching to initial second generation TKIs in all patients. We need improved OS and at least comparable safety data in long-term comparison studies with standard imatinib.

In CML-CP, patients deemed imatinib-resistant or -intolerable, nilotinib or dasatinib should be used. Clinical responses with second generation TKIs in these patients are high, except patients who present with various BCR-ABL mutations.

Accordingly, these patients should be followed vigilantly for progression. Initial disease responses can be lost and progression can occur due to the emergence of new mutations. The pros and cons of AlloHSCT should be clearly discussed with each patient. Obviously, the answer is complicated and must be individualized due to multiple factors such as age, co-morbid conditions, donor availability, HLA-match status, disease control status, donor related issues, BCR-ABL mutation status, and available clinical trials.

In patients with advanced CML (AP/BP) regardless of prior imatinib failure, nilotinib or dasatinib should be initiated. Concomitant chemotherapy use is at the discretion of prescribers. In choosing nilotinib or dasatinib, mutation analysis can be useful. If a patient carries a mutation that is well-described by clinical studies to confer resistance to a particular TKI, then another TKI with improved potency against BCR-ABL mutants should be initiated. Additionally, patient co-morbid conditions and the adverse effect profile of a particular TKI should be taken into consideration. In patients with T315I mutation, novel agents in clinical trials such as XL228 (Smith et al. 2009) or omacetaxine (Cortes-Franco et al. 2009) should be considered. Short responses are seen in advanced stage CML patients; therefore an alloHSCT should be pursued.

The development of TKIs has revolutionized the treatment of CML. However, TKIs have not produced "a cure" which has been sought from targeted therapy. Approximately, 15% of patients will fail on imatinib. Further, there are mechanisms by which CML can be resistant to imatinib and second generation TKIs. Through continued research targeting CML blasts and stem cells, as well as additional drug discovery, CML may one day be eradicated without an alloHSCT.

References

Abdelhalim A, Barcos M, Block AW et al (2007) Remission of Philadelphia chromosome-positive central nervous system leukemia after dasatinib therapy. Leuk Lymphoma 48:1053–1056

Aguayo A, Couban S (2009) State-of-the-art in the management of chronic myelogenous leukemia in the era of the tyrosine kinase inhibitors: evolutionary trends in diagnosis, monitoring and treatment. Leuk Lymphoma 50(Suppl 2):1–8

Altekruse SF, Kosary C, Krapcho M, Neyman N (2010) SEER cancer statistics review 1975–2007. National Cancer Institute, Bethesda, MD. http://seer.cancer.gov/csr/1975_2007/

Baccarani M, Cortes J, Pane F et al (2009) Chronic myeloid leukemia: an update of concepts and management recommendations of European LeukemiaNet. J Clin Oncol 27:6041–6051

Barnes DJ, Palaiologou D, Panousopoulou E et al (2005) Bcr-Abl expression levels determine the rate of development of resistance to imatinib mesylate in chronic myeloid leukemia. Cancer Res 65:8912–8919

Baskaran R, Wood LD, Whitaker LL et al (1997) Ataxia telangiectasia mutant protein activates c-Abl tyrosine kinase in response to ionizing radiation. Nature 387:516–519

Bergeron A, Rea D, Levy V et al (2007) Lung abnormalities after dasatinib treatment for chronic myeloid leukemia: a case series. Am J Respir Crit Care Med 176:814–818

Bhatia R, Holtz M, Niu N et al (2003) Persistence of malignant hematopoietic progenitors in chronic myelogenous leukemia patients in complete cytogenetic remission following imatinib mesylate treatment. Blood 101:4701–4707

Biggs JC, Szer J, Crilley P et al (1992) Treatment of chronic myeloid leukemia with allogeneic bone marrow transplantation after preparation with BuCy2. Blood 80:1352–1357

Bjork J, Albin M, Welinder H et al (2001) Are occupational, hobby, or lifestyle exposures associated with Philadelphia chromosome positive chronic myeloid leukaemia? Occup Environ Med 58:722–727

Boehm A, Walcherberger B, Sperr WR et al (2011) Improved outcome in patients with chronic myeloid leukemia after allogeneic hematopoietic stem cell transplantation over the past 25 years: a single center experience. Biol Blood Marrow Transplant 17:133–140

Borthakur G, Kantarjian H, Daley G et al (2006) Pilot study of lonafarnib, a farnesyl transferase inhibitor, in patients with chronic myeloid leukemia in the chronic or accelerated phase that is resistant or refractory to imatinib therapy. Cancer 106:346–352

Branford S, Rudzki Z, Parkinson I et al (2004) Real-time quantitative PCR analysis can be used as a primary screen to identify patients with CML treated with imatinib who have BCR-ABL kinase domain mutations. Blood 104:2926–2932

Breccia M, Alimena G (2009) Resistance to imatinib in chronic myeloid leukemia and therapeutic approaches to circumvent the problem. Cardiovasc Hematol Disord Drug Targets 9:21–28

Breccia M, Palandri F, Iori AP et al (2010) Second-generation tyrosine kinase inhibitors before allogeneic stem cell transplantation in patients with chronic myeloid leukemia resistant to imatinib. Leuk Res 34:143–147

Canitrot Y, Falinski R, Louat T et al (2003) p210 BCR/ABL kinase regulates nucleotide excision repair (NER) and resistance to UV radiation. Blood 102:2632–2637

Clift RA, Buckner CD, Thomas ED et al (1994) Marrow transplantation for patients in accelerated phase of chronic myeloid leukemia. Blood 84:4368–4373

Cohen MH, Williams G, Johnson JR et al (2002) Approval summary for imatinib mesylate capsules in the treatment of chronic myelogenous leukemia. Clin Cancer Res 8:935–942

Conchon M, Sanabani SS, Bendit I, Santos FM, Serpa M, Dorliac-Llacer PE (2009) Two successful pregnancies in a woman with chronic myeloid leukemia exposed to nilotinib during the first trimester of her second pregnancy: case study. J Hematol Oncol 2:42

Conchon M, Sanabani SS, Serpa M et al (2010) Successful pregnancy and delivery in a patient with chronic myeloid leukemia while on dasatinib therapy. Adv Hematol 2010:136252

Corso A, Lazzarino M, Morra E et al (1995) Chronic myelogenous leukemia and exposure to ionizing radiation–a retrospective study of 443 patients. Ann Hematol 70:79–82

Cortes J, O'Brien S, Kantarjian H (2004) Discontinuation of imatinib therapy after achieving a molecular response. Blood 104:2204–2205

Cortes J, Giles F, Salvado AJ, McDougall K (2006) Molecular responses in newly diagnosed chronic myelocytic leukemia (CML) patients treated with 800 mg imatinib daily. An update from the RIGHT trial study group. Blood 108:2149

Cortes J, Rousselot P, Kim DW et al (2007a) Dasatinib induces complete hematologic and cytogenetic responses in patients with imatinib-resistant or -intolerant chronic myeloid leukemia in blast crisis. Blood 109:3207–3213

Cortes J, Quintas-Cardama A, Garcia-Manero G et al (2007b) Phase 1 study of tipifarnib in combination with imatinib for patients with chronic myelogenous leukemia in chronic phase after imatinib failure. Cancer 110:2000–2006

Cortes JE, O'brien SM, Ferrajoli A, Borthakur G (2008) Efficacy of nilotinib (AMN107) in patients (Pts) with newly diagnosed, previously untreated philadelphia chromosome (Ph)- positive chronic myelogenous leukemia in early chronic phase (CML-CP). J Clin Oncol 26:7016

Cortes JE, Jones D, O'Brien S et al (2010a) Results of dasatinib therapy in patients with early chronic-phase chronic myeloid leukemia. J Clin Oncol 28:398–404

Cortes JE, Jones D, O'Brien S et al (2010b) Nilotinib as front-line treatment for patients with chronic myeloid leukemia in early chronic phase. J Clin Oncol 28:392–397

Cortes-Franco J, Raghunadharao D, Parikh S, Wetzler M (2009) Safety and efficacy of subcutaneous-administered omacetaxine mepesuccinate in chronic myeloid leukemia (CML) patients who are resistant or intolerant to two or more tyrosine kinase inhibitors – Results of a multicenter phase 2/3 study. Blood (ASH Annual Meeting Abstracts) 114:861

de Lavallade H, Punnialingam S, Milojkovic D et al (2008) Pleural effusions in patients with chronic myeloid leukaemia treated with dasatinib may have an immune-mediated pathogenesis. Br J Haematol 141:745–747
Deininger M, O'brien SG, Guilhot F (2009) International randomized study of interferon vs STI571 (IRIS) 8-year follow up: sustained survival and low risk for progression or events in patients with newly diagnosed chronic myeloid leukemia in chronic phase (CML-CP) treated with imatinib. Blood 114:1126
deKlein A, Van Kessel AG, Grosveld G, Bartram CR (1982) A cellular oncogene is translocated to the Philadelphia chromosome in chronic myelocytic leukaemia. Nature 300:765–767
Doyle LA, Ross DD (2003) Multidrug resistance mediated by the breast cancer resistance protein BCRP (ABCG2). Oncogene 22:7340–7358
Druker BJ, Tamura S, Buchdunger E et al (1996) Effects of a selective inhibitor of the Abl tyrosine kinase on the growth of Bcr-Abl positive cells. Nat Med 2:561–566
Druker BJ, Talpaz M, Resta DJ et al (2001a) Efficacy and safety of a specific inhibitor of the BCR-ABL tyrosine kinase in chronic myeloid leukemia. N Engl J Med 344:1031–1037
Druker BJ, Sawyers CL, Kantarjian H et al (2001b) Activity of a specific inhibitor of the BCR-ABL tyrosine kinase in the blast crisis of chronic myeloid leukemia and acute lymphoblastic leukemia with the Philadelphia chromosome. N Engl J Med 344:1038–1042
Druker BJ, Guilhot F, O'Brien SG et al (2006) Five-year follow-up of patients receiving imatinib for chronic myeloid leukemia. N Engl J Med 355:2408–2417
Drummond M, Lennard A, Brummendorf T, Holyoake T (2004) Telomere shortening correlates with prognostic score at diagnosis and proceeds rapidly during progression of chronic myeloid leukemia. Leuk Lymphoma 45:1775–1781
Fialkow PJ, Jacobson RJ, Papayannopoulou T (1977) Chronic myelocytic leukemia: clonal origin in a stem cell common to the granulocyte, erythrocyte, platelet and monocyte/macrophage. Am J Med 63:125–130
Fiskus W, Pranpat M, Bali P et al (2006a) Combined effects of novel tyrosine kinase inhibitor AMN107 and histone deacetylase inhibitor LBH589 against Bcr-Abl-expressing human leukemia cells. Blood 108:645–652
Fiskus W, Pranpat M, Balasis M et al (2006b) Cotreatment with vorinostat (suberoylanilide hydroxamic acid) enhances activity of dasatinib (BMS-354825) against imatinib mesylate-sensitive or imatinib mesylate-resistant chronic myelogenous leukemia cells. Clin Cancer Res 12:5869–5878
Frigeri F, Arcamone M, Luciano L, Di Francia R, Pane F, Pinto A (2009) Systemic dasatinib fails to prevent development of central nervous system progression in a patient with BCR-ABL unmutated Philadelphia chromosome-positive leukemia. Blood 113:5028–5029
Furstoss O, Dorey K, Simon V, Barila D, Superti-Furga G, Roche S (2002) c-Abl is an effector of Src for growth factor-induced c-myc expression and DNA synthesis. EMBO J 21:514–524
Garg RJ, Kantarjian H, O'Brien S et al (2009) The use of nilotinib or dasatinib after failure to 2 prior tyrosine kinase inhibitors: long-term follow-up. Blood 114:4361–4368
Garland P, Dazzi F, Marin D (2010) Dasatinib may not suppress the GVL effect of donor lymphocyte infusions for CML. Bone Marrow Transplant 45:395–396
Giles F, Le Coutre P, Bhalla K (2007) A phase II study of nilotinib administered to patients with imatinib resistant or intolerant chronic myelogenous leukemia (CML) in chronic phase, accelerated phase, or blast crisis who also failed dasatinib. J Clin Oncol 25:7038
Giles F, Larson R, Kantarjian H (2008) Nilotinib in chronic myelogenous leukemia in blast crisis (CML-BC) patients with imatinib resistance or intolerance: update of a phase 2 results. 13th Congress of the European Hematology Association (EHA), Copenhagen
Goff SP, Gilboa E, Witte ON, Baltimore D (1980) Structure of the Abelson murine leukemia virus genome and the homologous cellular gene: studies with cloned viral DNA. Cell 22:777–785
Goldblatt M, Huggins JT, Doelken P, Gurung P, Sahn SA (2009) Dasatinib-induced pleural effusions: a lymphatic network disorder? Am J Med Sci 338:414–417
Gorre ME, Mohammed M, Ellwood K et al (2001) Clinical resistance to STI-571 cancer therapy caused by BCR-ABL gene mutation or amplification. Science 293:876–880

Graham SM, Jorgensen HG, Allan E et al (2002) Primitive, quiescent, Philadelphia-positive stem cells from patients with chronic myeloid leukemia are insensitive to STI571 in vitro. Blood 99:319–325
Groffen J, Stephenson JR, Heisterkamp N, de Klein A, Bartram CR, Grosveld G (1984) Philadelphia chromosomal breakpoints are clustered within a limited region, bcr, on chromosome 22. Cell 36:93–99
Guilhot F, Apperley J, Kim DW et al (2007) Dasatinib induces significant hematologic and cytogenetic responses in patients with imatinib-resistant or -intolerant chronic myeloid leukemia in accelerated phase. Blood 109:4143–4150
Hayes RB, Yin SN, Dosemeci M et al (1997) Benzene and the dose-related incidence of hematologic neoplasms in China. Chinese Academy of Preventive Medicine–National Cancer Institute Benzene Study Group. J Natl Cancer Inst 89:1065–1071
He Y, Wertheim JA, Xu L et al (2002) The coiled-coil domain and Tyr177 of bcr are required to induce a murine chronic myelogenous leukemia-like disease by bcr/abl. Blood 99:2957–2968
He K, Lago MW, Iyer RA, Shyu WC, Humphreys WG, Christopher LJ (2008) Lacteal secretion, fetal and maternal tissue distribution of dasatinib in rats. Drug Metab Dispos 36:2564–2570
Hidaka H, Inagaki M, Kawamoto S, Sasaki Y (1984) Isoquinolinesulfonamides, novel and potent inhibitors of cyclic nucleotide dependent protein kinase and protein kinase C. Biochemistry 23:5036–5041
Hochhaus A, Kantarjian HM, Baccarani M et al (2007) Dasatinib induces notable hematologic and cytogenetic responses in chronic-phase chronic myeloid leukemia after failure of imatinib therapy. Blood 109:2303–2309
Hochhaus A, Muller MC, Radich J et al (2009) Dasatinib-associated major molecular responses in patients with chronic myeloid leukemia in chronic phase following imatinib failure: response dynamics and predictive value. Leukemia 23:1628–1633
Holtz MS, Forman SJ, Bhatia R (2005) Nonproliferating CML CD34+ progenitors are resistant to apoptosis induced by a wide range of proapoptotic stimuli. Leukemia 19:1034–1041
Hu Y, Swerdlow S, Duffy TM, Weinmann R, Lee FY, Li S (2006) Targeting multiple kinase pathways in leukemic progenitors and stem cells is essential for improved treatment of Ph+ leukemia in mice. Proc Natl Acad Sci USA 103:16870–16875
Hughes TP, Branford S (2009) Measuring minimal residual disease in chronic myeloid leukemia: fluorescence in situ hybridization and polymerase chain reaction. Clin Lymphoma Myeloma 9(Suppl 3):S266–271
Jabbour E, Kantarjian H, Jones D et al (2006a) Frequency and clinical significance of BCR-ABL mutations in patients with chronic myeloid leukemia treated with imatinib mesylate. Leukemia 20:1767–1773
Jabbour E, Kantarjian H, Giles F, O'Brien S (2006b) Treatment with nilotinib for patients with chronic myeloid leukemia (CML) who failed prior therapy with imatinib and dasatinib. Blood 108:2171
Jabbour E, Cortes J, Kantarjian H et al (2007) Novel tyrosine kinase inhibitor therapy before allogeneic stem cell transplantation in patients with chronic myeloid leukemia: no evidence for increased transplant-related toxicity. Cancer 110:340–344
Jain SK, Susa M, Keeler ML, Carlesso N, Druker B, Varticovski L (1996) PI 3-kinase activation in BCR/abl-transformed hematopoietic cells does not require interaction of p85 SH2 domains with p210 BCR/abl. Blood 88:1542–1550
Jemal A, Siegel R, Xu J, Ward E (2010) Cancer statistics. CA Cancer J Clin 60:277–300
Kadia TM, Yang H, Ferrajoli A et al (2010) A phase I study of vorinostat in combination with idarubicin in relapsed or refractory leukaemia. Br J Haematol 150:72–82
Kantarjian HM, Deisseroth A, Kurzrock R, Estrov Z, Talpaz M (1993) Chronic myelogenous leukemia: a concise update. Blood 82:691–703
Kantarjian H, Sawyers C, Hochhaus A et al (2002) Hematologic and cytogenetic responses to imatinib mesylate in chronic myelogenous leukemia. N Engl J Med 346:645–652
Kantarjian H, Giles F, Wunderle L et al (2006) Nilotinib in imatinib-resistant CML and Philadelphia chromosome-positive ALL. N Engl J Med 354:2542–2551

Kantarjian H, Pasquini R, Hamerschlak N et al (2007a) Dasatinib or high-dose imatinib for chronic-phase chronic myeloid leukemia after failure of first-line imatinib: a randomized phase 2 trial. Blood 109:5143–5150

Kantarjian HM, Giles F, Gattermann N et al (2007b) Nilotinib (formerly AMN107), a highly selective BCR-ABL tyrosine kinase inhibitor, is effective in patients with Philadelphia chromosome-positive chronic myelogenous leukemia in chronic phase following imatinib resistance and intolerance. Blood 110:3540–3546

Kantarjian H, Pasquini R, Levy V et al (2009a) Dasatinib or high-dose imatinib for chronic-phase chronic myeloid leukemia resistant to imatinib at a dose of 400 to 600 milligrams daily: two-year follow-up of a randomized phase 2 study (START-R). Cancer 115:4136–4147

Kantarjian H, Cortes J, Kim DW et al (2009b) Phase 3 study of dasatinib 140 mg once daily versus 70 mg twice daily in patients with chronic myeloid leukemia in accelerated phase resistant or intolerant to imatinib: 15-month median follow-up. Blood 113:6322–6329

Kantarjian HM, Giles FJ, Bhalla KN, Larson R (2009c) Update on imatinib-resistant chronic myeloid leukemia patients in chronic phase (CML-CP) on nilotinib therapy at 24 months: Clinical response, safety, and long-term outcomes. Blood 114:1129

Kantarjian H, Shah N, Hochhaus A, Cortes J (2010a) Dasatinib compared to imatinib(IM) in patients with newly diagnosed chronic phase chronic myelogenous leukemia (CML-CP): twleve month efficacy and safety from the phase III DASISION study. J Clin Oncol 28:LBA6500

Kantarjian H, Shah NP, Hochhaus A et al (2010b) Dasatinib versus imatinib in newly diagnosed chronic-phase chronic myeloid leukemia. N Engl J Med 362:2260–2270

Keller G, Brassat U, Braig M, Heim D, Wege H, Brummendorf TH (2009) Telomeres and telomerase in chronic myeloid leukaemia: impact for pathogenesis, disease progression and targeted therapy. Hematol Oncol 27:123–129

Kharas MG, Fruman DA (2005) ABL oncogenes and phosphoinositide 3-kinase: mechanism of activation and downstream effectors. Cancer Res 65:2047–2053

Kharas MG, Deane JA, Wong S et al (2004) Phosphoinositide 3-kinase signaling is essential for ABL oncogene-mediated transformation of B-lineage cells. Blood 103:4268–4275

Kheifets LI, Afifi AA, Buffler PA, Zhang ZW, Matkin CC (1997) Occupational electric and magnetic field exposure and leukemia. A meta-analysis. J Occup Environ Med 39:1074–1091

Klyuchnikov E, Kroger N, Brummendorf TH, Wiedemann B, Zander AR, Bacher U (2010) Current status and perspectives of tyrosine kinase inhibitor treatment in the posttransplant period in patients with chronic myelogenous leukemia (CML). Biol Blood Marrow Transplant 16:301–310

Kroger N, Bacher U, Bader P et al (2010) NCI first international workshop on the biology, prevention, and treatment of relapse after allogeneic hematopoietic stem cell transplantation: report from the committee on disease-specific methods and strategies for monitoring relapse following allogeneic stem cell transplantation. part II: chronic leukemias, myeloproliferative neoplasms, and lymphoid malignancies. Biol Blood Marrow Transplant 16:1325–1346

Kurzrock R, Gutterman JU, Talpaz M (1988) The molecular genetics of Philadelphia chromosome-positive leukemias. N Engl J Med 319:990–998

Lahaye T, Riehm B, Berger U et al (2005) Response and resistance in 300 patients with BCR-ABL-positive leukemias treated with imatinib in a single center: a 4.5-year follow-up. Cancer 103:1659–1669

Larson RA, Druker BJ, Guilhot F et al (2008) Imatinib pharmacokinetics and its correlation with response and safety in chronic-phase chronic myeloid leukemia: a subanalysis of the IRIS study. Blood 111:4022–4028

Laurent E, Talpaz M, Kantarjian H, Kurzrock R (2001) The BCR gene and philadelphia chromosome-positive leukemogenesis. Cancer Res 61:2343–2355

Le Coutre P, Ottmann OG, Giles F et al (2008a) Nilotinib (formerly AMN107), a highly selective BCR-ABL tyrosine kinase inhibitor, is active in patients with imatinib-resistant or -intolerant accelerated-phase chronic myelogenous leukemia. Blood 111:1834–1839

Le Coutre P, Giles F, Hochhaus A, Apperley J (2008b) Nilotinib in chronic myeloid leukemia patients in accelerated phase (CML-AP) with imatinib resistance or intolerance: 2 year follow-up results from a Phase 2 study. Blood 112:3229

Le Coutre P, O'Dwyer E, Mendes L (2008c) The occurence of pleural/pericardial effusions in Ph+CML patients failing prior tyrosine kinase inhibitors (TKI) before starting nilotinib-Analysis of data from compassionate use program. 13th Congress of the European Hematology Association (EHA), Copenhagen, p 0557

Liu J, Joha S, Idziorek T et al (2008) BCR-ABL mutants spread resistance to non-mutated cells through a paracrine mechanism. Leukemia 22:791–799

Mahon FX, Deininger MW, Schultheis B et al (2000) Selection and characterization of BCR-ABL positive cell lines with differential sensitivity to the tyrosine kinase inhibitor STI571: diverse mechanisms of resistance. Blood 96:1070–1079

Mahon FX, Belloc F, Lagarde V et al (2003) MDR1 gene overexpression confers resistance to imatinib mesylate in leukemia cell line models. Blood 101:2368–2373

Maru Y, Witte ON (1991) The BCR gene encodes a novel serine/threonine kinase activity within a single exon. Cell 67:459–468

Mattei D, Feola M, Orzan F, Mordini N, Rapezzi D, Gallamini A (2009) Reversible dasatinib-induced pulmonary arterial hypertension and right ventricle failure in a previously allografted CML patient. Bone Marrow Transplant 43:967–968

McWhirter JR, Galasso DL, Wang JY (1993) A coiled-coil oligomerization domain of Bcr is essential for the transforming function of Bcr-Abl oncoproteins. Mol Cell Biol 13:7587–7595

Melo JV (1996) The diversity of BCR-ABL fusion proteins and their relationship to leukemia phenotype. Blood 88:2375–2384

Melo JV, Deininger MW (2004) Biology of chronic myelogenous leukemia--signaling pathways of initiation and transformation. Hematol Oncol Clin North Am 18:545–568, vii–viii

Melo JV, Myint H, Galton DA, Goldman JM (1994) P190BCR-ABL chronic myeloid leukaemia: the missing link with chronic myelomonocytic leukaemia? Leukemia 8:208–211

Menzel H, von Bubnoff N, Hochhaus A, Haferlach C, Peschel C, Duyster J (2007) Successful allogeneic stem cell transplantation in second chronic-phase CML induced by the tyrosine kinase inhibitor nilotinib (AMN107) after blast crisis under imatinib. Bone Marrow Transplant 40:83–84

Merante S, Orlandi E, Bernasconi P, Calatroni S, Boni M, Lazzarino M (2005) Outcome of four patients with chronic myeloid leukemia after imatinib mesylate discontinuation. Haematologica 90:979–981

Merante S, Colombo AA, Calatroni S et al (2009) Nilotinib restores long-term full-donor chimerism in Ph-positive acute lymphoblastic leukemia relapsed after allogeneic transplantation. Bone Marrow Transplant 44:263–264

Misaghian N, Ligresti G, Steelman LS et al (2009) Targeting the leukemic stem cell: the Holy Grail of leukemia therapy. Leukemia 23:25–42

Nakahara F, Sakata-Yanagimoto M, Komeno Y et al (2010) Hes1 immortalizes committed progenitors and plays a role in blast crisis transition in chronic myelogenous leukemia. Blood 115:2872–2881

National Comprehensive Cancer Network clincal practice guidelines in oncology (2010) Chronic myelogenous leukemia, retrieved 8-07-2012 http://www.nccn.org/professionals/physician_gls/pdf/cml.pdf

Noens L, van Lierde MA, De Bock R et al (2009) Prevalence, determinants, and outcomes of nonadherence to imatinib therapy in patients with chronic myeloid leukemia: the ADAGIO study. Blood 113:5401–5411

Novartis Pharmaceuticals Corporation (2010) Package insert. Tasigna (nilotinib). Novartis Pharmaceuticals Corporation, East Hanover

Nowell PC, Hungerfold DA (1960) A minute chromosome in human chronic granulocytic leukemia. Science 132:1497–1501

O'Brien SG, Guilhot F, Larson RA et al (2003) Imatinib compared with interferon and low-dose cytarabine for newly diagnosed chronic-phase chronic myeloid leukemia. N Engl J Med 348:994–1004

Oweini H, Otrock ZK, Mahfouz RA, Bazarbachi A (2011) Successful pregnancy involving a man with chronic myeloid leukemia on dasatinib. Arch Gynecol Obstet 283:133–134

Pane F, Frigeri F, Sindona M et al (1996) Neutrophilic-chronic myeloid leukemia: a distinct disease with a specific molecular marker (BCR/ABL with C3/A2 junction). Blood 88:2410–2414
Pendergast AM (2002) The Abl family kinases: mechanisms of regulation and signaling. Adv Cancer Res 85:51–100
Pendergast AM, Muller AJ, Havlik MH, Maru Y, Witte ON (1991) BCR sequences essential for transformation by the BCR-ABL oncogene bind to the ABL SH2 regulatory domain in a non-phosphotyrosine-dependent manner. Cell 66:161–171
Pendergast AM, Quilliam LA, Cripe LD et al (1993) BCR-ABL-induced oncogenesis is mediated by direct interaction with the SH2 domain of the GRB-2 adaptor protein. Cell 75:175–185
Porkka K, Koskenvesa P, Lundan T et al (2008) Dasatinib crosses the blood–brain barrier and is an efficient therapy for central nervous system Philadelphia chromosome-positive leukemia. Blood 112:1005–1012
Press RD, Galderisi C, Yang R et al (2007) A half-log increase in BCR-ABL RNA predicts a higher risk of relapse in patients with chronic myeloid leukemia with an imatinib-induced complete cytogenetic response. Clin Cancer Res 13:6136–6143
Quintas-Cardama A, Cortes J (2009) Molecular biology of bcr-abl1-positive chronic myeloid leukemia. Blood 113:1619–1630
Quintas-Cardama A, Kantarjian H, Jones D et al (2007a) Dasatinib (BMS-354825) is active in Philadelphia chromosome-positive chronic myelogenous leukemia after imatinib and nilotinib (AMN107) therapy failure. Blood 109:497–499
Quintas-Cardama A, Kantarjian H, O'Brien S et al (2007b) Pleural effusion in patients with chronic myelogenous leukemia treated with dasatinib after imatinib failure. J Clin Oncol 25:3908–3914
Quintas-Cardama A, Kantarjian H, Cortes J (2007c) Flying under the radar: the new wave of BCR-ABL inhibitors. Nat Rev Drug Discov 6:834–848
Radich JP, Gooley T, Bensinger W et al (2003) HLA-matched related hematopoietic cell transplantation for chronic-phase CML using a targeted busulfan and cyclophosphamide preparative regimen. Blood 102:31–35
Rasheed W, Flaim B, Seymour JF (2009) Reversible severe pulmonary hypertension secondary to dasatinib in a patient with chronic myeloid leukemia. Leuk Res 33:861–864
Robinson DR, Wu YM, Lin SF (2000) The protein tyrosine kinase family of the human genome. Oncogene 19:5548–5557
Ross DM, Branford S, Moore S, Hughes TP (2006) Limited clinical value of regular bone marrow cytogenetic analysis in imatinib-treated chronic phase CML patients monitored by RQ-PCR for BCR-ABL. Leukemia 20:664–670
Rousselot P, Huguet F, Rea D et al (2007) Imatinib mesylate discontinuation in patients with chronic myelogenous leukemia in complete molecular remission for more than 2 years. Blood 109:58–60
Rowley JD (1973) A new consistent chromosome abnormality in chronic myelogenous leukemia identified by quinacrine fluorescence and Giemsa staining. Nature 243:209–213
Roy A, Shah N, Le Coutre P, Wang X (2010) Relationship between dasatinib exposure and long term clinical outcomes in chronic phase chronic myleoid leukemia (CML-CP) patients. J Clin Oncol 28:6518
Saglio G, Kim DW, Issaragrisil S et al (2010) Nilotinib versus imatinib for newly diagnosed chronic myeloid leukemia. N Engl J Med 362:2251–2259
Sattler M, Verma S, Shrikhande G et al (2000) The BCR/ABL tyrosine kinase induces production of reactive oxygen species in hematopoietic cells. J Biol Chem 275:24273–24278
Sattler M, Mohi MG, Pride YB et al (2002) Critical role for Gab2 in transformation by BCR/ABL. Cancer Cell 1:479–492
Savage DG, Szydlo RM, Chase A, Apperley JF, Goldman JM (1997) Bone marrow transplantation for chronic myeloid leukaemia: the effects of differing criteria for defining chronic phase on probabilities of survival and relapse. Br J Haematol 99:30–35
Sawyers CL, McLaughlin J, Goga A, Havlik M, Witte O (1994) The nuclear tyrosine kinase c-Abl negatively regulates cell growth. Cell 77:121–131

Sawyers CL, Hochhaus A, Feldman E et al (2002) Imatinib induces hematologic and cytogenetic responses in patients with chronic myelogenous leukemia in myeloid blast crisis: results of a phase II study. Blood 99:3530–3539

Schindler T, Bornmann W, Pellicena P, Miller WT, Clarkson B, Kuriyan J (2000) Structural mechanism for STI-571 inhibition of abelson tyrosine kinase. Science 289:1938–1942

Shah NP, Kim DW, Kantarjian H et al (2010a) Potent, transient inhibition of BCR-ABL with dasatinib 100 mg daily achieves rapid and durable cytogenetic responses and high transformation-free survival rates in chronic phase chronic myeloid leukemia patients with resistance, suboptimal response or intolerance to imatinib. Haematologica 95:232–240

Shah N, Cortes J, Schiffer A, Coutre P (2010b) Four year follow-up of patients with chronic phase chronic myeloid leukemia (CP-CML) receiving 100 mg of dasatinib once daily. J Clin Oncol 28:6512

Shimoni A, Leiba M, Schleuning M et al (2009) Prior treatment with the tyrosine kinase inhibitors dasatinib and nilotinib allows stem cell transplantation (SCT) in a less advanced disease phase and does not increase SCT Toxicity in patients with chronic myelogenous leukemia and philadelphia positive acute lymphoblastic leukemia. Leukemia 23:190–194

Sirard C, Lapidot T, Vormoor J et al (1996) Normal and leukemic SCID-repopulating cells (SRC) coexist in the bone marrow and peripheral blood from CML patients in chronic phase, whereas leukemic SRC are detected in blast crisis. Blood 87:1539–1548

Smith DC, Britten C, DON Clary LT (2009) A phase I study of XL228, a potent IGF1R/AURORA/SRC inhibitor, in patients with solid tumors or hematologic malignancies. J Clin Oncol 27:3512

Sokal JE, Baccarani M, Russo D, Tura S (1988) Staging and prognosis in chronic myelogenous leukemia. Semin Hematol 25:49–61

Spencer A, Prince M, Deangelo D (2007) Phase IA/II study of oral LBH589, a novel deactylase inhibitor (DACi), administered on 2 schedules, in patients with advanced hematologic malignancies. Blood 110:1041

Stoklosa T, Poplawski T, Koptyra M et al (2008) BCR/ABL inhibits mismatch repair to protect from apoptosis and induce point mutations. Cancer Res 68:2576–2580

Strom SS, Yamamura Y, Kantarijian HM, Cortes-Franco JE (2009) Obesity, weight gain, and risk of chronic myeloid leukemia. Cancer Epidemiol Biomarkers Prev 18:1501–1506

Talpaz M, Silver RT, Druker BJ et al (2002) Imatinib induces durable hematologic and cytogenetic responses in patients with accelerated phase chronic myeloid leukemia: results of a phase 2 study. Blood 99:1928–1937

Talpaz M, Shah NP, Kantarjian H et al (2006) Dasatinib in imatinib-resistant Philadelphia chromosome-positive leukemias. N Engl J Med 354:2531–2541

Varticovski L, Daley GQ, Jackson P, Baltimore D, Cantley LC (1991) Activation of phosphatidylinositol 3-kinase in cells expressing abl oncogene variants. Mol Cell Biol 11:1107–1113

Velev N, Cortes J, Champlin R et al (2010) Stem cell transplantation for patients with chronic myeloid leukemia resistant to tyrosine kinase inhibitors with BCR-ABL kinase domain mutation T315I. Cancer 116(15):3631–3637

Wang L, Knight K, Lucas C, Clark RE (2006) The role of serial BCR-ABL transcript monitoring in predicting the emergence of BCR-ABL kinase mutations in imatinib-treated patients with chronic myeloid leukemia. Haematologica 91:235–239

Welch PJ, Wang JY (1993) A C-terminal protein-binding domain in the retinoblastoma protein regulates nuclear c-Abl tyrosine kinase in the cell cycle. Cell 75:779–790

White DL, Saunders VA, Dang P et al (2007) Most CML patients who have a suboptimal response to imatinib have low OCT-1 activity: higher doses of imatinib may overcome the negative impact of low OCT-1 activity. Blood 110:4064–4072

Yaish P, Gazit A, Gilon C, Levitzki A (1988) Blocking of EGF-dependent cell proliferation by EGF receptor kinase inhibitors. Science 242:933–935

Zhao C, Chen A, Jamieson CH et al (2009) Hedgehog signalling is essential for maintenance of cancer stem cells in myeloid leukaemia. Nature 458:776–779

Novel Targeted Therapeutics for Acute Myeloid Leukemia

Vu Duong and Jeffrey Lancet

Introduction

Acute myeloid leukemia (AML) is a heterogenous group of malignant hematopoietic disorders characterized by uncontrolled proliferation and accumulation of clonal immature myeloid cells in the bone marrow, ultimately leading to hematopoietic failure. Despite some advances in therapy for patients with adult acute myeloid leukemia, the overall prognosis remains dismal. The backbone of induction therapy has not changed in decades, and usually includes a combination of an anthracycline and cytarabine. While the complete remission (CR) rate after induction is approximately 60–80% for patients younger than the age of 60, the overall cure rate remains only 25–30%, owing to a high rate of relapse despite postremission therapy. Outcomes are even worse in older patients, patients who evolve from previous myelodysplastic syndromes, and patients whose disease is linked to environmental and occupational exposures (Appelbaum et al. 2006; Kantarjian et al. 2006a). These patients frequently relapse, and the remission duration is generally short. A minority of patients are eligible for allogeneic hematopoietic stem cell transplant (HSCT), which offers the possibility of cure but is associated with high morbidity and mortality, even in highly selected populations (Yanada et al. 2005a; Cornelissen et al. 2007; Koreth et al. 2009).

With such poor outcomes, the therapeutic limits with conventional cytotoxic chemotherapy are clear, thus underscoring the need for newer, targeted therapies. Conventional chemotherapy has the added disadvantage of being nonselective, and induction therapy is therefore associated with high rates of adverse events, most

V. Duong, M.D.
Division of Hematology/Oncology, University of Maryland, Greenebaum Cancer Center, 22 S. Greene St, 21201, Baltimore, MD, USA
e-mail: vduong@umm.edu

J. Lancet, M.D. (✉)
Department of Hematologic Malignancies, H. Lee Moffitt Cancer Center and Research Institute, 12902 Magnolia Drive, 33612, Tampa, FL, USA
e-mail: Jeffrey.lancet@moffitt.org

J. Tao and E. Sotomayor (eds.), *Hematologic Cancers: From Molecular Pathobiology to Targeted Therapeutics*, Cancer Growth and Progression 14, DOI 10.1007/978-94-007-5028-9_14, © Springer Science+Business Media Dordrecht 2012

notably severe pancytopenia and neutropenic infections. The prototype for selective, targeted therapy can be seen in the treatment of chronic myelogenous leukemia, in which a single fusion protein is responsible for pathogenesis in the vast majority of cases. Tyrosine kinases directed against the fusion product of bcr-abl has resulted in a dramatic improvement in the therapy and prognosis of these patients (Saglio et al. 2010; Kantarjian et al. 2010; Hochhaus et al. 2009; O'Brien et al. 2003). However, other leukemias including AML are much less likely to see such success owing to their heterogeneous nature and the involvement of multiple cell-signaling cascades in disease pathogenesis. Research in the last decade has uncovered new intracellular signals and mechanisms that govern AML and promote proliferation and survival of leukemic stem cell. Despite progress in uncovering potential targets for therapy, many of these principles have yet to bear a clinical impact on our treatment of the disease.

This chapter will first review some of the major concepts and molecular pathways involved in leukemogenesis, followed by a focused discussion on some of the emerging therapeutic agents targeting these changes. It should be noted that what follows is, by no means, an all-inclusive list of potential targets and agents, but rather a discussion of some of the more heavily-researched topics.

Leukemia Stem Cells (LSCs) in AML

The concept of leukemia stem cells has been hypothesized for decades, but proof of this immature population is a relatively recent discovery in AML (Bonnet and Dick 1997). Dick and Bonnet showed that a subpopulation of CD34+ and CD38- cells, when transplanted to non-obese diabetic with severe combined immunodeficient (NOD-SCID) mice, were capable of initiating acute myeloid leukemia (Bonnet and Dick 1997). It was demonstrated from these experiments that AML is arranged as a cellular hierarchy, with only a small percentage of cells at the apex capable of producing leukemic blasts, thus giving them the ability to both initiate and maintain the leukemic clone. These leukemia stem cells (LSCs) possess the requisite properties of self-renewal, differentiation, and proliferation. Furthermore, while rapidly proliferating leukemic blasts are particularly susceptible to conventional cytotoxic chemotherapy, LSCs spend the majority of their time in the G_0 phase of the cell cycle, rendering them naturally resistant to cell-cycle specific chemotherapeutic agents including cytarabine and anthracyclines (Guzman et al. 2001). The persistence of these cells may explain the disparity between relatively high rates of complete remission after induction chemotherapy and low rates of cure. Allogeneic HSCT is the most effective form of consolidation therapy and likely offers the only true chance of cure, particularly in those with high-risk features (Cornelissen et al. 2007; Koreth et al. 2009). While this procedure enables the use of high-dose chemotherapy, its success and failure likely rests upon the graft vs. leukemia effect on quiescent leukemia stem cells, mediated by donor T cells.

Studies have shown that larger LSC populations at diagnosis are correlated with increased minimal residual disease and poorer overall survival, and the ability to engraft readily in NOD/SCID mice is a marker of poor prognosis, thus highlighting that eradication of these cells is paramount in determining outcomes (van Rhenen et al. 2005; Pearce et al. 2006).

Based on the similar surface phenotype, the high rate of quiescence, and similar levels of telomerase (Morrison et al. 1995), it has been postulated that LSCs arise from HSCs. HSCs possess a prolonged lifespan with an increased number of cell divisions, increasing the chances of random mutation. As the development of AML is a multi-step process requiring multiple mutations, HSCs are logical candidates for leukemic transformation into AML LSCs. On the other hand, several X-inactivating studies based on G6PD expression by Fialkow and colleagues suggest that AML may arise from more committed myeloid progenitors (Fialkow et al. 1987, 1981). More recent studies appear to confirm this hypothesis, based upon the identification of myeloid antigens (CD13, CD33, and CD123) upon leukemia repopulating cells within the CD34-positive/CD38-negative compartment (Taussig et al. 2005). Other studies have reported transformation of multipotent progenitors (MPP) or myeloid-restricted progenitors that lack self-renewal capacity (Misaghian et al. 2009). The ability to distinguish normal HSCs from LSCs may also be aided by differential expression of cell surface antigens such as CD90, CD117, and CD123 (Guzman and Jordan 2004; Blair et al. 1998). The issue of the precise cell targeted for leukemic transformation continues to be debated, and the ability to identify and discriminate AML-LSCs clinically could have enormous implications for therapeutic and monitoring strategies for this disease.

A variety of signaling pathways are also dysregulated in AML LSCs, and may be potential targets for future therapy, as indicated in Fig. 1.

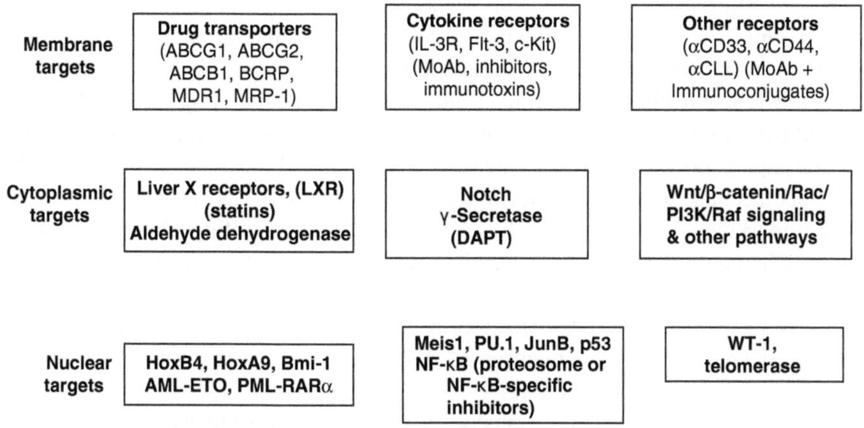

Fig. 1 Potential therapeutic targets on leukemic stem cells. Adapted with permission from Misaghian et al. (2009)

Wnt/β-Catenin Signaling

The Wnt/β-catenin signaling pathway is critical in the self-renewal ability of normal hematopoietic stem cells and is requisite for self-renewal in leukemia stem cells (Wang et al. 2010).

Multiple components of this pathway are aberrantly expressed and it has been shown to be constitutively activated in the majority of AML cases, thus implicating it in the pathogenesis of AML (Simon et al. 2005). This may arise from mutations in FLT-3 and balanced translocation products such as AML1-ETO, PML-RARα, and PLZF-RARα (Muller-Tidow et al. 2004). Furthermore, high levels of β-catenin have been linked with poorer event-free and overall survival (Ysebaert et al. 2006).

In the absence of Wnt, β-catenin is phosphorylated by multiple kinases, targeting it for ubiquitin-dependant proteasomal degradation. On the other hand, binding of Wnt ligand to Frizzled and low-density lipoprotein receptor–related protein family (LRP5 and LRP6) blocks glycogen synthase kinase-3-beta (GSK3-beta) and inhibits phosphorylation of β-catenin, stabilizing it. This results in translocation of non-phosphorylated β-catenin to the nucleus, where it interacts with T-cell transcription factor (TCF) and lymphoid enhancer factor (LEF) resulting in the transcription of growth-promoting genes including c-Myc and cyclin D1 (He et al. 1998; Tetsu and McCormick 1999). As β-catenin is not required for self-renewal of normal hematopoietic stem cells and actually exhausts the long-term stem cell pool, it is an attractive target in AML.

Clustered HOX Genes

Clustered homeobox (HOX) proteins are a family of 39 polypeptides characterized by a 60 amino acid DNA-binding motif known as the homeodomain. Hox genes of the A, B, and C (but not D) clusters are required for hematopoietic development and HSC self-renewal (Sauvageau et al. 1994). Their expression is tightly regulated in normal hematopoiesis, with preferential expression on HSCs and early progenitors but subsequent downregulation as the cells differentiate (Pineault et al. 2002). While the exact mechanism by which they influence hematopoiesis has yet to be elucidated, their importance is underscored by experiments demonstrating abnormalities of multiple hematopoietic cell lines in mice that either overexpress or are deficient in HOX (Frohling et al. 2007).

These genes likely play a major role in LSC self-renewal and are known to be globally dysregulated in the majority of AML cases. In the majority of cases, the upstream mutations responsible for HOX dysregulation have yet to be elucidated, but some involve recurrent gene translocation products involving MLL (Rawat et al. 2008; Andreeff et al. 2008), NUP98 (Palmqvist et al. 2007), and the less common CALM/AF10 (Okada et al. 2006). MLL is a known regulator of HOX gene expression, and partial tandem mutations may also result in upregulation in a small minority of AML cases. NPM1 mutations are also known to be associated with HOX gene dysregulation (Rawat et al. 2008; Alcalay et al. 2005).

Another family involved in HOX gene regulation is the caudal-related homeobox genes, or CDX family (Frohling et al. 2007). Multiple HOX genes have a promoter region containing binding sites for the 3 Cdx homologs CDX1, CDX2, and CDX4. CDX2 has both tumor suppressor function and proto-oncogene activity in the gastrointestinal system, and aberrant expression is associated with multiple GI malignancies. CDX2 is not expressed in normal adult hematopoeisis yet almost 90% of patients with AML aberrantly express it (Chan and Huntly 2008). This over-expression has been shown to alter HOX gene expression in murine progenitors (Frohling et al. 2007; Rawat et al. 2008; Chan and Huntly 2008; Scholl et al. 2007). CDX4 is upregulated in approximately 25% of patients with AML, and had been shown to induce HOX gene dysregulation in both in vitro and in vivo experiments. While our understanding of the upstream and downstream mechanisms of HOX genes is incomplete, growing evidence supports their importance in the self-renewal abilities of myeloid leukemia stem cells.

Bmi-1 Protein

B-cell-specific Moloney murine leukemia vius integration site 1 (Bmi-1) is an essential regulator of HSC and LSC self-renewal. Bmi-1 belongs to the family of Polycomb group (PcG) of genes, which are responsible for the preservation of gene-silencing and thus the maintenance of cell identity. Its expression is limited to immature HSC and the CD34+ fraction of AML patients (Raaphorst 2003; Park et al. 2003). The importance of this gene was demonstrated in experiments with Bmi-1 knockout mice, in which progressive loss of all hematopoietic cell lines was seen. When Bmi-$1^{-/-}$ cells were transplanted into lethally radiated mice, partial reconstitution of myeloid cells was seen, but this effect was short-lived due to the lack of self-renewal capabilities (Park et al. 2003). To study the effect in leukemia stem cells, a murine model of HSCs transfected with Meis1a and Hoxa9 was used. It was previously demonstrated that Meisa1a and Hoxa9 are sufficient to produce AML. When Bmi-$1^{-/-}$ mice were transfected with these two genes, AML was induced but with lower populations of leukemic cells in the blood. Furthermore, complementation studies showed that Bmi-1 restored the ability of leukemic cells to repopulate, thus showing that Bmi-1 is essential in the self-renewal of L-HSCs (Lessard and Sauvageau 2003).

PI3K/PTEN/Akt Pathway

The phosphoinositide 3-kinase(PI3K)/Akt signaling network is critical in multiple cellular processes including cell cycle progression, differentiation, transcription, translation, and apoptosis (Martelli et al. 2010). The family of PI3K lipid kinases can be further organized into classes I, II, and III, of which class I PI3K are the most well-characterized. These kinases are composed of heterodimers of an adaptor/regulatory and a catalytic subunit. The adaptor/regulatory subunits help localize PI3K to the plasma membrane and stabilize the catalytic subunit. The lipid products of PI3K at the plasma membrane interact with NH_2-terminal of Akt, resulting in a conformational change and phosphorylation of Akt. Akt is then directed through the cytoplasm into the nucleus, where it exerts its actions on gene regulation (Martelli et al. 2010, 2006). One of its downstream targets includes the mammalian target of rapamycin (mTOR), a serine/threonine kinase that regulates translation of key molecules such as c-Myc, cyclin D1 and ribosomal proteins in response to growth factors (Martelli et al. 2006). The interplay between Akt and mTOR is complex, but Akt can either directly phosphorylate mTOR or it can activate mTOR by modulating tuberous sclerosis 2 (TSC2), which in turn affects Rheb and the mTOR/Raptor/mLST8 complex. Furthermore, mTOR can inhibit Akt in a classic negative-feedback loop (Martelli et al. 2006).

In states of cellular stress such as heat shock, ischemia, hypoxia, and oxidative stress, this pathway is upregulated as a compensatory cellular protective mechanism (Martelli et al. 2006). As many chemotherapeutic agents act through reactive oxygen species, the PI3K/Akt pathway is often upregulated in response to chemotherapy. This pathway is consistently elevated in both AML blasts and in LSCs transplanted in NOD-SCID mice, where it suppresses apoptosis (Martelli et al. 2010; Zhao et al. 2004; Xu et al. 2003, 2005).

Aberrant activation of this cascade results in dysregulated cell proliferation and apoptosis, and upregulation is associated with drug resistance (Tazzari et al. 2007). Other causes of upregulation are numerous and include mutations of FLT-3 and c-kit, N- or K-ras mutations, PI3K p110β and/or δ overexpression, CXCR4/CXCL12 interactions and autocrine/paracrine secretion of growth factors such as VEGF or IGF-1 (Martelli et al. 2010). In addition, the PI3K/Akt pathway appears to be activated in response to leukemic cell adherence to fibronectin via VLA-4, leading to increased survival and chemotherapy resistance (Matsunaga et al. 2003). Given the emerging understanding of the vital importance of LSC interactions with their environmental niche for survival and maintenance of stem cell properties, this pathway may be of increasing importance for therapeutic targeting (Konopleva and Jordan 2011). Prognostically, some studies have suggested that upregulation of this pathway in AML is associated with poor outcome, a prospective study by Tamburini and colleagues suggest it actually be a favorable prognostic factor in de novo cases, even after adjustment for FLT3-ITD (Tamburini et al. 2007).

NF-κB

Nuclear factor kappa-light chain enhancer of B cells (NF-κB) is a heterodimer composed of p50 and p65 subunits, and is normally bound to IκBα in its inactive state. IκBα, when phosphorylated by the IκB kinase (IKK) complex, becomes targeted by the 26S proteasome complex for ubiquitination and degradation. Thus, when IκBα is degraded, NFκB is liberated and translates to the nucleus, where it is involved in the regulation of many genes that code for mediators of immune responses, inflammatory responses, and cell survival (Nakanishi and Toi 2005). NF-κB is known to be constitutively active in AML blasts and stem cells (Guzman et al. 2001), but not in normal HSCs. Furthermore, inhibition of NF-κB with the proteasome inhibitor MG-132 resulted in rapid LSC death with sparing of normal HSCs, suggesting a central role for NF-κB in LSC survival (Guzman et al. 2001).

Other New Stem Cell Targets

Within the last 2 years, several other molecular targets within leukemic stem cells have been elucidated, each playing roles in maintenance and/or regulation of leukemic hematopoiesis. The role of the Hedgehog (Hh) signaling has become evident in chronic myelogenous leukemia, whereby investigations have demonstrated that the Smo protein (an essential component of Hh signaling), is critical for the initiation, maintenance, and propagation of the disease in a mouse model (Zhao et al. 2009). The Mushashi (Msi) family of RNA-binding proteins has been found to play a key role in somatic stem cell function, including in adult and fetal neural stem cells (Okano et al. 2005). Comparatively high expression of Msi2 has been observed in hematopoietic stem cells, and its depletion has been associated with failure of engraftment. Within AML cell lines, Msi2 is overexpressed, and its expression in primary AML appears to be independently associated with adverse prognostic markers such as monosomy 7 and *FLT3* ITD mutations, suggesting its relevance both as a prognostic indicator and target for therapy (Okano et al. 2005). Finally, hypoxia-inducible factors (HIF), which mediate cellular response to hypoxia, may also regulate cancer stem cell function (Keith and Simon 2007). In AML, HIF1α transcript levels were found to be highest in the stem cell subset ($CD34^+CD38^-$) compared to other populations. In addition, inhibition of *HIF1a* expression was found to inhibit colony forming unit activity in leukemic stem cells, further highlighting the importance of HIF1α as a potential therapeutic target (Wang et al. 2011).

Epigenetic Modifications

In contrast to genetic aberrations that lead to structural DNA changes, epigenetic alterations involve the loss of gene function without direct modification of the DNA sequence. DNA promoter-region methylation and histone deacetylation represent the best-studied intracellular processes that mediate epigenetic alterations. In AML, a host of genes may be hypermethylated and subsequently silenced, including tumor-suppressor, DNA repair, and apoptosis genes (Agrawal et al. 2007; Issa et al. 1997; Melki et al. 2000; Seedhouse et al. 2003; Voso et al. 2004). More globally, genomic scanning has suggested that up to 8% of all human CpG islands may be aberrantly methylated in AML (Rush et al. 2001), suggesting a potentially important role for pharmacologic reversal of gene silencing in the treatment of these diseases.

More specifically, in various subtypes of AML, there are several well-described epigenetic phenomenon that appear to have an important pathophysiologic role, some of which can be targeted in a way to reverse gene silencing and restore a normal cellular phenotype.

PML-RAR

Acute promyelocytic leukemia (APL) is characterized by a reciprocal translocation between chromosomes 15 and 17, which ultimately generates the onco-fusion protein PML-RAR, an aberrant retinoic acid receptor with altered transcriptional regulatory function. This fusion protein is able to bind promoter regions of retinoic acid receptor α target genes, thereafter recruiting DNA methyltransferases and histone deacetylases to inhibit transcription of these target genes, resulting in a block of normal retinoic-acid induced myelopoietic differentiation and an accumulation of immature myeloid progenitors (promyelocytes), which characterizes this disease (Di Croce et al. 2002; Plass et al. 2008; Lin et al. 1998). Therefore, pharmacologic interruption of the gene silencing process induced by the PML-RAR fusion protein could restore normal myeloid differentiation.

AML1-ETO

The translocation t(8;21)(q22;q22) is a common karyotypic abnormality in AML that results in the creation of a chimeric protein, AML1-ETO (also known as RUNX1/MTG8), which functions as an aberrant transcription factor for several hematopoietic-specific target genes, including *GM-CSF* and *interleukin-3* (Plass et al. 2008). The AML1-ETO fusion protein recruits a transcriptional repressor complex that includes both histone deacetylases and DNA methyltransferases, thereby silencing RUNX1 target genes (Gelmetti et al. 1998; Liu et al. 2005). Pharmacologic inhibition of DNA methyltransferase and histone deacetylation allows for restoration of target gene expression in an AML1-ETO model (Klisovic et al. 2003), suggesting that epigenetic therapies may have a clinical role in AML, particularly those subtypes characterized by DNMT or HDAC-mediated gene silencing.

Selected Aberrant Signaling Pathways

FLT3

Fms-like tyrosine kinase 3 (FLT3), also known as fetal liver kinase-2 (FLK-2) or stem cell kinase 1 (STK-1), is a member of the class III subfamily of receptor tyrosine kinases (RTK). It is composed of an extracellular ligand-binding domain, a trans-membrane domain, a juxtamembrane domain, and a cytoplasmic tyrosine kinase domain interrupted by a kinase insert. Primarily expressed on myeloid and lymphoid progenitors including CD34+ cells (Rosnet et al. 1996), its expression is lost as the cell differentiates (Knapper 2007). FLT3 regulates many intracellular processes involved in normal hematopoiesis and cellular growth, including phospholipid metabolism, transcription, proliferation, and apoptosis (Meshinchi and Appelbaum 2009). Upon binding with FLT3 ligand (FL), FLT3 undergoes a conformational change allowing receptor-receptor dimerization and autophosphorylation, which activates cell signaling through multiple downstream regulators including β-catenin, phosphatidylinositol-3 kinase (PI3)/AKT, and RAS/mitogen-activated protein kinase (MAPK), ultimately leading to the promotion of cell differentiation, proliferation and survival, and inhibition of apoptosis.

FLT3 is expressed at various levels in 80–100% of patients across all subtypes of AML (Rosnet et al. 1996; Drexler 1996). In addition, two broad categories of mutations have been described, namely the internal tandem duplications (FLT3-ITD) and the activation loop mutations in the tyrosine kinase domain (FLT3-TKD). FLT3-ITDs are found in approximately 25% of AML patients overall, and their prevalence is highly age dependent, ranging from 5–10% in children to >35% in elderly patients (Stirewalt et al. 2001; Meshinchi et al. 2006). FLT3-ITD mutations are most frequently seen in acute promyelocytic leukemia and in those with normal cytogenetics. FLT3-TKD mutations are have been reported to account for 7–10% of AML cases (Yamamoto et al. 2001), and the prevalence does not appear to be age dependent (Knapper 2007). Taken together, FLT3 mutations are the most common molecular abnormality found in AML. Although these two mututations were originally thought to be mutually exclusive, so-called dual mutations in which the same patient harbors both mutations have been described and account for approximately 1% of cases (Chen et al. 2005).

First described by Nakao and colleagues in 1996 (Nakao et al. 1996), FLT3-ITDs are in-frame duplications of variable lengths in the juxtamembrane domain-encoding sequence of exons 14 and 15. These mutations promote receptor-receptor dimerization even in the absence of FLT ligand (FL), leading to constitutive activation of multiple downstream targets including the PI3K/Akt and RAS/MAPK pathway. FLT3-TKDs are most often caused by a single missense mutation within the activating loop domain. The majority of these mutations occur in codon 835, in which a tyrosine is substituted for aspartic acid. Similar to FLT3-ITDs, FLT3-TKDs also promote autophosphorylation of the receptor, but studies suggest that different downstream regulators are involved leading to different biological effects.

Studies regarding the prognostic significance of FLT3-ITD have been uniform in linking these mutations to adverse outcomes in intermediate-risk cytogenetics. In contrast, it does not seem to have prognostic significance in patients with poor-risk or low-risk cytogenetics (including APL) (Santos et al. 2009). Kiyoi and colleagues first showed the presence of FLT3/ITD was the strongest prognostic factor for overall survival in younger (>60 years old) patients (Kiyoi et al. 1999). FLT3-ITD is also associated with a higher peripheral white blood cell count, higher bone marrow blast percentage, lower complete remission (CR) rates, higher relapse rate, and shorter remission duration, especially in patients less than 60 years of age (Kiyoi et al.

1999; Ozeki et al. 2004; Yanada et al. 2005b; Moreno et al. 2003; Thiede et al. 2002; Frohling et al. 2002; Kottaridis et al. 2001). The negative prognostic impact persists in patients in first relapse, as patients harboring the mutation are less likely to achieve a second CR and have shorter survival (Ravandi et al. 2010a). The size of the duplication may also have prognostic relevance, with one study showing a decrease in overall survival and relapse-free survival with increasing ITD size (Stirewalt et al. 2006). Similarly, as newly diagnosed patients with FLT3-ITD are usually heterozygous for the mutation, the ratio between mutant and wild-type alleles may also be confer prognostic significance. Those patients with high mutant to wt ratios may have shorter disease-free and overall survival (Thiede et al. 2002; Kottaridis et al. 2001; Schnittger et al. 2002). With respect to FLT-3 inhibitors, wild-type alleles may reduce the efficacy, but may also protect normal hematopoietic precursors from further marrow suppression (Mori et al. 2009; Levis et al. 2009).

On the other hand, there are conflicting studies regarding the prognostic impact of FLT3-TKD and there continues to be controversy. While some studies have shown an association with adverse outcomes comparable to that of ITDs (Yanada et al. 2005b), others have even suggested a survival advantage over FLT3-wild type (Mead et al. 2007). Similarly, some have shown an association with a higher peripheral white blood cell count while others have not. It appears to have less impact upon survival, even in intermediate-risk patients (Santos et al. 2009; Mead et al. 2007).

RAS

The rat sarcoma (Ras) protein family includes Kras, Nras, and Hras, which lay downstream of receptor tyrosine kinases such as FLT3 and mediate a variety of cell signaling pathways involving survival, proliferation, and differentiation (Downward 2003). The first step in activation of Ras is the transfer of a 15 carbon moiety at a cysteine near the carboxyl terminus of a polypeptide, a process catalyzed by farnesyltransferase (FTase). Farnesylated Ras is transported to the endoplasmic reticulum, where it undergoes further processing that allows incorporation into the phospholipid plasma membrane. Once membrane-bound, Ras fluctuates between an active GTP bound form and an inactive GDP bound form. This in turn regulates several downstream effectors, including the MAPK/ERK pathway, and the PI3K/Akt pathway, and the Rac/Rho pathway, all of which are integral to cell regulation of apoptosis, cell cycling, and cell survival (Braun and Fenaux 2008; Rowinsky et al. 1999).

Mutations in RAS protein, particularly N-Ras, occur in approximately 10–30% of AML cases, resulting in constitutive activation of the GTP-bound form (Kiyoi et al. 1999; Bowen et al. 2005; Neubauer et al. 2008; Ritter et al. 2004). However, it may be insufficient to produce AML by itself, suggesting that it may only be a secondary mutation in leukemogenesis. Its frequency varies with cytogenetics and FAB subtype, occurring more frequently with inv(16)/t(16;16) and inv(3)/t(3;3), and rarely in t(15;17) and MLL/11q23 (Bowen et al. 2005; Bacher et al. 2006; Callens et al. 2005). It appears to confer sensitivity to cytarabine by inducing p53-dependent differentiation (Neubauer et al. 2008; Meyer et al. 2009), but multiple studies have reported conflicting associations with outcome (Bowen et al. 2005; Ritter et al. 2004; Illmer et al. 2005).

Multidrug Resistance and P-gp

The multidrug resistance (MDR) family of ATP-binding cassette transporters comprises a multitude of structurally and functionally related proteins that regulate the movement of specific substrates across the cell membrane (Mahadevan and List 2004). Pertaining to cytoprotective effect against chemotoxins, MDR1 (also known as P-glycoprotein), a 170 k-Da adenosine triphosphate-dependent drug transporter that is encoded by the *MDR1* (*ABCB1*) gene, is the best characterized of the MDR transport proteins, being constitutively expressed in many normal tissues as well as malignant cells. In normal cells, MDR1 may be upregulated in response to stress induced from a variety of sources, including reactive oxygen species and DNA damage (Hirsch-Ernst et al. 1998; Fardel et al. 1997; Thévenod et al. 2000). In leukemia, MDR1-expression may enhance cellular survival not only via substrate drug/toxin efflux, but also via suppression of caspase activation (Ruefli et al. 2002; Smyth et al. 1998). Lipophilic or positively charged compounds such as anthracyclines are particularly susceptible to the actions of MDR1.

As a clinical therapeutic target, MDR1 appears to be especially relevant in AML, its expression being strongly linked to factors such as older age, presence of antecedent hematologic disease, unfavorable cytogenetic profile, and CD34 expression (Leith et al. 1997, 1999). Specifically, studies by the Southwest Oncology Group (SWOG) demonstrated a much higher rate of MDR1 expression in leukemic cells taken from older AML patients compared with younger patients (Leith et al. 1997, 1999). In addition, MDR1 appears to be much more frequently expressed in CD34-positive compared with CD34-negative leukemic cell populations (Leith et al. 1999), suggesting a predilection for a more functionally immature progenitor leukemic cells, and thereby an avenue, perhaps, by which to target leukemic stem cells. To this point, it has been shown that pharmacologic inhibition of MDR1 by cyclosporine results in delayed engraftment of MDR1 expressing leukemic cells in a mouse xenograft model (Lehne et al. 2002). Prognostically, MDR1 expression has been reproducibly and independently associated with adverse outcomes, hence fortifying its candidacy as a target for pharmacologic inhibition.

Leukemic Stem Cell Agents

Targeted agents against leukemic stem cells may either take advantage of the unique surface phenotype or one of the many aberrant pathways described above. One strategy is the targeting of CD123, a receptor for IL-3 and a unique marker in AML

LSCs (Jordan et al. 2000). A diphtheria toxin conjugate specific for this cell surface marker has shown promise in preclinical models, showing a selective reductions in LSCs and AML blasts, a decrease in the ability to engraft in NOD/SCID mice, and better disease-free survival in mice transfected with AML (Feuring-Buske et al. 2002; Frankel et al. 2008, 2000; Kim et al. 2010; Liu et al. 2004; Su et al. 2010; Black et al. 2003). Quantitation of the IL-3 receptor also correlates with response to these fusion proteins (Testa et al. 2005; Yalcintepe et al. 2006). In one phase I trial, the diphtheria-IL3 fusion was successful in inducing 1 complete remission and 3 partial remissions in AML patients (Frankel et al. 2008).

As discussed previously, evidence suggests that AML-LSCs with repopulating ability express the myeloid antigen CD33 (Taussig et al. 2005). Gemtuzumab ozogamicin (GO) is a recombinant humanized monoclonal antibody directed against CD33. In addition to being expressed on LSCs, approximately 90% of blasts in patients with AML express CD33 (Larson et al. 2005). Approximately 50% of the CD33 monoclonal antibody is loaded with calicheamicin, an antitumor antibiotic. Binding of gemtuzumab to CD33 results in an antibody-antigen complex, which becomes internalized, wherby the calicheamicin derivative is then activated and released inside the lysozomes due to the acidic pH. Calicheamicin then translocates to the nucleus, where it binds to the minor groove of DNA, resulting in double strand breaks and cell death. Resistance to gemtuzumab ozogamicin has been linked to the multidrug resistance (MDR) phenotype, and high levels of CD33 in the peripheral blood may also reduce efficacy likely due to peripheral consumption (Stasi et al. 2008).

In three separate phase II trials that enrolled a total of 277 patients, gemtuzumab ozogamicin was given at a dose of 9 mg/m^2 over 2 h with 14–28 days in between doses (Larson et al. 2002, 2005; Sievers et al. 2001). Inclusion criteria differed slightly between the three trials, but all had CD33+ AML in first relapse. Twenty-six percent of patients achieved an overall remission, approximately half of which were CR and half of which were CRp. Amongst the 157 patients above the age of 60, 24% achieved overall remission (Larson et al. 2002, 2005; Sievers et al. 2001). As expected, infusion reactions such as fever and chills were common. Prolonged neutropenia and thrombocytopenia is almost universal with the use of this agent and lasts 7–8 weeks on average, and is likely due to the expression of CD33 on hematopoietic precursors (Larson et al. 2005). Transient and reversible hyperbilirubinemia and elevated liver transaminases is relatively common, and the unique side effect of hepatic venoocclusive disease (VOD) has also been reported (Giles et al. 2001). In May 2000, based on the interim results of these three studies, gemtuzumab was approved by the FDA for treatment of patients 60 years or older with CD33+ acute myeloid leukemia in first relapse and who are not candidates for other cytotoxic chemotherapy (Bross et al. 2001).

Gemtuzumab was then studied in patients with newly diagnosed AML, but the results were poor overall. In elderly patients who were ineligible for cytotoxic chemotherapy, gemtuzumab monotherapy at the same dose used in previous trials yielded a disappointing 17% CR rate (Amadori et al. 2005). When added as maintenance in postremission therapy in elderly patients in CR1, there was no clinical benefit in terms of disease-free survival or overall survival (Lowenberg et al. 2010). Finally, in a large, international phase III study conducted by the Southwest Oncology Group, gemtuzumab ozogamicin was added to daunorubicin and cytarabine induction and showed no benefit in response rates, CR rates or relapse-free survival, and was also associated with a statistically significant increase in fatal adverse events. Based on these results, SWOG recommended closure of both arms of the study on August 2009 (Petersdorf et al. 2009) and gemtuzumab was voluntarily withdrawn from the market in June 2010. Despite its recent withdrawal, several studies involving gemtuzumab ozogamicin in AML are still actively recruiting.

Other potential cell-surface markers on LSCs include CD44, CD47, and CD123. CD44 mediates the interactions between the LSC and the microenvironment, and high levels have been associated with poor prognosis (Legras et al. 1998). Using anti-CD44 monoclonal antibodies, differentiation and selective eradication of AML stem cells were observed in engrafted NOD/SCID mice (Jin et al. 2006). CD47 serves as the ligand for signal regulatory protein alpha (SIRPα), is highly expressed on AML LSC compared to HSC and may portend poorer overall survival. Activation of SIRPα ultimately leads to inhibition of phagocytosis, an effect that was blocked by anti-CD47 antibodies in vivo. The use of these antibodies resulted in decreased leukemia cells in the bone marrow and decreased engraftment in mice (Legras et al. 1998; Majeti et al. 2009). Testing of CD44 and CD47 antibodies in human trials has not yet begun. An early phase trial of CSL360, a monoclonal anti-CD123 antibody, recently completed phase 1 testing, with downregulation of CD123 but without evidence of single-agent clinical activity in advanced AML (Roberts 2010).

Bortezomib

Owing to its well-documented success in the treatment of multiple myeloma, bortezomib (Velcade, PS-341) is the most widely-used proteasome inhibitor. It is a boronic acid dipeptide that reversibly inhibits the 26S proteasome by way of binding to the 20S complex, and has been shown to have very high response rates and prolongs survival in multiple myeloma (San Miguel et al. 2008; Richardson et al. 2005). The side-effect profile includes peripheral sensory neurotoxicity, cytopenias, and hypotension. While bortezomib currently has FDA-approved indications for the treatment of multiple myeloma and mantle cell lymphoma only, it has also been studied in refractory acute leukemias. Multiple preclinical studies have validated its proapoptotic and anti-proliferative activity in AML (Matondo et al. 2010; Conticello et al. 2008; Colado et al. 2008; Stapnes et al. 2007; Riccioni et al. 2007; Gil et al. 2007; Horton et al. 2006; Servida et al. 2005), and it was well-tolerated in a phase I study in acute leukemias and MDS that also demonstrated proteasomal inhibition. The maximum-tolerated dose of bortezomib was found to be 1.25 mg/m^2 administered twice weekly for 4 weeks every 6 weeks. Dose-limiting toxicities included orthostatic hypotension, nausea, diarrhea, and edema. Four out of 15 patients had a hematological improvement in this heavily pretreated population, but the effect was transient, suggesting it has limited activity as monotherapy (Cortes et al. 2004). Dosing was similar to a previous phase II study in patients with refractory multiple myeloma in which bortezomib was dosed at 1.3 mg/m^2 on days 1, 4, 8, and 11 in a 21-day cycle (Richardson et al. 2004).

Interestingly, anthracyclines have been shown to increase degradation of IκBα and induce rapid expression of NF-κB activity, perhaps as an escape mechanism for the toxic effects of the drug (Laurent and Jaffrezou 2001; Boland et al. 1997). Thus, bortezomib was paired with both anthracyclines and cytarabine in vitro with good results (Horton et al. 2006; Guzman et al. 2002; Pigneux et al. 2007), and one phase I study has shown that it adds little additional toxicity to idarubicin and cytarabine in AML patients (Attar et al. 2008). The authors also demonstrated good tolerability at all doses up to 1.5 mg/m^2 and a CR rate of 61% (19 patients) (Attar et al. 2008). Bortezomib has also been paired with pegylated liposomal doxorubicin in a study involving patients with advanced hematologic malignancies, of which most were multiple myeloma. Of the 42 patients enrolled, five had AML, and the only two evaluable patients from this group both had a PR. High-grade toxicities were not surprising and mostly included cytopenias, neutropenic fever, and peripheral neuropathy. Most recently, a phase I/II study of bortezomib and fludarabine, cytarabine and idarubicin in patients with relapsed/refractory AML. Of the 19 evaluable patients, five had a response after one cycle, including three CRs. One patient who achieved a PR converted to CR with bortezomib consolidation, and one patient converted from MRD positive to negative. Peripheral neuropathy was seen in 10% of patients and suggests that cytarabine may enhance the neurotoxicity of bortezomib (Mateos et al. 2009).

Parthenolide

Parthenolide (PTL) is a sesquiterpene lactone found naturally in *Tanacetum parthenium*, commonly known as Feverfew, which is used to treat migraine headaches and rheumatoid arthritis (Jordan 2007). It has been found to have other biological properties including antitumor activity, inhibition of DNA synthesis, and inhibition of cell proliferation in a variety of cancer cell lines, and sensitization of cancer cells to chemotherapy agents (Patel et al. 2000; Zhang et al. 2004; Ross et al. 1999; Wen et al. 2002). It has potent inhibitory activity against NF-κB by direct binding to IκB-kinase, thereby increasing the pool of IκB available to bind NF-κB, and also been shown to modify of the p50 and p65 subunits of NF-κB, generate reactive oxygen species (ROS), block signal transducers and activators of transcription 3 (STAT3) phosphorylation on Tyr705, and sustain c-Jun N-terminal kinase (JNK) activation (Zhang et al. 2004; Wen et al. 2002; Guzman et al. 2005). An initial in vitro study showed preferential targeting of human AML LSCs with relative sparing of normal HSCs in a murine model. This study also compared results to cytarabine, which was found to be more toxic to normal HSCs with little activity against LSCs. There was a strong association with NF-κB inhibition, activation of p53 leading to apoptosis, and generation of ROS. Futhermore, the effect of PTL on LSCs was abrogated by the antioxidant N-acetylcysteine, thus the authors proposed that NF-κB inhibition may make LSCs more sensitive to ROS-related cell death (Guzman et al. 2005).

Despite its impressive activity in animal models, PTL is limited by poor aqueous solubility, leading to the development of a dimethylamino analog of parthenolide (DMAPT). This compound has approximately 70% oral bioavailability and greater than 1,000 times the water solubility compared to parthenolide. In both a murine and canine model, it also demonstrated inhibition of NF-κB, induction of ROS, upregulation of p53, and preferential eradication of LSCs (Guzman et al. 2007).

Epigenetic Modifiers

Although the classification of pharmacologic epigenetic modifiers as targeted therapies may be questioned, given the generalized lack of knowledge in AML as to the precise genes being targeted for down regulation, their ability to modify gene expression is well-documented, and has spurred the clinical development of such agents.

Perhaps the most classic (and successful) example of a clinically active epigenetic modifier is All-transretinoic acid (ATRA), used in the treatment of acute promyelocytic leukemia (APML). As described previously, in APML, the PML-RARα fusion oncoprotein serves to inhibit transcription of retinoic acid receptor-responsive genes. Since supra-physiologic (pharmacologic) levels of retinoic acid can induce RAR signaling, restoring transcription of target genes (Chomienne et al. 1991; Raelson et al. 1996), the pharmacologic use of retinoic acid, in the form of ATRA, has been implemented. Earlier studies of single-agent ATRA in APML demonstrated very high rates of complete response in patients with previously untreated or relapsed disease, along with acceptable toxicity (Chomienne et al. 1991; Huang et al. 1988; Castaigne et al. 1990; Kanamaru et al. 1995). In addition to its ability to promote myleopoietic differentiation, ATRA may also trigger clearance of leukemia-initiating cells, suggesting a dual mechanism for its beneficial long-term clinical benefit (Nasr et al. 2008). More recently, its use has been refined in a manner that allows for combination with cytotoxic drugs, leading to very high rates of long-term remission (Ades et al. 2008; Sanz et al. 2004; Tallman et al. 1997). The novel strategy of combining ATRA with arsenic trioxide (a highly active agent in relapsed APML with both apoptosis and differentiation-inducing properties), has also been demonstrated to induce high rates of durable response, leading many investigators to anticipate the eventual possibility of curative therapy for APML without the need for cytotoxic chemotherapy (Ravandi et al. 2009).

DNA methyltransferase inhibitors represent another class of epigenetic modulators currently in use for the treatment of AML. While 5-azacitidine (azacitidine) and 5-aza-2′-deoxycytidine (decitabine) are nucleoside analogues, developed originally as a cytotoxic agents (Saiki et al. 1978), these drugs also function as DNA methyltransferase inhibitors following incorporation into DNA, where they covalently bind and sequester DNA methyltransferases (DNMT), leading to decreased promoter hypermethylation and expression of silenced genes at doses much lower than the maximum tolerated doses (MTD) (Goffin and Eisenhauer 2002). Examples of genes in AML whose expression may be restored following exposure to DNMT inhibitors include differentiation and tumor suppressor genes, such as *E-Cadherin* and $P15^{INK4B}$ (Kantarjian et al. 2007; Lakshmikuttyamma et al. 2009).

Clinically, both azacitidine and decitabine have received FDA approval for the treatment of myelodysplastic syndromes (MDS), based upon studies demonstrating leukemia-free survival advantages compared with best supportive care (Kantarjian et al. 2006b; Silverman et al. 2002). In a more recent phase three trial, azacitidine significantly improved overall survival in higher-risk MDS patients (compared with conventional care regimens), demonstrating for the first time that a drug therapy could impart a disease-modifying effect in MDS (Fenaux et al. 2009). In a subgroup of 113 patients in this study who had AML by World Health Organization criteria (20–29% blasts), azacitidine significantly improved overall survival, demonstrating a potential disease-modifying effect in AML (Fenaux et al. 2010). Similarly, single-agent decitabine has documented activity in AML. In a phase I study that included patients with a variety of hematologic malignancies, 5 of the 37 patients with AML had a CR with varying doses of decitabine (Issa et al. 2004). Decitabine has also been studied in elderly patients as first-line therapy in a phase II trial. At a dose of 20 mg/m^2 for 5 days of a 4-week cycle, the overall response rate was 25% with acceptable toxicity (Cashen et al. 2010). Of the 34 patients with cytogenetic abnormalities, five achieved a complete cytogenetic response (Cashen 2009). Despite the approval and relative wide-scale use of these agents in both AML and higher-risk MDS, the precise mechanism through which hypomethylating agents induce clinical response remains unclear.

Early phase studies have also investigated the use of hypomethylating agents in combination with other therapies in AML, including cytotoxic chemotherapy. Combinations of azacitidine with cytarabine, thalidomide, ATRA and gemtuzumab have demonstrated clear clinical activity, although the benefit of the combination beyond more conventional or single-agent therapy will need to be confirmed in larger randomized studies (Borthakur et al. 2009, 2010; Raza et al. 2008; Lubbert et al. 2009; Chowdhury et al. 2009).

Histone deacetylase inhibitors have also been tested in clinical trials, but with limited success as single agents. Depsipeptide, vorinostat (suberoylanilide hydroxamic acid, SAHA), MS-275, and MGCD0103 have all been tested in phase I trials involving AML patients (Gojo et al. 2007; Byrd et al. 2005; Garcia-Manero et al. 2008a, b). In these four trials, a total of 111 AML/MDS patients were treated with HDAC inhibitors, 7 having achieved a CR (4 treated with vorinostat and 3 patients treated with MGCD0103) (Garcia-Manero et al. 2008a, b). It should be noted, however, that correlative studies did show increased histone acetylation and some patients did have blast count reductions.

While single-agent inhibition of either DNA methyltransferase or histone deacetylase can augment gene expression, it appears that combining this approach leads to a higher degree and more sustained reversal of gene silencing (Cameron et al. 1999). The ability to more comprehensively reverse gene silencing by dual, non-overlapping mechanisms of epigenetic modification has led to interest in the therapeutic targeting both hypermethylation and histone deacetylase in AML. A 14% CR rate in 29 patients with AML/MDS was seen in patients treated with azacitidine and sodium phenylbutyrate. Decitabine and valproic acid have been paired in AML/MDS, resulting in a 54% CR rate in one phase I study. In another phase I study, the addition of valproic acid to decitibine was associated with increased risk of encephalopathy but with similar response rates, with demonstration of ER promoter demethylation and histone H3 and H4 acetylation (Garcia-Manero et al. 2006).

Although much evidence supports a potential pathophysiologic and prognostic role of several silenced genes in these diseases, clinical/translational studies that have attempted to correlate clinical response with treatment-induced changes in DNA promoter methylation status or gene expression have yielded mixed results. In one study of combined DNMT inhibitor (azacitidine) with an HDAC inhibitor (phenylbutyrate), responding patients clearly demonstrated inhibited much greater reversal of *P15* methylation within bone marrow cells compared with non-responding patients; however, a relationship between actual upregulation of surrogate gene expression and response was not seen (Gore et al. 2006). Other studies have failed to demonstrate any correlation between clinical response and down-regulation of DNA methylation or increased histone acetylation (Kantarjian et al. 2007; Garcia-Manero et al. 2006, 2008a; Blum et al. 2007). Clearly, further investigations into the precise targets and mediators of response to epigenetic modifiers will be necessary to augment their role in the treatment of AML and MDS.

Signal Transduction Inhibitors

Targeted signaling pathway inhibition with small molecules is emerging as an important therapeutic modality in many cancers. In AML, small molecule inhibitors of targeted pathways are also being clinically developed, some of which are described below.

FLT3 Inhibitors

The high frequency of *FLT3* mutations in AML, along with its important prognostic relevance, has brought forth over the past 10 years an enormous effort to therapeutically target the product of this mutation, such that the clinical development of inhibitors of FLT3 has far outpaced that of any other target in AML. FLT3 inhibitors are all are heterocyclic compounds that directly inhibit FLT3, containing structural similarity to the purine analog ring of adenosine, thus competing with ATP for the ATP-binding pocket of the FLT3 kinase domain (Knapper 2007). All differ in their

Table 1 Status of FLT3 inhibitors in clinical development in AML

Inhibitor	Company	Route	FLT3 inhib in vitro potency	FLT3 inhib in vivo potency	Selectivity	Oral bioav	Status
Sunitinib	Pfizer	Oral	+++	+++	++	++	Discont'd
Sorafenib	Bayer	Oral	+++	+++	+++	++++	Ph 2
Midostaurin	Novartis	Oral	+++	++	++	++	Ph 3
Lestaurtinib	Cephalon	Oral	+++	++	++	+	Ph 3
Tandutinib	Millennium	Oral	++	++	++++	+	Discont'd
AC220	Ambit	Oral	++++	++++	++++	+++	Ph 2
KW2449	Kyowa	Oral	++	+++	++	+++	Discont'd

affinity for FLT3 due to changes in the tertiary structure of the binding pocket. Specific inhibitors (some of which have had development discontinued) brought into the clinic over the years include Sunitinib, Sorafenib, Midostaurin, Lestaurtinib, Tandutinib, AC220, and KW2449. Basic pharmacokinetic and pharmacodynamic features of these compounds are summarized in Table 1.

Sorafenib is an orally bioavailable, multi-kinase inhibitor currently approved for hepatocellular carcinoma and renal cell carcinoma. It has also shown strong FLT3 inhibitory activity, appearing more potent against FLT3-ITD mutants compared to FLT3-TKD mutants (Zhang et al. 2008). Data on sorafenib monotherapy in FLT3-ITD patients are limited, but have been associated with hematologic, bone marrow, and even complete molecular responses in some patients, making it somewhat unique in its activity as a single agent as compared to most other FLT3 inhibitors. In phase I study of 16 patients with relapsed/refractory AML, the use of sorafenib resulted in a significant decrease in the number of peripheral and bone marrow blasts in patients, but only in patients with FLT3-ITD (Zhang et al. 2008). The drug has been well-tolerated in these patients, but resistance may develop after prolonged use (Metzelder et al. 2009a; Schroeder et al. 2009). In another study, six patients with FLT3-ITD positive AML were treated with sorafenib before and after allogeneic hematopoietic stem cell transplant (HSCT) with rapid and durable responses observed in all six patients (Metzelder et al. 2009b). Most recently, a phase I/II study with sorafenib (400 mg twice daily × 7 days) combined with idarubicin and cytarabine in younger patients with AML revealed a 75% CR rate, including a 93% CR rate in the FLT3-ITD patients (Ravandi et al. 2010b).

Midostaurin (PKC412) is an indolocarbazole multitargeted kinase inhibitor that has activity against both FLT3-mutant and FLT3-wt AML (Odgerel et al. 2008; Barry et al. 2007; Weisberg et al. 2002). It was first studied by Stone and colleagues as monotherapy in relapsed/refractory AML and MDS, where 14 of 20 patients achieved a transient decrease in peripheral blasts by at least 50% and concomitant decreases in FLT3 autophosphorylation (Stone et al. 2005). Several years later, the same group published results of a phase II study of young patients with newly diagnosed AML, receiving induction chemotherapy with daunorubicin and cytarabine, followed by cytarabine postremission therapy plus midostaurin. A dose of 50 mg twice daily was best tolerated (higher doses were associated with unacceptable rates of nausea and vomiting), and overall survival at 1 and 2 years was found to be comparable in both the FLT3-mutant and FLT3-wt subgroups, suggesting that addition of a FLT3 inhibitor to cytotoxic chemotherapy may preclude the need for allogeneic stem cell transplant in FLT3-mutated AML patients in CR1 (Stone et al. 2009). Based on this study, the Cancer and Leukemia Group B (CALGB) is currently conducting a phase III study of cytarabine and daunorubicin induction followed by cytarabine postremission therapy plus midostaurin or placebo given comcomitantly with all cycles.

Lestaurtinib (CEP701) is another orally bioavailable indolocarbazole tyrosine kinase inhibitor that has been shown to selectively inhibit phosphorylation of both FLT3-mutated and FLT3-wt cell lines, although with seemingly the most potency against FLT3_ITD mutants (Levis et al. 2002; Mead et al. 2008). In two early trials lestaurtinib was well-tolerated when given as monotherapy and resulted in transient reductions in blast percentage in both the bone marrow and peripheral blood (Smith et al. 2004; Knapper et al. 2006). This drug has also demonstrated synergy with standard induction chemotherapy agents in vitro, but may antagonize the effects if given prior to chemotherapy due to induction of cell-cycle arrest (Levis et al. 2004). A randomized, open-label study comparing mitoxantrone, etoposide, and cytarabine (MEC) alone or MEC followed by lestaurtinib in AML patients with activating FLT3 mutations in first relapse was recently published, demonstrating no improvement in response rates or survival with the combination, a disappointing result possibly explained by the fact that only 42% of patients achieved sustained target (FLT3) inhibition (Levis et al. 2011).

AC220 is an orally available, potent, selective second generation RTK inhibitor optimized for FLT3 inhibition. Preclinical studies in a mouse model showed survival extension in doses as low as 1 mg/kg once daily, and it has been shown to be highly selective for both WT and mutant FLT3 (Zarrinkar et al. 2009). When tested in a phase I study in predominantly relapsed or refractory AML patients unselected for FLT3 mutations, the agent was well-tolerated, with the most common side effects being GI events, peripheral edema, and dysguesia, which were grade 2 or less. The maximum tolerated dose (MTD) was found to be 200 mg daily for 28 days and most responses occurred within one cycle. Intriguingly, 9 (12%) of 76 patients achieved a CR/CRi and 18% achieved a PR. Of note, 4 out of 6 patients with FLT3-ITD patients responded, of which two had failed prior treatment with sorafenib (Cortes et al. 2009). There are also preclinical data indicating that AC220 combined with cytarabine or azacitadine leads to increased efficacy (Belli et al. 2009). A phase 2 monotherapy efficacy study with AC220 in relapsed/refractory FLT3-ITD AML is currently underway.

Farnesyl Transferase Inhibitors

Farnesyl transferase inhibitors (FTIs), functional inhibitors of cytosolic Ras activation (and its subsequent downstream signaling pathways), were originally developed on the basis of a high frequency of *Ras* mutations in human cancers (Rowinsky et al. 1999; End 1999; Lancet and Karp 2003). In addition to Ras pathway modification, preclinical studies have shown the ability of FTIs to disrupt a number of other intracellular proteins including Rho-B, Rac, and Lamin proteins (Rowinsky et al. 1999; End et al. 2001). Available compounds tested in clinical studies include tipifarnib (R115777), lonafarnib (SCH-66336), and BMS-214662, of which tipifarnib has undergone the most development.

Tipifarnib (Zarnestra, Johnson & Johnson) is an orally available selective inhibitor of farnesyl transferase. Across several early phase studies in AML utilizing tipifarnib as monotherapy, both partial and complete responses were documented, even in heavily pretreated populations (Karp et al. 2001; Lancet et al. 2007; Harousseau et al. 2007). Dose-limiting toxicities included transient neurologic deficits, diarrhea, enterocololitis, arrhythmias, and delayed hematologic recovery after consolidation (Karp et al. 2001; Brandwein et al. 2009). Disappointingly, a pivotal phase III trial did not demonstrate a survival advantage for elderly patients with newly diagnosed AML treated with tipifarnib versus best supportive care, indicating the need to study this class of compounds in more selected patient populations or in combination (Harousseau et al. 2009). To this end, more recent data have indicated that the presence of a specified 2-gene ratio (*RASGRP1* and *APTX*) within leukemic blasts was predictive of response and survival in tipifarnib-treated patients across two large studies (Raponi et al. 2008). These findings have led to the development of a new trial testing tipifarnib prospectively in preselected patients, in order to validate this 2-gene ratio as a potential biomarker for efficacy.

HSP 90 Inhibitors

Heat shock proteins are a group of structurally unrelated chaperone proteins that regulate folding and refolding of polypeptides and also play a role in protein degradation via the proteasome complex. Heat shock protein 90 (Hsp90) has two isoforms, α and β, that when activated, form a multiprotein complex that has ATPase activity. Importantly in AML, Hsp90 clients include c-KIT, FLT3, Akt, and Bcr/abl, which play vital roles in a variety of cell-signaling pathways and likely leukemogenesis. In fact, the functionality and intracellular maintenance of mutated FLT3 and BCR-ABL appear dependent upon Hsp90 binding (An et al. 2000; Minami et al. 2002). In addition, Hsp90 appears to be highly expressed in primary AML cells and has been associated with a poor prognosis (Flandrin et al. 2008; Thomas et al. 2005).

The clinical experience of Hsp90 inhibitors in acute leukemias is limited to date. However, a recent phase I study evaluating the MTD of the 2nd generation geldanamycin analogue alvespimicin, as monotherapy in primarily advanced AM,L demonstrated good tolerance, with 3 out of 17 patients achieving CR with incomplete blood count recovery. Common adverse events were neutropenic fever, fatigue, nausea, and diarrhea. Elevated troponins and myocardial infarctions were seen at the doses higher than 24 mg/m^2 twice weekly (Lancet et al. 2010). Based upon the importance of many Hsp90 client proteins in the leukemogenic process, further investigation of this class of agents is warranted with newer, more potent and selective inhibitors of Hsp90.

MDR Inhibitors

Given the highly significant impact of MDR1 expression upon outcome in patients with AML, along with the knowledge that commonly used antileukemic therapies were substrates of MDR1, the clinical application of MDR1 inhibitors in AML was once greeted with great enthusiasm as a prototypical approach to targeted therapy. Initial efforts at P-gp inhibition involved a variety of pharmacologically diverse agents including verapamil, tamoxifen, cyclosporine, and quinine. As these agents were originally developed for other therapeutic uses, most early studies were met with unacceptable toxicity and conflicting data. The UK medical research council added cyclosporine A to cytarabine, daunorubicin, and etoposide, and found no difference in CR or OS rates, except in patients over the age of 60, who actually fared worse (Liu Yin et al. 2001). On the other hand, SWOG conducted a study assessing the benefit of cyclosporine added to daunorubicin and cytarabine in poor-risk AML patients. While CR rates did not improve significantly, RFS (34% vs. 9% at 2 years) and OS (22% vs. 12%) both improved with the addition of cyclosporine over placebo (List et al. 2001). In another small randomized study of elderly patients with AML arising from MDS, comparing idarubicin plus cytarabine with or without oral cyclosporine, the authors observed a lower rate of treatment failure rates and improved leukemia-free survival in the cyclosporine arm (Matsouka et al. 2006). Similarly, the addition of quinine to conventional cytotoxic chemotherapy did not improve outcomes for acute leukemias in two separate studies (Solary et al. 2003, 1996), yet in another phase III trial was shown to improve CR and OS survival rates in P-gp positive high-risk MDS or related leukemia patients when combined with mitoxantrone and cytarabine (Wattel et al. 1998).

The next agent tested was valspodar (PSC-833), a cyclosporine D analog with ten-fold great potency for P-gp blockade than its predecessor, without the propensity for renal toxicity or immunosuppression, but with the undesirable effect of reducing hepatic elimination of concurrently administered anthracycline compounds (Mahadevan and List 2004). Phase III studies adding valspodar to induction chemotherapy have been uniformly disappointing. Two large phase 3 cooperative group trials were closed early due to either a lack of benefit (Greenberg et al. 2004) or excessive induction mortality in the PSC-833 arms (Baer et al. 2002). Two later trials showed no benefit in either younger or elderly patients, even in patients demonstrating inhibitable MDR1 efflux (van der Holt et al. 2005; Kolitz et al. 2010). As such, further development of PSC-833 was halted.

Newer, highly potent and selective MDR1 inhibitors, lacking the unfavorable pharmacokinetic interactions with anthracyclines and epipodophyllotoxins, were viewed as the next generation of logical compounds for targeting MDR1 in a clinically effective manner. Among these newer compounds, only zosuquidar (LY335979) has been clinically developed in AML. In two phase I trials, the combination of daunorubicin and cytarabine with zosuquidar was shown to be well-tolerated and P-gp inhibition was demonstrated (Lancet et al. 2009; Gerrard et al. 2004). Most recently, however, a randomized, double-blind ECOG study of elderly patients with AML compared induction chemotherapy with daunorubicin and cytarabine and with either zosuquidar or placebo was published. Disappointingly, there was no demonstrable difference in overall survival, progression free survival, or CR rates with zosuquidar. Median and 2-year overall survival, while not statistically significant, actually favored placebo (7.2 months and 20% on zosuquidar and 9.4 months and 23% on placebo, respectively, $p = .281$) (Cripe et al. 2010).

The general failure of MDR1 inhibitors in combination with chemotherapeutic drugs to successfully treat AML likely has many explanations. One simple, but plausible argument is that most of the negative randomized studies did not focus upon a targeted population of patients with functional MDR1 expression such that the ability of the compound to exert a therapeutic effect on the most relevant patient population was diminished. Another argument might be that the negative pharmacokinetic interactions with chemotherapeutic agents promoted excessive toxicity that diluted any potential clinical benefit. Finally, the case can be made that selective inhibition of a limited number of these proteins within a very large family of transporters will be insufficient to overcome the overall multidrug resistance phenotype (Mahadevan and List 2004; Ross 2000).

Conclusions

AML remains largely an incurable illness, defined by a rapid disease course and a complex array of molecular events that adversely affect outcomes for the majority of patients. However, through fastidious investigation of unique tumor markers, gene expression, and signaling pathways, along with the rapid development of targeted therapies, it will likely become possible to escape the paradigm of 'one size fits all" and instead to target specific patients, based upon unique molecular features, to the precise therapies which will most likely affect the biologic properties that drive the malignant clone within that individual. AML provides us with a comprehensive network of molecular anomalies and genetic alterations that has expanded exponentially in the past few years and upon which to focus research efforts, and while much remains to be learned about the pathophysiologic relevance and "targetability" of each potential molecular target and anomaly, will allow for the design of more rationally based clinical trials, with higher expectations of response, lower likelihoods of toxicity, and faster time to approval of critically needed new agents.

References

Ades L, Sanz MA, Chevret S et al (2008) Treatment of newly diagnosed acute promyelocytic leukemia (APL): a comparison of French-Belgian-Swiss and PETHEMA results. Blood 111:1078–1084

Agrawal S, Hofmann W-K, Tidow N et al (2007) The C/EBP{delta} tumor suppressor is silenced by hypermethylation in acute myeloid leukemia. Blood 109:3895–3905

Alcalay M, Tiacci E, Bergomas R et al (2005) Acute myeloid leukemia bearing cytoplasmic nucleophosmin (NPMc+AML) shows a distinct gene expression profile characterized by up-regulation of genes involved in stem-cell maintenance. Blood 106:899–902

Amadori S, Suciu S, Stasi R et al (2005) Gemtuzumab ozogamicin (Mylotarg) as single-agent treatment for frail patients 61 years of age and older with acute myeloid leukemia: final results of AML-15B, a phase 2 study of the European organisation for research and treatment of cancer and Gruppo Italiano Malattie Ematologiche dell'Adulto leukemia groups. Leukemia 19:1768–1773

An WG, Schulte TW, Neckers LM (2000) The heat shock protein 90 antagonist geldanamycin alters chaperone association with p210bcr-abl and v-src proteins before their degradation by the proteasome. Cell Growth Differ 11:355–360

Andreeff M, Ruvolo V, Gadgil S et al (2008) HOX expression patterns identify a common signature for favorable AML. Leukemia 22:2041–2047

Appelbaum FR, Gundacker H, Head DR et al (2006) Age and acute myeloid leukemia. Blood 107:3481–3485

Attar EC, De Angelo DJ, Supko JG et al (2008) Phase I and pharmacokinetic study of bortezomib in combination with idarubicin and cytarabine in patients with acute myelogenous leukemia. Clin Cancer Res 14:1446–1454

Bacher U, Haferlach T, Schoch C, Kern W, Schnittger S (2006) Implications of NRAS mutations in AML: a study of 2502 patients. Blood 107:3847–3853

Baer MR, George SL, Dodge RK et al (2002) Phase 3 study of the multidrug resistance modulator PSC-833 in previously untreated patients 60 years of age and older with acute myeloid leukemia: cancer and leukemia group B study 9720. Blood 100:1224–1232

Barry EV, Clark JJ, Cools J, Roesel J, Gilliland DG (2007) Uniform sensitivity of FLT3 activation loop mutants to the tyrosine kinase inhibitor midostaurin. Blood 110:4476–4479

Belli BA, Dao A, Bhagwat S, Wierenga W, Armstrong RC (2009) AC220, a potent and specific FLT3 inhibitor, enhances the cytotoxic effects of chemotherapeutic agents in cell culture and in mouse tumor xenografts. Blood (ASH Annual Meeting Abstracts) 114:2052

Black JH, McCubrey JA, Willingham MC, Ramage J, Hogge DE, Frankel AE (2003) Diphtheria toxin-interleukin-3 fusion protein (DT(388)IL3) prolongs disease-free survival of leukemic immunocompromised mice. Leukemia 17:155–159

Blair A, Hogge DE, Sutherland HJ (1998) Most acute myeloid leukemia progenitor cells with long-term proliferative ability in vitro and in vivo have the phenotype CD34(+)/CD71(−)/HLA-DR. Blood 92:4325–4335

Blum W, Klisovic RB, Hackanson B et al (2007) Phase I study of decitabine alone or in combination with valproic acid in acute myeloid leukemia. J Clin Oncol 25:3884–3891

Boland MP, Foster SJ, O'Neill LA (1997) Daunorubicin activates NFkappaB and induces kappaB-dependent gene expression in HL-60 promyelocytic and Jurkat T lymphoma cells. J Biol Chem 272:12952–12960

Bonnet D, Dick JE (1997) Human acute myeloid leukemia is organized as a hierarchy that originates from a primitive hematopoietic cell. Nat Med 3:730–737

Borthakur G, Garcia-Manero G, Estrov Z et al (2009) Phase 2 study of decitabine and gemtuzumab ozogamicin in acute myelogenous leukemia and high-risk myelodysplastic syndrome- outcome in previously untreated patients. ASH Annual Meeting Abstracts 114:1053

Borthakur G, Huang X, Kantarjian H et al (2010) Report of a phase 1/2 study of a combination of azacitidine and cytarabine in acute myelogenous leukemia and high-risk myelodysplastic syndromes. Leuk Lymphoma 51:73–78

Bowen DT, Frew ME, Hills R et al (2005) RAS mutation in acute myeloid leukemia is associated with distinct cytogenetic subgroups but does not influence outcome in patients younger than 60 years. Blood 106:2113–2119

Brandwein JM, Leber BF, Howson-Jan K et al (2009) A phase I study of tipifarnib combined with conventional induction and consolidation therapy for previously untreated patients with acute myeloid leukemia aged 60 years and over. Leukemia 23:631–634

Braun T, Fenaux P (2008) Farnesyltransferase inhibitors and their potential role in therapy for myelodysplastic syndromes and acute myeloid leukaemia. Br J Haematol 141:576–586

Bross PF, Beitz J, Chen G et al (2001) Approval summary: gemtuzumab ozogamicin in relapsed acute myeloid leukemia. Clin Cancer Res 7:1490–1496

Byrd JC, Marcucci G, Parthun MR et al (2005) A phase 1 and pharmacodynamic study of depsipeptide (FK228) in chronic lymphocytic leukemia and acute myeloid leukemia. Blood 105:959–967

Callens C, Chevret S, Cayuela JM et al (2005) Prognostic implication of FLT3 and Ras gene mutations in patients with acute promyelocytic leukemia (APL): a retrospective study from the European APL Group. Leukemia 19:1153–1160

Cameron EE, Bachman KE, Myohanen S et al (1999) Synergy of demethylation and histone deacetylase inhibition in the re-expression of genes silenced in cancer. Nat Genet 21:103–107

Cashen AF (2009) Decitabine can induce complete cytogenetic responses when used as initial therapy in older AML patients. ASH Annual Meeting Abstracts 114:4123

Cashen AF, Schiller GJ, O'Donnell MR, DiPersio JF (2010) Multicenter, phase II study of decitabine for the first-line treatment of older patients with acute myeloid leukemia. J Clin Oncol 28:556–561

Castaigne S, Chomienne C, Daniel M et al (1990) All-trans retinoic acid as a differentiation therapy for acute promyelocytic leukemia. I. Clinical results [see comments]. Blood 76:1704–1709

Chan WI, Huntly BJ (2008) Leukemia stem cells in acute myeloid leukemia. Semin Oncol 35:326–335

Chen W, Jones D, Medeiros LJ, Luthra R, Lin P (2005) Acute myeloid leukaemia with FLT3 gene mutations of both internal tandem duplication and point mutation type. Br J Haematol 130:726–728

Chomienne C, Balitrand N, Ballerini P, Castaigne S, de Thé H, Degos L (1991) All-trans retinoic acid modulates the retinoic acid receptor-alpha in promyelocytic cells. J Clin Invest 88:2150–2154

Chowdhury S, Seropian S, Marks PW (2009) Decitabine combined with fractionated gemtuzumab ozogamicin therapy in patients with relapsed or refractory acute myeloid leukemia. Am J Hematol 84:599–600

Colado E, Alvarez-Fernandez S, Maiso P et al (2008) The effect of the proteasome inhibitor bortezomib on acute myeloid leukemia cells and drug resistance associated with the CD34+ immature phenotype. Haematologica 93:57–66

Conticello C, Adamo L, Vicari L et al (2008) Antitumor activity of bortezomib alone and in combination with TRAIL in human acute myeloid leukemia. Acta Haematol 120:19–30

Cornelissen JJ, van Putten WL, Verdonck LF et al (2007) Results of a HOVON/SAKK donor versus no-donor analysis of myeloablative HLA-identical sibling stem cell transplantation in first remission acute myeloid leukemia in young and middle-aged adults: benefits for whom? Blood 109:3658–3666

Cortes J, Thomas D, Koller C et al (2004) Phase I study of bortezomib in refractory or relapsed acute leukemias. Clin Cancer Res 10:3371–3376

Cortes J, Foran J, Ghirdaladze D et al (2009) AC220, a potent, selective, second generation FLT3 Receptor Tyrosine Kinase (RTK) Inhibitor, in a First-in-Human (FIH) phase 1 AML study. Blood (ASH Annual Meeting Abstracts) 114:636

Cripe LD, Uno H, Paietta EM et al (2010) Zosuquidar, a novel modulator of P-glycoprotein, does not improve the outcome of older patients with newly diagnosed acute myeloid leukemia: a randomized, placebo-controlled, trial of the Eastern Cooperative Oncology Group (ECOG 3999). Blood 116(20):4077–4085

Di Croce L, Raker VA, Corsaro M et al (2002) Methyltransferase recruitment and DNA hypermethylation of target promoters by an oncogenic transcription factor. Science 295:1079–1082

Downward J (2003) Targeting RAS signalling pathways in cancer therapy. Nat Rev Cancer 3:11–22

Drexler HG (1996) Expression of FLT3 receptor and response to FLT3 ligand by leukemic cells. Leukemia 10:588–599

End DW (1999) Farnesyl protein transferase inhibitors and other therapies targeting the Ras signal transduction pathway. Invest New Drugs 17:241–258

End DW, Smets G, Todd AV et al (2001) Characterization of the antitumor effects of the selective farnesyl protein transferase inhibitor R115777 in vivo and in vitro. Cancer Res 61:131–137

Fardel O, Lecureur V, Daval S, Corlu A, Guillouzo A (1997) Up-regulation of P-glycoprotein expression in Rat liver cells by acute doxorubicin treatment. Eur J Biochem 246:186–192
Fenaux P, Mufti GJ, Hellstrom-Lindberg E et al (2009) Efficacy of azacitidine compared with that of conventional care regimens in the treatment of higher-risk myelodysplastic syndromes: a randomised, open-label, phase III study. Lancet Oncol 10:223–232
Fenaux P, Mufti GJ, Hellstrom-Lindberg E et al (2010) Azacitidine prolongs overall survival compared with conventional care regimens in elderly patients with low bone marrow blast count acute myeloid leukemia. J Clin Oncol 28:562–569
Feuring-Buske M, Frankel AE, Alexander RL, Gerhard B, Hogge DE (2002) A diphtheria toxin-interleukin 3 fusion protein is cytotoxic to primitive acute myeloid leukemia progenitors but spares normal progenitors. Cancer Res 62:1730–1736
Fialkow PJ, Singer JW, Adamson JW et al (1981) Acute nonlymphocytic leukemia: heterogeneity of stem cell origin. Blood 57:1068–1073
Fialkow PJ, Singer JW, Raskind WH et al (1987) Clonal development, stem-cell differentiation, and clinical remissions in acute nonlymphocytic leukemia. N Engl J Med 317:468–473
Flandrin P, Guyotat D, Duval A et al (2008) Significance of heat-shock protein (HSP) 90 expression in acute myeloid leukemia cells. Cell Stress Chaperones 13:357–364
Frankel AE, McCubrey JA, Miller MS et al (2000) Diphtheria toxin fused to human interleukin-3 is toxic to blasts from patients with myeloid leukemias. Leukemia 14:576–585
Frankel A, Liu JS, Rizzieri D, Hogge D (2008) Phase I clinical study of diphtheria toxin-interleukin 3 fusion protein in patients with acute myeloid leukemia and myelodysplasia. Leuk Lymphoma 49:543–553
Frohling S, Schlenk RF, Breitruck J et al (2002) Prognostic significance of activating FLT3 mutations in younger adults (16 to 60 years) with acute myeloid leukemia and normal cytogenetics: a study of the AML study group Ulm. Blood 100:4372–4380
Frohling S, Scholl C, Bansal D, Huntly BJ (2007) HOX gene regulation in acute myeloid leukemia: CDX marks the spot? Cell Cycle 6:2241–2245
Garcia-Manero G, Kantarjian HM, Sanchez-Gonzalez B et al (2006) Phase 1/2 study of the combination of 5-aza-2′-deoxycytidine with valproic acid in patients with leukemia. Blood 108:3271–3279
Garcia-Manero G, Yang H, Bueso-Ramos C et al (2008a) Phase 1 study of the histone deacetylase inhibitor vorinostat (suberoylanilide hydroxamic acid [SAHA]) in patients with advanced leukemias and myelodysplastic syndromes. Blood 111:1060–1066
Garcia-Manero G, Assouline S, Cortes J et al (2008b) Phase 1 study of the oral isotype specific histone deacetylase inhibitor MGCD0103 in leukemia. Blood 112:981–989
Gelmetti V, Zhang J, Fanelli M, Minucci S, Pelicci PG, Lazar MA (1998) Aberrant Recruitment of the Nuclear Receptor Corepressor-Histone Deacetylase Complex by the Acute Myeloid Leukemia Fusion Partner ETO. Mol Cell Biol 18:7185–7191
Gerrard G, Payne E, Baker RJ et al (2004) Clinical effects and P-glycoprotein inhibition in patients with acute myeloid leukemia treated with zosuquidar trihydrochloride, daunorubicin and cytarabine. Haematologica 89:782–790
Gil L, Styczynski J, Dytfeld D et al (2007) Activity of bortezomib in adult de novo and relapsed acute myeloid leukemia. Anticancer Res 27:4021–4025
Giles FJ, Kantarjian HM, Kornblau SM et al (2001) Mylotarg (gemtuzumab ozogamicin) therapy is associated with hepatic venoocclusive disease in patients who have not received stem cell transplantation. Cancer 92:406–413
Goffin J, Eisenhauer E (2002) DNA methyltransferase inhibitorsâ€'state of the art. Ann Oncol 13:1699–1716
Gojo I, Jiemjit A, Trepel JB et al (2007) Phase 1 and pharmacologic study of MS-275, a histone deacetylase inhibitor, in adults with refractory and relapsed acute leukemias. Blood 109:2781–2790
Gore SD, Baylin S, Sugar E et al (2006) Combined DNA methyltransferase and histone deacetylase inhibition in the treatment of myeloid neoplasms. Cancer Res 66:6361–6369

Greenberg PL, Lee SJ, Advani R et al (2004) Mitoxantrone, etoposide, and cytarabine with or without valspodar in patients with relapsed or refractory acute myeloid leukemia and high-risk myelodysplastic syndrome: a phase III trial (E2995). J Clin Oncol 22:1078–1086

Guzman ML, Jordan CT (2004) Considerations for targeting malignant stem cells in leukemia. Cancer Control 11:97–104

Guzman ML, Neering SJ, Upchurch D et al (2001) Nuclear factor-kappaB is constitutively activated in primitive human acute myelogenous leukemia cells. Blood 98:2301–2307

Guzman ML, Swiderski CF, Howard DS et al (2002) Preferential induction of apoptosis for primary human leukemic stem cells. Proc Natl Acad Sci U S A 99:16220–16225

Guzman ML, Rossi RM, Karnischky L et al (2005) The sesquiterpene lactone parthenolide induces apoptosis of human acute myelogenous leukemia stem and progenitor cells. Blood 105: 4163–4169

Guzman ML, Rossi RM, Neelakantan S et al (2007) An orally bioavailable parthenolide analog selectively eradicates acute myelogenous leukemia stem and progenitor cells. Blood 110:4427–4435

Harousseau JL, Lancet JE, Reiffers J et al (2007) A phase 2 study of the oral farnesyltransferase inhibitor tipifarnib in patients with refractory or relapsed acute myeloid leukemia. Blood 109:5151–5156

Harousseau JL, Martinelli G, Jedrzejczak WW et al (2009) A randomized phase 3 study of tipifarnib compared with best supportive care, including hydroxyurea, in the treatment of newly diagnosed acute myeloid leukemia in patients 70 years or older. Blood 114:1166–1173

He TC, Sparks AB, Rago C et al (1998) Identification of c-MYC as a target of the APC pathway. Science 281:1509–1512

Hirsch-Ernst KI, Ziemann C, Foth H, Kozian D, Schmitz-Salue C, Kahl GF (1998) Induction of mdr1b mRNA and P-glycoprotein expression by tumor necrosis factor alpha in primary rat hepatocyte cultures. J Cell Physiol 176:506–515

Hochhaus A, O'Brien SG, Guilhot F et al (2009) Six-year follow-up of patients receiving imatinib for the first-line treatment of chronic myeloid leukemia. Leukemia 23:1054–1061

Horton TM, Gannavarapu A, Blaney SM, D'Argenio DZ, Plon SE, Berg SL (2006) Bortezomib interactions with chemotherapy agents in acute leukemia in vitro. Cancer Chemother Pharmacol 58:13–23

Huang M, Ye Y, Chen S et al (1988) Use of all-trans retinoic acid in the treatment of acute promyelocytic leukemia. Blood 72:567–572

Illmer T, Thiede C, Fredersdorf A et al (2005) Activation of the RAS pathway is predictive for a chemosensitive phenotype of acute myelogenous leukemia blasts. Clin Cancer Res 11:3217–3224

Issa J-PJ, Zehnbauer BA, Kaufmann SH, Biel MA, Baylin SB (1997) HIC1 Hypermethylation is a late event in hematopoietic neoplasms. Cancer Res 57:1678–1681

Issa JP, Garcia-Manero G, Giles FJ et al (2004) Phase 1 study of low-dose prolonged exposure schedules of the hypomethylating agent 5-aza-2′-deoxycytidine (decitabine) in hematopoietic malignancies. Blood 103:1635–1640

Jin L, Hope KJ, Zhai Q, Smadja-Joffe F, Dick JE (2006) Targeting of CD44 eradicates human acute myeloid leukemic stem cells. Nat Med 12:1167–1174

Jordan CT (2007) The leukemic stem cell. Best Pract Res Clin Haematol 20:13–18

Jordan CT, Upchurch D, Szilvassy SJ et al (2000) The interleukin-3 receptor alpha chain is a unique marker for human acute myelogenous leukemia stem cells. Leukemia 14:1777–1784

Kanamaru A, Takemoto Y, Tanimoto M et al (1995) All-trans retinoic acid for the treatment of newly diagnosed acute promyelocytic leukemia. Japan Adult Leukemia Study Group [see comments]. Blood 85:1202–1206

Kantarjian H, O'Brien S, Cortes J et al (2006a) Results of intensive chemotherapy in 998 patients age 65 years or older with acute myeloid leukemia or high-risk myelodysplastic syndrome: predictive prognostic models for outcome. Cancer 106:1090–1098

Kantarjian H, Issa J-PJ, Rosenfeld CS et al (2006b) Decitabine improves patient outcomes in myelodysplastic syndromes. In: Wiley Subscription Services, Inc., A Wiley Company, pp 1794–803

Kantarjian H, Oki Y, Garcia-Manero G et al (2007) Results of a randomized study of 3 schedules of low-dose decitabine in higher-risk myelodysplastic syndrome and chronic myelomonocytic leukemia. Blood 109:52–57

Kantarjian H, Shah NP, Hochhaus A et al (2010) Dasatinib versus imatinib in newly diagnosed chronic-phase chronic myeloid leukemia. N Engl J Med 362:2260–2270

Karp JE, Lancet JE, Kaufmann SH et al (2001) Clinical and biologic activity of the farnesyltransferase inhibitor R115777 in adults with refractory and relapsed acute leukemias: a phase 1 clinical-laboratory correlative trial. Blood 97:3361–3369

Keith B, Simon MC (2007) Hypoxia-inducible factors, stem cells, and cancer. Cell 129:465–472

Kim HP, Frankel AE, Hogge DE (2010) A diphtheria toxin interleukin-3 fusion protein synergizes with tyrosine kinase inhibitors in killing leukemic progenitors from BCR/ABL positive acute leukemia. Leuk Res 34:1035–1042

Kiyoi H, Naoe T, Nakano Y et al (1999) Prognostic implication of FLT3 and N-RAS gene mutations in acute myeloid leukemia. Blood 93:3074–3080

Klisovic MI, Maghraby EA, Parthun MR et al (2003) Depsipeptide (FR 901228) promotes histone acetylation, gene transcription, apoptosis and its activity is enhanced by DNA methyltransferase inhibitors in AML1//ETO-positive leukemic cells. Leukemia 17:350–358

Knapper S (2007) FLT3 inhibition in acute myeloid leukaemia. Br J Haematol 138:687–699

Knapper S, Burnett AK, Littlewood T et al (2006) A phase 2 trial of the FLT3 inhibitor lestaurtinib (CEP701) as first-line treatment for older patients with acute myeloid leukemia not considered fit for intensive chemotherapy. Blood 108:3262–3270

Kolitz JE, George SL, Marcucci G et al (2010) P-glycoprotein inhibition using valspodar (PSC-833) does not improve outcomes for patients under age 60 years with newly diagnosed acute myeloid leukemia: cancer and leukemia group B study 19808. Blood 116(9):1413–1421

Konopleva MY, Jordan CT (2011) Leukemia stem cells and microenvironment: biology and therapeutic targeting. J Clin Oncol 29:591–599

Koreth J, Schlenk R, Kopecky KJ et al (2009) Allogeneic stem cell transplantation for acute myeloid leukemia in first complete remission: systematic review and meta-analysis of prospective clinical trials. JAMA 301:2349–2361

Kottaridis PD, Gale RE, Frew ME et al (2001) The presence of a FLT3 internal tandem duplication in patients with acute myeloid leukemia (AML) adds important prognostic information to cytogenetic risk group and response to the first cycle of chemotherapy: analysis of 854 patients from the united kingdom medical research council AML 10 and 12 trials. Blood 98:1752–1759

Lakshmikuttyamma A, Scott SA, DeCoteau JF, Geyer CR (2009) Reexpression of epigenetically silenced AML tumor suppressor genes by SUV39H1 inhibition. Oncogene 29:576–588

Lancet JE, Karp JE (2003) Farnesyltransferase inhibitors in hematologic malignancies: new horizons in therapy. Blood 102:3880–3889

Lancet JE, Gojo I, Gotlib J et al (2007) A phase 2 study of the farnesyltransferase inhibitor tipifarnib in poor-risk and elderly patients with previously untreated acute myelogenous leukemia. Blood 109:1387–1394

Lancet JE, Baer MR, Duran GE et al (2009) A phase I trial of continuous infusion of the multidrug resistance inhibitor zosuquidar with daunorubicin and cytarabine in acute myeloid leukemia. Leuk Res 33:1055–1061

Lancet JE, Gojo I, Burton M et al (2010) Phase I study of the heat shock protein 90 inhibitor alvespimycin (KOS-1022, 17-DMAG) administered intravenously twice weekly to patients with acute myeloid leukemia. Leukemia 24:699–705

Larson RA, Boogaerts M, Estey E et al (2002) Antibody-targeted chemotherapy of older patients with acute myeloid leukemia in first relapse using Mylotarg (gemtuzumab ozogamicin). Leukemia 16:1627–1636

Larson RA, Sievers EL, Stadtmauer EA et al (2005) Final report of the efficacy and safety of gemtuzumab ozogamicin (Mylotarg) in patients with CD33-positive acute myeloid leukemia in first recurrence. Cancer 104:1442–1452

Laurent G, Jaffrezou JP (2001) Signaling pathways activated by daunorubicin. Blood 98:913–924

Legras S, Gunthert U, Stauder R et al (1998) A strong expression of CD44-6v correlates with shorter survival of patients with acute myeloid leukemia. Blood 91:3401–3413

Lehne G, Sorensen D, Tjonnfjord G et al (2002) The cyclosporin PSC 833 increases survival and delays engraftment of human multidrug-resistant leukemia cells in xenotransplanted NOD-SCID mice. Leukemia 16:2388–2394

Leith CP, Kopecky KJ, Godwin J et al (1997) Acute myeloid leukemia in the elderly: assessment of multidrug resistance (MDR1) and cytogenetics distinguishes biologic subgroups with remarkably distinct responses to standard chemotherapy. A southwest oncology group study. Blood 89:3323–3329

Leith CP, Kopecky KJ, Chen IM et al (1999) Frequency and clinical significance of the expression of the multidrug resistance proteins MDR1/P-glycoprotein, MRP1, and LRP in acute myeloid leukemia: a southwest oncology group study. Blood 94:1086–1099

Lessard J, Sauvageau G (2003) Bmi-1 determines the proliferative capacity of normal and leukaemic stem cells. Nature 423:255–260

Levis M, Allebach J, Tse KF et al (2002) A FLT3-targeted tyrosine kinase inhibitor is cytotoxic to leukemia cells in vitro and in vivo. Blood 99:3885–3891

Levis M, Pham R, Smith BD, Small D (2004) In vitro studies of a FLT3 inhibitor combined with chemotherapy: sequence of administration is important to achieve synergistic cytotoxic effects. Blood 104:1145–1150

Levis M, Sato T, Murphy K, Rajkhowa T, Pratz K (2009) FLT3 mutant to wild type allelic ratio and clinical status are predictive of response to FLT3 inhibitors in AML. ASH Annual Meeting Abstracts 114:1716

Levis M, Ravandi F, Wang ES et al (2011) Results from a randomized trial of salvage chemotherapy followed by lestaurtinib for patients with FLT3 mutant AML in first relapse. Blood 117:3294–3301

Lin RJ, Nagy L, Inoue S, Shao W, Miller WH, Evans RM (1998) Role of the histone deacetylase complex in acute promyelocytic leukaemia. Nature 391:811–814

List AF, Kopecky KJ, Willman CL et al (2001) Benefit of cyclosporine modulation of drug resistance in patients with poor-risk acute myeloid leukemia: a southwest oncology group study. Blood 98:3212–3220

Liu Yin JA, Wheatley K, Rees JK, Burnett AK (2001) Comparison of 'sequential' versus 'standard' chemotherapy as re-induction treatment, with or without cyclosporine, in refractory/relapsed acute myeloid leukaemia (AML): results of the UK medical research council AML-R trial. Br J Haematol 113:713–726

Liu TF, Urieto JO, Moore JE et al (2004) Diphtheria toxin fused to variant interleukin-3 provides enhanced binding to the interleukin-3 receptor and more potent leukemia cell cytotoxicity. Exp Hematol 32:277–281

Liu S, Shen T, Huynh L et al (2005) Interplay of RUNX1/MTG8 and DNA methyltransferase 1 in acute myeloid leukemia. Cancer Res 65:1277–1284

Lowenberg B, Beck J, Graux C et al (2010) Gemtuzumab ozogamicin as postremission treatment in AML at 60 years of age or more: results of a multicenter phase 3 study. Blood 115: 2586–2591

Lubbert M, Ruter BH, Schmid M et al (2009) Low-Dose Decitabine (DAC), alone or in combination with All-Trans Retinoic Acid (ATRA), is an active first-line treatment in older AML patients of all cytogenetic risk groups: final results of the FR00331 multicenter phase II study. ASH Annual Meeting Abstracts 114:4141

Mahadevan D, List AF (2004) Targeting the multidrug resistance-1 transporter in AML: molecular regulation and therapeutic strategies. Blood 104:1940–1951

Majeti R, Chao MP, Alizadeh AA et al (2009) CD47 is an adverse prognostic factor and therapeutic antibody target on human acute myeloid leukemia stem cells. Cell 138:286–299

Martelli AM, Nyakern M, Tabellini G et al (2006) Phosphoinositide 3-kinase/Akt signaling pathway and its therapeutical implications for human acute myeloid leukemia. Leukemia 20:911–928

Martelli AM, Evangelisti C, Chiarini F, McCubrey JA (2010) The phosphatidylinositol 3-kinase/Akt/mTOR signaling network as a therapeutic target in acute myelogenous leukemia patients. Oncotarget 1:89–103

Mateos M-V, Martinez R, Brunet S et al (2009) A phase I/II, national, multicenter, open label study of bortezomib (Velcade) and fludarabine, cytarabine and idarubicin (Flag-Ida) (V-Flag-Ida) in Pts with relapsed and/or refractory Acute Myeloblastic Leukemia (AML). Blood (ASH Annual Meeting Abstracts) 114:4140
Matondo M, Bousquet-Dubouch MP, Gallay N et al (2010) Proteasome inhibitor-induced apoptosis in acute myeloid leukemia: a correlation with the proteasome status. Leuk Res 34:498–506
Matsouka P, Pagoni M, Zikos P et al (2006) Addition of cyclosporin-A to chemotherapy in secondary (post-MDS) AML in the elderly. A multicenter randomized trial of the Leukemia Working Group of the Hellenic Society of Hematology. Ann Hematol 85:250–256
Matsunaga T, Takemoto N, Sato T et al (2003) Interaction between leukemic-cell VLA-4 and stromal fibronectin is a decisive factor for minimal residual disease of acute myelogenous leukemia. Nat Med 9:1158–1165
Mead AJ, Linch DC, Hills RK, Wheatley K, Burnett AK, Gale RE (2007) FLT3 tyrosine kinase domain mutations are biologically distinct from and have a significantly more favorable prognosis than FLT3 internal tandem duplications in patients with acute myeloid leukemia. Blood 110:1262–1270
Mead AJ, Gale RE, Kottaridis PD, Matsuda S, Khwaja A, Linch DC (2008) Acute myeloid leukaemia blast cells with a tyrosine kinase domain mutation of FLT3 are less sensitive to lestaurtinib than those with a FLT3 internal tandem duplication. Br J Haematol 141:454–460
Melki JR, Vincent PC, Brown RD, Clark SJ (2000) Hypermethylation of E-cadherin in leukemia. Blood 95:3208–3213
Meshinchi S, Appelbaum FR (2009) Structural and functional alterations of FLT3 in acute myeloid leukemia. Clin Cancer Res 15:4263–4269
Meshinchi S, Alonzo TA, Stirewalt DL et al (2006) Clinical implications of FLT3 mutations in pediatric AML. Blood 108:3654–3661
Metzelder S, Scholl S, Matthias K et al (2009a) Compassionate use of sorafenib in relapsed and refractory Flt3-ITD positive acute myeloid leukemia. Blood (ASH Annual Meeting Abstracts) 114:2060
Metzelder S, Wang Y, Wollmer E et al (2009b) Compassionate use of sorafenib in FLT3-ITD-positive acute myeloid leukemia: sustained regression before and after allogeneic stem cell transplantation. Blood 113:6567–6571
Meyer M, Rubsamen D, Slany R et al (2009) Oncogenic RAS enables DNA damage- and p53-dependent differentiation of acute myeloid leukemia cells in response to chemotherapy. PLoS One 4:e7768
Minami YF, Kiyoi H, Kiyoi HF, Yamamoto Y, Yamamoto YF, Yamamoto K et al (2002) Selective apoptosis of tandemly duplicated FLT3-transformed leukemia cells by Hsp90 inhibitors. Leukemia 16:1535–1540
Misaghian N, Ligresti G, Steelman LS et al (2009) Targeting the leukemic stem cell: the Holy Grail of leukemia therapy. Leukemia 23:25–42
Moreno I, Martin G, Bolufer P et al (2003) Incidence and prognostic value of FLT3 internal tandem duplication and D835 mutations in acute myeloid leukemia. Haematologica 88:19–24
Mori Y, Kiyoi H, Ishikawa Y, Naoe T (2009) FL-dependent wild-type FLT3 signals reduce the inhibitory effects of FLT3 inhibitors on wild-type and mutant FLT3 co-expressing cells. ASH Annual Meeting Abstracts 2009;114:2067
Morrison SJ, Prowse KR, Ho P, Weissman IL (1995) Telomerase activity in hematopoietic cells is associated with self-renewal potential. Immunity 5:207–216
Muller-Tidow C, Steffen B, Cauvet T et al (2004) Translocation products in acute myeloid leukemia activate the Wnt signaling pathway in hematopoietic cells. Mol Cell Biol 24:2890–2904
Nakanishi C, Toi M (2005) Nuclear factor-kappaB inhibitors as sensitizers to anticancer drugs. Nat Rev Cancer 5:297–309
Nakao M, Yokota S, Iwai T et al (1996) Internal tandem duplication of the flt3 gene found in acute myeloid leukemia. Leukemia 10:1911–1918
Nasr R, Guillemin M-C, Ferhi O et al (2008) Eradication of acute promyelocytic leukemia-initiating cells through PML-RARA degradation. Nat Med 14:1333–1342

Neubauer A, Maharry K, Mrozek K et al (2008) Patients with acute myeloid leukemia and RAS mutations benefit most from postremission high-dose cytarabine: a cancer and leukemia group B study. J Clin Oncol 26:4603–4609
O'Brien SG, Guilhot F, Larson RA et al (2003) Imatinib compared with interferon and low-dose cytarabine for newly diagnosed chronic-phase chronic myeloid leukemia. N Engl J Med 348:994–1004
Odgerel T, Kikuchi J, Wada T et al (2008) The FLT3 inhibitor PKC412 exerts differential cell cycle effects on leukemic cells depending on the presence of FLT3 mutations. Oncogene 27:3102–3110
Okada Y, Jiang Q, Lemieux M, Jeannotte L, Su L, Zhang Y (2006) Leukaemic transformation by CALM-AF10 involves upregulation of Hoxa5 by hDOT1L. Nat Cell Biol 8:1017–1024
Okano H, Kawahara H, Toriya M, Nakao K, Shibata S, Imai T (2005) Function of RNA-binding protein Musashi-1 in stem cells. Exp Cell Res 306:349–356
Ozeki K, Kiyoi H, Hirose Y et al (2004) Biologic and clinical significance of the FLT3 transcript level in acute myeloid leukemia. Blood 103:1901–1908
Palmqvist L, Pineault N, Wasslavik C, Humphries RK (2007) Candidate genes for expansion and transformation of hematopoietic stem cells by NUP98-HOX fusion genes. PLoS One 2:e768
Park IK, Qian D, Kiel M et al (2003) Bmi-1 is required for maintenance of adult self-renewing haematopoietic stem cells. Nature 423:302–305
Patel NM, Nozaki S, Shortle NH et al (2000) Paclitaxel sensitivity of breast cancer cells with constitutively active NF-kappaB is enhanced by IkappaBalpha super-repressor and parthenolide. Oncogene 19:4159–4169
Pearce DJ, Taussig D, Zibara K et al (2006) AML engraftment in the NOD/SCID assay reflects the outcome of AML: implications for our understanding of the heterogeneity of AML. Blood 107:1166–1173
Petersdorf S, Kopecky K, Stuart RK et al (2009) Preliminary results of southwest oncology group study S0106: an international intergroup phase 3 randomized trial comparing the addition of gemtuzumab ozogamicin to standard induction therapy versus standard induction therapy followed by a second randomization to post-consolidation gemtuzumab ozogamicin versus no additional therapy for previously untreated acute myeloid leukemia. ASH Annual Meeting Abstracts 114:790
Pigneux A, Mahon FX, Moreau-Gaudry F et al (2007) Proteasome inhibition specifically sensitizes leukemic cells to anthracyclin-induced apoptosis through the accumulation of Bim and Bax pro-apoptotic proteins. Cancer Biol Ther 6:603–611
Pineault N, Helgason CD, Lawrence HJ, Humphries RK (2002) Differential expression of Hox, Meis1, and Pbx1 genes in primitive cells throughout murine hematopoietic ontogeny. Exp Hematol 30:49–57
Plass C, Oakes C, Blum W, Marcucci G (2008) Epigenetics in acute myeloid leukemia. Semin Oncol 35:378–387
Raaphorst FM (2003) Self-renewal of hematopoietic and leukemic stem cells: a central role for the Polycomb-group gene Bmi-1. Trends Immunol 24:522–524
Raelson J, Nervi C, Rosenauer A et al (1996) The PML/RAR alpha oncoprotein is a direct molecular target of retinoic acid in acute promyelocytic leukemia cells. Blood 88:2826–2832
Raponi M, Lancet JE, Fan H et al (2008) A 2-gene classifier for predicting response to the farnesyltransferase inhibitor tipifarnib in acute myeloid leukemia. Blood 111:2589–2596
Ravandi F, Estey E, Jones D et al (2009) Effective treatment of acute promyelocytic leukemia with all-trans-retinoic acid, arsenic trioxide, and gemtuzumab ozogamicin. J Clin Oncol 27:504–510
Ravandi F, Kantarjian H, Faderl S et al (2010a) Outcome of patients with FLT3-mutated acute myeloid leukemia in first relapse. Leuk Res 34:752–756
Ravandi F, Cortes JE, Jones D et al (2010b) Phase I/II study of combination therapy with sorafenib, idarubicin, and cytarabine in younger patients with acute myeloid leukemia. J Clin Oncol 28:1856–1862
Rawat VP, Thoene S, Naidu VM et al (2008) Overexpression of CDX2 perturbs HOX gene expression in murine progenitors depending on its N-terminal domain and is closely correlated with deregulated HOX gene expression in human acute myeloid leukemia. Blood 111:309–319

Raza A, Mehdi M, Mumtaz M, Ali F, Lascher S, Galili N (2008) Combination of 5-azacytidine and thalidomide for the treatment of myelodysplastic syndromes and acute myeloid leukemia. Cancer 113:1596–1604

Riccioni R, Senese M, Diverio D et al (2007) M4 and M5 acute myeloid leukaemias display a high sensitivity to Bortezomib-mediated apoptosis. Br J Haematol 139:194–205

Richardson PG, Hideshima T, Mitsiades C, Anderson K (2004) Proteasome inhibition in hematologic malignancies. Ann Med 36:304–314

Richardson PG, Sonneveld P, Schuster MW et al (2005) Bortezomib or high-dose dexamethasone for relapsed multiple myeloma. N Engl J Med 352:2487–2498

Ritter M, Kim TD, Lisske P, Thiede C, Schaich M, Neubauer A (2004) Prognostic significance of N-RAS and K-RAS mutations in 232 patients with acute myeloid leukemia. Haematologica 89:1397–1399

Roberts AW, He S, Ritchie D et al (2010) A phase I study of anti-CD123 monoclonal antibody (mAb) CSL360 targeting leukemia stem cells (LSC) in AML. ASCO Meeting Abstracts;28:e13012

Rosnet O, Buhring HJ, Marchetto S et al (1996) Human FLT3/FLK2 receptor tyrosine kinase is expressed at the surface of normal and malignant hematopoietic cells. Leukemia 10:238–248

Ross DD (2000) Novel mechanisms of drug resistance in leukemia. Leukemia 14:467–473

Ross JJ, Arnason JT, Birnboim HC (1999) Low concentrations of the feverfew component parthenolide inhibit in vitro growth of tumor lines in a cytostatic fashion. Planta Med 65:126–129

Rowinsky EK, Windle JJ, Von Hoff DD (1999) Ras protein farnesyltransferase: a strategic target for anticancer therapeutic development. J Clin Oncol 17:3631–3652

Ruefli AATK, Darcy PK, Smyth MJ, Johnstone RW (2002) P-glycoprotein inhibits caspase-8 activation but not formation of the death inducing signal complex (disc) following Fas ligation. Cell Death Differ 9:1266–1272

Rush LJ, Dai Z, Smiraglia DJ et al (2001) Novel methylation targets in de novo acute myeloid leukemia with prevalence of chromosome 11 loci. Blood 97:3226–3233

Saglio G, Kim DW, Issaragrisil S et al (2010) Nilotinib versus imatinib for newly diagnosed chronic myeloid leukemia. N Engl J Med 362:2251–2259

Saiki JH, McCredie KB, Vietti TJ et al (1978) 5-azacytidine in acute leukemia. Cancer 42:2111–2114

San Miguel JF, Schlag R, Khuageva NK et al (2008) Bortezomib plus melphalan and prednisone for initial treatment of multiple myeloma. N Engl J Med 359:906–917

Santos FP, Qiao W, Cortes JE et al (2009) Prognostic value of FLT3 mutations among different cytogenetic subgroups in acute myeloid leukemia (AML). J Clin Oncol (Meeting Abstracts) 27:7015

Sanz MA, Martín G, González M et al (2004) Risk-adapted treatment of acute promyelocytic leukemia with all-trans-retinoic acid and anthracycline monochemotherapy:a multicenter study by the PETHEMA group. Blood 103:1237–1243

Sauvageau G, Lansdorp PM, Eaves CJ et al (1994) Differential expression of homeobox genes in functionally distinct CD34+ subpopulations of human bone marrow cells. Proc Natl Acad Sci U S A 91:12223–12227

Schnittger S, Schoch C, Dugas M et al (2002) Analysis of FLT3 length mutations in 1003 patients with acute myeloid leukemia: correlation to cytogenetics, FAB subtype, and prognosis in the AMLCG study and usefulness as a marker for the detection of minimal residual disease. Blood 100:59–66

Scholl C, Bansal D, Dohner K et al (2007) The homeobox gene CDX2 is aberrantly expressed in most cases of acute myeloid leukemia and promotes leukemogenesis. J Clin Invest 117:1037–1048

Schroeder T, Saure C, Bruns I et al (2009) Clinical efficacy of sorafenib in patients with Acute Myeloid Leukemia (AML) and activating FLT3-mutations. ASH Annual Meeting Abstracts 114:2057

Seedhouse CH, Das-Gupta EP, Russell NH (2003) Methylation of the hMLH1 promoter and its association with microsatellite instability in acute myeloid leukemia. Leukemia 17:83–88

Servida F, Soligo D, Delia D et al (2005) Sensitivity of human multiple myelomas and myeloid leukemias to the proteasome inhibitor I. Leukemia 19:2324–2331

Sievers EL, Larson RA, Stadtmauer EA et al (2001) Efficacy and safety of gemtuzumab ozogamicin in patients with CD33-positive acute myeloid leukemia in first relapse. J Clin Oncol 19:3244–3254
Silverman LR, Demakos EP, Peterson BL et al (2002) Randomized controlled trial of azacitidine in patients with the myelodysplastic syndrome: a study of the cancer and leukemia group B. J Clin Oncol 20:2429–2440
Simon M, Grandage VL, Linch DC, Khwaja A (2005) Constitutive activation of the Wnt/beta-catenin signalling pathway in acute myeloid leukaemia. Oncogene 24:2410–2420
Smith BD, Levis M, Beran M et al (2004) Single-agent CEP-701, a novel FLT3 inhibitor, shows biologic and clinical activity in patients with relapsed or refractory acute myeloid leukemia. Blood 103:3669–3676
Smyth MJ, Krasovskis E, Sutton VR, Johnstone RW (1998) The drug efflux protein, P-glycoprotein, additionally protects drug-resistant tumor cells from multiple forms of caspase-dependent apoptosis. Proc Natl Acad Sci U S A 95:7024–7029
Solary E, Witz B, Caillot D et al (1996) Combination of quinine as a potential reversing agent with mitoxantrone and cytarabine for the treatment of acute leukemias: a randomized multicenter study. Blood 88:1198–1205
Solary E, Drenou B, Campos L et al (2003) Quinine as a multidrug resistance inhibitor: a phase 3 multicentric randomized study in adult de novo acute myelogenous leukemia. Blood 102:1202–1210
Stapnes C, Doskeland AP, Hatfield K et al (2007) The proteasome inhibitors bortezomib and PR-171 have antiproliferative and proapoptotic effects on primary human acute myeloid leukaemia cells. Br J Haematol 136:814–828
Stasi R, Evangelista ML, Buccisano F, Venditti A, Amadori S (2008) Gemtuzumab ozogamicin in the treatment of acute myeloid leukemia. Cancer Treat Rev 34:49–60
Stirewalt DL, Kopecky KJ, Meshinchi S et al (2001) FLT3, RAS, and TP53 mutations in elderly patients with acute myeloid leukemia. Blood 97:3589–3595
Stirewalt DL, Kopecky KJ, Meshinchi S et al (2006) Size of FLT3 internal tandem duplication has prognostic significance in patients with acute myeloid leukemia. Blood 107:3724–3726
Stone RM, DeAngelo DJ, Klimek V et al (2005) Patients with acute myeloid leukemia and an activating mutation in FLT3 respond to a small-molecule FLT3 tyrosine kinase inhibitor, PKC412. Blood 105:54–60
Stone RM, Fischer T, Paquette R et al (2009) A phase 1b study of midostaurin (PKC412) in combination with daunorubicin and cytarabine induction and high-dose cytarabine consolidation in patients under age 61 with newly diagnosed de novo acute myeloid leukemia: overall survival of patients whose blasts have FLT3 mutations is similar to those with wild-type FLT3. ASH Annual Meeting Abstracts 114:634
Su Y, Li SY, Ghosh S, Ortiz J, Hogge DE, Frankel AE (2010) Characterization of variant diphtheria toxin-interleukin-3 fusion protein, DTIL3K116W, for phase I clinical trials. Biologicals 38:144–149
Tallman MS, Andersen JW, Schiffer CA et al (1997) All-trans-retinoic acid in acute promyelocytic leukemia. N Eng J Med 337:1021–1028
Tamburini J, Elie C, Bardet V et al (2007) Constitutive phosphoinositide 3-kinase/Akt activation represents a favorable prognostic factor in de novo acute myelogenous leukemia patients. Blood 110:1025–1028
Taussig DC, Pearce DJ, Simpson C et al (2005) Hematopoietic stem cells express multiple myeloid markers: implications for the origin and targeted therapy of acute myeloid leukemia. Blood 106:4086–4092
Tazzari PL, Cappellini A, Ricci F et al (2007) Multidrug resistance-associated protein 1 expression is under the control of the phosphoinositide 3 kinase/Akt signal transduction network in human acute myelogenous leukemia blasts. Leukemia 21:427–438
Testa U, Riccioni R, Biffoni M et al (2005) Diphtheria toxin fused to variant human interleukin-3 induces cytotoxicity of blasts from patients with acute myeloid leukemia according to the level of interleukin-3 receptor expression. Blood 106:2527–2529

Tetsu O, McCormick F (1999) Beta-catenin regulates expression of cyclin D1 in colon carcinoma cells. Nature 398:422–426

Thévenod F, Friedmann JM, Katsen AD, Hauser IA (2000) Up-regulation of Multidrug Resistance P-glycoprotein via Nuclear Factor-Î°B Activation Protects Kidney Proximal Tubule Cells from Cadmium- and Reactive Oxygen Species-induced Apoptosis. J Biol Chem 275:1887–1896

Thiede C, Steudel C, Mohr B et al (2002) Analysis of FLT3-activating mutations in 979 patients with acute myelogenous leukemia: association with FAB subtypes and identification of subgroups with poor prognosis. Blood 99:4326–4335

Thomas X, Campos L, Mounier C et al (2005) Expression of heat-shock proteins is associated with major adverse prognostic factors in acute myeloid leukemia. Leuk Res 29:1049–1058

van der Holt B, Lowenberg B, Burnett AK et al (2005) The value of the MDR1 reversal agent PSC-833 in addition to daunorubicin and cytarabine in the treatment of elderly patients with previously untreated acute myeloid leukemia (AML), in relation to MDR1 status at diagnosis. Blood 106:2646–2654

van Rhenen A, Feller N, Kelder A et al (2005) High stem cell frequency in acute myeloid leukemia at diagnosis predicts high minimal residual disease and poor survival. Clin Cancer Res 11:6520–6527

Voso MT, Scardocci A, Guidi F et al (2004) Aberrant methylation of DAP-kinase in therapy-related acute myeloid leukemia and myelodysplastic syndromes. Blood 103:698–700

Wang Y, Krivtsov AV, Sinha AU et al (2010) The Wnt/beta-catenin pathway is required for the development of leukemia stem cells in AML. Science 327:1650–1653

Wang Y, Liu Y, Malek Sami N, Zheng P, Liu Y (2011) Targeting HIF1[alpha] eliminates cancer stem cells in hematological malignancies. Cell Stem Cell 8:399–411

Wattel E, Solary E, Hecquet B et al (1998) Quinine improves the results of intensive chemotherapy in myelodysplastic syndromes expressing P glycoprotein: results of a randomized study. Br J Haematol 102:1015–1024

Weisberg E, Boulton C, Kelly LM et al (2002) Inhibition of mutant FLT3 receptors in leukemia cells by the small molecule tyrosine kinase inhibitor PKC412. Cancer Cell 1:433–443

Wen J, You KR, Lee SY, Song CH, Kim DG (2002) Oxidative stress-mediated apoptosis. The anticancer effect of the sesquiterpene lactone parthenolide. J Biol Chem 277:38954–38964

Xu Q, Simpson SE, Scialla TJ, Bagg A, Carroll M (2003) Survival of acute myeloid leukemia cells requires PI3 kinase activation. Blood 102:972–980

Xu Q, Thompson JE, Carroll M (2005) mTOR regulates cell survival after etoposide treatment in primary AML cells. Blood 106:4261–4268

Yalcintepe L, Frankel AE, Hogge DE (2006) Expression of interleukin-3 receptor subunits on defined subpopulations of acute myeloid leukemia blasts predicts the cytotoxicity of diphtheria toxin interleukin-3 fusion protein against malignant progenitors that engraft in immunodeficient mice. Blood 108:3530–3537

Yamamoto Y, Kiyoi H, Nakano Y et al (2001) Activating mutation of D835 within the activation loop of FLT3 in human hematologic malignancies. Blood 97:2434–2439

Yanada M, Matsuo K, Emi N, Naoe T (2005a) Efficacy of allogeneic hematopoietic stem cell transplantation depends on cytogenetic risk for acute myeloid leukemia in first disease remission: a metaanalysis. Cancer 103:1652–1658

Yanada M, Matsuo K, Suzuki T, Kiyoi H, Naoe T (2005b) Prognostic significance of FLT3 internal tandem duplication and tyrosine kinase domain mutations for acute myeloid leukemia: a meta-analysis. Leukemia 19:1345–1349

Ysebaert L, Chicanne G, Demur C et al (2006) Expression of beta-catenin by acute myeloid leukemia cells predicts enhanced clonogenic capacities and poor prognosis. Leukemia 20:1211–1216

Zarrinkar PP, Gunawardane RN, Cramer MD et al (2009) AC220 is a uniquely potent and selective inhibitor of FLT3 for the treatment of acute myeloid leukemia (AML). Blood 114:2984–2992

Zhang S, Ong CN, Shen HM (2004) Critical roles of intracellular thiols and calcium in parthenolide-induced apoptosis in human colorectal cancer cells. Cancer Lett 208:143–153

Zhang W, Konopleva M, Shi YX et al (2008) Mutant FLT3: a direct target of sorafenib in acute myelogenous leukemia. J Natl Cancer Inst 100:184–198
Zhao S, Konopleva M, Cabreira-Hansen M et al (2004) Inhibition of phosphatidylinositol 3-kinase dephosphorylates BAD and promotes apoptosis in myeloid leukemias. Leukemia 18:267–275
Zhao C, Chen A, Jamieson CH et al (2009) Hedgehog signalling is essential for maintenance of cancer stem cells in myeloid leukaemia. Nature 458:776–779

Novel Targeted Therapeutics for Peripheral T-Cell Lymphoma

Owen O. Connor, Salvia Jain, and Jasmine Zain

Introduction

The peripheral T-cell lymphomas (PTCL) are a group of aggressive mature T-cell and natural killer cell (NK) neoplasm's that present with great morphological and molecular heterogeneity (Anderson et al. 1998). They are relatively rare diseases, constituting <15% of all NHL's (Groves et al. 2000). The present 2008 WHO classification recognizes over 20 sub-types of mature T-cell and NK-cell malignancies (WHO 2008). These sub-types are listed in Table 1 along with the annual incidence rates and 3 year survival rates as reported in the SEER database (Seer Cancer Statistic 1975). The most common histologies include: PTCL, not otherwise specified (PTCL-NOS); anaplastic large cell lymphoma (ALCL) and angioimmunoblastic T-cell lymphoma (AITL). Similar to the B-cell neoplasm's, the T-cell lymphomas can be broadly classified as aggressive or indolent malignancies. The most aggressive histologies include PTCL-NOS, hepatosplenic T-cell lymphoma, the gamma/delta T-cell malignancies and extranodal NK/T-cell nasal type lymphoma. PTCL-NOS is often viewed as a "wastebasket" category for those diseases that do not fit cleanly into the other sub-types. Most cases of PTCL lack distinct genetic or biological alterations and prognostic models have largely relied on clinical features or simple biological factors such as proliferation. Despite a relatively poor understanding of the molecular pathogenesis of these diseases, significant progress has

O.O. Connor, M.D., Ph.D. • S. Jain, M.D.
Division of Hematology and Medical Oncology, NYU Cancer Institute,
NYU Langone Medical Center, New York, NY, USA

J. Zain, M.D. (✉)
Department of Medicine, Division of Hematology and medical Oncology,
NYU Cancer Insitue, NYU Langone Medical Center,
7th Floor, 160 East 34th Street, New York, NY 10016, USA
e-mail: jasmine.zain@numc.org

J. Tao and E. Sotomayor (eds.), *Hematologic Cancers: From Molecular Pathobiology to Targeted Therapeutics*, Cancer Growth and Progression 14,
DOI 10.1007/978-94-007-5028-9_15, © Springer Science+Business Media Dordrecht 2012

Table 1 WHO 2008: the mature T-cell and NK-cell neoplasm's classification along with incidence and survival rates as reported in the SEER data-base

Subtype	Incidence rates: all ages, both sexes (01–07)	Incidence rates: all ages, males (01–07)	Incidence rates: all ages, females (01–07)	3 year relative survival: all ages, both sexes (01–06)	3 year relative survival: all ages, males (01–06)	3 year relative survival: all ages, females (01–06)
T-cell prolymphocytic leukemia	0	0	0	29.9	33.2	26.5
T-cell LGL leukemia	0	0	0	76.7	71.8	81
Adult T-cell leukemia/lymphoma	0	0	0	31.6	34.8	27.8
Extranodal NK/T-cell lymphoma, nasal type	0.1	0.1	0	35.4	33.1	39.3
Enteropathy T-cell lymphoma	0	0	0	21.2	16.2	–[a]
Hepatosplenic T-cell lymphoma	0	0	–[a]	19.6	–[a]	–[a]
Panniculitis-like T-cell lymphoma	0	0	0	60.6	–[a]	65.2
Mycosis fungoides	0.5	0.6	0.4	93.3	93.5	93.2
Sezary syndrome	0	0	0	71.3	–[a]	–[a]
Primary cutaneous ALCL	0.1	0.1	0.1	90.4	87.8	93.1
PTCL NOS	1.1	0.5	0.3	42.8	44.2	40.5
AITL lymphoma	0.1	0.1	0.1	46.2	44.8	47.8
ALCL Alk +ve	0.2	0.3	0.1	58.6	57.6	60.2
ALCL Alk –ve	0.1	0.1	0.1	30	28.4	32.5

[a] (-) Incidence/survival rate not provided in the SEER data-base due to <25 cases reported

been made in the understanding of many PTCL entities. For example, ALK positive anaplastic large cell lymphoma (ALCL) is considered a distinct disease entity which is distinguished from the provisional entity of ALK negative ALCL, due to the distinct molecular pathogenesis, relatively younger age group in which it presents and better prognosis. Based on the recent publications from the 'The International T-cell Lymphoma Project', an international effort of over 22 centers worldwide which collected data from over 1,314 cases of T-cell and NK -cell lymphoma, we now recognize that PTCL-NOS is a distinct entity from ALK negative ALCL as the former is associated with a markedly inferior prognosis (Vose et al. 2008) Table 2.

Advanced disease stage, high International Prognostic Index at presentation (Gallamini et al. 2004) and inherent chemo-resistance (Jillella et al. 2000) contribute to dismal outcome in most sub-types of T-cell lymphoma. Long term survival at 5 years remains at 10–30% with traditional combination chemotherapy regimens compared with 65–70% in aggressive B-cell NHL's (Coiffier et al. 1990; Gisselbrecht et al. 1998). Increasing intensity of induction chemotherapy has not been shown to improve survival as effectiveness may be offset by increased toxicity although autologous stem cell transplantation in first remission is likely to be beneficial for those patients who attain a complete remission (CR) with induction chemotherapy.

Present Treatment Strategies for Patients with Peripheral T-Cell Lymphoma

Presently there is no standard therapy for patients with PTCL which has produced a great deal of variability in clinical practice for management of both the upfront and relapsed settings. This is primarily due to a paucity of multi-center clinical trials with sufficient number of patients which is no doubt due to the rarity of the condition, morphologic heterogeneity and numerous diagnostic challenges. Nonetheless, most types of PTCL and NK-cell lymphomas are initially treated with an anthracycline containing regimen such as CHOP which usually produces complete response rates of 50–70%. The median progression free survival (PFS) for patients with PTCL following standard CHOP- based chemotherapy is only 12–14 months. However for patients with ALK pos ALCL and PTCL-NOS with low/low-intermediate IPI scores and localized disease, 70–80% and 25–30% of patients respectively, remained disease free at 5 years. Because of the poor benefit from induction chemotherapy alone, several groups have explored the role of high-dose chemotherapy followed by autologous stem cell transplantation (ASCT) in first remission. Autologous stem cell transplantation seems to offer long-term disease control in over 50–70% of cases if performed in CR or near CR in first remission (Philip et al. 1995). For patients not eligible for ASCT due to progressive or refractory disease or extensive bone marrow involvement at the end of induction chemotherapy an allogeneic stem cell transplantation may be able to offer long term disease control in over 50% of cases through a graft versus lymphoma effect (Kim et al. 2006; Le Gouill et al. 2008; Corradini et al. 2004).

Table 2 Novel classes of drugs and their response in patients with PTCL and CTCL

Agent	Target	ORR (%)	CR (%)	PR (%)	Median DOR (months)	No. of patients	Citation
Pralatrexate	RFC-1	29	10	17	9.4	115	O'Connor et al. (2009a)
Vorinostat	Histone & non-H proteins	24.4	None	24	3.7	33	Duvic et al. (2007a)
Romidepsin	Histone & non-H proteins	33	11	22	9	46	Piekarz et al. (2009b)
Belinostat	Histone & non-H proteins	17	17	None	2.5	12	Brad et al. (2009)
Panobiostat	Histone & non-H proteins	60	20	40	5.9	10	Ellis et al. (2008)
Bortezomib	Various	61.5 (with CHOP)	61.5	None	Unknown	13	Lee et al. (2008)
Gemcitabine	Purine synthesis	51	30	25	34	20	Zinzani et al. (2010)
Lenalidamide	Various	34	None	30	3.2	24	Dueck et al. (2009)
Everolimus	mTOR	47	5	42	7.2	19	Johnston et al. (2010)
Alemtuzumab	CD52	36	21.4	14.2	4	14	Enblad et al. (2004)

Targeted Therapeutic Agents in T-Cell Lymphoma

While the lack of insight into the biology of T-cell lymphomas has hindered the development of true targeted therapies, there is now an abundance of new drugs which have shown potentially significant activity either alone or in combination with conventional agents.

Anti-folate

Pralatrexate (Fotolyn)

Pralatrexate (PDX, 10-propargyl 10-deazaaminopterin), is a novel antifolate designed to have high affinity for the reduced folate carrier, RFC. It became the first drug ever approved for the treatment of relapsed or refractory PTCL in September 2009. Pralatrexate has been rationally designed to have a higher affinity (approximately tenfold greater than methotrexate) for the one carbon- reduced folate carrier (RFC-1) (DeGraw et al. 1993). Pralatrexate is also a more effective substrate for polyglutamation by FPGS so it is more efficiently polyglutamated as compared to methotrexate. Longer chain polyglutamates have been demonstrated after only a 3-h exposure compared to other analogs (Wang et al. 2003), which subsequently leads to higher intracellular accumulation of the drug compared to methotrexate which also enhances its affinity for dihydrofolate reductase (DHFR). Initial cytotoxicity assays revealed that pralatrexate was more potent compared to methotrexate across a variety of cell lines including human breast, non small cell lung cancer, mesothelioma and human lymphoma cell lines (Izbicka and Diaz 2009). Preclinical animal data using xenograft models of lymphoma have confirmed the superior efficacy of pralatrexate compared to methotrexate in inducing tumor responses and complete remissions (Sirotnak et al. 1998). Early clinical data established by O'Connor et al. demonstrated that pralatrexate exhibited marked activity in patients with relapsed or refractory T-cell lymphoma (Krug et al. 2000; O'Connor et al. 2009b). Population pharmacokinetic (PK) and pharmacodynamic (PD) modeling based on pretreatment levels of homocysteine and methylmalonic acid revealed that elevated nutritional covariates (especially MMA) and higher area under the curve of pralatrexate exposure (AUC) was associated with a significantly higher incidence of stomatitits. These data also demonstrated that normalization of the pretreatment homocysteine and methylmalonic levels with folic acid and Vitamin B12 supplementation markedly reduced the severity and incidence of mucositis in patients receiving pralatrexate (Azzoli et al. 2007). Based on the encouraging results seen in the single institution single arm phase II experience, pralatrexate was studied in an international multicenter registration directed clinical trial called PROPEL.

PROPEL (O'Connor et al. 2009a) (Pralatrexate in Relapsed or Refractory Peripheral T-cell Lymphoma) is the largest prospective clinical trial ever conducted in patients with relapsed or refractory PTCL. Most aggressive histologies of T-cell lymphoma were eligible including PTCL, transformed MF, HTLV-1 ATLL, angio-immunoblastic, gamma-delta T-cell lymphoma and enteropathy associated T-cell lymphoma. One-hundred and fifteen patients were enrolled between August 2006 and April 2008 of which 109 were evaluable. This was a heavily pretreated population with a median of 3 (1–12) prior treatment regimens. Importantly, in the population eligible for efficacy, 24% (n=26) of patients never demonstrated any response to any prior therapy, while 63% (n=69) did not have evidence of response to their most recent prior therapy. Sixteen percent (n=18) of patients relapsed following ASCT and the median time from diagnosis to study entry was 15.6 months. The treatment schedule consisted of pralatrexate given at 30 mg/m^2 weekly for 6 weeks followed by 1 week of rest in a 7 week cycle. In the evaluable patient population (n =109), the ORR was 29% (95% CI: 21–39%) as assessed by independent central review and 39% when assessed by investigator. Twelve (11%) patients achieved CR/CRu, 20 (18%) achieved PR, and 21 (19%) experienced stable disease (SD). Of the 69 patients who did not have any evidence of response to their most recent prior therapy, 17 (25%) responded to pralatrexate. Of the 26 patients who did not have evidence of response to any prior conventional therapy, 5 (19%) responded to pralatrexate. When IWC was supplemented with PET scans, the response rate was 26% (n=28), (14% CR and 12% PR). The majority of responding patients attained response quickly, 63% of all responses occurred within the first cycle of pralatrexate, but responses were observed as late as cycle 7. Interestingly the median duration of response was 10.1 months while the median PFS and OS were 3.5 months and 14.5 months respectively. Median follow-up time for all patients still alive at the time of the analysis was 18 months. Overall the drug was well tolerated with the most common AEs being mucositis, nausea, thrombocytopenia, and fatigue. The most common grade 3 or 4 AEs were thrombocytopenia, mucositis, neutropenia, and anemia.

Recently Marchi et al. (2010) reported that pralatrexate is synergistic with bortezomib in various T-cell lymphoma cell lines and in a severe combined immunodeficient-beige xenograft mouse model of transformed cutaneous T-cell lymphoma. Cytotoxicity studies showed that the combination was not more toxic than either single agent thereby suggesting that this could represent a novel platform for treatment of T-cell malignancies. Pralatrexate is also being studied in CTCL as a single agent (Horwitz et al. 2009a) and in combination with other therapies including gemcitabine (Horwitz et al. 2009b) and HDAC inhibitors.

Histone Deacetylase Inhibitors (HDACi)

Histone deacetylase inhibitors (HDACis) are epigenetic agents found to be active in the treatment of T-cell lymphomas. Two agents of this class, vorinostat (Mann et al. 2007) (Zolinza) and romidepsin (Demierre et al. 2009) (Istodax) are approved for the treatment of relapsed or refractory CTCL in the US.

HDACis work through a myriad of different mechanisms, including: (1) upregulation of cell dependent kinase (cdk) inhibitors like p21/p27 and down regulation of CyclinD$_1$; (2) acetylation of non- histone proteins including STAT-3, RelA/p65; p53, HIF-1alpha, Bcl-6, Hsp 90 in a way that modulate their activity and thus their ability to regulate cell growth and survival; and (3) direct activation of apoptotic pathways by affecting the balance between the antiapoptotoc proteins like Bcl-2 and the proapapototic proteins like Bax and Bak (Bhalla 2005; Piekarz and Bates 2009). Despite these many pleiotropic effects, it has been difficult to assign a precise mechanistic basis to any one or more of these drugs in any particular tumor type, let alone T-cell lymphoma. It is peculiar that these drugs appear to have a class effect in T-cell lymphomas in contrast to other sub-types of lymphoproliferative neoplasm's. Gene expression profiling on paired tissue samples and studies of select biomarkers including gene activation with HDACis has shown that up to 5–10% of the genome can be affected by HDAC inhibitors. In one study the genes that were consistently affected included genes affecting cell cycle (CCNDI, IGFI) apoptosis (septin10, TEF, SORBBS2), angiogenesis (GUCY1A1, ANGPT1) and immune modulation (LAIR1). A brief description of the various HDACis currently in clinical use or trials for T cell malignancies are listed below.

Suberoylanilide Hydroxamic Acid (Vorinostat)

Vorinostat was the first HDAC inhibitor approved by the U.S. FDA for the treatment of cancer. Vorinostat was approved in October 2006 for the treatment of patients with relapsed or refractory cutaneous T-cell lymphoma (CTCL). Based on the original phase 1 study (O'Connor et al.; Kelly et al.), the maximum tolerated dose of vorinostat is 400 mg (PO) per day. Duvic et al. (2007a) conducted a phase II trial of 33 advanced, heavily pre-treated CTCL patients who received three different dosages and schedules of vorinostat including 400 mg daily, 300 mg twice daily for 3 days followed by 4 days of rest and 300 mg twice daily for 14 days with a week of rest followed by 200 mg twice daily. Treatment was continued until the patients showed signs of progression or toxicity. Strict criteria based on Severity Weighted Assessment Tool (SWAT) for skin care were used to determine responses. The overall response rate was 24.4% with eight patients having partial remission (PR) including four patients with Sezary Syndrome (SS). More importantly, 14 of the 33 patients (42%) reported significant relief from pruritus. The median time to response was 11.9 weeks, while the median overall response duration was 15.1 weeks, thus confirming that the 400 mg per day dosage was optimal for this patient population. Subsequently the registration directed phase II study of 74 patients reported by Olsen et al. (2007) in CTCL showed an overall response rate of 29.7%, with one patient CR. The median time to objective response was 56 days, though some patients took up to 6 months to respond. Vorinostat is a very promising agent in the treatment of T-cell lymphoma. As HDACis can modulate a variety of survival proteins, combinations of HDACis with other targeted agents and conventional chemotherapy should be investigated.

Romidepsin

Romidepsin (Depsipeptide, FK228) is cyclic peptide originally isolated from the broth culture of *Chromobacterium violaceum* is presently approved for the treatment of CTCL patients who have failed at least 1 prior systemic therapy. Two phase II trials of romidepsin in patients with relapsed PTCL have shown impressive activity. The NCI sponsored study (Piekarz et al. 2009b) with 46 evaluable patients showed an overall response rate of 33% with 5 (11%) CR's assessed by Cheson Criteria. Patients who received than 2 cycles of therapy had a response rate of 44%. The median duration of response was 9 months with a time to progression of 12 months. Responses were seen across all histologic subtypes of PTCL, though again, AITL appeared to be associated with a slightly lower overall response rate. Romidepsin was administered at a dose of 14 mg/m^2 on days 1,8,15 of a 28 day cycle. The results of a registration directed phase 2 study in patients with relapsed or refractory PTCL are still awaited. The most common side effects associated with the drug appear to be nausea, fatigue, anorexia, anemia and leukopeina. Serious drug related adverse effects seen in approximately 2% of patients included supraventricular tachycardia, infection and neutropenia. In patients with CTCL, pooled analysis from 2 phase II trials established an overall response rate of 41% with a CR rate of 7% with duration of response of 14.9 months. The response rate was 58% in patients with SS. Relief of pruritus was seen in over 60% of the patients. These results are quite exciting and hopefully further investigation will establish its role in patients with PTCL either alone or in combination with other agents.

Belinostat

Belinostat (PXD101) is a hydroxamate pan-histone deacetylase inhibitor which demonstrates broad anti-neoplastic activity in vitro and in vivo. A multicenter phase II trial (Brad et al. 2009) evaluated the activity of Belinostat in patients with PTCL or CTCL who had failed 1 prior systemic therapy. Belinostat (1,000 mg/m^2) was administered as a 30-min IV infusion on days 1–5 of a 3 week cycle. In the PTCL arm, 12 patients were enrolled; 7 with PTCL-NOS, 2 with anaplastic large cell lymphoma, 2 with AITL and 1 with NK/T-cell lymphoma and were treated for a median of 2 cycles (range 1–8). Of the 11 evaluable patients: 2 achieved a CR and 5 patients experienced stable disease (SD). The two CR (11+ and 15+ wks) were both in patients with PTCL-NOS. A third patient with PTCL-NOS had stable disease lasting for 14 weeks and discontinued treatment due to an adverse event (AE). Both patients with ALCL had durable SD (20+ and 14 weeks). In the CTCL arm, 16 patients were enrolled (including 9 with mycosis fungoides (MF), 5 with SS and 2 with primary cutaneous ALCL) and treated for a median of 2 cycles (range 1–6). Four responses were observed (25%): 1 CR (ALCL) and 3 PR (Mycosis Fungoides and Sezary Syndrome). The median duration of the response was 10 weeks (range 7–39+ weeks). Four additional patients had a 25–50% decrease in their SWAT score. Improvement in pruritus score by three on a visual analog scale was reported in 5/6 patients with significant pruritus at baseline. Overall in both study arms the drug was well-tolerated and most adverse events were only grade 1/2 (nausea, fatigue, constipation, diarrhea, and vomiting). Grade3 AE attributed to the study drug was reported in seven patients and included peripheral edema, apraxia, adynamic ileus, pruritus, rash, and infections. Only 1 treatment related grade 4 AE was noted which was thrombocytopenia. One patient with progressive PTCL died 6 days after their last belinostat dose with ventricular fibrillation and drug causality could not be excluded.

Panobinostat

Oral panobinostat, also known as LBH589, is a pan-HDACI belonging to the hydroxamic acid group that has shown remarkable activity in patients with CTCL in a phase I/II trial (Ellis et al. 2008). The dose administered was 20 mg given every three times per week. Six out of 10 heavily pretreated patients showed a CR and 4 showed PR. Duvic et al. (2008) conducted an open label multicenter phase II trial with the primary objective of establishing the efficacy and safety of panobinostat for patients with relapsed/refractory CTCL with Stage IB–IVA mycosis fungoides or sezary syndrome. Ninety five patients (70 with MF and 25 with SS) were enrolled who had received 3–4 median prior treatment regimens and mostly were ≥ Stage IIB at study entry. They received 1–17+ (median = 3) treatment cycles of panobinostat. In grade 1 patients, 11 out of 62 pts had confirmed skin responses by SWAT, including two complete skin responses. In grade 2, 4 out of 33 pts had confirmed skin and CT scan responses. Common adverse events (>20%; all grades regardless of causality) included diarrhea, thrombocytopenia, nausea etc. Recently Dickinson et al. (2009) showed activity of panobinostat in patients with refractory Hodgkin's Lymphoma (HL). Thirteen HL patients were treated with escalating doses of panobinostat in a phase IA/II multicentre study. A computed tomography partial response was achieved in 5 out of 13(38%), and a metabolic response by ^{18}F-fluoro-2-deoxy-d-glucose positron emission tomography scanning in 7 out of 12 (58%) evaluable patients. Thus panobinostat has demonstrated efficacy with good safety profile in patients with T-cell lymphoma. Panobinostat-based combination studies are currently being conducted, including a phase I/II trial of panobinostat and everolimus in patients with relapsed Hodgkin's disease and NHL.

Proteosome Inhibitors

Bortezomib (Velcade)

Bortezomib is the first agent of this class to be approved for the treatment of cancer now with indications in multiple myeloma and mantle cell lymphoma [O'Connor et al.; Goy et al.; Richardson et al.]. The ubiquitin-proteosome pathway is critical for maintaining the balance of intracellular proteins which it does by eliminating proteins no longer required by the cell. Typically these proteins are involved in cell cycle regulation, survival and apoptosis. Bortezomib targets the catalytic 20S core of the proteosome with a multitude of downstream effects including inhibition of cellular protesosome, accumulation of the cell cycle-dependent kinase inhibitors such as p27/p21, activation of p53, inhibition of NF-kB and accumulation of pro-apoptotic proteins like Noxa which inactivates anti-apoptotic proteins like McL-1. The net effect of proteasome inhibition involves induction of cell death via the intrinsic apoptotic pathway. Zinzani et al. (2007) conducted a phase II trial of bortezomib (1.3 mg/m^2 on days 1,4, 8, and 11 of a 21-day cycle) in 15 patients with relapsed/refractory T-cell lymphoma, including PTCL (2 patients) or CTCL (10 patients) with a reported overall response rate (ORR) of 67% (with 2 CR's and 6 PR's). Of the two patients with PTCL, one attained a CR. The responses were durable, lasting on average 7–14 months. A phase I study (Lee et al. 2008) evaluated the use of standard dose CHOP (cyclophosphamide, doxorubicin, vincristine, prednisone) plus bortezomib in 13 patients with advanced, aggressive T-cell or NK/T-cell lymphoma. No dose limiting toxicities were observed up to the maximal dose of bortezomib of 1.6 mg/m^2. The CR rate was 62%. No data were provided for progression free or overall survival. It has also been demonstrated *in vitro* that expression of PRDM1beta leads to resistance to chemotherapeutic agents in T-cell lymphomas which could be down-regulated by the proteasome inhibitor bortezomib, thereby providing another rationale for therapeutic interference by bortezomib.

PR-171 (Carfilzomib)

PR-171 is a novel epoxyketone-based irreversible proteosome inhibitor that is currently in clinical development. Compared to bortezomib, PR-171 exhibits equal potency but greater selectivity for the chymotrypsin-like activity of the proteosome and is considered to be an irreversible inhibitor of the proteasome.

In cell culture PR-171 is more cytotoxic than bortezomib with demonstrable accumulation of proteosome substrates and induction of cell cycle arrest and/or apoptosis (Susan et al. 2007). Recent phase I studies have explored a number of different schedules, including a weekly × 5 day schedule, and a Day 1 and 2 schedule (O'Connor et al. 2010). While there is essentially no experience with carfilzomib in T-cell lymphoma, the lack of apparent neurotoxicity, coupled with the activity already seen with bortezomib, may make it a promising agent to explore in this disease setting.

Nucleoside Analogs

Gemcitabine

Gemcitabine is an analog of deoxycitadine that is deaminated and inactivated by deoxycytidine deaminase which then gets incorporated into DNA resulting in chain termination and inhibition of DNA synthesis. In addition, it also inhibits ribonucleoside reductase, an enzyme that is required for DNA synthesis. Gemcitabine has been studied as a single agent in patients with relapsed or refractory PTCL by Zinzani et al. (2010) demonstrating an overall response rate was 51% (20 of 39 patients); complete response (CR) and partial response (PR) rates were 23% (9 of 39 patients) and 28% (11 of 39 patients), respectively. Patients with MF had a CR rate of 16% and a PR rate of 32% compared with a CR rate of 30% and a PR rate of 25% of PTCLU patients. Duration of response was reported to be atleast 15 months when the data was published. Marchi et al. (2005) have studied single agent gemcitabine in patients with CTCL (n=32) and reported a response rate of 75% with 22% CR's and a median duration of response lasting 10 months. It is largely these data that have supported a role for the use of gemcitabine in T-cell malignancies either alone or incombination with other agents.

Forodesine

Forodesine is a transition state purine nucleoside phophoryalse inhibitor (PNP). Inhibiton of this enzyme leads to apoptosis and cell cycle arrest in lymphocytes. Forodesine has shown in vitro activity against CLL, many subtypes of NHL, T-ALL and B-ALL as well as synergy with other anti-lymphoma targeted agents. Furman et al. (2006) treated 34 patients with T-ALL using oral forodesine and demonstrated a 32.4% response rate (11 patients; 7 [20.6%] CR, 0 CRi [0%] and 4 [11.8%] PR). Two patients with CR proceeded to hematopoietic stem cell transplantation following remission. The restoration of normal hematopoiesis observed in patients while receiving therapy indicates a specificity of forodesine for leukemic cell populations. A total of 18 patients were dose escalated to 90 mg/m^2 of forodesine after cycle 2 due to a lack of response. None of these patients demonstrated an improved response with dose escalation suggesting that the 40 mg/m^2 should become the recommended phase II dose in T-ALL. An open-label dose-escalation study of forodesine, 40–320 mg/m^2 every day for 4 weeks, was undertaken by Duvic et al. (2007b) to evaluate the safety and PK profile of oral forodesine. This was followed by an investigation of the expansion of the optimal biologic dose (dose with maximum PNP inhibition and elevation of plasma deoxyguanosine levels) to assess efficacy in patients with refractory CTCL. Overall, 37 patients were treated, including 14 patients in the dose-escalation portion of the study. The overall response rate was 53.6% (15 out of 28 patients; 2 [7.1%] with a CR and 13 [46.4%] with a PR). The most common adverse events classified as grade2 or less, without regard to causality, were nausea (30%), dizziness (22%), pruritus (22%), fatigue (19%), headache (19%), peripheral edema (19%), and pyrexia (16%) thereby suggesting an encouraging role of forodesine in CTCL. Forodesine definitely shows encouraging response in T-cell lymphoma and is overall well tolerated.

Clofarabine

Clofarabine blocks DNA synthesis via inhibition of ribonucleotide reductase and DNA polymerases and induces apoptosis. Clofarabine has been shown to have increased stability relative to fludarabine or cladribine because it exhibits increased resistance to deamination and phosphorolysis. Clofarabine is approved for the treatment of pediatric ALL. Though early, results of a phase I/ II trial of this agent have been reported by Horwitz et al. (2008). Twenty nine patients with refractory or relapsed T-cell or NK-cell lymphomas were treated with a dose of 4 mg/m^2 IV for 3 consecutive days every 3 weeks. Seventeen percent (5/29) of all patients and 21% (5/24) of evaluable patients responded. Two patients had a CR, 3 had a PR, 7 patients had stable disease and 12 patients progressed. Two out of 4 pts with AITL had a CR. Although early there may be a modest signal response. Future studies exploring combinations with other nucleoside analogs or even antifolates have a strong rationale, and are likely to be pursued in the near future.

Nelarabine

Nelarabine (compound 506U78), a novel purine nucleoside, is a soluble pro-drug of 9-beta-D-arabinofuranosylguanine (ara-G). Nelarabine is rapidly demethoxylated in blood by adenosine deaminase to ara-G. Pre-clinical and clinical studies have demonstrated the selective cytotoxicity of ara-G in T-lineage derived cells. CALGB Protocol 59901 (Czuczman et al. 2007) was a phase II study of nelarabine in patients with systemically untreated cutaneous T-cell lymphoma (CTCL) or relapsed/refractory systemic peripheral T-cell lymphoma (STCL). The objectives were to determine response rate, remission duration and safety profile associated with nelarabine given at 1.5 gm/m^2 per day on days 1, 3 and 5 as an IV infusion every 21 days for a minimum of two cycles with the intent to continue up to two cycles beyond CR for a maximum of eight cycles. Nineteen patients were enrolled in the study: Eleven CTCL and eight STCL patients. Grade 3 or 4 adverse events were documented in 50% and 28% of patients respectively. In particular, 33% of patients experienced grade3 or 4 neurologic toxicities. There were two partial remissions lasting 3 months and 5.5 months each. Median event-free survival was 1.2 months and median overall survival was 3 months. Due to lack of efficacy and excessive toxicity, nelarabine is not recommended as monotherapy in adult patients with CTCL and STCL at this dose schedule. Preliminary results from another phase II trial of nelarabine (Thompson et al. 2005) (1.5 gm/m^2 on days 1, 3, and 5 of a 28-day cycle) in refractory B- or T-cell non-Hodgkin lymphoma reported an overall response rate of 44% (CR 22%) with a median time to progression of 8 months. The principle side effects were reversible neurotoxicity and myelosuppression. This group of investigators suggested that this agent was worthy of further study in PTCL. However at this moment its role in treatment of PTCL needs to be defined and cannot be routinely recommended.

Immunomodulatory Drugs

Lenalidomide (Revlimid)

Lenalidomide (CC-5013) a second generation analog of thalidomide, is an immunomodualtory agent (IMiD) with biological effects that include activation of natural killer (NK) cells and T cells, modulation of various cytokines such as TNF-alpha, Interleukin-12 and Interferon-gamma in tumor microenvironment and inhibition of angiogenesis. The IMiDs have also shown to have direct antitumor activity (Lentzsch et al. 2003; Richardson et al. 2002). A phase II trial of lenalidomide in PTCL was reported by Dueck et al. (2009). This study to date has enrolled 24 patients with mature T cell lymphomas. The ORR was 34% with no CRs. Major side effects included dose related myelosuppression fatigue, pruritis and rash. Deep vein thrombosis is rarely reported in trials where lenalidomide is used as a monotherapy, but the incidence is much higher when combined with dexamethasone. While lenalidomide may not have significant enough single agent activity to be useful in aggressive chemotherapy resistant diseases like PTCL, it is possible that such agents may have the greatest utility as a maintenance therapy for these patients. Such studies are now ongoing in mantle cell lymphoma, though no such data exists at present in T-cell lymphomas using this strategy.

Mammalian Target of Rapamycin (mTOR) Inhibitors

mTOR is a critical mediator of signals that are essential for mRNA translation and protein synthesis in mammalian cells . mTOR is a downstream effector of the phosphatidylinositol 3-kinase (PI3′K) and the AKT kinase pathway, and mediates signals critical for regulating cell survival and growth . Among the various downstream effectors of the PI3′K/mTOR pathway are the p70 S6 kinase and its targets including S6 ribosomal protein (rpS6), eukaryotic initiation factor 4B (eIF4B) and the translational repressor 4E-BP1. When eIF4B is phosphorylated in an mTOR-dependent manner it dissociates from the eukaryotic initiation factor 4E (eIF4E), so cap-dependent mRNA translation can proceed. In addition it has been shown that mTOR plays the role of a rheostat capable of integrating extra-cellular, plasma membrane-associated and intracellular signals with relevance to T-cell priming and tolerance. Hirase et al. (2009) reported their studies on the role of the mTOR pathway in A-TLL. They evaluated the effects of the mTOR inhibitor, rapamycin, on the proliferation of HTLV-1 positive T cell lines and were able to demonstrate significant growth inhibition in HTLV-1 positive T cell lines after 24 and 48 h of treatment with rapamycin. On the other hand, such growth inhibition was not seen in HTLV-1 negative cells lines or normal peripheral blood lymphocytes, raising the prospect of some degree of specific activation of mTOR in HTLV-1-transformed lines. Nevertheless, there were also some effects seen in the HTLV-1 negative MOLT-4 cell line, suggesting the suppressive effects of mTOR inhibitors may extend in some non-HTLV-I infected T cell leukemia cells as well. A recent phase II trial of an oral mTOR inhibitor everolimus showed promising activity in patients with relapsed or refractory Hodgkin's lymphoma (Johnston et al. 2010). Nineteen patients were enrolled who had received a median of six prior therapies (range, 3–14) and 84% had undergone prior autologous stem cell transplant. The ORR was 47% with eight patients achieving a PR and one patient achieving a CR. The median time to progression was 7.2 months. Patients received a median of seven cycles of therapy. Of the 19 patients, one remained on therapy at 36 months; while the others went off study because of progressive disease (16), toxicity (1), and death from infection (1). Four patients experienced a Grade 3 or higher pulmonary toxicity. This suggested that everolimus has single-agent activity in relapsed/refractory HL and provided proof-of-concept that targeting the mTOR pathway in HL is clinically relevant. Other agents including temsirolimus are currently being tested in clinical trials alone and in combination with other biologic agents.

Agents Targeting Apoptotic Pathways

The apoptotic pathways are intricately controlled by a balance between proapoptotic (Bax, Bak, Mtd/Bok) and antiapoptotic proteins (Bcl-2, Bcl-xL, Mcl-1, Bcl-w). The Bcl-2 family members are overexpressed in several hematologic malignancies. Small molecules that can target antiapoptotic Bcl-2 family members represent a new opportunity to affect this biology directly (Kluck et al. 1997; Cory et al. 2003). Small molecules such as the gossypol analogue AT-101, ABT-263 and obatoclax represent agents that can target antiapoptotic Bcl-2 family members. They have exhibited inhibitory properties in carcinoma cell lines and tumor xenograft models. ABT –737 is a BH-3 only mimetic capable of binding with high affinity to the prosurvival proteins Bcl-xL, Bcl-2 and Bcl-w, inducing Bax/Bak-dependent killing (Van Delft et al. 2006). ABT-263, a potent orally bioavailable analog of ABT-737 is designed to be Bad-like BH3 mimetic. In human tumor cells, ABT-263 induces Bax translocation, cytochrome *c* release, and subsequent apoptosis. This agent is now in clinical trials in NHL where patients are dosed on days 1–14 of a 21 day cycle (10–440 mg). One patient with NK/T-cell lymphoma demonstrated a 75% reduction of his skin lesions after cycle 2. The main toxicities so far include thrombocytopenia, elevation of LFT's. Paoluzzi et al. (2008a) showed that ABT-737 showed synergism when combined with the proteasome inhibitors bortezomib or carfilzomib in select lymphoma cell lines and induced potent mitochondrial membrane depolarization and apoptosis when combined with either. ABT-737 plus bortezomib also induced significant apoptosis in primary samples of MCL, DLBCL, and chronic lymphocytic leukemia (CLL) but no significant cytotoxic effect was observed in peripheral blood mononuclear cells from healthy donors. In severe combined immunodeficient beige mouse models of MCL, the addition of ABT-737 to bortezomib enhanced efficacy compared with either drug alone and with the control suggesting that ABT-737 alone or in combination with a proteasome inhibitor represents a novel and potentially important platform for the treatment of B-cell malignancies. Paoluzzi et al. (2008b) also showed that AT-101 has potent anti-tumor activity both in vitro and in vivo in mantle cell lymphoma cell lines and SCID beige mouse model of drug resistant B-cell lymphoma. The addition of AT-101 to cyclophosphamide and rituximab in a schedule- dependent manner enhanced the efficacy of the conventional agents and was safe.

Oblimersen sodium (Genasense) a 16-mer directed at the open reading frame of Bcl-2 mRNA induces apoptosis of CLL and lymphoma cells in vitro and exhibits synergy with other drugs. Despite modest single agent activity in CLL, a phase III randomized trial by O'Brien et al. (2007) showed that when it was combined with cyclophosphamide and fludarabine, major response rate and duration of response was found to be superior than with-out oblimersen. Seventeen percent CR was reported which correlated with extended time to progression and improved survival. Activity has also been reported in patients with mantle cell lymphoma.

Monoclonal Antibodies

Anti-CD52 (Alemtuzumab)

Alemtuzumab or Campath-1 H is a recombinant DNA- derived humanized monoclonal antibody hat is directed against CD52, a 21–28 kDa cell surface glycoprotein, which is expressed on mature lymphocytes. Ginaldi et al. (1998) investigated the intensity of expression of CD52 in a study of 45 cases of lymphoid leukemia and showed that normal T lymphocytes expressed higher CD-42 expression than B lymphocytes and that the antigen was also significantly higher in T-prolymphocytic leukemia (T-PLL) compared with CLL. Moreover the differences in CD52 expression were somewhat higher in campath-IH treated patients who responded than in non-responders. Campath-1 H is administered as an infusion over 2 h. The dose has to be escalated to 30 mg and is to be given 3 times per week for a total of 12 weeks. It can be given as a subcutaneous injection, but the data in T-cell prolymphocytic leukemia indicate decreased efficacy if given by subcutaneous route. Dearden et al. (2001) evaluated the efficacy of Campath-IH in 39 patients with T-prolymphocytic leukemia and reported an ORR of 76% including 60% CR rate. The median disease free survival was 7 months. Survival was significantly prolonged in patients achieving CR compared with those in PR or with no response. A pilot study by Enblad et al. (2004) in 14 patients with PTCL found an ORR of 36% including 3 CR's and 2 PR's lasting upto 12 months. A phase II study by Zinzani et al. (2005) assessed the impact of a reduced dose (10 mg three times per week for four weeks) schedule of campath-IH in ten patients with pretreated cutaneous/peripheral T-cell lymphoma. Of these, six had nodal PTCLU and four had MF. All MF patients were in stageT3 or T4, N0, M0 of the TNM classification. All PTCLU patients were in stage III-IV according to the Ann Arbor system. Among PTCLU patients, four presented with 4 involved nodal sites and three had 3 involved nodal sites with bulky disease. The median number of prior treatments was 3 (range, 2–4), and the median time from original diagnosis was 13 months (range, 6–15). The overall response rate (ORR) was 60%, with 2 (20%) patients achieving complete responses (CR), and with 4 (40%) obtaining partial response (PR). In the MF subset, the best response was PR (3/4, 75%). However, in the PTCLU subset, there were 2 (33%) CR's which lasted 3 and 8 months as well as 1 (17%) PR. The median duration of response was 7 months. Alemtuzumab is now being studied in combination with CHOP therapy as upfront therapy for PTCL. In one study of 24 patients by Gallamini et al. (2007), 17 CR's and 1 PR were reported. Six patients had stable and progressive disease with median duration of response of 11 months. The major toxicity is immunosuppression and increased risk of CMV reactivation. Hematologic toxicity can occur and blood counts need to be monitored closely.

Anti-CD30

CD30 is a cell surface antigen glycoprotein of 120 kDa that shows sequence homology to members of the TNF receptor superfamily. It is expressed on mitogen activated B and T lymphocytes but not on resting monocytes and lymphocytes. It is also expressed on malignant hematopoetic cells including Hodgkin's disease, ALCL, primary cutaneous ALCL, lymphomatoid papulosis and certain cases of transformed MF. There are several monoclonal antibodies that target CD30. SGN-30 developed by Seattle Genetics has shown promising activity against ALCL in phase I trial (Bartlett et al. 2008) with a 17% partial response with no toxicity. The agent is now being tested in a larger phase II trial in ALCL. However a CALGB trial (Cancer and Leukemia Group B) combining this agent with systemic chemotherapy (vinorelbine, liposomal doxorubicin and gemcitabine) in patients with relapsed Hodgkin's disease showed increased pulmonary toxicity (Blum et al. 2008) in patients with an FcgammaRIIIa V/F polymorphism. A phase II trial of single agent SGN-30 in cutaneous ALCL, lymphomatoid papulosis and MF has shown promising activity (CR plus PR plus stable disease) of 87% (Duvic et al. 2009). The median duration of CR and PRs was 84 days. The dose was 4 mg/kg given every 3 weeks, which was then increased to 12 mg/kg. Side-effects were minimal. This agent is now being tested in a broader multicenter trial in patients with ALCL.

SGN-35

SGN-35 is a drug-antibody conjugate in which SGN-30 is linked to Auristatin. Multicenter phase I dose escalation trial was conducted by Younes et al. (2008) in 45 patients with refractory or recurrent CD30-positive hematologic malignancies, including HL ($n = 42$), systemic anaplastic large cell lymphoma (sALCL; $n = 2$), and angioimmunoblastic T cell lymphoma ($n = 1$). Approximately 75% of patients reporting B- symptoms at baseline experienced symptom resolution on study. Most patients (86%) had reductions in target lesion size. Among 28 evaluable patients treated at doses ≥1.2 mg/kg every 3 week, the objective response rate (CR + PR) was 46% (n = 13) and the complete remission rate was 25% (n = 7). Two additional PRs were observed in the 0.6 mg/kg cohort. Median response duration to date was 22 weeks (range, 0.1–to 38+ weeks). Thirteen patients continued to remain on the study. The most common adverse events (occurring in ≥20% of patients) were fatigue, pyrexia, nausea, and diarrhea. To assess if more frequent dosing might maximize anti-tumor activity with acceptable tolerability, Bartlett et al. (2009) conducted a multicenter, phase 1, weekly dosing, dose-escalation study (3 + 3 design) in pts with refractory or recurrent HL or systemic ALCL. SGN-35 was administered weekly at doses of 0.4–1 mg/kg (2-h IV infusions). Multiple CRs were observed at higher doses (1 patient with sALCL and 2 with HL achieved CR at the dose of 0.8 mg/kg, 1 patient with systemic ALCL and 3 patients with HL achieved CR at the dose of 1 mg/kg). Two independent phase II single-arm open-label multicenter clinical trials to evaluate the efficacy and safety of SGN-35 as a single agent in patients with relapsed or refractory Hodgkin lymphoma and relapsed or refractory systemic ALCL are ongoing. The ATHERA trial which is a randomized, double-blind, placebo-controlled, multicenter phase 3 trial to evaluate the efficacy and safety of brentuximab vedotin (SGN-35) and best supportive care (BSC) compared to placebo and BSC in treatment of residual HL following ASCT is currently recruiting patients.

Anti-CD4

CD4 belongs to the immunoglobulin superfamily and acts as a co-receptor of TCR (T-cell receptor). It is normally expressed in helper Tcells, regulatory Tcells, macrophages, monocytes and dendritic cells and highly expressed by T-cell malignancies including PTCL's and CTCL's. Different monoclonal antibodies against T-cell antigens have been developed and tried clinically in T-cell lymphomas without much success. Knox et al. reported results with SK3, a chimeric anti-CD4 antibody, in seven patients with MF with some effect and a good safety profile. The trial was limited by the development of antichimeric antibodies. M-T412, an anti-CD4 chimeric antibody directed against a different epitope of the CD4 molecule, demonstrated a higher affinity and was able to induce CD4 positive lymphocyte depletition through an Fc-mediated mechanism. This was studied in MF and showed an ORR of 88% with freedom from progression lasting 25 weeks.

Zanolimumab (HuMax)

This is a fully humanized anti-CD4 monoclonal antibody under active investigation for the treatment of CD4 positive malignancies, mainly CTCL in early and advanced stages and other non-cutaneous PTCL's. A phase II trial in refractory CTCL was reported by Obitz et al. (Kim et al. 2007) with a response of 40% and no adverse effects.

D'Amore et al. (2007) presented data from a phase II trial of HuMax in PTCL demonstrating an ORR of 62.5% in the first eight patients enrolled in the trial and only one related case of febrile neutropenia. Two phase II clinical trials are evaluating the role of this drug in early and late stage CTCL. A blinded randomized phase III trial comparing two different dosing of Zanolimumab (8 v.s 14 mg/m^2) in previously treated MF is ongoing.

Targeting Angiogenesis

Bevacizumab (Avastin)

Bevacizumab, a humanized monoclonal antibody directed against vascular endothelial growth factor (VEGF-A), is approved in combination with chemotherapy in lung, colorectal, breast and brain cancers. VEGF-A expression is reported in all lymphoma subtypes, with strongest expression in PTCL. A recent ECOG 2404 phase II clinical trial (Advani et al. 2009) evaluated bevacizumab with CHOP as upfront therapy in patients with PTCL. Patients received bevacizumab 15 mg/ kg IV and CHOP chemotherapy on day 1 for 6–8 cycles. Patients with CR/PR/SD received maintenance therapy with bevacizumab 15 mg/ kg IV q 3 weeks for 4 cycles. Among 33 pts, 22 received at least 4 cycles of A-CHOP. Results from the trial were not mentioned. However one or more designated cardiac toxicities were recorded for 5 of these 22 pts. Congestive heart failure was recorded in 4 patients who received cumulative doxorubicin doses of 300–400 mg/m^2. It was also felt that bevacizumab might potentiate the adverse cardiac effects of anthracyclines, resulting in clinically significant toxicity and EF in such patients will need to be very carefully monitored if this drug is used in future trials in conjunction with an anthracycline.

Plitidepsin (Aplidin)

Plitidepsin is a cyclic depsipeptide isolated from the marine tunicate *Aplidium albicans*. Plitidepsin displays a broad spectrum of antitumor activities, inducing apoptosis by triggering mitochondrial cytochrome c release, initiating the Fas/ DC95, JNK pathway and activating caspase 3 activation. This agent also inhibits elongation factor 1-a, thereby interfering with protein synthesis, and induces G1 arrest and G2 blockade, thereby inhibiting tumor cell growth. A phase II prospective open-label study (Ferme et al. 2008) to evaluate its activity in patients with relapsed/ refractory non-cutaneous PTCL was conducted in 14 patients. They were treated with plitidepsin 3.2 mg/m^2 IV infusion over 1-h on days 1, 8 and 15 every 4 weeks. Eleven patients were evaluable when preliminary results were reported, one was too early and two were non-evaluable. One confirmed CR and 3 PR's were observed with a 36% objective response rate. Median overall survival was 11 months.

Plitidepsin was tolerable in this heavily pretreated population with low hematological toxicity. Transient but reversible ALT elevation was noticed in 7 patients.

Future trials will establish its role in T-cell lymphoma.

Conclusion

At present there are many promising agents with activity against T-cell lymphomas. However it is imperative that insight into the biology of T-cell lymphomagenesis guide the development of novel agents. Major challenges still persist for this field such as: (1) What is the optimal way to combine currently available agents and what are the optimal agents to combine? ; (2) Given the rarity and heterogeneity of this disease what is best strategy to accrue patients to clinical trials to answer these questions? Designing and implementing well designed clinical trials with careful thought to correlative studies could answer some of these questions. This will not only help move the field forward towards a cure of these diseases but may also change the treatment paradigm to more targeted and tailored approach to individual tumors in cancer patients.

References

Adams J (2004) The development of proteasome inhibitors as anticancer drugs. Cancer Cell 5:417–421

Advani RH, Hong F, Kristen N et al (2009) Cardiac toxicity associated with the anti-VEGF monoclonal antibody bevacizumab (Avastin) in combination with CHOP (A-CHOP) chemotherapy for Peripheral T Cell Lymphoma (PTCL):the ECOG 2404 Trial. Blood (ASH Annual Meeting Abstracts), Nov 114:1671

Anderson JR, Armitage JO, Weisenburger DD (1998) Epidemiology of the non-Hodgkin's lymphomas: distributions of the major subtypes differ by geographic locations. Non-Hodgkin's lymphoma classification project. Ann Oncol 9:717–720

Azzoli CG, Krug L, Miller V et al (2007) Phase I study of the antifolate pralatrexate given with vitamin B12 and folic acid supplementation in patients (pts) with advanced non-small-cell lung cancer. J Clin Oncol 25:608s ((suppl) Abstract 13006)

Bartlett NL, Younes A, Carabasi MH (2008) A phase 1 multidose study of SGN-30 immunotherapy in patients with refractory or recurrent CD30? hematologic malignancies. Blood 111:1848–1854

Bartlett N, Forero-Torres A, Rosenblatt J et al (2009) Complete remissions with weekly dosing of SGN-35, a novel antibody-drug conjugate (ADC) targeting CD30, in a phase I dose-escalation study in patients with relapsed or refractory Hodgkin lymphoma (HL) or systemic anaplastic large cell lymphoma. J Clin Oncol 27:15s, (suppl; Abstract 8500)

Bhalla KN (2005) Epigenetic and chromatin modifiers as targeted therapy of hematologic malignancies. J Clin Oncol 23:3971–3993

Blum KA, Johnson JL, Jung S-H (2008) Serious pulmonary toxicity with SGN-30 and gemcitabine, vinorelbine, and liposomaldoxorubicin in patients with relapsed/refractory Hodgkin lymphoma (HL). Cancer Leuk B (CALGB).2008; 50:502 (Blood; 112:92–93 (Abstract 232))

Cheson BD, Horning SJ, Coiffier B et al (1999) Report of an international workshop to standardize response criteria for non-Hodgkin's lymphomas. NCI sponsored international working group. J Clin Oncol 17:1244

Coiffier B, Brousse N, Peuchmaur M et al (1990) Peripheral T-cell lymphomas have a worse prognosis than B-cell lymphomas: a prospective study of 361 immunophenotyped patients treated with the LNH-84 regimen. The GELA (Groupe d'Etude des Lymphomes Agressives). Ann Oncol 1:45–50

Corradini P, Dodero A, Zallio F et al (2004) Graft-versus-lymphoma effect in relapsed peripheral T-cell non-Hodgkin's lymphomas after reduced-intensity conditioning followed by allogeneic transplantation of hematopoietic cells. J Clin Oncol 22:2172–2176

Cory S, Huang DC, Adams JM (2003) The Bcl-2 family: roles in cell survival and oncogenesis. Oncogene 22:8590–8607

Czuczman MS, Porcu P, Johnson J et al (2007) Results of a phase II study of 506U78 in cutaneous T-cell lymphoma and peripheral T-cell lymphoma: CALGB 59901. Leuk Lymphoma 48(1):97–103

D'Amore F, Radford J, Jerkeman M (2007) Zanolimumag (HuMax- CD4), a fully human monoclonal antibody: efficacy and safety in patients with relapsed or treatment-refractory noncutaneous CD4? T-cell lymphoma. Blood 110:999a (Abstract#3409) 2007

De Coninck EC, Kim YH, Varghese A et al (2001) Clinical characteristics and outcome of patients with extracutaneous mycosis fungoides. J Clin Oncol 19:779–784

Dearden CE, Matutes E, Cazin B et al (2001) High remission rate in T-cell prolymphocytic leukemia with Campath-1 H. Blood 98(6):1721–1726, 15 September

DeGraw J, Colwell W, Sirotnik FM (1993) Synthesis and antitumor activity of 10- Propargyl-10-deazaminoterin. J Med Chem 1936:2228–2231

Demierre M, Whittaker S, Kim Y, Kim E, Piekarz R, Prince M, Nichols J, Balser J, Prentice A, Bates S (2009) Pooled analyses of two international, multicenter clinical studies of romidepsin in 167 patients with cutaneous T-cell lymphoma (CTCL). J Clin Oncol 27:Abstract 8546

Dickinson M, Ritchie D, DeAngelo DJ et al (2009) Preliminary evidence of disease response to the pan deacetylase inhibitor panobinostat (LBH589) in refractory Hodgkin lymphoma. Br J Haematol 147(1):97–101

Dueck GS, Chua N, Prasad A, Stewart D (2009) Activity of lenalidomide in a phase II trial for T-cell lymphoma: report on the first 24 cases. J Clin Oncol 27:15s (suppl; Abstract 8524)

Duvic M, Talpur R, Ni X et al (2007a) Phase 2 trial of oral vorinostat (suberoylanilide hydroxamic acid, SAHA) for refractory cutaneous T-cell lymphoma (CTCL). Blood 109:31–39

Duvic M, Forero-Torres A, Foss F (2007b) Response to oral forodesine in refractory cutaneous T-cell lymphoma: interim results of a phase I/II study. Blood 110 (Abstract #122)

Duvic M, Becker JC, Dalle S et al (2008) Phase II trial of oral panobinostat in patients with refractory Cutaneous T-cell Lymphoma, Blood (ASH Annual Meeting Abstracts), Nov 112:1005

Duvic M, Reddy S, Pinter-Brown L (2009) A phase II study of SGN- 30 in cutaneous anaplastic large cell lymphoma and related lymphoproliferative disorders. Clin Cancer Res 15(9): 6217–6224

Ellis L, Pan Y, Smyth GK et al (2008) Histone deacetylase inhibitor panobinostat induces clinical responses with associated alterations in gene expression profiles in cutaneous T-cell lymphoma. Clin Cancer Res 14:4500–4510

Enblad G, Hagberg H, Erlanson M (2004) A pilot study of alemtuzumab (anti-CD52 monoclonal antibody) therapy for patients with relapsed or chemotherapy-refractory peripheral T-cell lymphomas. Blood 103:2920–2924

Escalon MP, Liu NS, Yang Y et al (2005) Prognostic factors and treatment of patients with T-cell non-Hodgkin lymphoma: the M. D. Anderson cancer center experience. Cancer 103:2091–2098

Ferme C, Mateos MV, Szyldergemajn S et al (2008) Plitidepsin is active in PTCL: a subset analysis from an ongoing multicenter phase II trial. Blood (ASH Annual Meeting Abstracts) 112:Abstract 1566

Furman RR, Gore L, Ravandi F (2006) Forodesine IV (Bcx-1777) is clinically active in relapsed/refractory T-cell leukemia: results of a phase II study (interim report) Blood 108:524a (Abstract #1851)

Gallamini A, Stelitano C, Calvi R et al (2004) Peripheral T-cell lymphoma unspecified (PTCL-U): a new prognostic model from a retrospective multicentric clinical study. Blood 103:2474–2479

Gallamini A, Zaja F, Patti C (2007) Alemtuzumab (Campath-1 H) and CHOP chemotherapy as first-line treatment of peripheral T-cell lymphoma: results of a GITIL (Gruppo Italiano Terapie Innovative nei Linfomi) prospective multicenter trial. Blood 110:2316–2323

Gimsing P, Hansen M, Knudsen LM et al (2008) A phase I clinical trial of the histone deacetylase inhibitor belinostat in patients with advanced hematological neoplasia. Eur J Haematol 81:170–176

Ginaldi L, De Massimo M, Matutes E (1998) Levels of expression of CD52 in normal and leukemic B and T cells: correlation with *in vivo* therapeutic responses to campath-1 H. Leuk Res 22(2):185–191, February

Gisselbrecht C, Gaulard P, Lepage E et al (1998) Prognostic significance of T-cell phenotype in aggressive non-Hodgkin's lymphomas. Groupe d'Etudes des Lymphomes de l'Adulte (GELA). Blood 92:76–82

Glaser KB, Staver MJ, Waring JF et al (2003) Gene expression profiling of multiple histone deacetylase (HDAC) inhibitors: defining common gene set produced by HDAC inhibition in T24 and MDA carcinoma cell lines. Mol Cancer Ther 2:151–163

Groves FD, Linet MS, Travis LB, Devesa SS et al (2000) Cancer surveillance series: non-Hodgkin's lymphoma incidence by histologic subtype in the United States from 1978 through 1995. J Natl Cancer Inst 92:1240–1251

Hirase C, Maeda Y, Yamaguchi T, Miyatake J-I, Kanamaru A (2009) mTOR inhibition and adult T-cell leukemia. Leuk Lymphoma 50:645–647

Horwitz S, Mulford D, Paul H (2008) Clofarabine is active in peripheral T-cell lymphomas: results of the phase I portion of a phase I/II study. Ann Oncol 19 (Suppl 4):iv157 (Abstract #244)

Horwitz S, Zain J, O'Connor O (2009a) Pralatrexate (PDX) is active in cutaneous T-Cell lymphoma: preliminary results of a multi-center dose-finding trial. ASH (Abstract 1569)

Horwitz S, Vose J, O'Connor O (2009b) A phase 1/2A open label study of pralatrexate and emcitiabine in patients with relapsed or refractory lymphoproliferative malignancies. ASH (poster 1570)

Izbicka E, Diaz A, Saunders M (2009) Distince mechansistic activity profile of pralatrexate in comparison to other antifolates in vitro and in vivo molds of human cancers. Cancer Chemother Pharmacol 64:993–999

Jagasia M, Morgan D, Goodman S et al (2004) Histology impacts the outcome of peripheral T-cell lymphomas after high dose chemotherapy and stem cell transplant. Leuk Lymphoma 45:2261–2267

Jillella AP, Murren JR, Hamid KK et al (2000) P-glycoprotein expression and multidrug resistance in cutaneous T-cell lymphoma. Cancer Invest 18:609–613

Johnston PB, Inwards DJ, Colgan JP et al (2010) A phase II trial of the oral mTOR inhibitor everolimus in relapsed Hodgkin's lymphoma. Am J Hematol 85(5):320–324, May

Johnstone RW, Licht JD (2003) Histone deacetylase inhibitors in cancer therapy: is transcription the primary target? Cancer Cell 4:13–18

Kelly WK, Richon VM, O'Connor O (2003) Phase I clinical trial of histone deacetylase inhibitor: suberoylanilide hydroxamic acid administered intravenously. Clin Cancer Res 9:3578–3588

Kim YH, Liu HL, Mraz-Gernhard S et al (2003) Long-term outcome of 525 patients with mycosis fungoides and Sezary syndrome: clinical prognostic factors and risk for disease progression. Arch Dermatol 139:857–866

Kim SW, Tanimoto TE, Hirabayashi N et al (2006) Myeloablative allogeneic hematopoietic stem cell transplantation for non-Hodgkin lymphoma: a nationwide survey in Japan. Blood 108:382–389

Kim Y, Duvic M, Obitz E (2007) Clinical efficacy of zanolimumab (HuMax-CD4): two phase 2 studies in refractory cutaneous T-cell lymphoma. Blood 109:4655–4662

Kluck RM, Bossy-Wetzel E, Green DR (1997) The release of cytochrome c from mitochondria: a primary site for Bcl-2 regulation of apoptosis. Science 275:1132–1136

Krug LM, Ng KK, Kris MG et al (2000) Phase I and pharmacokinetic study of 10-propargyl-10-deazaaminopterin, a new antifolate. Clin Cancer Res 6:3493–3498

Le Gouill S, Milpied N, Buzyn A et al (2008) Graft-versus-lymphoma effect for aggressive T-cell lymphomas in adults: a study by the Societe Francaise de Greffe de Moelle et de Therapie Cellulaire. J Clin Oncol 26:2264–2271

Lee J, Suh C, Kang HJ (2008) Phase I study of proteasome inhibitor bortezomib plus CHOP in patients with advanced, aggressive T-cell or NK/T-cell lymphoma. Ann Oncol 19(12):2079–2083

Lentzsch S, LeBlanc R, Podar K (2003) Immunomodulatory analogs of thalidomide inhibit growth of Hs Sultan cells and angiogenesis in vivo. Leukemia 17:41–44

Mann BS, Johnson JR, Cohen MH et al (2007) FDA approval summary: vorinostat for treatment of advanced primary cutaneous T-cell lymphoma. Oncologist 12:1247–1252

Marchi E, Paoluzzi L, Scotto L et al (©2010 AACR) Pralatrexate is synergistic with the proteasome inhibitor bortezomib in *in vitro* and *in vivo* models of T-cell lymphoid malignancies. Clin Cancer Res 16(14):3648–3658

Marchi E, Alinari L, Tani M, Stefoni V (2005) Gemcitabine as frontline treatment for cutaneous T-cell lymphoma: phase II study of 32 patients. Cancer 104(11):2437–2441

Matsuoka H, Unami A, Fujimura T (2007) Mechanisms of HDAC inhibitor-induced thrombocytopenia. Eur J Pharmacol 571(2–3):88–96

Mounier N, Gisselbrecht C, Briere J et al (2004) All aggressive lymphoma subtypes do not share similar outcome after front-line autotransplantation: a matched-control analysis by the Groupe d'Etude des Lymphomes de l'Adulte (GELA). Ann Oncol 15:1790–1797

O'Brien S, Moore JO, Boyd TE (2007) Randomized phase III trial of fludarabine plus cyclophosphamide with or without oblimersen sodium [Bcl-2 antisense] in patients with relapsed or refractory chronic lymphocytic leukemia. J Clin Oncol 25(9):1114–1120, March 20

O'Connor OA, Wright J, Moskowitz C et al (2005) Phase II clinical experience with the novel proteasome inhibitor bortezomib inpatients with indolent non-Hodgkin's lymphoma and mantle celllymphoma. J Clin Oncol 23:676–684

O'Connor OA, Heaney ML, Schwartz L (2006) Clinical experience with intravenous and oral formulations of the novel histone deacetylase inhibitor suberoylanilide hydroxamic acid in patients with advanced hematologic malignancies. J Clin Oncol 24:166–173

O'Connor O, Pro B, Pinter-Brown LL (2009a) Results of the pivotal, multicenter, phase II study of pralatrexate in patients with relapsed or refractory peripheral T-cell lymphoma (PTCL). J Clin Oncol 27:15s (suppl; Abstract 8561)

O'Connor O, Horwitz S, Zelenetz A (2009b) Phase II–I–II study for two different doses and schedules of Pralatrexate, a high affinity substrate for the reduced folate carrier in patients with relapsed or refractory lymphoma reveals marked activity in T cell malignancies. J Clin Oncol 27(26):4357–4364

Olsen EA, Kim YH, Kuzel TM et al (2007) Phase IIb multicenter trial of vorinostat in patients with persistent, progressive, or treatment refractory cutaneous T-cell lymphoma. J Clin Oncol 25:3109–3115

Paoluzzi L, Gonen M, Bhagat G (2008a) The BH3-only mimetic ABT- 737 synergizes the antineoplastic activity of proteasome inhibitors in lymphoid malignancies. Blood 112(7):2906–2916

Paoluzzi L, Gonen M, Gardner JR (2008b) Targeting Bcl-2 family members with the BH-3 mimetic AT-101 markedly enhances the therapeutic effects of chemotherapeutic agents in in vitro and vivo models of B-cell lymphoma. Blood 111:5350–5358

Philip T, Guglielmi C, Hagenbeek A et al (1995) Autologous bone marrow transplantation as compared with salvage chemotherapy in relapses of chemotherapy-sensitive non-Hodgkin's lymphoma. N Engl J Med 333:1540–1545

Piekarz R, Bates S (2009) Epigenetic modifiers: basic understanding and clinical development. Clin Cancer Res 15(12):3918–3926

Piekarz RL, Robey R, Sandor V et al (2001) Inhibitor of histone deacetylation, depsipeptide (FR901228), in the treatment of peripheral and cutaneous T-cell lymphoma: a case report. Blood 98:2865–2868

Piekarz RL, Frye AR, Wright JJ (2006) Cardiac studies in patients treated with depsipeptide, FK228, in a phase II trial for T-cell lymphoma. Clin Cancer Res 12(12):3762–3773

Piekarz RL, Robin F, Maria T (2009a) A multi-institutional phase II trial of t he HDAC inhibitor romidepsin as monotherapy for patients with cutaneous t-cell lymphoma. J Clin Oncol 27(32):5410–5417

Piekarz et al (2009b) Phase II trial of single agent romidepsin in relapsed peripheral T cell lymphoma ASH (Abstract 14157)

Pohlman B, Advani R, Duvic M et al (2009) Final results of a phase II trial of Belinostat (PXD-101) in patients with recurrent ore refractory peripheral or cutaneous T-cell lymphoma. Blood (ASH Annual Meeting Abstracts) 114:Abstract 920

Querfeld C, Kuzel TM, Guitart J, Rosen ST (2009) Lenalidomide (Revlimid_) patients with cutaneous T-cell lymphoma. Hematol Meet Rep 3(1):103–105

Ri M, Iida S, Ishida T (2009) Bortezomib-induced apoptosis in mature T-cell lymphoma cells partially depends on upregulation of Noxa and functional repression of Mcl-1. Cancer Sci 100:341–348

Richardson PG, Schlossman RL, Weller E (2002) Immunomodulatory drug CC-5013 overcomes drug resistance and is well tolerated in patients with relapsed multiple myeloma. Blood 100:3063–3067

Richardson PG, Sonneveld P, Schuster MW et al (2005) Bortezomib or high-dose dexamethasone for relapsed multiple myeloma. N Engl J Med 352:2487–2498

Richardson PG, Mitsiades C, Hideshima T, Anderson KC (2006) Bortezomib: proteasome inhibition as an effective anticancer therapy. Annu Rev Med 57:33–47

Rudiger T, Weisenburger DD, Anderson JR et al (2002) Peripheral T-cell lymphoma (excluding anaplastic large-cell lymphoma): results from the non-Hodgkin's lymphoma classification project. Ann Oncol 13:140–149

Savage KJ, Chhanabhai M, Gascoyne RD et al (2004) Characterization of peripheral T-cell lymphomas in a single North American institution by the WHO classification. Ann Oncol 15:1467–1475

Seer Cancer Statistic Review 1975–2007

Sirotnak FM, DeGraw JI, Moccio DM et al (1984) New folate analogs of the 10-deaza-aminopterin series: basis for structural design and biochemical and pharmacologic properties. Cancer Chemother Pharmacol 12:18–25

Sirotnak FM, DeGraw JI, Colwell WT et al (1998) A new analogue of 10-deazaaminopterin with markedly enhanced curative effects against human tumor xenografts in mice. Cancer Chemother Pharmacol 42:313–318

Susan DD, Christopher JK, Monette A (2007) Antitumor activity of PR-171, a novel irreversible inhibitor of the proteasome. Cancer Res 67:6383–6391

Thompson, MA, Pro, B, Sarris, A et al (2005) Results of a phase II study of 506U78 (Nelarabine) in refractory indolent B-cell or peripheral T-cell lymphoma. American Society of Hematology Annual Meeting, Abstract 2681

Tse C, Shoemaker AR, Adickes J, Anderson MG (2008) ABT-263: a potent and orally bioavailable Bcl-2 family inhibitor. Cancer Res 68(9):3421–3428

Van Delft MF, Wei AH, Mason KD (2006) The BH3 mimetic ABT-737 targets selective Bcl-2 proteins and efficiently induces apoptosis via Bak/Bax if Mcl-1 is neutralized. Cancer Cell 10:389–399

Vose J, Armitage J, Weisenburger D (2008) International peripheral T-cell and natural killer/T-cell lymphoma study: pathology findings and clinical outcomes. J Clin Oncol 26:4124–4130

Wang ES, O'Connor O, She Y et al (2003) Activity of a novel antifolate (PDX, 10-propargyl 10-deazaaminopterin) against human lymphoma is superior to methotrexate and correlates with tumor RFC-1 gene expression. Leuk Lymphoma 44:1027–1035

WHO (2008) Classification of tumours of haematopoietic and lymphoid tissues, 4th edn. WHO Press, Geneva

Willemze R, Jaffe ES, Burg G et al (2005) WHO-EORTC classification for cutaneous lymphomas. Blood 105:3768–3785

Younes A, Forero-Torres A, Bartlett NL et al (2008) Multiple complete responses in a phase I dose escalation study of the antibody drug conjugate SGN-35 in patients with relapsed or refractory CD-30 positive lymphomas. Blood (ASH Annual Meeting Abstracts) 112:Abstract 1006

Yuan R, Kay A, Berg WJ, Lebwohl D (2009) Targeting tumorigenesis: development and use of mTOR inhibitors in cancer therapy. J Hematol Oncol 2(45)

Zinzani PL, Alinari L, Tani M et al (2005) Preliminary observations of a phase II study of reduced-dose alemtuzumab treatment in patients with pretreated T-cell lymphoma. Haematologica 90(5):702–703, May

Zinzani PL, Musuraca G, Tani M (2007) Phase II trial of proteasome inhibitor bortezomib in patients with relapsed or refractory cutaneous T-cell lymphoma. J Clin Oncol 25(27):4293–4297

Zinzani PL, Venturini F, Stefoni V et al (2010) Gemcitabine as single agent in pre-treated T-cell lymphoma patients: evaluation of the long term outcome. Ann Oncol 21(4):860–863

Large Granular Lymphocyte Leukemia – From Molecular Pathogenesis to Targeted Therapy

Mithun Vinod Shah and Thomas P. Loughran Jr.

Introduction

Large Granular Lymphocytes (LGL) are part of the immune system normally comprising 10–15% of adult peripheral blood mononuclear cells (PBMC). LGL actively survey for virus-infected or transformed cells in the body. Phenotypically, LGL contain two distinct subpopulations – cytotoxic T-lymphocytes (CTL) and Natural Killer (NK) cells. Despite important differences in origin and functions, CTL share several characteristics with NK-cells (Smyth et al. 2001).

LGL leukemia is a chronic lymphoproliferative disorder of cytotoxic lymphocytes. LGL leukemia can be of cytotoxic T-cells (known as T-cell LGL leukemia or T-LGL leukemia) or that of NK cells (NK-cell LGL leukemia or NK-LGL leukemia). In T-LGL leukemia, the expansion of clonal CTL is seen as the expansion of CD3+ CD8+/CD57+ T-cell receptor (TCR)-$\alpha\beta$+CTL (Loughran 1993). Leukemic T-LGL are CD45RA+CD62L-, a phenotype consistent with effector-memory RA T-cells (T_{EMRA}). Thus, leukemic T-LGL can be considered as malignant expansion of T_{EMRA} cells (Loughran 1993; Yang et al. 2008).

NK-LGL leukemia is characterized by CD3- CD56+ and/or CD16+ cells (Loughran 1993). In the absence of clonal TCR gene it is more difficult to establish the clonality of leukemic NK-LGL. However, recent reports studying NK-receptor

M.V. Shah
Department of Internal Medicine, Penn State Milton S. Hershey Medical Center, 500 University Drive, Hershey, PA, USA

T.P. Loughran Jr. (✉)
Department of Internal Medicine, Penn State Milton S. Hershey Medical Center, 500 University Drive, Hershey, PA, USA

Penn State Hershey Cancer Institute, Penn State Milton S. Hershey Medical Center, 500 University Drive, Hershey, PA, USA
e-mail: tloughran@psu.edu

J. Tao and E. Sotomayor (eds.), *Hematologic Cancers: From Molecular Pathobiology to Targeted Therapeutics*, Cancer Growth and Progression 14, DOI 10.1007/978-94-007-5028-9_16, © Springer Science+Business Media Dordrecht 2012

repertoire suggest clonal expansion of NK cells in NK-LGL leukemia patients (Loughran 1993; Epling-Burnette et al. 2004a).

Following recognition of target cells, LGL are powerful executers of cell-mediated cytotoxicity. Upon finding their targets, LGLs make brief contact with target cells and induce apoptosis in the latter. Apoptosis induction in the target cells is achieved either by death receptor mediated pathway or by delivering cytotoxins in the target cells. The death receptor mediated signaling depends on the activation of 'death receptors' such as Fas by their ligands. LGL contain various cytotoxins such as perforin (pore forming protein) and granzyme B (GrB) in their azurophilic granules. Perforin punches pores in target cells facilitating the delivery of cytotoxins such as GrB into the target cells. Once in target cells, GrB cleaves and activates various proteins such as caspases, filamin, nuclear poly(ADP-ribose) polymerase (PARP) and Bid leading to cell death in both caspase-dependent and -independent manners (Smyth et al. 2001; Krammer 2000).

Clinical Features of LGL Leukemia

LGL leukemia is a disorder of middle-aged individuals with the median age around 55 years. Both T- and NK-LGL leukemia may manifest as an indolent disorder or an aggressive leukemia. T-LGL leukemia is far more common than NK-LGL leukemia, with more than 85% LGL leukemia manifesting as indolent T-cell disease. Indolent T-LGL leukemia is a chronic disease with median survival of around 14.5 years. On the other hand, aggressive NK-LGL leukemia is one of the most aggressive tumors known, with median survival of a few months following the diagnosis (Loughran 1993; Sokol and Loughran 2006).

About one third of patients with T-LGL leukemia are asymptomatic and are diagnosed coincidentally. Some patients present with B-symptoms such as fever, unexplained weight loss and night sweats. T-LGL leukemia has significant overlap with a variety of hematological and autoimmune diseases. Rheumatoid arthritis (RA) is the most common autoimmune condition seen in the patients with T-LGL leukemia. Recurrent infections due to coexistent neutropenia is a common feature of LGL leukemia. Other hematologic conditions such as hemolytic anemia, pure red cell aplasia, cyclic neutropenia and aplastic anemia also coexist in patients with T-LGL leukemia. While the exact pathogenesis is not understood, it is believed that leukemic T-LGL infiltrate into normal tissues where they mediate direct destruction of normal tissue (Sokol and Loughran 2006; Lamy and Loughran 1998; Zhang et al. 2009).

Treatment of Patients with LGL Leukemia

Asymptomatic LGL leukemia patients with indolent course usually do not require any treatment. Indications for therapy include symptomatic or severe anemia (such as transfusion dependent anemia) or neutropenia (absolute neutrophil count or ANC of <500/μl). The goals of therapy are (i) eradication of clone (ii) correction of cytopenias and (iii) symptomatic relief. The mainstay of treatment in LGL leukemia is immunosuppressive therapy such as methotrexate (MTX), cyclophosphamide, cyclosporine A (CSA) (Sokol and Loughran 2006; Lamy and Loughran 2003). MTX is one of the most commonly used drugs for the treatment of indolent T-LGL leukemia. Generally, low-dose (~10 mg/m^2), oral treatment is given on weekly basis. The overall response rate (ORR) is about 50%. It may take 3–4 months to observe response. Steroids such as prednisone are sometimes used as adjuvant to hasten the clinical response and symptomatic relief. Due to absence of cross-resistance among MTX, CSA, and cyclophosphamide, trial of other drugs is considered in the event of non-response to the initial treatment. Treatment options such as fludarabine, alemtuzumab (anti-CD52 monoclonal antibody), anti-thymocyte globulin (ATG), pentostatin or splenectomy are considered second-line options in the cases of failure with the first-line treatment (Sokol and Loughran 2006; Osuji et al. 2006).

Pathophysiology of CTL Homeostasis

Activation-Induced Cell Death and Its Dysregulation in LGL Leukemia

In periphery, antigen encounter by an antigen-specific naïve T-cell leads to proliferation of that T-cell. Within days, vigorous proliferation leads to increase in antigen-specific T-cells by about 50,000-fold accompanied by acquisition of effector functions (Thome and Tschopp 2001). Unchecked proliferation and persistent cytotoxicity of CTL is undesirable due to the risk of developing autoimmunity or malignancy. Thus, most of these activated cells are selectively eliminated following antigen clearance (Thome and Tschopp 2001). This process, known as activation-induced cell death (AICD), is paramount for maintenance of T-cell homeostasis and peripheral immune tolerance (Zhang et al. 2004). In the periphery, AICD is achieved by contribution of both granule-mediated pathway and death receptor mediated signaling.

The activation of CTL upregulates the surface expression of both Fas and FasL, ensuring effective elimination of activated CTL via Fas-FasL mediated apoptosis either in autocrine or paracrine fashion (Krammer 2000; Zhang et al. 2004). Fas is a member of tumor necrosis factor receptor (TNFR) family of proteins that plays role in CTL-mediated apoptosis of target cells including other activated CTL (Thome and Tschopp 2001; Matiba et al. 1997). Interaction of FasL with its receptor leads to trimerization of Fas. The cytosolic portion of the trimerized receptor complex binds to an adaptor protein known as Fas-associated death domain (FADD). Collectively, this complex is called death-inducing signaling complex (DISC) (Matiba et al. 1997). Formation of DISC allows for binding and activation of a zymogen known as procaspase-8. Recruitment of procaspase-8 to DISC leads to its cleavage and formation of heterotetramers of two p10 and p18 subunits each (Lavrik et al. 2003). Activated caspase-8 is the key-initiator of death receptor-mediated apoptosis activating other caspases. The activation of caspase cascade eventually culminates in cell death. DISC formation and the activation of caspase-8 are, in part, negatively regulated by cellular FADD-like IL1-converting enzyme (FLICE)-inhibitory protein (c-FLIP) (Thome and Tschopp 2001).

Dysregulation of Fas-Mediated Apoptosis in Leukemic LGL

Like activated normal CTL, leukemic LGL also express abundant Fas and FasL on their surfaces. However, while normal activated CTL readily undergo Fas-FasL mediated apoptosis, leukemic LGL are resistant to FasL-mediated apoptosis (Lamy et al. 1998).

Gene expression profiling carried out using microarray technique showed a unique gene expression signature in LGL leukemia PBMC. Leukemic LGL show expression pattern in agreement with acquisition of effector functions, but with severe dysregulation in apoptotic machinery. Various genes known to have pro-apoptotic function were downregulated while those with known anti-apoptotic functions were upregulated. Thus, in leukemic LGL the processes of activation and apoptosis that are normally tightly coupled is uncoupled, leading to the inhibition of AICD (Shah et al. 2008).

It is proposed that although leukemic LGL are capable of undergoing Fas-mediated apoptosis, various survival signals keep them from doing so. LGL leukemia patients do not carry any known mutation in Fas or FasL (Liu et al. 2002a). *In vitro* treatment with interleukin-2 (IL2), phytohemagglutinin and IL2, or ceramide sensitize leukemic LGL to Fas-mediated apoptosis (Yang et al. 2008; Epling-Burnette et al. 2001). Inhibitors of various survival signaling pathways restore Fas-sensitivity in leukemic LGL suggesting intact Fas-FasL apoptotic machinery (Shah et al. 2008; Epling-Burnette et al. 2001).

Role of Soluble Fas and FasL in LGL Leukemia

Various isoforms of both Fas and FasL have been found in the sera of LGL leukemia patients. The sera from LGL leukemia patients contained elevated levels of soluble form of Fas receptor (sFas) compared to their healthy counterparts. It is believed that sFas may work as a decoy for FasL, resulting in Fas-resistance phenotype of leukemic LGL (Liu et al. 2002a). In support of this hypothesis, the source of these sFas variants in serum was traced to leukemic LGL. Leukemic LGL were found to express alternative spliced Fas variants not seen in naïve or activated PBMC from healthy controls. When overexpressed, these variants were secreted in supernatant. The supernatant containing these sFas variants blocked Fas-mediated apoptosis of leukemic LGL (Liu et al. 2002a).

Therapeutic Implications

LGL leukemia patients have high amounts of sFasL in serum whereas serum from normal donors does not contain any detectable levels of sFasL. The soluble form of FasL (sFasL) is produced either by alternative splicing or by proteolytic cleavage of membrane bound FasL by matrix metalloproteinase (MMP) family of enzymes (Liu et al. 2002a; Tanaka et al. 1996).

While systemic treatment with Fas agonists such as anti-Fas antibodies or multimeric recombinant FasL seem as a possibility, severe systemic (mainly hepatic) toxicities, preclude their therapeutic use. On the other hand, sFasL is biologically inactive and has activity only when aggregated secondarily by cross linking antibodies. Thus, sFasL can be directed to the target cells by a tumor marker-specific antibody reducing the systemic toxicity while maintaining anti-tumor activity. This approach has shown promise in various leukemia as well as solid tumors (Schrama et al. 2006; Bremer et al. 2008).

MMP inhibitors could theoretically be proposed for the treatment of LGL leukemia, based on the rationale of the inhibition of cleavage of membrane bound FasL resulting in decreased sFasL in the serum. More than fifty MMP inhibitors such as marimastat have been pursued in clinical trials. It would be of interest to see if MMP inhibitors can be used for treatment of LGL leukemia.

Abnormal DISC Formation

DISC formation is the immediate downstream event of Fas-FasL ligation in Fas-mediated apoptosis. Both short and long forms of FLIP (known as c-FLIP$_S$ and c-FLIP$_L$ respectively) contain caspase-homologous regions, enabling them to be recruited to DISC. However, FLIP lack proteolytic capabilities of caspase-8. Thus, recruitment of FLIP to DISC inhibits the execution of Fas-induced apoptosis signals mediated by caspases. FLIP not only inhibit Fas-mediated apoptosis, but is also known to induce NF-κB- and Erk-mediated proliferation in T-cells (Thome and Tschopp 2001). Downregulation of FLIP is seen towards the end of CTL response and correlates with their increased sensitivity to Fas-mediated apoptosis. In contrast, leukemic LGL express higher basal levels of both isoforms of FLIP that may contribute to Fas-resistant phenotype of leukemic LGL (Yang et al. 2008).

Deregulation of Jak-Stat Pathway in LGL Leukemia

Janus kinase-signal transducers and activators of transcription (Jak-Stat) signaling cascade plays role in conferring survival in various tumors. It is also known to be activated following T-cell activation. JAK proteins are kinases that phosphorylate STAT proteins. STAT proteins are latent transcription factors that, upon activation, transcribe various known anti-apoptotic genes. Jak-Stat pathway plays an essential role following cytokine signaling. Four members of Jak family and seven members of Stat family are known in humans. Stat family members may form homo- or heterodimers resulting in various combinations that may have overlapping but distinct transcription profiles (Murray 2007; Levy and Lee 2002).

Upon activation of cytokine or growth factor receptor, aggregation of receptors leads to transphosphorylation of JAK leading to its activation. Activated JAK can then bind to STAT. STAT proteins contain an N-terminus dimerization domain, a central DNA-binding domain and a C-terminus transactivation domain. Phosphorylation of tyrosine and threonine (believed to be mediated by ERK) of the dimerization domain allows STAT proteins to form homo- or heterodimers. Dimerized STAT proteins then translocate to nucleus where they exert their transcriptional activities by binding to enhancer regions of the target genes including Bcl-xL, myeloid cell leukemia sequence 1 (Mcl-1), IAP-family of protein survivin (BIRC5), cell-cycle regulator cyclin D1, c-Myc and vascular endothelial growth factor (VEGF) (Levy and Inghirami 2006). While physiological activation of STATs last for a few minutes to a few hours, constitutively activated STATs are frequently found in a wide variety of human tumors (Steelman et al. 2008). Transformations mediated by oncogenes such as v-*src*, v-*abl*, v-*fps*, v-*fes*, v-*eyk* and Gα_{12} have been shown to be mediated by STAT especially by STAT3 (Buettner et al. 2002; Kumar et al. 2005).

Leukemic LGL harbor constitutively activated STAT1 and/or STAT3 but not STAT5. STAT3 and/or STAT1 dimers from LGL leukemia patients' PBMC showed DNA binding activity equivalent to that of *in vitro* activated normal PBMC, suggesting that leukemic LGL are activated *in vivo*. The inhibition of JAK2/3 using small molecular tyrosine kinase inhibitor AG490 induced apoptosis in leukemic LGL as well as restored Fas-sensitivity. Specific inhibition of STAT3 using antisense to *STAT3* induced significant apoptosis as well as restored Fas-sensitivity in leukemic LGL.

The promoter region of human *MCL1* – a BCL2 family member important for maintaining mitochondria integrity – contains a STAT3 binding site. In leukemic LGL, STAT3 binds to this site and induces expression of *MCL1*. The induction of apoptosis through STAT3 inhibition correlates with decreased MCL1 expression indicating a role of anti-apoptotic protein MCL1 in survival of leukemic LGL (Epling-Burnette et al. 2001).

Therapeutic Implications

Recently, there has been a great interest in finding specific methods to inhibit Stat3 signaling. Small molecules inhibitors (JSI124 and platinum (IV) compounds such as CPA-7) or peptide-based inhibitors specific to STAT3 (Iwamaru et al. 2006; Schust et al. 2006; Tan et al. 2006; Turkson et al. 2004) have shown promising results in specifically inhibiting Stat3-mediated signaling.

Stat3 decoy is a double-stranded 15-mer oligonucleotide, corresponding closely to the STAT3 response element (SRE) within the c-fos promoter. Stat3 decoy oligonucleotide binds specifically to activated STAT3 and blocks binding of STAT3 to SREs, acting as a functional antagonist of STAT3 activity in cells. The inhibition of STAT3-mediated transcription is the key to anti-tumor activity. STAT3 decoy has shown to have potent anti-tumor activity *in vitro*, as well as in animal models (Xi et al. 2004). A clinical trial evaluating the efficacy of STAT3 decoy in head and neck cancer is currently ongoing (STAT3 DECOY in Head and Neck Cancer).

Specific STAT3 peptide inhibitors have been developed by fusing the STAT3 SH2 binding domain to a membrane-translocating sequence, which allows delivery of the active peptide in the cells. The STAT3 peptide inhibitor competitively inhibits STAT3 dimerization *in vitro* by direct interaction with STAT3 monomers. Since dimerization of STAT3 is essential for its activity as a transcription factor, these peptides inhibit STAT3 activity by functioning as decoys (Christine et al. 2005).

Small molecular inhibitor OPB-31121, an oral agent, is being evaluated in clinical trials for non-hodgkin's lymphoma, multiple myeloma, and in various advanced solid tumors. OPB-31121 inhibits interleukin-6 dependent phosphorylation of STAT3 thereby exerting its anti-tumor activity (STAT3 Inhibitor for Solid Tumors).

Lestarutinib (CEP 701), an orally available JAK2 inhibitor, is being evaluated in hematopoietic malignancies such as acute myelogenous leukemia and myeloproliferative disorders as well as in prostate cancer, either as a single agent or in combination with other chemotherapeutic agents. Interestingly it is also shown to have activity in autoimmune conditions such as psoriasis (Santos et al. 2010; http://clinicaltrials.gov/ct2/results?term=CEP701).

Trisenox (Arsenic Trioxide, ATO, As2O3) is a highly effective treatment for patients with acute promyelocytic leukemia (APL). It is shown to improve event-free survival and disease-free survival in patients with APL. It is shown to inhibit STAT3 activity in dose-dependent manner by inhibiting the activity of protein tyrosine kinase such as JAK (http://clinicaltrial.gov; Powell et al. 2010; Wetzler et al. 2006).

Given the role of Jak-Stat pathway in survival of leukemic LGL, it will be interesting to see if any of these novel agents would have therapeutic activity in LGL leukemia patients.

Deregulation of Ras-Mek-Erk Signaling in LGL Leukemia

Ras has a well-established role in tumor biology. Mutations of Ras occur in about 30% of all human cancers. Ras family of proteins belongs to guanidine triphosphatase (GTPase). In its active (GTP-bound) form, Ras engages various downstream effector pathways that play essential role in cellular responses such as survival, proliferation and differentiation (Schubbert et al. 2007).

Ras is activated following signal delivered to receptor tyrosine kinases (RTK). Ras signaling can be activated by Src-family kinases (SFK), platelet-derived growth factor (PDGF), or sphingosine-1-phosphate (S1P) through $G\alpha_{12}$. Ras cascade is the major pro-survival regulator following T-cell activation. In T-cells, immunoreceptor tyrosine-based activation motifs (ITAM) of the TCR translate the signal of antigen engagement to Ras (Samelson 2002; Veillette et al. 2002; Pyne and Pyne 2002; Heldin and Westermark 1999; Mor and Philips 2006).

Ras activity is also regulated by post-transcriptional regulations. Prenyl transferases are a class of enzymes that include farnesyl transferases (FT) and geranylgeranyl transferases (GGT). Prenyl-transferases modify Ras activity by adding either one or two hydrophobic moieties on C-terminus of RAS. This modification is essential to anchor Ras on cytosolic leaflet of cellular membranes which is a pre-requisite for activation (Schubbert et al. 2007; Mor and Philips 2006).

Once activated, Ras phosphorylates and activates Raf-1. Raf-1 phosphorylates MAPK-extracellular regulated kinase (ERK) kinase (MEK) resulting in its activation. Activated MEK in turn phosphorylates ERK resulting in its translocation to nucleus. In the nucleus, ERK activates various transcription factors, including Fos and Jun of Ets family, resulting in proliferation, differentiation or survival of cells (Steelman et al. 2008; Schubbert et al. 2007; Mor and Philips 2006). Ras signaling also promotes survival by directly promoting the transcription of FLIP and MCL1 (Budd et al. 2006; Huntington et al. 2007). In Jurkat T-cells MAPK/ERK activity is inversely proportional to Fas-sensitivity and anti-apoptotic activity of MAPK/ERK signaling overrides Fas-mediated apoptotic signals (Holmström et al. 2000).

LGL leukemia patients harbor constitutively active form of Ras (H-Ras-GTP) (Epling-Burnette et al. 2004b). Ras-Mek-Erk signaling was found to be constitutively activated in leukemic LGL. The inhibition of RAS, MEK or ERK induced apoptosis in as well as restored Fas-sensitivity in leukemic LGL (Epling-Burnette et al. 2004b). The inhibition of Ras either using chemical inhibitor FTI2153 or by overexpressing dominant negative form of Ras, induced apoptosis in leukemic LGL by inhibiting ERK activity. Similar results were obtained using MAPK inhibitors (PD98059 or U0216) that induced apoptosis in leukemic LGL and restored Fas-sensitivity in Erk-dependent manner. These results suggest that overactive RAS and MEK lies upstream to ERK in mediating survival signals in leukemic LGL (Epling-Burnette et al. 2004b).

Therapeutic Implications

R1150777 (Zarnestra, Tipifarnib) is a farnesyl transferase inhibitor (FTI) designed to inhibit Ras pathway. Zarnestra is being investigated as a potential treatment in various tumors including leukemia (Armand et al. 2007). Since Ras isoforms such as H-Ras requires farnesylation for malignant transformation activity (Zhu et al. 2005), it was hypothesized that by inhibiting farnesylation of Ras, Zarnestra would inhibit Ras-mediated signaling in LGL leukemia. A clinical trial was conducted on eight

LGL leukemia patients using this drug. While none of the patients achieved clinical response, interesting biological responses were observed in most of these patients.

In a patient with NK-LGL leukemia with coexisting primary pulmonary hypertension (PPH), improvement in the symptoms and signs of pulmonary hypertension following treatment with Zarnestra was observed (Epling-Burnette et al. 2008). Treatment with Zarnestra also resulted in decreased marrow LGL, improved hematopoiesis using *in vitro* cultures, and increased normal marrow hematopoiesis (unpublished observations). It was proposed that activating NKR (or TCR in T-LGL leukemia) acts through adaptor proteins to activate Ras-Mek-Erk and PI3k-Akt signaling. This activation ultimately culminates into Fas-resistance phenotype, granule redistribution, and cytotoxicity. Disruption of Ras signaling using FTI may prevent these downstream effects.

Deregulation of PI3k-Akt Pathway LGL Leukemia

Among downstream effector pathways of Ras cascade, phosphoinositide-3-kinase (PI3k) -v-akt murine thymoma viral oncogene homolog (Akt) mediated signaling has a well established role in metabolism, survival and proliferation – and hence in tumor formation. This deregulation is commonly caused by mutation or amplification of PI3K-α, or deletion or mutation of a negative regulator of PI3k-Akt signaling, phosphatase and tensin homolog (PTEN) (Vivanco and Sawyers 2002).

In T-cells, PI3k-Akt signaling plays pivotal role in TCR-mediated activation and proliferation. PI3K is activated upon membrane relocation. This relocation is mediated by the interaction of PI3K with RAS or SFK, or through the direct interaction of PI3K with cytokine receptors at the SH2-domain binding site (Steelman et al. 2008; Vivanco and Sawyers 2002; Schade et al. 2006). The most studied component downstream of PI3K is a serine/threonine kinase – AKT (or protein kinase B, PKB). The activation of AKT accounts for many of the biological functions of PI3K. To promote proliferation, AKT acts through upregulating cyclin D1, and mammalian target of rapamycin (mTOR) pathway, while downregulating forkhead box class O (FOXO) transcription factors and cyclin-dependent kinase inhibitors (CKI) such as $p21^{WAF}$ and $p27^{KIP1}$.

One of the main targets downstream to Akt in pro-survival signaling is NF-κB signaling. Further, AKT phosphorylates Bcl-2 antagonist of cell death (BAD), preventing its interaction with anti-apoptotic factor Bcl-x_L. This leaves Bcl-x_L to exert its anti-apoptotic functions. AKT phosphorylates and inhibits caspase-9, thus inhibiting apoptosis. Another anti-apoptotic phosphorylation target of AKT is MDM2 – a negative regulator of tumor suppressor p53. Phosphorylated MDM2 binds to p53, expediting its degradation and interfering with tumor suppression effects of the latter (Wymann and Schneiter 2008; Mayo and Donner 2001).

In LGL leukemia PBMC, SFK maintain PI3K in its constitutively active form as assessed by phosphorylated state of AKT and glycogen synthase kinase-3 (GSK3). Following activation, PI3k-Akt signaling can enhance the survival of normal T-cells through the inhibition of Fas clustering and DISC (Jones et al. 2002), a phenotype consistent with leukemic LGL. Further, the inhibition of SFK or PI3K induced apoptosis in leukemic LGL. The inhibition of SFK or PIK3 was accompanied by the inhibition of ERK1/2 activity, placing Akt upstream to Erk in LGL survival signaling. In leukemic LGL, SFK-mediated activation of PI3k-Akt pathway results in constitutive ERK activity. The inhibition of this pathway at any level – namely the inhibition of SFK, AKT or ERK –interferes with the survival of leukemic LGL resulting in induction of apoptosis (Epling-Burnette et al. 2004b; Schade et al. 2006).

Therapeutic Implications

Due to their role in promoting proliferation and survival as well as potent anti-angiogenic effects, PI3k and Akt are well established candidates for anti-tumor therapy. Currently, many pan-PI3K inhibitors are being studied in pre-clinical and clinical trials. The strategies to use PI3k-Akt signaling inhibitors are either to use them as single agents or in combination with conventional chemotherapy or radio-therapy. Among many isoforms, PI3Kα is the most commonly involved isoform in cancer. Hence, as a strategy to minimize side effects, efforts are underway to develop a PI3Kα specific inhibitor. Currently, inhibitor of PI3K such as PI-103, NVP-BEZ235, SF1126, PX-866, and ZSTK474 (all pan-PI3K isoform inhibitors), have shown promise as anti-tumor agents as well in the treatment for autoimmune diseases. Given this dual effect, it would be of interest to see if these novel agents possess therapeutic potential in treatment of LGL leukemia (Wymann and Schneiter 2008; Kong and Yamori 2008).

Deregulation of NF-κB Signaling in LGL Leukemia

NF-κB plays an essential role in hematopoiesis, inflammation, as well as survival and proliferation of adaptive immune system cells. Thus, NF-κB is now recognized as a critical player in almost all the aspects of immune responses (Sen and Baltimore 1986; Hayden et al. 2006). In inactivated state, NF-κB is in cytoplasm as a complex with the inhibitor of NF-κB (IκB). This complex keeps NF-κB from both entering to nucleus and binding to DNA, thus depriving its transcriptional functions. This inhibition is void once IκB is phosphorylated by (IκB)-kinase complex (IKK). Phsophorylation of IκB, leads to its ubiquitination and proteasomal degradation. IKK is a known AKT substrate. In summary, phosphorylation of IKK by AKT leads to the activation of the IKK, which in turn leads to phosphorylation and degradation of IκB, leading to NF-κB activation (Wymann and Schneiter 2008).

The activation of NF-κB is one of the most characterized pathways in antigen-receptor signaling in T-cells. The activation of NF-κB downstream of TCR ligation facilitates proliferation and acquisition of effector functions. NF-κB orchestrates T-cell activation by providing a milieu to proliferate (such as inducing IL2 production) and acquire effector functions (such as promoting transcription of RANTES, Fas and FasL) (Hoffmann and Baltimore 2006).

An important NF-κB function is to protect T-cells against AICD (Rivera-Walsh et al. 2000). This function is executed through promoting the expression of pro-survival Bcl-2 family member and inhibitor of apoptosis (IAPs) proteins (Hoffmann and Baltimore 2006). Deficiency or inhibition of NF-κB activity results in failure of activation, either due to lack of proliferation or premature onset of apoptosis suggesting critical role of NF-κB signaling in mounting effective T-cell response (Hayden et al. 2006; Kontgen et al. 1995; Jeremias et al. 1998; Wan and DeGregori 2003).

Due to its pivotal role in CD8+ T-cell activation and survival along with an established role in various aspects of tumorigenesis as well as inflammation, NF-κB is an interesting candidate to study in LGL leukemia. Gene expression data of leukemic LGL showed that c-Rel was overexpressed in leukemic LGL (Shah et al. 2008). Further, leukemic LGL showed constitutively active NF-κB. The inhibition of NF-κB using a specific inhibitor BAY 11-7082 resulted in apoptosis of leukemic LGL. The inhibition of Akt led to the inhibition of NF-κB activity; whereas, the inhibition of NF-κB activity did not affect Akt phosphorylation. This suggests that NF-κB acts downstream of PI3k-Akt pathway in leukemic LGL. We also found that NF-κB maintains expression of Mcl-1 independent of STAT3 activity (Zhang et al. 2008).

Therapeutic Implications

NF-κB is identified as a major player in cell growth, proliferation, immunity and transformation, making it an attractive target treatment of various diseases. However, due to its widespread role in cell biology, the inhibition of NF-κB signaling is also prone to significant toxicities. Bardoxolone methyl (CDDO-Me, RTA402), a synthetic triterpenoid, is an oral inhibitor that is known to inhibit two key signaling pathways implicated in survival of leukemic LGL: NF-κB- and Jak/Stat- pathways. CDDO-Me inhibits tumor necrosis factor-α (TNFα)-induced phosphorylation of IKKβ (a component of IKK complex) thus inhibiting translocation of NF-κB to the nucleus thus inhibiting NF-κB signaling. CDDO-Me also blocks interleukin-6 (IL6)-induced and constitutive STAT3. CDDO-Me binds directly to STAT3inhibiting the formation of STAT3 dimers (Ahmad et al. 2006, 2008). CDDO-Me is being tested in clinical trials for the treatment of both lymphoid malignancies and solid tumors (RTA-402).

Deregulation of the Sphingolipid Rheostat in LGL Leukemia

Sphingolipids are biologically active lysophospholipids that act either directly or as second messengers to regulate diverse biological functions such as survival, proliferation, and migration. Ceramide (N-acyl sphingosine), a pro-apoptotic sphingolipid, can be synthesized either *de novo* (by condensation of serine and palmytoyl-CoA) or by catabolic pathway (by breaking down sphingomyelin). Ceramide is synthesized in cells following wide variety of stress or death signals, including Fas-FasL interaction. Ceramide can be deacylated into sphingosine by ceramidases such as acid ceramidase (ASAH1) or phosphorylated to ceramide-1-phosphate (C1P) by ceramide kinase. Sphingosine, then, can be phosphorylated by one of the two sphingosine kinases (SPHK1 or SPHK2) into sphingosine-1-phosphate (S1P). Though structurally related to ceramide, S1P is a pro-survival molecule. Given that pro-apoptotic (such as ceramide and sphingosine) and anti-apoptotic (such as S1P and C1P) sphingolipids exist in a rapidly exchanging equilibrium, it has been proposed that the relative amounts rather than absolute quantities of these molecules, determine the cell fate – this equilibrium is known as the sphingolipid rheostat (Wymann and Schneiter 2008; Rosen and Goetzl 2005; Spiegel and Milstien 2002).

Role of sphingolipids, especially that of S1P, in oncogenesis, metastasis and angiogenesis is well established (Visentin et al. 2006). S1P acts in autocrine or paracrine manner, either intracellularly or through one of the five highly specific S1P receptors (S1PR) on the cell surface – $S1P_1$ through $S1P_5$ (Spiegel and Milstien 2002). S1PRs belong to a class of receptors known as G-protein coupled receptors (GPCR). GPCR are coupled with G-proteins that act as relay junction for transmitting extracellular signals to various signaling pathways. One such G-protein relevant to S1P biology is $G\alpha_{12}$. The roles of three components of sphingolipid rheostat are well studied – sphingosine kinase, acid ceramidase, and G-protein $G\alpha_{12}$. It was shown that sphingolipid rheostat is dysregulated in leukemic LGL. This altered sphingolipid rheostat is believed to confer survival on leukemic LGL. It was shown that ASAH1 and SPHK were upregulated in leukemic LGL. It was also shown that Galpha12-mediated signaling was upregulated in leukemic LGL. Further, leukemic LGL expressed different pattern of S1PR compared to their healthy counterparts. Finally, disruption of the sphingolipid rheostat induced apoptosis in leukemic LGL but not in LGL derived from healthy controls.

SPHK is overexpressed in variety of tumors and is considered an oncogene (Spiegel et al. 1998; Milstien and Spiegel 2006). The activation of various survival signaling pathways implicated in tumorigenesis such as platelet-derived growth factor (PDGF) or vascular endothelial growth factor (VEGF) result in the activation of SPHK. SPHK in turn activates many of these pathways constituting a positive feedback loop. Persistent elevation of phosphatidylinositol (3,4,5) trisphosphate results in the activation of SPHK linking PI3k-Akt cascade to sphingolipid signaling. Similarly, ERK activation leads to the activation of SPHK linking Ras-Mek-Erk signaling to sphingolipid rheostat. Activated SPHK feeds into several proliferative and pro-survival pathways by activating ERK, PI3K and NF-κB. SPHK promotes cell survival by increasing cellular concentration of S1P while reducing concentrations of ceramide and sphingosine. This puts SPHK in unique position where various survival signaling pathways converge, rendering it an interesting candidate for anti-tumor therapy.

Acid ceramidase functions upstream to SPHK in sphingolipid metabolism. Upregulation of ceramidase is a survival mechanism used by variety of human tumors to combat ceramide production that normally follows various apoptosis-inducing insults (Spiegel et al. 1998; Kolesnick 2002; Park and Schuchman 2006). N-oleoylethanolamine (NOE), a chemical inhibitor of acid ceramidase, selectively induces apoptosis in various types of tumor cells (Morales et al. 2006).

Gα_{12} is coupled with S1PRs S1P$_1$ and S1P$_5$. Constitutive active form of Gα_{12} is sufficient to induce transformation that is at least partially mediated by activating STAT3. The components believed to play a role in Gα_{12}-mediated transformation include JAK3, PDGFα, and PI3K (Kumar et al. 2005). All these components are known to be deregulated in LGL leukemia suggesting involvement of Gα_{12} in survival of leukemic LGL (Shah et al. 2008; Epling-Burnette et al. 2001; Schade et al. 2006; Zhang et al. 2008).

S1P receptor S1P$_1$ is the most predominant S1PR in human naïve CD8+ T-cells, while other receptors are expressed at very low levels. Following activation, S1P$_1$ and S1P$_5$ are further downregulated in normal CD8+ cells (Shah et al. 2008). In contrast to normal CD8+ T-cells, S1P$_5$ is the predominant S1PR on leukemic LGL. S1P$_1$ is constitutively downregulated in LGL leukemia PBMC suggesting that leukemic LGL are activated *in vivo* (Shah et al. 2008; Brinkmann 2007).

In T-cells, S1P plays a role by protecting cells against ceramide and Fas-FasL mediated apoptosis (Wymann and Schneiter 2008; Cuvillier et al. 1998; Goetzl et al. 1999). S1P protects T-cells against Fas-mediated apoptosis in various T-cell lines as well as in human PBMC (Shah et al. 2008; Cuvillier et al. 1998; Goetzl et al. 1999). Microarray analysis showed that sphingolipid- and Gα_{12}-mediated signaling was enriched in leukemic LGL. *ASAH1* was identified as a core enriched component in leukemic LGL. In agreement with Fas-sensitive phenotype of activated normal PBMC, ASAH1 is downregulated to undetectable levels following activation in normal PBMC. In contrast, leukemic LGL express abundant ASAH1 presumably facilitating the breakdown of ceramide resulting in the Fas-resistant phenotype observed in leukemic LGL (Lamy et al. 1998; Shah et al. 2008).

As discussed above, sphingolipid rheostat is dysregulated in leukemic LGL. Hence, it was hypothesized that the disruption of the sphingolipid rheostat in a way that tilts the balance in the favor of pro-apoptotic molecules (such as ceramide) and away from anti-apoptotic molecules (such as S1P) should lead to induction of apoptosis in leukemic LGL. Indeed, the inhibition of ASAH1 using its chemical inhibitor NOE induced significant apoptosis in leukemic LGL (Shah et al. 2008). Similarly, the inhibition of SPHK using chemical inhibitors (SKI-I and SKI-II) led to induction of apoptosis in leukemic LGL (Zhang et al. 2008). Further, the inhibition of S1P-mediated signaling using FTY720 led to induction of apoptosis in leukemic LGL but not in LGL isolated from healthy controls. FTY720 treatment also restored Fas-sensitivity in leukemic LGL suggesting a role of S1P-mediated signaling in protection against AICD (Shah et al. 2008). Collectively, these results suggest the role of sphingolipid rheostat in the pathogenesis of LGL leukemia.

Therapeutic Opportunities

FTY720 or Fingolimod is a novel compound in clinical trials as immunomodulator in post-renal transplant patients, in patients with allergic conditions such as asthma, and in those with autoimmune conditions such as multiple sclerosis (MS). FTY720 is structurally similar to S1P and following phosphorylation, can bind to four out of five S1PR (except $S1P_2$) acting as a functional antagonist (Brinkmann 2007). Given that FTY720 selectively induces apoptosis of leukemic LGL and that LGL leukemia patients often have coexisting autoimmune diseases, it is possible that FTY720 may have a therapeutic role in these patients.

Sphingosine kinase inhibitors, SKI-I and SKI-II, are known to have anti-tumor activity and have proven their anti-tumor activity in mouse models (French et al. 2006). Development of newer, more selective inhibitors targeting SPHK offer exciting therapeutic opportunities (Paugh et al. 2008).

Neutralizing antibody to S1P has been proposed as a candidate for blocking survival and angiogenic signaling in various tumors. It works as a 'sponge' to absorb extracellular S1P – leading to abolition of S1P-mediated protection against apoptosis (Milstien and Spiegel 2006). However, there are no reports of neutralizing antibody to S1P as a potential in therapeutic in leukemia. It remains to be seen if it might have activity in LGL leukemia.

Ceramide is a well established mediator of apoptosis *in vitro* and *in vivo*. However, the use of ceramide as a therapeutic agent is stymied by its insolubility. Liposome-based drug delivery system has been proposed as a mean to overcome this therapeutic barrier. It was recently shown that treatment with ceramide containing nanoliposomes resulted in complete remission in rat model of aggressive NK-LGL leukemia. Polyethylene glycol coated liposomes, called pegylated liposomes, were shown to facilitate entry of ceramide into the cells. Nanoliposomes that are formulated at 80 ± 15 nm in size and contain 30 mol% cell-permeable ceramide were shown to be much less toxic than the delivery of 'free' ceramide both *in vitro* and *in vivo*.

Following delivery in nanoliposomal form, ceramide accumulated in mitochondria of leukemic NK cells resulting in caspase-dependent apoptosis of leukemic NK-cells both *in vitro* and *in vivo*. This was accompanied by the inhibition of ERK activity and downregulation of survivin expression.

To test the therapeutic benefit of nanolipsomal ceramide *in vivo*, Fischer F344 rat LGL leukemia model were treated either with ghost or C6-ceramide nanoliposomes. It was shown that treatment with ceramide containing nanoliposomes conferred significantly prolonged survival in addition to normalization of LGL counts in the blood, bone marrow, lymph nodes and lungs. Given the safety profile in various animal models, ceramide containing nanoliposomes may be an important therapeutic modality in patients with LGL leukemia. (Liu et al., *in press*) With the availability of technique to coat nanoliposomes with antibodies, it is tempting to predict the advent of targeted nanoliposomes such as those targeting only CD3+ cells. Such a treatment modality would be expected to achieve greater efficacy at lower dose with fewer side effects.

The Role of Interleukin-15 Signaling in LGL Leukemia

Recently, network theory approach was used to identify the most important regulators in the survival of leukemic LGL. First, a survival signaling network was constructed by integrating signaling pathways involved in normal CTL activation and the known deregulations of survival signaling in leukemic T-LGL. Then, the network was translated into a predictive discrete dynamic Boolean model. By simulating node deregulations corresponding to known signaling deregulations, it was concluded that the constitutive presence of interleukin-15 (IL15), the presence of PDGF, and the initial T-cell activation signal are sufficient to reproduce all known deregulations in leukemic T-LGL. IL15 and PDGF were shown to play a crucial roles in the survival of leukemic LGL, experimentally validating predictions made using network modeling approach (Zhang et al. 2008).

The generation of memory lymphocytes depends on antigenic stimulation. The survival of these memory lymphocytes requires cytokines such as IL2 and IL15. Even though IL2 and IL15 share two subunits of their heterotrimeric receptors, their role in CTL homeostasis is contrasting – while IL2 facilitates AICD, IL-15 promotes proliferation and long-term survival of memory phenotype CD8+ T-cells in an antigen-independent fashion. Il15 also promotes survival of NK cells *in vivo* (Waldmann 2006; Liu et al. 2002b). Given its role in survival of both memory T-cells and NK-cells it was not unexpected that IL15 has a pivotal role in the pathogenesis of both forms of LGL leukemia. The inhibition of IL15 signaling (by blocking IL15 or the unique subunit of IL15 receptor complex, IL15Rα) induces apoptosis of leukemic LGL and increases the levels of BH3-interacting domain death agonist (Bid) in leukemic LGL, suggesting role that Bid plays in abnormal survival of leukemic LGL (Hodge et al. 2009).

Bid is a BH3-only member of Bcl-2 family of proteins. Bid is a proteolytic substrate for activated caspases -3 and -8 in death receptor signaling pathway as well as for granzyme B. Following cleavage, truncated Bid (tBid) translocates to the outer mitochondrial membrane to participate in mitochondrial permeabilization, thus acting as a sentinel for protease-mediated death signaling. This places Bid in a unique position connecting death-receptor signaling pathway to mitochondrial pathway of cell death (Billen et al. 2008).

In normal NK cells IL15 upregulates E3 ligase HDM2. HDM2 directly interacts with Bid and significantly reduces Bid accumulation by enhancing proteasomal degradation of the latter. Bid levels are low in both leukemic T- and NK-LGL compared to their normal counterparts. It was further shown that forced overexpression of Bid using transduction approach leads to enhanced apoptosis of T-LGL leukemia cells (Hodge et al. 2009).

Therapeutic Implications

Bortezomib (PS341, Velcade) is a proteasomal inhibitor that is used in the treatment of various cancers such as multiple myeloma and mantle cell lymphoma. Bortezomib is an inhibitor of 26 S proteasomal complex, a large multi-subunit complex that degrades ubiquinated proteins in eukaryotic nucleus and cytosol. Bortezomib induces apoptosis in a wide variety of cancer cell lines with relatively few toxic effects on their normal counterparts. By inhibiting proteasomal pathway, Bortezomib interferes with numerous signaling pathways (Mitchell 2003).

The treatment of leukemic T- or NK-LGL with Brotezomib led to induction of apoptosis in these cells. Bortezomib also induced the expression of Bid. It was proposed that Bortezomib induces the expression of Bid which is followed by cleavage of Bid into tBid. tBid, then, proceeds with its pro-apoptotic activity. Bortezomib also induced Fas or TRAIL-independent apoptosis in leukemic LGL, leading to a hypothesis that Bid induced apoptosis may be explained by its role in DNA damage and repair (Hodge et al. 2009).

Deregulation of PDGF Signaling in LGL Leukemia

PDGF is a major growth factor and mitogen for various cell types. A molecule of PDGF is a dimer made up of structurally similar A, B, C, and D polypeptide chains, which combine to four homodimeric (PDGF-AA, PDGF-BB, PDGF-CC and PDGF-DD) and one heterodimeric PDGF-AB proteins. PDGF is synthesized by many different cell types, and its expression is broad. The PDGF isoforms exert their cellular effects by binding to two structurally related receptor subunits, denoted the α-receptor and the β-receptor. Three PDGF receptor (PDGFR) complexes PDGF-αα, PDGF-ββ and PDGF-αβ mediate overlapping but not identical signaling. PDGF-α and PDGF-β belong to type III–receptor tyrosine kinase (RTK) families and are characterized by five immunoglobulin-like domains in the extracellular region, a transmembrane domain, an ATP-binding site, and a hydrophilic kinase insert domain in the cytosol. The activation of PDGFR complex can activate many major signal transduction pathways, including the Jak-Stat, PI3k-Akt, and Ras-Mek-Erk signaling pathways. The activation of PDGFR thus results to downstream effects such as stimulation of cell growth, DNA replication, angiogenesis, and wound healing.

The *sis* oncogene of simian sarcoma virus (SSV) was found to encode the B-chain of PDGF more than two decades ago. SSV transformation involves autocrine stimulation by a PDGF-like molecule. Since then, dysregulation of PDGF signaling has been implicated in various malignancies including leukemia, gastrointestinal stromal tumors (GIST), glioblastoma multiforme (GBM), dermatofibrosarcoma protruberans, and in myeloproliferative diseases such as chronic myelomonocytic leukemia (CMML). In addition, overproduction of PDGF is implicated in autocrine and paracrine growth stimulation of various tumors (Heldin and Westermark 1999; Alvarez et al. 2006).

As described above, the presence of enhanced IL15 and PDGF signaling is sufficient to reproduce all known deregulations in T-LGL leukemia (Zhang et al. 2008). In this sense, PDGF can be considered one of the master switches in orchestrating survival signaling in T-LGL leukemia. Platelet-poor plasma from patients from both forms of LGL leukemia had about threefold higher levels of circulating PDGF-BB compared to their normal counterparts. The source of PDGF-BB was traced back to leukemic LGL. Interestingly, PDGF signaling operates in autocrine nature as leukemic LGL overexpress PDGFR-β transcripts compared to their normal counterparts. Sera from patient with LGL leukemia results in enhanced autophosphorylation of PDGF-RTK in leukemic LGL. Enhanced PDGF signaling results in constitutive activation of Sfk-PI3k-Akt and Mek-Erk signaling pathways in leukemic LGL. The inhibition of SFK activity (using chemical inhibitor PP2) or PI3K activity (using chemical inhibitor LY294002) inhibited PDGF-BB-induced activation of AKT and ERK in both the forms of LGL leukemia, resulting in the induction of apoptosis. Finally, the inhibition of PDGF-β-RTK (using chemical inhibitor AG1296) induces apoptosis in leukemic LGL (Zhang et al. 2008; Yang et al. 2010).

The direct effect of PDGF signaling on proliferation of leukemic LGL was demonstrated when the addition of PDGF-BB led to sixfold increase in cell numbers compared to sera from normal counterparts. This PDGF-BB dependent proliferation of leukemic LGL could be blocked by inhibition of PDGF-β-RTK, SFK, or PI3K. Neutralizing antibody to PDGF-BB inhibited cell proliferation in the leukemic LGL in PI3k-Akt dependent manner, placing PDGF signaling upstream to PI3k-Akt in LGL leukemia survival signaling (Yang et al. 2010).

Therapeutic Implications

Imatinib (STI-571, Gleevec) is a RTK-inhibitor that inhibits tyrosine kinases such as PDGFR, ABL and KIT at therapeutic concentrations *in vivo*. Imatinib has been successfully tried in patients with myeloproliferative disease involving dysregulation of PDGFR signaling (Apperley et al. 2002). While the anti-neoplastic role of imatinib is well established, recent publications suggest its usefulness in chronic inflammation and autoimmune conditions such as scleroderma and arthritis (Paniagua et al. 2006; Chung et al. 2009). While the exact mechanism of action of imatinib in autoimmune conditions is not known, it may be hypothesized that the inhibition of PDGF signaling as occurs in leukemic LGL may play a role. Given the safety profile of imatinib and the fact that PDGF-β is implicated in the survival of leukemic LGL, it would be of interest to see if imatinib can be used as a potential therapeutic strategy for patients with both T- and NK-LGL leukemia.

Concluding Remarks

LGL leukemia is a rare lymphoproliferative disorder of cytotoxic cells. The mainstay of treatment for LGL leukemia as discussed above is immunosupression. While drugs such as MTX and CSA are effective in many patients, the current modalities of treatment are not curative. The problems with such conventional therapies include the long duration of treatment with many patients receiving the treatment indefinitely, side effects including the possibility of second malignancy, and frequent recurrence. Further, in most studies complete hematological response (CHR) was achieved far less common than partial response (PR) indicating residual disease. Given these problems, it is imperative that more targeted therapy for LGL leukemia be developed.

Given the cytotoxic potential of CTL and NK cells, it is hardly any surprise that LGL leukemia is associated with various autoimmune disorders. Due to this close association with variety of autoimmune diseases, it is proposed that therapy aimed at leukemic LGL may also be of use in patients with autoimmune diseases. Since its first description, many survival mechanisms in leukemic LGL are unraveled. With development of more specific small molecular inhibitors, oligonucleotides, peptides, and monoclonal antibodies, it will be of great interest to see evolution of the therapeutic armamentarium for patients with LGL leukemia.

References

Ahmad R et al (2006) Triterpenoid CDDO-Me blocks the NF-kappaB pathway by direct inhibition of IKKbeta on Cys-179. J Biol Chem 281(47):35764–35769

Ahmad R et al (2008) Triterpenoid CDDO-methyl ester inhibits the janus-activated kinase-1 (JAK1) à signal transducer and activator of transcription-3 (STAT3) pathway by direct inhibition of JAK1 and STAT3. Cancer Res 68(8):2920–2926

Alvarez RH, Kantarjian HM, Cortes JE (2006) Biology of platelet-derived growth factor and its involvement in disease. Mayo Clin Proc 81(9):1241–1257

Apperley JF et al (2002) Response to imatinib mesylate in patients with chronic myeloproliferative diseases with rearrangements of the platelet-derived growth factor receptor beta. N Eng J Med 347(7):481–487

Armand J-P et al (2007) The emerging role of targeted therapy for hematologic malignancies: update on bortezomib and tipifarnib. Oncologist 12(3):281–290

Billen LP, Shamas-Din A, Andrews DW (2008) Bid: a Bax-like BH3 protein. Oncogene 27(S1):S93–S104

Bremer E et al (2008) Superior activity of fusion protein scFvRit:sFasL over cotreatment with rituximab and Fas agonists. Cancer Res 68(2):597–604

Brinkmann V (2007) Sphingosine 1-phosphate receptors in health and disease: mechanistic insights from gene deletion studies and reverse pharmacology. Pharmacol Ther 115(1):84–105

Budd RC, Yeh W-C, Tschopp J (2006) cFLIP regulation of lymphocyte activation and development. Nat Rev Immunol 6(3):196–204

Buettner R, Mora LB, Jove R (2002) Activated STAT signaling in human tumors provides novel molecular targets for therapeutic intervention. Clin Cancer Res 8(4):945–954

Christine R et al (2005) Implication of STAT3 signaling in human colonic cancer cells during intestinal trefoil factor 3 (TFF3) â€“ and vascular endothelial growth factorâ€”mediated cellular invasion and tumor growth. Cancer Res 65(1):195–202

Chung L et al (2009) Molecular framework for response to imatinib mesylate in systemic sclerosis. Arthritis Rheum 60(2):584–591

Cuvillier O et al (1998) Sphingosine 1-phosphate inhibits activation of caspases that cleave poly(ADP-ribose) polymerase and lamins during Fas- and ceramide-mediated apoptosis in jurkat T lymphocytes. J Biol Chem 273(5):2910–2916

Epling-Burnette PK, Liu JH, Catlett Falcone R, Turkson J, Oshiro M, Kothapalli R, Li Y, Wang JM, Yang-Yen HF, Karras J, Jove R, Loughran TP Jr (2001) Inhibition of STAT3 signaling leads to apoptosis of leukemic large granular lymphocytes and decreased Mcl-1 expression. J Clin Invest 107(3):351–362

Epling-Burnette PK et al (2004a) Dysregulated NK receptor expression in patients with lymphoproliferative disease of granular lymphocytes. Blood 103(9):3431–3439

Epling-Burnette PK et al (2004b) ERK couples chronic survival of NK cells to constitutively activated Ras in lymphoproliferative disease of granular lymphocytes (LDGL). Oncogene 23(57):9220–9229

Epling-Burnette PK et al (2008) Clinical improvement by farnesyltransferase inhibition in NK large granular lymphocyte leukemia associated with imbalanced NK receptor signaling. Blood 112(12):4694–4698

French KJ et al (2006) Antitumor activity of sphingosine kinase inhibitors. J Pharmacol Exp Ther 318(2):596–603

Goetzl EJ, Kong Y, Mei B (1999) Lysophosphatidic acid and sphingosine 1-phosphate protection of T cells from apoptosis in association with suppression of Bax. J Immunol 162(4): 2049–2056

Hayden MS, West AP, Ghosh S (2006) NF-[kappa]B and the immune response. Oncogene 25(51):6758–6780

Heldin C-H, Westermark B (1999) Mechanism of action and in vivo role of platelet-derived growth factor. Physiol Rev 79(4):1283–1316

Hodge DL et al (2009) Interleukin-15 enhances proteasomal degradation of bid in normal lymphocytes: implications for large granular lymphocyte leukemias. Cancer Res 69(9):3986–3994

Hoffmann A, Baltimore D (2006) Circuitry of nuclear factor{kappa}B signaling. Immunol Rev 210(1):171–186

Holmström TH et al (2000) MAPK/ERK signaling in activated T cells inhibits CD95/Fas-mediated apoptosis downstream of DISC assembly. EMBO J 19(20):5418–5428

Huntington ND et al (2007) Interleukin 15-mediated survival of natural killer cells is determined by interactions among Bim, Noxa and Mcl-1. Nat Immunol 8(8):856–863

Iwamaru A et al (2006) A novel inhibitor of the STAT3 pathway induces apoptosis in malignant glioma cells both in vitro and in vivo. Oncogene 26(17):2435–2444

Jeremias I et al (1998) Inhibition of nuclear factor kappa B activation attenuates apoptosis resistance in lymphoid cells. Blood 91(12):4624–4631

Jones RG et al (2002) CD28-Dependent activation of protein kinase B/Akt blocks Fas-mediated apoptosis by preventing death-inducing signaling complex assembly. J Exp Med 196(3):335–348

Kolesnick R (2002) The therapeutic potential of modulating the ceramide/sphingomyelin pathway. J Clin Invest 110(1):3–8

Kong D, Yamori T (2008) Phosphatidylinositol 3-kinase inhibitors: promising drug candidates for cancer therapy. Cancer Sci 99(9):1734–1740

Kontgen F et al (1995) Mice lacking the c-rel proto-oncogene exhibit defects in lymphocyte proliferation, humoral immunity, and interleukin-2 expression. Genes Dev 9(16):1965–1977

Krammer PH (2000) CD95′s deadly mission in the immune system. Nature 407(6805):789–795

Kumar RN, Shore SK, Dhanasekaran N (2005) Neoplastic transformation by the gep oncogene, G[alpha]12, involves signaling by STAT3. Oncogene 25(6):899–906

Lamy T, Loughran TP (1998) Large granular lymphocyte leukemia. Cancer Control 5(1):25–33

Lamy T, Loughran TP Jr (2003) Clinical features of large granular lymphocyte leukemia. Semin Hematol 40(3):185–195

Lamy T et al (1998) Dysregulation of CD95/CD95 ligand-apoptotic pathway in CD3(+) large granular lymphocyte leukemia. Blood 92(12):4771–4777

Lavrik I et al (2003) The active caspase-8 heterotetramer is formed at the CD95 DISC. Cell Death Differ 10(1):144–145

Levy DE, Inghirami G (2006) STAT3: a multifaceted oncogene. Proc Natl Acad Sci U S A 103(27):10151–10152

Levy DE, Lee C-k (2002) What does Stat3 do? J Clin Invest 109(9):1143–1148

Liu JH et al (2002a) Blockade of Fas-dependent apoptosis by soluble Fas in LGL leukemia. Blood 100(4):1449–1453

Liu K et al (2002b) IL-15 mimics T cell receptor crosslinking in the induction of cellular proliferation, gene expression, and cytotoxicity in CD8+ memory T cells. Proc Natl Acad Sci U S A 99(9):6192–6197

Loughran TP Jr (1993) Clonal diseases of large granular lymphocytes. Blood 82(1):1–14

Matiba B, Mariani SM, Krammer PH (1997) The CD95 system and the death of a lymphocyte. Semin Immunol 9(1):59–68
Mayo LD, Donner DB (2001) A phosphatidylinositol 3-kinase/Akt pathway promotes translocation of Mdm2 from the cytoplasm to the nucleus. Proc Natl Acad Sci U S A 98(20):11598–11603
Milstien S, Spiegel S (2006) Targeting sphingosine-1-phosphate: a novel avenue for cancer therapeutics. Cancer Cell 9(3):148–150
Mitchell BS (2003) The proteasome – an emerging therapeutic target in cancer. N Eng J Med 348(26):2597–2598
Mor A, Philips MR (2006) Compartmentalized Ras/Mapk signaling. Annu Rev Immunol 24(1):771–800
Morales A et al (2006) Pharmacological inhibition or small interfering RNA targeting acid ceramidase sensitizes hepatoma cells to chemotherapy and reduces tumor growth in vivo. Oncogene 26(6):905–916
Murray PJ (2007) The JAK-STAT signaling pathway: input and output integration. J Immunol 178(5):2623–2629
Osuji N et al (2006) T-cell large granular lymphocyte leukemia. Cancer 107(3):570–578
Paniagua RT et al (2006) Selective tyrosine kinase inhibition by imatinib mesylate for the treatment of autoimmune arthritis. J Clin Invest 116:2633–2642
Park J-H, Schuchman EH (2006) Acid ceramidase and human disease. Biochimica et Biophysica Acta (BBA) Biomembr 1758(12):2133–2138
Paugh SW et al (2008) A selective sphingosine kinase 1 inhibitor integrates multiple molecular therapeutic targets in human leukemia. Blood 112(4):1382–1391
Powell BL et al (2010) Arsenic trioxide improves event-free and over-all survival for adults with acute promyelocytic leukemia: north American leukemia intergroup study C9710. Blood 116(19):3751–3757. doi:10.1182/blood-2010-02-269621
Pyne S, Pyne NJ (2002) Sphingosine 1-phosphate signalling and termination at lipid phosphate receptors. Biochimica et Biophysica Acta (BBA) Mol Cell Biol Lipids 1582(1–3):121–131
Rivera-Walsh I et al (2000) The NF-kappa B signaling pathway is Not required for Fas ligand gene induction but mediates protection from activation-induced cell death. J Biol Chem 275(33):25222–25230
Rosen H, Goetzl EJ (2005) Sphingosine 1-phosphate and its receptors: an autocrine and paracrine network. Nat Rev Immunol 5(7):560–570
RTA-402; http://clinicaltrials.gov/ct2/results?term=cddo-me
Samelson LE (2002) Signal transduction mediated by the T cell antigen receptor: the role of adapter proteins. Annu Rev Immunol 20(1):371–394
Santos FPS et al (2010) Phase 2 study of CEP-701, an orally available JAK2 inhibitor, in patients with primary or post-polycythemia vera/essential thrombocythemia myelofibrosis. Blood 115(6):1131–1136
Schade AE et al (2006) Phosphatidylinositol-3-phosphate kinase pathway activation protects leukemic large granular lymphocytes from undergoing homeostatic apoptosis. Blood 107(12):4834–4840
Schrama D, Reisfeld RA, Becker JC (2006) Antibody targeted drugs as cancer therapeutics. Nat Rev Drug Discov 5(2):147–159
Schubbert S, Shannon K, Bollag G (2007) Hyperactive Ras in developmental disorders and cancer. Nat Rev Cancer 7(4):295–308
Schust J et al (2006) Stattic: a small-molecule inhibitor of STAT3 activation and dimerization. Chem Biol 13(11):1235–1242
Sen R, Baltimore D (1986) Multiple nuclear factors interact with the immunoglobulin enhancer sequences. Cell 46(5):705–716
Shah MV et al (2008) Molecular profiling of LGL leukemia reveals role of sphingolipid signaling in survival of cytotoxic lymphocytes. Blood 112(3):770–781
Smyth MJ et al (2001) Unlocking the secrets of cytotoxic granule proteins. J Leukoc Biol 70(1):18–29

Sokol L, Loughran TP Jr (2006) Large granular lymphocyte leukemia. Oncologist 11(3):263–273
Spiegel S, Milstien S (2002) Sphingosine 1-phosphate, a key cell signaling molecule. J Biol Chem 277(29):25851–25854
Spiegel S et al (1998) Sphingosine-1-phosphate in cell growth and cell death. Ann N Y Acad Sci 845(1):11–18
STAT3 DECOY in head and neck cancer; clinicalTrials.gov identifier:NCT00696176, http://clinicaltrials.gov/ct2/show/NCT00696176
STAT3 Inhibitor for solid tumors; clinicalTrials.gov identifier: NCT00955812; http://clinicaltrials.gov/ct2/show
Steelman LS et al (2008) Contributions of the Raf/MEK/ERK, PI3K/PTEN/Akt/mTOR and Jak/STAT pathways to leukemia. Leukemia 22(4):686–707
Tan M et al (2006) Selective inhibition of ErbB2-overexpressing breast cancer in vivo by a novel TAT-based ErbB2-targeting signal transducers and activators of transcription 3-blocking peptide. Cancer Res 66(7):3764–3772
Tanaka M et al (1996) Fas ligand in human serum. Nat Med 2(3):317–322
Thome M, Tschopp J (2001) Regulation of lymphocyte proliferation and death by FLIP. Nat Rev Immunol 1(1):50–58
Tretinoin, cytarabine, and daunorubicin with or without arsenic trioxide followed by tretinoin with or without mercaptopurine and methotrexate in treating patients with acute promyelocytic leukemia; NCT00003934; http://clinicaltrial.gov
Turkson J et al (2004) Inhibition of constitutive signal transducer and activator of transcription 3 activation by novel platinum complexes with potent antitumor activity. Mol Cancer Ther 3(12):1533–1542
Veillette A, Latour S, Davidson D (2002) Negative regulation of immunoreceptor signaling. Annu Rev Immunol 20(1):669–707
Visentin B et al (2006) Validation of an anti-sphingosine-1-phosphate antibody as a potential therapeutic in reducing growth, invasion, and angiogenesis in multiple tumor lineages. Cancer Cell 9(3):225–238
Vivanco I, Sawyers CL (2002) The phosphatidylinositol 3-Kinase-AKT pathway in human cancer. Nat Rev Cancer 2(7):489–501
Waldmann TA (2006) The biology of interleukin-2 and interleukin-15: implications for cancer therapy and vaccine design. Nat Rev Immunol 6(8):595–601
Wan YY, DeGregori J (2003) The survival of antigen-stimulated T cells requires NF[kappa] B-mediated inhibition of p73 expression. Immunity 18(3):331–342
Wetzler M et al (2006) Arsenic trioxide affects signal transducer and activator of transcription proteins through alteration of protein tyrosine kinase phosphorylation. Clin Cancer Res 12(22):6817–6825
Wymann MP, Schneiter R (2008) Lipid signalling in disease. Nat Rev Mol Cell Biol 9(2):162–176
Xi S, Gooding WE, Grandis JR (2004) In vivo antitumor efficacy of STAT3 blockade using a transcription factor decoy approach: implications for cancer therapy. Oncogene 24(6):970–979
Yang J et al (2008) Antigen activation and impaired Fas-induced death-inducing signaling complex formation in T-large-granular lymphocyte leukemia. Blood 111(3):1610–1616
Yang J et al (2010) Platelet-derived growth factor mediates survival of leukemic large granular ymphocytes via an autocrine regulatory pathway. Blood 115(1). doi:51
Zhang J, Xu X, Liu Y (2004) Activation-induced cell death in T cells and autoimmunity. Cell Mol Immunol 1(3):186–192
Zhang R et al (2008) Network model of survival signaling in large granular lymphocyte leukemia. Proc Natl Acad Sci U S A 105(42):16308–16313
Zhang R, Shah MV, Loughran TP Jr (2009) The root of many evils: indolent large granular lymphocyte leukaemia and associated disorders. Hematol Oncol 28(3):105–117
Zhu K et al (2005) Farnesyltransferase inhibitor R115777 (Zarnestra, Tipifarnib) synergizes with paclitaxel to induce apoptosis and mitotic arrest and to inhibit tumor growth of multiple myeloma cells. Blood 105(12):4759–4766

Epigenetic Regulation and Therapy in Lymphoid Malignancies

Yizhuo Zhang, Shanqi Guo, and Haifeng Zhao

Introduction

Lymphoid malignancies usually represent malignant counterparts of lymphocytes at discrete stages of normal lymphocyte differentiation; this disease is characterized by uncontrolled proliferation and survival of clonal neoplastic lymphoid cells. Major breakthroughs in the past have contributed to our understanding of the genetic failures and the changed biology in lymphoma cells that underlie the initiation and progression of the disease. It is now recognized that not only genetic but also epigenetic alterations are similarly important in this process. Enormous evidence has accumulated in the past decades that establishes the importance of epigenetic modifications in cancer and has resulted in shifting the focus from entirely genetic-based studies to integrated studies involving both genetic and epigenetic alterations. Since these alterations do not change the DNA sequences and are pharmacologically reversible, they have been regarded as optimal targets for what is now known as epigenetic therapy. In this review, we will discuss our current understanding of normal epigenetic processes, outline our knowledge of epigenetic alterations, discuss advances in the understanding of post-translational histone modifications (DNA acetylation and methylation) and microRNA in normal and in lymphoma cells, and explore novel therapies. We will also discuss how this information is being used to improve current therapy of this disease.

Y. Zhang, M.D., Ph.D. (✉) • S. Guo • H. Zhao
Department of Hematology and Blood and Marrow Transplantation, Tianjin Medical University Cancer Institute and Hospital and Tianjin Key Laboratory of Cancer Prevention and Therapy, Huan-Hu-Xi Road, Ti-Yuan-Bei, He Xi District, Tianjin, 300060, P.R. China
e-mail: aprilzyz@yahoo.com; shanqi_guo@yahoo.com; zhaohaifeng457@yahoo.cn

J. Tao and E. Sotomayor (eds.), *Hematologic Cancers: From Molecular Pathobiology to Targeted Therapeutics*, Cancer Growth and Progression 14,
DOI 10.1007/978-94-007-5028-9_17,

Targeting Histone Deactylation in the Treatment of Lymphoma

Histone acetylation plays a critical role in the regulation of gene expression. In general, hyperacetylated histones are associated with transcriptionally active genes, whereas hypoacetylated histones are associated with suppressed transcription. Histone acetyltransferase and histone deacetylase (HDAC) controls these changes and the dynamic balance between chromatin structure and gene expression of proteins involved in regulation of a variety of functions (Sharma et al. 2010). Therefore, HDAC is considered closely associated with a variety of tumors (Hagelkruys et al. 2011). As a promising new generation of targeted anti-cancer drugs, HDAC inhibitors (HDACi) have a higher response rate and have been highly focused in recent years. The anti-tumor activity of HDACi has been confirmed from a number of experiments, and I/ II clinical trials have shown that HDACi can have an effect on refractory cutaneous T cell lymphomas (CTCL), Hodgkin lymphoma (HL), myeloid tumors, and solid tumors (Mehnert and Kelly 2007). Presently, vorinostat and romidepsin have been approved by the US FDA for the treatment of relapsed CTCL.

Concept of HDAC

Studies have shown that histone acetylation and deacetylation modifications play important roles in the regulation of gene expression and tumor development. There are two types of enzymes that determine the degree of histone acetylation: histone acetyltransferase and HDAC.

HDACs are enzymes involved in the remodeling of chromatin and have a key role in the epigenetic regulation of gene expression. They catalyze the removal of acetyl groups from lysine residues in histone amino termini, leading to chromatin condensation and transcriptional repression. In addition, the activity of non-histone proteins can be regulated through HDAC-mediated hypo-acetylation (Fig. 1) (Hagelkruys et al. 2011).

Classification of HDAC

To date, eighteen HDACs have been identified in humans. They have been grouped into two major categories: zinc-dependent HDACs and NAD-dependent HDACs. These have been further classified into four major classes based on their homology to yeast HDACs, their subcellular localization, and their enzymatic activities. The class I HDACs (1, 2, 3, and 8) are homologous to the yeast RPD3 protein, can generally be detected in the nucleus, and show ubiquitous expression in various human cell lines and tissues. Class II HDACs (4, 5, 6, 7, 9, and 10) share homologies with

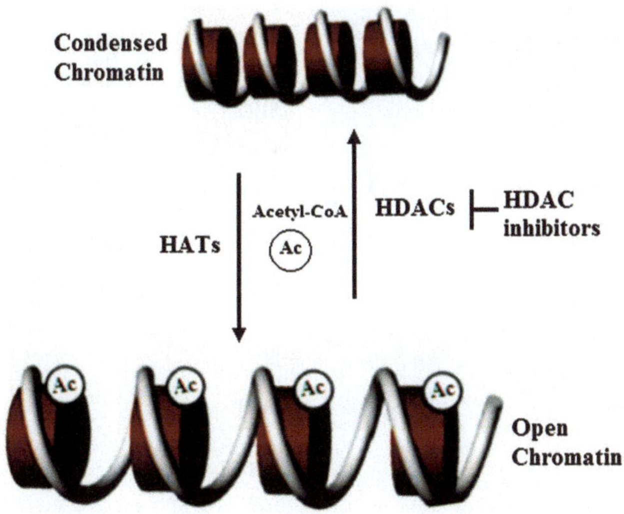

Fig. 1 Histone acetyltransferase (HAT) and histone deacetylase (HDAC) determine the degree of histone acetylation. The extent of acetylation and deacetylation on different positions of core histones is determined by the antagonistic activity of histone acetylases (HAT) and histone deacetylases (HDAC) and alters the nucleosomal conformation of both transformed and non-transformed cells

the yeast Hda1 protein and can shuttle between the nucleus and the cytoplasm. The class IIb HDACs, HDAC6 and 10, are found in the cytoplasm and contain two deacetylase domains. HDAC6 has unique substrate specificity with a tubulin deacetylase domain specific for the cytoskeletal protein tubulin. The class III HDACs (SIRT1, 2, 3, 4, 5, 6, and 7) are homologues of the yeast protein Sir2 and require NAD+ for their activity to regulate gene expression in response to changes in the cellular redox status. SIRT1 has been shown to interact with p53, then deacetylate it, resulting in repression of p53-mediated transcriptional activity. SIRT1 can negatively regulate multiple pathways, including both tumor suppressors (p53, FOXO) and oncogenic proteins (survivin, β-catenin, NF-κB). HDAC11 is the sole member of the class IV HDACs. There is similarity to the catalytic core regions of both class 1 and 2 enzymes; however, HCAD11 does not have a strong enough identity to be placed in either class (Fig. 2) (Copeland et al. 2010).

Role and Classification of HADC Inhibitors (HDACi)

HDACi alters gene transcription by inhibiting histone deacetylase activity and remodeling chromatin architecture. The rapid development in understanding the biology of lymphoid cells has led to the development of HDACi as a promising class of mechanism-based agents for the treatment of lymphoma. All currently

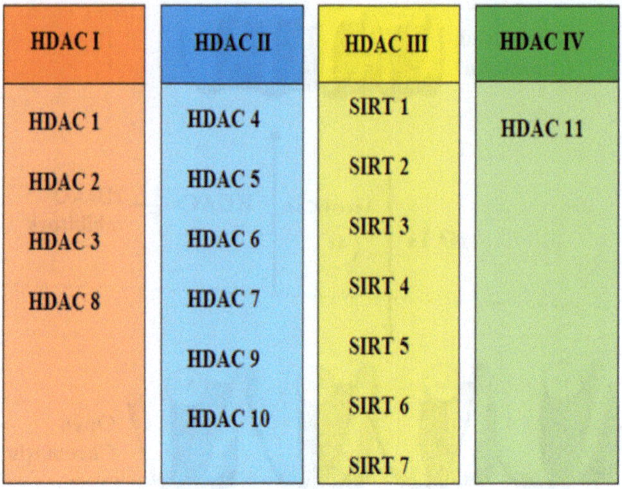

Fig. 2 Histone deacetylase (HDAC) classes based on function and DNA sequence similarity

Table 1 Classification of histone deacetyalse inhibitors

HDAC inhibitors	Compounds	Pharmacologic profile
Short chain fatty acids	Valproic acid, phenylbutyrate	Short plasma half life, rapid metabolism, non-specific mode of action
Hydroxamic acids	TrichostatinA, vorinostat (SAHA), PXD-101 (belinostat), LBH589 (panobinostat), ITF2357, pyroxamide, oxamflatin, scriptaid	Potent HDAC I/II/IV inhibitors, also called pan-HDAC inhibitors
Benzamides	MGCD0103, MS-275, CI-994, SK-7041	MGCD0103 is a class I specific HDAC inhibitor
Cyclic peptides	Romidepsin, etinostat (MS-275)	Very potent HDAC inhibitor with IC50s in low nanomolar range
Situin inhibitors	Niacinamide, sirtinol	Class II specific HDAC inhibitors

available HDACis mainly inhibit class 1 and 2 HDACs. Many structurally diverse compounds can bind to HDACs and inhibit their enzymatic activity. These compounds include hydroxamates, cyclic peptides, aliphatic acids, benzamides, and electrophilic ketones.

HDACi can be divided into the following five types based on their chemical structure (Table 1):

1. Hydroxamic acids, including trichostatin A, vorinostat (SAHA), panobinostat (LBH-589), and others. They are the earliest discovered and most extensively studied HDAC inhibitors so far. Hydroxamic acids have strong activity and simple structure, with major effects on the ClassI and Class II HDACs.

2. Short-chain fatty acids, such as butyric acid, benzene acid, and valproic acid. Compounds of this type possess a relatively simple structure, less selectivity, low inhibitory activity and bioavailability, and fast metabolism in vivo.
3. Cyclic tetrapeptide classes, including trapoxin and depsipeptide (FK228). This class can be further divided into two subclasses: Aoe-containing inhibitors and that without Aoe moiety. Moreover, Aoe-containing cyclic peptide could irreversibly bind and inhibit HDACs, while the other subclass without Aoe moiety refers to reversible inhibitors.
4. Benzamide comprises MS-275, MGCD103 and CI-994 and so on. The activity of these inhibitors is lower than that of the corresponding hydroxamic acids and cyclic peptide compounds, but this type has a high selectivity with class I HDAC.
5. Electrophilic ketones include a variety of trifluoromethyl ketones. When trifluoromethyl ketone carbonyl is reduced to hydroxyl groups, then it loses its activity.

Mechanisms of Action of HDACis

HDAC inhibitors exert a myriad of biological effects, including induction of differentiation/apoptosis, cell-cycle/mitotic arrest and induction of autophagic cell death. HDACIs can modify gene expression and altering the acetylation status of transcription factors and other proteins involved in transcription, and can lead to cell death by activating apoptosis by both the intrinsic and the extrinsic pathways, also, HDACis can cause DNA damage through production of ROS, and blocking the activity of the chaperone protein Hsp90.

Abnormal regulation of cell cycle is an important mechanism of tumorigenesis. Studies have shown that a variety of HDACi can induce cyclin-dependent kinase inhibitor p21 and activate the expression of apoptosis precursor protein. p21 is an important differentiation-related genes, and plays an important role for growth arrest by blocking cyclin-dependent kinase activity. Richon et al found that cell cycle kinase inhibitor p21 protein levels were significantly increased after giving HDACi for 6 ~ 24 hours, while the levels of cyclin A and D1 were decreased. Researches also indicate that HDACi can also induce p16, cyclin E, thioredoxin binding protein 2 (TBP2) and other cell cycle regulators.

It is noteworthy that though HDACi can cause high degree of histone acetylation, these genes in the activation of transcription are less than 2%, so we conclude that the HDACi on gene expression is selective.

HDACi can also regulate non-histone acetylation. The substrate proteins of HS0 involved in tumor cell proliferation and apoptosis and related to anti-tumor properties of the regulation. Yang et al show that by inducing acetylation of HSP90, Panobinostat induce tumor cell growth arrest and apoptosis. HDACi-induced HSP90 acetylation achieved by inhibiting the activity of HDAC6 .

HDACi mainly through the death receptor pathway or the mitochondrial pathway activate caspase and induces tumor cell apoptosis. Bokelmann reports suggest that valproic acid and vorinostat can activate pro-apoptotic factors Bid and Bim, while inhibiting anti-apoptotic Bcl-2 factor. Zhao and others found that excessive expression of E2F1 which is regulated by the pro-apoptotic factor Bim leading to cell apoptosis. Interestingly, Rosato et al have shown that HDACi can still induce cell death even deal all the cells with inhibitors of apoptosis. So we believe HDACi can also cause non-apoptotic cell-death. At present, most reports consider this is related to autophagy , but the specific mechanism remains unclear. Some believe that the mechanisms of HDACi-induced cell death also including the expression of Bcl-2 family with BH-3 region,P53 pathway, and ROS regulation (Mehnert and Kelly 2007; Cotto et al. 2010; Copeland et al. 2010).

Clinical Application of HDACis in Lymphoma

The rapid development in understanding the biology of lymphoid cells has led to HDACi developed as a promising class of mechanism-based agents for the treatment of lymphoma (Zain and O'Connor 2010).

All currently available HDACis mainly inhibit classIand II HDACs. Many structurally diverse compounds can bind to HDACs and inhibit their enzymatic activity, including hydroxamates, cyclic peptides, aliphatic acids, benzamides, and electrophilic ketones. Presently, vorinostat and romidepsin have been approved in the United States for the treatment of relapsed and refractory CTCL (Shakovich and Melnick 2011).

Hodgkin Lymphoma

With current therapies available, such as chemotherapy, radiotherapy, and hematopoietic stem cell transplantation, the cure rate of Hodgkin lymphoma (HL) is higher than 85%. However, for patients without complete remission, the prognosis is relatively poor and the median survival time is about 3 years. The latest clinical trials have shown that HDACi has a promising effect on the relapsed or refractory HL (Buglio and Younes 2010).

A phase II clinical trial of oral panobinostat (LBH589), conducted in 28 patients with relapsed or refractory HL, reported clinical efficacy in 2 cases (complete response; CR) and 24 cases of progressive response (PR), with maximum tolerated dose (MTD) defined as 40 mg/day. Subsequently, in another following phase II clinical trial in which panobinostat was dosed at MTD, of all 61 relapsed or refractory HL patients enrolled, one patient achieved CR, ten patients achieved PR, and 31 patients showed stable disease (SD). The common toxicity was thrombocytopenia, which was reversible by decreasing the dose. Panobinostat demonstrated an encouraging clinical activity in relapsed or refractory HL with fewer side effects (Sureda et al. 2010).

Mocetinostat (MGCD0103) is an amino phenylbenzamide inhibitor specific for HDAC classesIand IV. Researches have demonstrated that mocetinostat has a potent anti-proliferative activity in HL cell lines. Furthermore, mocetinostat can induce tumor necrosis factor α (TNF-α) expression and secretion. In a phase II clinical trial in relapsed or refractory HL, mocetinostat was administrated at a dose of 110 mg three times per week. Among the enrolled 20 patients, 7 had CR. However, as a result of an intolerable side effect, an additional ten patients were enrolled at a dose of 85 mg three times per week .The disease control rates in the 85 mg and 100 mg cohorts were 35% and 25%. Of four patients who died, all in the 100 mg cohort, two cases may have resulted from the pharmaceutical administration. However, mocetinostat, at 85 mg three times per week, was shown to be an effective single agent for relapsed or refractory HL with a manageable safety profile (Buglio et al. 2010).

B Cell Non-Hodgkin Lymphoma

B-cell non-Hodgkin lymphoma (NHL) is the most common type of lymphoma and has been readily classified into aggressive subtypes and indolent diseases. Diffuse large B-cell lymphoma (DLBCL) and follicular lymphoma (FL) are common types of NHL. Because of frequent relapse in B-cell lymphomas, novel effective treatments are clearly needed. At present, there are no HDACis approved specifically for the treatment of B-cell lymphomas, but a strong rational and clinical trials of some HDACis have shown potent activity for B-cell NHL (Crump et al. 2008a).

Vorinostat (SAHA), a pan-HDACi of class Iand II, is reported to be the first clinically effective HDACi in B-cell lymphomas, because it helped a transformed B-cell lymphoma patient achieve a CR. In an oral vorinostat phase I clinical trial in ten patients with B-cell malignancy, four patients achieved CR/PR (3 with follicular and 1 with mantle cell lymphoma), with MTD of oral vorinostat defined as 400 mg daily. The common related adverse events were anorexia, hyperlipidemia, albuminurea, fatigue, nausea, diarrhea, hypocalcemia, ana abnormal hematology (thrombocytopenia, anemia, leukopenia). However, adverse hematology reactions were reversible, with patients recovering from these events through a period of rest. In a phase II trial of vorinostat in the DLBCL, 1 of 18 patients enrolled achieved CR, with the time to remission more than 468 days. Thus, vorinostat, as a single agent, does not show the expecting activity in the treatment of B-cell lymphoma (Watanabe et al. 2010; Crump et al. 2008b).

Belinostat is another hydroxamic acid derivative, without preferential activity versus a special class of HDACs. In a phase I clinical trial of belinostat for treatment of advanced B cell tumor, with belinostat administered by shirt infusion at different doses (600, 900, and 1,000 mg/m^2/day) in 16 patients with relapse or refractory B cell lymphoma, five patients, including two patients with DLBCL, two with chronic lymphocytic leukemia (CLL), and one with multiple myeloma achieved disease stabilization, The adverse events reported were nausea, vomiting, fatigue, and diarrhea. Importantly, there were two patients, both with multiple myeloma, who suffered renal failure, the cause of which was not clear and needed further research.

Table 2 Clinical trials of HDACi in the CLTL

Agent	Phase	Sponsor or author	Number of cases	ORR	mDOR	mTTP	mTTR
Vorinostat	I/II	Duvic M	33	8%	15.1w	30.2w	11.9w
Vorinostat	II	Olsen	74	29.7%	>185d	4.9 m	55d
Romidepsin	II	NCI	71	34% (4CR)	13.7 m	15.1 m	2 m
Romidepsin	IIb	Gloucester Pharmaceutical company	96	34% (6CR)	15 m	8 m	2 m

Data in another phaseI clinical trial of oral belinostat in recurrent refractory lymphoma also showed an acceptable safety profile and more importantly signs of clinical efficiency in terms of stable disease. Particularly, clinical responses were reported in all patients with mantle cell lymphoma (Gimsing et al. 2008; Zain et al. 2009).

T Cell Non-Hodgkin's Lymphoma

T cells and NK cell tumors account for about 12% of NHL, with 15–20% being aggressive lymphomas. Presently, the 5-year survival rate for these patients was only 10–30%. Relapsed or refractory lymphoma is still a significant clinical problem to be solved (Zain and O'Connor 2010).

In a phaseI/II clinical trial of oral vorinostat, in which a total of 33 patients with relapsed or refractory CTCL were enrolled, the overall response rate (ORR) was 24.4%, the medium time to respond (mTTR) was 11.9 weeks, and the medium duration of response (mDOR) was 15.1 weeks. Consequently, the 400 mg/day dose was considered optimum in terms of response-toxicity profile for evaluation in a phase II multicenter trial. The related dose-limited toxicities included gastrointestinal disorders, anorexia, dehydration, fatigue, and bone marrow depression.

In a subsequent phase II study conducted in 74 relapsed or refractory CTCL patients, the ORR was 29.7% (including 1 CR), the mTTR, by far shorter, was just 8 weeks, and the mDOR was more than 185 days. The most common toxicities were diarrhea, fatigue, nausea, and anorexia. Compared with other drugs, vorinostat has the advantage of short time to respond, a good tolerance, and convenient oral administration, The agent has already been approved by the US FDA for treatment of relapsed or refractory CTCL in 2006 (Duvic et al. 2007; Olsen et al. 2007).

In a phase II study conducted by the National Cancer Institute, 71 patients with relapsed or refractory CTCL were given intravenous romidepsin at 14 mg/m^2. The results showed that the ORR was 34%, a CR was observed in four patients, mTTR was 2 months, and mDOR was 13.7 months. The related side effects included fatigue, anorexia, and hematologic abnormalities.

The activity of romidepsin in CTCL was defined further in another phase IIb trial conducted by Gloucester Pharmaceuticals. The results showed that romidepsin was associated with an ORR of 34% in the overall population, with four patients achieving a CR. The mDOR was 13 months, resulting in romidepsin being regarded as an important drug in treatment of relapsed or refractory CTCL. However, this drug should be carefully used in patients with heart disease because the cause of cardiac side effects is not yet exactly known. The agent was approved for the treatment of relapsed or refractory CTCL by the US FDA in 2009 (Table 2) (Piekarz et al. 2009; Bates et al. 2008).

Conclusion

As a new class of targeted antitumor drugs in epigenetics, HDACi regulates gene expression mainly through changing the histone acetylation and modulating chromatin structure. Compared with other chemotherapy agents, HDACi is found to have higher selectivity to tumor cells than normal ones. They can effectively target tumor cells to growth arrest, differentiation, and apoptosis, while have little effect on normal cells. HDACi has become the focus of clinical research; at present, many HDACis have shown promising activity in the clinical research malignant lymphoma treatment. Among them, vorinostat and romidepsin were approved for the treatment of relapsed or refractory CTCL. Other studies, such as romidepsin used in peripheral T-cell lymphoma (PTCL) and MGCD0103 in HL, have demonstrated remarkable and prospecting clinical effect.

However, there are still many problems to be solved, such as the specific anticancer mechanism and the relationship between concentration and duration of HDACi with the clinical effect. In addition, because of the short half-life and fast metabolism rate of HDACi, endeavors to find the proper delivery methods and dosage and methods to improve pharmacodynamic stability are also needed.

Histone Methylation in Lymphoma

As a significant epigenetic regulation mechanism, histone methylation plays an important role in many biological processes. In cells, there are various histone methyltransferases and histone demethylases working cooperatively to regulate the histone methylation state. Upon histone modification, effector proteins recognize modification sites specifically and affect gene transcriptional process. Histone lysine methylation generally happens in lysine (K) residues and arginine residues. Histone H3 K4, K9, K27, K36, K79, and H4 K20 can be methylated. Post-translational modification of chromatin and regulation of gene transcription through amino-terminal residues of histone is an important part of epigenetics. Enhancer of zests homolog 2 (EZH2) is the catalytic component of a polycomb group protein that mediates repression of gene transcription, silencing of the target genes, and leading to tumorigenesis through its histone 3 lysine 27 methyltransferase activity. Because EZH2 gene is highly expressed in human malignancies and is capable of promoting cell proliferation, and facilitating malignant tumor cell diffusion, the progression of lymphoma has been shown to be closely associated with EZH2 dysregulation.

Histone Methylation

Histone lysine methylation is catalyzed by different specific histone lysine methyltransferases (histone lysine (K) methyltransferases). Histone lysine methylation carries histone H3 and H4; for both, there are a total of six protein lysine residue methylation sites. The lysine side chain could be methylated, dimethylated, and trimethylated.

Histone arginine methylation plays a key role in the regulation of gene transcription, with such modification mediated by protein arginine methyltransferases (PRMTs). The methylation modification of PRMT1 and PRMT4 are related to activation of gene transcription, whereas methylation modifications of PRMT5 and PRMT6 are associated with inhibition of gene transcription. Histone arginine methylation is a dynamic reversible process. The arginine residues are related to a variety of covalent modifications, and these modifications are present in multiple sites, with transcription of genes having different effects and which could further regulate other physiological functions within cell, such as cell development, dynamic chromatin regulation, and DNA replication and repair (Wang et al. 2008).

Histone Demethylation

Histone demethylases mainly contains PADI4, LSD1, and JmjC domain-containing proteins.

PADI4 (peptidyl arginine deiminase, type IV) is a member of a gene family that encodes enzymes responsible for the conversion of arginine residues to citrulline residues. This gene may play a role in granulocyte and macrophage development, leading to inflammation and immune response. Multiple arginine sites mainly on histone 3 and histone 4 can be catalysed by PADI4 (Liu et al. 2011).

LSD1 (lysine-specific demethylase 1), a nuclear homolog of amine oxidases, functions as a histone demethylase and transcriptional corepressor, represses transcription via histone demethylation, and specifically demethylates histone H3 lysine 4. LSD1 is a kind of protein with the functions of demethylation and transcriptional repressor. Experiments confirmed that the inhibition of LSD1 by interfering RNA (RNAi) can lead to increased H3-K4 methylation, so that the target gene is expressed and LSD1 achieves transcriptional repression by demethylation. In 2004, LSD1 was identified as the first histone demethylase. It contains a C-terminal amine oxidase domain that is responsible for the demethylase activity through a FAD-dependent mechanism. An N-terminal SWIRM domain that is important for the stability and chromatin targeting of LSD1 was also found in other chromatin regulators. The mechanism of LSD1-mediated demethylation requires a protonated nitrogen for the reaction to proceed, the substrate specificity of it is limited to mono- or dimethylated lysine residues. LSD1 regulates gene expression mainly through three aspects: (1) through the integration of the SANT2 structure domain of CoREST and the target gene, causing demethylation of H3K4 and then gene transcription repression; (2) through binding of LSD1 and androgen/estrogen receptor can cause the demethylation of H3K9, leading to activation of hormone receptor-dependent transcription; and (3) through demethylation of H3K4, DNMT3L, which is the positive regulator of DNA methylation transferase (DNMTs) and could combine with unmethylated K4, promoting expression of DNMTs and resulting in DNA re-methylation (de novo methylation) and the inhibition of gene transcription (Escoubet-Lozach et al. 2009).

Unlike LSD1, the JmjC domain-containing proteins that have been tested do not require a protonated nitrogen and are able to reverse all three states of lysine methylation. In vivo, the overexpression of it can reduce dimethylation of H3-K36 level. We could consider JmjC domain protein as a demethylase. When compared with LSD1, first, they have different reaction mechanisms; second, the LSD1 protein family only has a small part of the proteins associated with demethylation, whereas the JmjC domain protein is a large family with the potential ability to regulate and inhibit different histone methylation. Finally, the LSD1 demethylation reaction requires protonated nitrogen, thus limiting its three methylated histone methylation capacity (Pfau et al. 2008).

Role of EZH2 in Lymphoma

A recently identified human gene (EZH2) is the human homologous genes of the zeste gene enhancer of *Drosophila*. It is an important member of the polycomb gene family. EZH2 is highly expressed in a variety of tumors and can promote cell proliferation and facilitate malignant tumor cell differentiation (Chase and Cross 2011). EZH2's role in the tumor has become the focus of current research.

EZH2 is a subunit of PRC2, and its SET domain catalyzes trimethylation of H3K27, a histone modification associated with transcriptional silencing. H3K27me3 helps recruit PRC1 to chromatin; it is thought that PRC1 is the effector of PcG-mediated silencing and long-term epigenetic memory.

In the PcG family, EZH2 plays a central role. PcG and TrxG (trithorax group) are widely conserved genes and play an important role in cellular memory system to prevent cell identity changes. The system is established from the early embryonic development, throughout the development process, and until adulthood. Through the formation of different components of the PcG and TrxG protein complex, they regulate the chromatin "active" or "inactive" status, maintaining corresponding gene transcription or inhibition. As an important PcG protein, EZH2 participates in the formation of chromatin structure, gene expression regulation, and growth control. EZH2 affects the formation of PcG and TrxG protein complexes, development and differentiation of hematopoietic cells, X chromosome inactivation, and even the process of tumorigenesis (Wang et al. 2004).

The Role of EZH2 in Cancer

The EZH2 gene is expressed in all fetal and adult tissues but is significantly reduced in the adult heart, brain, and kidney. There are many reports that the EZH2 gene is highly expressed in human malignancies. High expression of the EZH2 gene was first found in hematological malignancies, including bone marrow cancer, Hodgkin disease, non-Hodgkin lymphoma, and mantle cell lymphoma. Subsequently, prostate cancer, breast cancer, bladder cancer, liver cancer, colorectal cancer, and gastric cancer were found to have high expression of the EZH2 gene, while EZH2 in the corresponding normal tissues is not expressed or shows only a small amount of expression (Chase and Cross 2011).

Deregulation of PcG gene expression in experimental model systems has clearly been linked to oncogenesis. Because of the crucial effects of EZH2 in individual development and cell growth, especially in normal lymphoid organs, B-cell differentiation, and proliferation of lymphoid tissue, EZH2 was first studied in hematopoietic malignances. BMI-1 and EZH2 are involved in two different types of PcG complexes that regulate specific binding of target gene expression in normal B lymphocytes. Germinal center (GC) B lymphocyte development at different stages is related to different types of PcG gene expression. Because the PcG complex could affect pluripotent hematopoietic stem cells (HSCs) and directed differentiation potential through chromatin modifications, the imbalanced expression of EZH2 may lead to lymphomas. Research has also indicated that EZH2 may be involved in the malignant disease process. BMI-1 gene in mice shows the imbalance in the gene leads to B-cell and T-cell lymphoma, and its absence causes serious developmental defects in the hematopoietic system (Chase and Cross 2011; Wang et al. 2004; Velichutina et al. 2010).

EZH2 and Lymphoma

Lymphoma represents a group of malignancies that originate in the lymph nodes or other lymphoid tissues. There is increasing evidence that the up-regulated EZH2 in normal GC B cells as well as the deregulation of PcG expression are related to the formation of lymphomas (Velichutina et al. 2010).

The overexpression of EZH2 has been reported in mantle cell lymphoma and adult T-cell leukemia/lymphoma cells. In Burkitt lymphoma, the overexpression of c-Myc contributes to stimulate EZH2 expression. Hodgkin/Reed-Sternberg (H/RS) cells co-express BMI-1 and EZH2 (Dukers et al. 2004). Within the B-cell lineage, it was shown that EZH2 is highly expressed in lymphoid progenitors, and EZH2 deficiency induces defects in early lymphopoiesis (Sneeringer et al. 2010). Moreover, EZH2 mutations also exert influence on lymphomagenesis. A missense mutation in the EZH2 SET domain was discovered in a sizeable fraction of DLBCLs, and a truncation mutation of EZH2 was found in only 2 of 221 other tumors, suggesting that mutation of EZH2 is a lesion specific to lymphoma.

There is increasing evidence that the deregulation of PcG expression is related to the formation of lymphomas. Also, the irregular expression profile of BMI-1 and EZH2 in B-NHL suggests that the distinct balance between the BMI-1 and EZH2-containing PcG complex is disturbed in these lymphomas. Furthermore, the extent of irregular PcG expression is correlated with the type of lymphoma and therefore the clinical behavior (Park et al. 2011).

The expression of PcG complexes during GC reaction is linked to the differentiation status of follicular B cells and a mutually exclusive pattern of BMI-1/RING1 and EZH2/EED. PcG proteins are in reactive centroblasts and centrocytes. These observations suggest that expression of PcG complexes is highly regulated during GC reaction and that PcG proteins may contribute to antigen-specific B-cell maturation. Research has also shown that adult T-cell leukemia/lymphoma cells (Sasaki et al. 2011) have deregulated polycomb repressive complex 2 with over-expressed EZH2 and that there is the possibility of a new therapeutic strategy targeting histone methylation in this disease.

On the basis of the discovery that EZH2 is a mutant oncogene in germinal B cell lymphomas and the fact that the extent of abnormal PcG expression correlated with the type of lymphoma and therefore the clinical behavior, a new therapeutic strategy against EZH2 may exist for mechanism-based epigenetic treatment of patients with these lymphomas. Important steps toward this goal include further delineation of the lymphoma types in which this mutation occurs and development of assays amenable to detection of these mutations in routine clinical specimens.

The Mechanism of Action of EZH2

Beside PcG, with the function of controlling expression of Hox genes, a growing number of studies have revealed the function of PcG in stem cell function, differentiation, development, and cell cycle control. PcG proteins have been identified as important proteins in tumorigenesis due to their potential to repress tumor suppressor genes. Through the repression of the INK4/ARF locus, which encodes both p16INK4a, which prevents inactivation of the tumor suppressor RB, and p19ARF, which stabilizes the tumor suppressor p53, PcG plays a central role in cell proliferation. Genome-wide mapping by chromatin immunoprecipitation experiments of PcG members in human and murine stem cells has shown that several genes involved in regulation of cell development and more importantly, the localization and expression patterns of PcG genes change during the stages of differentiation. The PcG proteins maintain cellular identity by preserving chromatin states against chromatin-disrupting processes during cell cycle transitions. For example, during the process of DNA replication, H3K27-methylated histones recruit PRC2 and PRC1 complexes to nucleosomes of the nascent DNA strand to continue gene silencing, then PRC1 and PRC2 provoke changes in chromatin structure. At present, PRC1 is believed to recognize the PRC2-methyl mark and the PRC2 complex is considered to be required for the repressive capacity of PRC1 complex. Furthermore, PcG

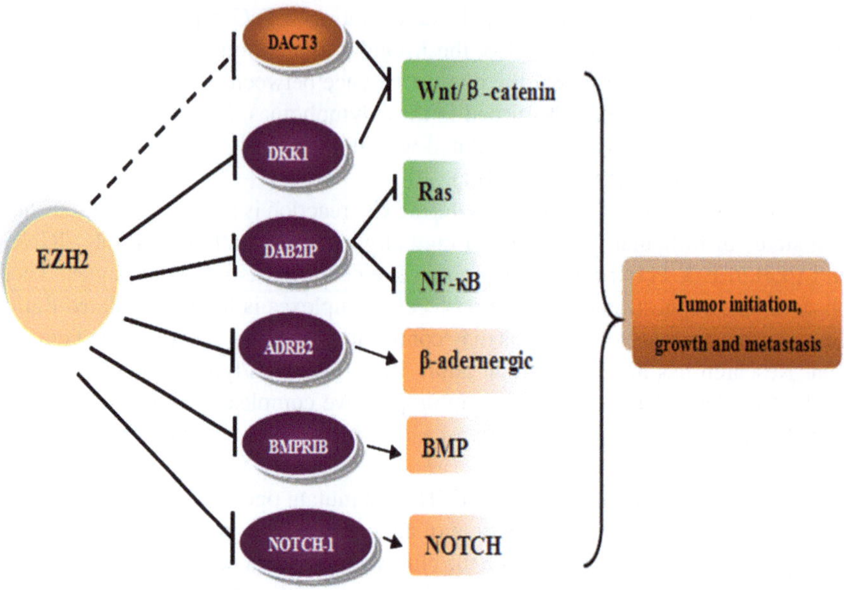

Fig. 3 EZH2-mediated epigenetic regulation of signaling pathways.EZH2 activates oncogenic signaling cascades and inhibits pro-differentiation pathways through epigenetic silencing of the negative regulators and positive effectors, respectively, thereby establishing an oncogene-tumor suppressorsignaling paradigm in driving tumor initiation, growth and progression. As DACT3 is repressed via H3K27me3, EZH2 is implicated to be its epigenetic regulator

genes play a major role in regulating hematopoietic function, especially for HSCs. For instance, BMI-1-deficient mice have severely impaired HSC self-renewal, and bone marrow progenitors lacking BMI-1 have a restricted proliferative potential; overexpression of EZH2 confers long-term repopulating potential on HSCs, preventing its exhaustion after replicative stress; and expression of one null EED allele or 2 hypomorphic EED alleles results in greater lymphoproliferation and a greater risk of developing hematologic tumors. Additionally, a gene expression signature of PRC2 target genes predicts poor outcome in patients with breast and prostate cancer, implying that there might exist a window to improve therapeutic efficiency for neoplasms (Tsang and Cheng 2011; Bracken et al. 2003; Yap et al. 2011).

Genome-wide searches of PRC2 target genes in mammalian cells including fibroblasts, ES, and cancer cells also revealed cohorts of signaling components that control cell differentiation. Emerging data indicate that EZH2 has a master regulatory function in controlling key signaling pathways in cancers by transcriptional repression of signaling molecules. EZH2 activates oncogenic signaling cascades and inhibits pro-differentiation pathways through epigenetic silencing of the negative regulators and positive effectors, respectively, thereby establishing an oncogene-tumor suppressor signaling paradigm in driving tumor initiation, growth and progression.

Conclusion

In summary, the roles for EZH2 in normal GC B-cells are to favor cellular proliferation by directly repressing several tumor suppressor genes; to create a repression state similar to that found in stem cells that might foster self-renewal and prevent premature differentiation; and to maintain and stabilize a transcriptional repression program (as shown in Fig. 3).

Preclinical data and results suggest that therapeutic targeting of EZH2 might have significant antilymphoma effects and support the rationale for development of inhibitors of the EZH2 SET domain (Hang et al. 2011).

Role of miRNAs in Lymphoma

MicroRNAs (miRNAs) are an abundant class of short regulatory (~22 nt) noncoding RNAs widely expressed in all metazoan eukaryotes and highly conserved throughout different organisms. Biogenesis of miRNAs is a complex process starting in the nucleus and ending in the cytoplasm (Li et al. 2011). miRNAs are encoded by genes that are presumably transcribed into single or clustered primary transcripts, which are processed to produce the mature miRNAs. These, in turn, are incorporated into a ribonucleoprotein complex called RNA-induced silencing complex (RISC) and guide the RISC to the target mRNA. RISC plus miRNA mediate target gene downregulation by mRNA cleavage or translational repression, mainly through binding to 3′UTR of target mRNA. miRNAs are involved in critical biologic processes, including development, cell differentiation, apoptosis, and proliferation. Recently, a growing body of evidence has implicated specific miRNAs in the pathogenesis of lymphoma.

MiRNAs in B Cell Lymphomas

Chronic Lymphoid Leukemia (CLL)

It was in CLL that a role of miRNAs in cancer was first reported by Calin et al. in 2002. The miR-15a/16-1 cluster at chromosome 13q14 within the most commonly deleted region in CLL was found to be down-regulated in 68% of cases with the deletion. Subsequently, these miRNAs have been shown to target Bcl-2 and remove the suppressor influence on anti-apoptotic Bcl-2, which is overexpressed consistently in CLL. Unique miRNA signatures have been generated dependent on VH gene mutation and ZAP70 status. The miR-29 family and miR-223 have been found consistently to be down-regulated in association with unmutated VH genes and are decreased dramatically in patients with poor prognosis independently of the prognostic marker classification used.

Possibly of as much importance in CLL associated with the very poor prognostic genotype, 17p13/TP53 deletion is consistently associated with down-regulation of miR-34a (Merkel et al. 2010). Members of the miR-34 family have been discovered to be direct p53 targets and mediate some of the p53-dependent effects. While miR-34b/c are not expressed in CLL cells, low miR-34a levels are a sensitive indicator of the activity of the p53 axis in CLL (Moussay et al. 2010) and are associated with p53 inactivation and chemotherapy-refractory disease irrespective of 17p deletion/TP53 mutation status. Among patients harboring a 17p deletion, low expression of the miR-223, miR-29c, miR-29b, and miR-181 family were associated strongly with disease progression, confirming the adverse effects of low levels of miR-29 and miR-223 in CLL progression. In a separate study, the 17p-deleted miRNA signature comprised high levels of miR-15a, miR-21, and miR-155 and low levels of miR-34a and miR-181b. In addition, miR-21 expression stratified survival in patients with 17p-deleted CLL and among cohorts of patients with CLL with a variety of chromosomal aberrations. Among both the good- and poor-prognosis subgroups, miR-181b levels identified patients who required early therapy and who thus need closer follow-up. Furthermore, studies have identified miRNAs that may predict clinical resistance of CLL to fludarabine and a similar expression profile emerges with miR-29a, miR-181a and miR-221 among those being expressed differentially (Moussay et al. 2010).

Diffuse Large B-Cell Lymphoma (DLBCL)

miR-155 was increased in several types of B cell lymphomas. SHIP1 has been identified as a target of miR-155 and is down-regulated by miR-155 in DLBCL as a consequence of autocrine stimulation by TNF-α. Another target of miR-155 is SMAD, which is suppressed by high levels of miR-155 and protects DLBCL cells from the growth-inhibitory effects of transforming growth factor-β (Rai et al. 2010). It is reported that miR-96, miR-182, miR-589, and miR-25 were shown to be significantly up-regulated in 7q+DLBCL, which were associated with lower death rate and better overall survival in patients treated with R-CHOP (Chigrinova et al. 2011). The expression of miR-18a correlated with overall survival, whereas the expression of miR-181a and miR-222 correlated with progression-free survival, suggesting that the expression of specific miRNAs may be useful for DLBCL survival prediction (Alencar et al. 2011). Myc-mediated repression of miRNA-34a promotes high-grade transformation of B-cell lymphoma by dysregulation of FoxP1 gastric marginal zone B-cell lymphoma of MALT type (MALT lymphoma). In addition, the miR-34a showed the strongest anti-proliferative properties through the targeted gene FoxP1, which promotes DLBCL proliferation, elucidating a novel Myc- and FoxP1-dependent pathway of malignant transformation and suggesting miR-34a replacement therapy as a promising strategy in lymphoma treatment (Craig et al. 2011a).

Mantle Cell Lymphoma (MCL)

miR-16-1 regulates the expression of cyclin D1 (CCND1) in MCL, which is associated with a poor prognosis, whereby miR-16-1 binding sites within the CCND1 mRNA 3′UTR are deleted and miRNA regulation of CCND1 is altered. miR-17-5p/miR-20a from the miR-17_92 cluster were expressed at high levels concomittantly with MYC and associated with poor disease prognosis. Subsequently, Zhao et al. (2010) found that 18 miRNAs were down-regulated and 21 were up-regulated in MCL compared with normal B lymphocytes, with the most frequent alterations being decreased miR-29, miR-142 and miR-150 and increased miR-124a and miR-155. Patients with significantly down-regulated miR-29 had a shorter survival compared with those who expressed relatively high levels of miR-29. Down-regulation of the miR-29 family was associated with CDK6 overexpression, and CDK6 was found to be a direct target of miR-29. Zhang et al. demonstrated that Myc acts as a repressor of miRNA-15a/16 by recruiting HDAC3, by the way that both c-Myc and HDAC3 co-localized to the two promoters of the miR-15a/16-1 cluster gene, DLEU2, suggesting the role of HDAC3 in Myc-induced miR-15a/16-1 changes and reveal novel mechanisms for c-Myc-driven miRNA suppression and malignant transformation in aggressive B-cell malignancies (Zhang et al. 2011). Follicular dendritic cells protect B-cell lymphoma cells against apoptosis, in part through activation of a miR-181a-dependent mechanism involving down-regulation of Bim expression, indicating that cell-cell contact controls tumor cell survival and apoptosis via miRNA in mantle cell and other non-Hodgkin lymphomas (Lwin et al. 2010).

Mucosa-Associated Lymphoid Tissue (MALT)

miR-203 is decreased in MALT lymphoma, which results from extensive promoter hypermethylation of the miR-203 and coincides with dysregulation of the miR-203 target ABL1. Reexpression of miR-203 can prevent tumor cell proliferation, and ABL inhibitors prevent tumor cell growth, supporting that the transformation from gastritis to MALT lymphoma is epigenetically regulated by miR-203 promoter methylation and identifying ABL1 as a novel target for the treatment of this malignancy (Craig et al. 2011b).

Follicular Lymphoma (FL)

The miR-155, miR-221 and miR-21 are overexpressed similarly in FL as they are in DLBCL, with no significant differences between expression levels in these different lymphomas. A signature of four miRNAs (miR-330, -17-5p, -106A, -210) has been proposed as a diagnostic classifier to discriminate FL, DLBCL, and non-neoplastic lymph node, but this has not been confirmed by other series.

Burkitt's Lymphoma (BL)

It had been reported that the down-regulation of hsa-miR-9* seems to specifically identify a particular subset of BL cases lacking MYC translocation and determined that the hsa-miR-9* is able to modulate E2F1 and c-Myc expression, suggesting it as a promising novel candidate for tumor cell marker (Onnis et al. 2010). The expression of hsa-miR-127 was shown to be strongly upregulated in EBV-positive BL samples compared with the EBV-negative cases. In addition, it was shown that hsa-miR-127 is involved in a B-cell differentiation process through posttranscriptional regulation of BLIMP1 and XBP1. These findings suggest that the overexpression of miR-127 may represent a key event in the lymphomagenesis of EBV-positive BL, likely by blocking the B-cell differentiation process (Leucci et al. 2010). miR-26a, repressed by MYC, was also found to be deregulated in primary human Burkitt lymphoma samples, and expression of miR-26a influenced cell cycle progression by targeting the oncogene EZH2, suggesting that MYC modulates genes important to lymphomagenesis via deregulation of miRNAs. Overexpression of let-7a decreased Myc mRNA and protein. Down-regulation of Myc protein and mRNA using siRNA MYC also elevated let-7a miRNA and decreased Myc gene expression. These findings with let-7a add to the complexity of MYC regulation and suggest that dysregulation of these miRNAs participates in the genesis and maintenance of the lymphoma phenotype in Burkitt lymphoma cells.

Others

MiR-21, miR-19, and miR-92a levels in cerebrospinal fluid (CSF) collected from patients with primary central nervous system lymphoma (PCNSL) indicated a significant diagnostic value, and, more importantly, combined miRNA analyses resulted in an increased diagnostic accuracy with 95.7% sensitivity and 96.7% specificity. This shows that the CSF miRNAs are potentially useful tools as novel noninvasive biomarkers for diagnosis of PCNSL (Baraniskin et al. 2011). It is shown that the plasma miR-92a values in NHL were extremely low when compared with healthy subjects, and the very low plasma level of miR-92a increased in the complete response (CR) phase but did not reach the normal range, with plasma levels lower again in the relapse phase, indicating that the plasma miR-92a value could be a novel biomarker not only for diagnosis but also for monitoring lymphoma patients after chemotherapy (Ohyashiki et al. 2011).

MicroRNAs in T-Cell lymphomas

NK/T Cell Lymphoma

It was observed that circulating miR-221 was decreased in patients with NK/T cell lymphoma and a reverse correlation with performance status and the overall survival after treatment, indicating that plasma miR221 may be a diagnostic and prognostic marker for NK/T cell lymphoma (Guo et al. 2010). miR-21 and miR-155 expression levels were significantly greater in NK-cell lymphoma. Reducing expression of miR-21 or miR-155 led to up-regulation of phosphatase and tensin homologue (PTEN), programmed cell death 4 (PDCD4), or Src homology-2 domain-containing inositol 5-phosphatase 1 (SHIP1). Moreover, transduction with either miR-21 or miR-155 led to down-regulation of PTEN and PDCD4 or SHIP1 with up-regulation of phosphorylated AKTser473. These results provide important new insight into the pathogenesis of NK-cell lymphoma/leukemia and suggest targeting miR-21 and/or miR-155 may represent a useful approach to treating NK-cell lymphoma/leukemia. It was shown that low miR-146a expression is an independent poor prognostic factors for NK/T cell lymphoma. miR-146a overexpression can inhibit nuclear factor-kB (NF-kB) activity, suppress cell proliferation, induce apoptosis, and enhance chemosensitivity. TNF receptor–associated factor 6, a target of miR-146a and a known NF-kB activator, was down-regulated by miR-146a. Promoter methylation of miR-146a gene was observed, as well as in NK/T-cell lymphoma tissues with low miR-146a expression, and miR-146a expression was induced by the conversion of methylation status with a demethylating agent. These results suggest that miR-146a might function as a potent tumor suppressor in NK/T-cell lymphomas and be useful for patient assessment and therapeutic targeting (Paik et al. 2011).

Mycosis Fungodes (MF)/Sezary Syndrome

The most significant differentially expressed microRNA were miR-155 and miR-92a (both up-regulated), while miR-93 showed the highest up-regulation in tumor stage mycosis fungoides (MF) by the miRNA microarray, unfortunately, when these miRNAs were validated by miRNA-qPCR on additional test groups. None of the miRNAs was up-regulated in tumor stage MF of those previously shown to be up-regulated in SS, and only 2 of the 19 miRNAs down-regulated in tumor stage MF were also down-regulated in SS (van Kester et al. 2011). miR-21 expression is increased in Sézary cells when compared with CD4+ T cells from healthy donors. Silencing of miR-21 in Sézary cells results in increased apoptosis, suggesting a functional role for miR-21 in the leukomogenic process (van der Fits et al. 2011).

It showed that expression of miR-223 can distinguish SzS samples from healthy controls and patients with mycosis fungoide. The miR-342 plays a role in the pathogenesis of Sézary syndrome by inhibiting apoptosis, thus describing a novel mechanism of regulation for this miRNA via binding of miR-199a* to its host gene. Moreover, it also provides the first in vivo evidence for down-regulation of the miR-17-92 cluster in malignancy and demonstrates that ectopic miR-17-5p expression increases apoptosis and decreases cell proliferation in Sézary syndrome cells (Ballabio et al. 2010). A total of 45 miRNAs have been shown to be differentially expressed between Sézary syndrome and controls. Using predictive analysis, 19 miRNAs, including miR-21, miR-214, miR-486, miR-18a, miR-342, miR-31, and let-7 members, were also found. Moreover, a signature of 14 miRNAs, including again miR-21, was identified, which may be able to discriminate patients with unfavorable and favorable outcomes (Narducci et al. 2011).

Anaplastic Large Cell Lymphoma (ALCL)

It was reported that the ALK(+) ALCL cell lines and biopsy specimens express a low level of miR-29a and that this down-modulation requires an active NPM-ALK kinase and that the low expression of miR-29a, probably through methylation repression, plays an important regulatory role in MCL-1 overexpression that could promote tumor cell survival by inhibiting apoptosis. Enforced miR-29a expression was found to modulate apoptosis through inhibition of MCL-1 expression in ALCL cell lines and in a xenografted model, with a concomitant tumor growth reduction. Thus, synthetic miR-29a represents a potential new tool to affect tumorigenesis in these lymphomas (Desjobert et al. 2011).

Cutaneous T Cell Lymphomas (CTCL)

Data have shown that the most induced (miR-326, miR-663b, and miR-711) and repressed (miR-203 and miR-205) miRNAs distinguish CTCL from benign skin diseases with >90% accuracy. A qRT-PCR-based classifier consisting of miR-155, miR-203, and miR-205 distinguishes CTCL from benign disorders with high specificity and sensitivity and with a classification accuracy of 95%, indicating that miRNAs have a high diagnostic potential in CTCL (Ralfkiaer et al. 2011).

Conclusion and Perspective

In summary, although it is clear that the functional importance of miRNAs in lymphoma is gaining momentum rapidly, we still have much to learn from these tiny molecules and further research on their target genes, specific transcription factors and their interactions is needed to integrate miRNAs into the regulatory network in the genesis of lymphoma. Recent researches have described remarkably effective inhibition of miRNAs in vivo, thus providing an entry point into the promising new discipline of microRNA therapeutics. Undoubtedly, the coming years will bring possibilities of designing miRNAs mediated targeting therapies and make miRNAs specifically attractive as new tools for the diagnosis, prognosis, and therapy of lymphoma. It is to be hoped that some of these will be efficient and will benefit future lymphoma patients.

References

Alencar AJ, Malumbres R, Kozloski GA, Advani R, Talreja N, Chinichian S, Briones J, Natkunam Y, Sehn LH, Gascoyne RD, Tibshirani R, Lossos IS (2011) MicroRNAs are independent predictors of outcome in diffuse large B-cell lymphoma patients treated with R-CHOP. Clin Cancer Res 17(12):4125–4135

Ballabio E, Mitchell T, van Kester MS, Taylor S, Dunlop HM, Chi J, Tosi I, Vermeer MH, Tramonti D, Saunders NJ, Boultwood J, Wainscoat JS, Pezzella F, Whittaker SJ, Tensen CP, Hatton CS, Lawrie CH (2010) MicroRNA expression in Sezary syndrome: identification, function, and diagnostic potential. Blood 116(7):1105–1113

Baraniskin A, Kuhnhenn J, Schlegel U, Chan A, Deckert M, Gold R, Maghnouj A, Zöllner H, Reinacher-Schick A, Schmiegel W, Hahn SA, Schroers R (2011) Identification of microRNAs in the cerebrospinal fluid as marker for primary diffuse large B-cell lymphoma of the central nervous system. Blood 117(11):3140–3146

Bates S, Piekarz R, Wright J et al (2008) Final clinical results of a phase 2 NCI multicenter study of Romidepsin in recurrent cutaneous T-cell lymphoma (molecular analyses included). Blood (ASH annual meeting abstracts) 112:Abstract 1568

Bracken AP, Pasini D, Capra M et al (2003) EZH2 is downstream of the pRB-E2F pathway, essential for proliferation and amplified in cancer. EMBO J 22(20):5323–5335

Buglio D, Younes A (2010) Histone deacetylase inhibitors in Hodgkin lymphoma. Invest New Drugs 28(Suppl 1):S21–S27, Dec Epub 2010 Dec 3

D Buglio, Vidya M, Noor M et al (2010) The class-I HDAC inhibitor MGCD0103 induces apoptosis in Hodgkin lymphoma cell lines and synergizes with proteasome inhibitors by an HDAC6-independent mechanism. Br J Haematol 151(4):387–396

Chase A, Cross NC (2011) Aberrations of EZH2 in cancer. Clin Cancer Res 17(9):2613–2618

Chigrinova E, Mian M, Shen Y, Greiner TC, Chan WC, Vose JM, Inghirami G, Chiappella A, Baldini L, Ponzoni M, Ferreri AJ, Franceschetti S, Gaidano G, Tucci A, Facchetti F, Lazure T, Lambotte O, Montes-Moreno S, Piris MA, Zucca E, Kwee I, Bertoni F (2011) Integrated profiling of diffuse large B-cell lymphoma with 7q gain. Br J Haematol 153(4):499–503

Copeland A, Buglio D, Younes A (2010) Histone deacetylase inhibitors in lymphoma. Curr Opin Oncol 22(5):431–436

Cotto M, Cabanillas F, Tirado M, García MV, Pacheco E (2010) Epigenetic therapy of lymphoma using histone deacetylase inhibitors. Clin Transl Oncol 12(6):401–409

Craig VJ, Cogliatti SB, Imig J, Renner C, Neuenschwander S, Rehrauer H, Schlapbach R, Dirnhofer S, Tzankov A, Müller A (2011a) Myc-mediated repression of microRNA-34a promotes high-grade transformation of B-cell lymphoma by dysregulation of FoxP1. Blood 117(23):6227–6236

Craig VJ, Cogliatti SB, Rehrauer H, Wündisch T, Müller A (2011b) Epigenetic silencing of microRNA-203 dysregulates ABL1 expression and drives Helicobacter-associated gastric lymphomagenesis. Cancer Res 71(10):3616–3624

Crump M, Andreadis C, Assouline S et al (2008a) Treatment of relapsed or refractory non-hodgkin lymphoma with the oral isotype-selective histone deacetylase inhibitor MGCD0103: interim results from a phase II study. J Clin Oncol 26:Abstract 8528

Crump M, Coiffier B, Jacobsen ED et al (2008b) Phase II trial of oral vorinostat (suberoylanilide hydroxamic acid) in relapsed diffuse large-B-cell lymphoma. Ann Oncol 19:964–969

Desjobert C, Renalier MH, Bergalet J, Dejean E, Joseph N, Kruczynski A, Soulier J, Espinos E, Meggetto F, Cavaillé J, Delsol G, Lamant L (2011) MiR-29a down-regulation in ALK-positive anaplastic large cell lymphomas contributes to apoptosis blockade through MCL-1 overexpression. Blood 117(24):6627–6637

Dukers DF, van Galen JC, Giroth C et al (2004) Unique polycomb gene expression pattern in Hodgkin's lymphoma and Hodgkin's lymphoma-derived cell lines. Am J Pathol 164(3):873–881

Duvic M, Talpur R, Ni X et al (2007) Phase 2 trial of oral vorinostat (suberoylanilide hydroxamic acid, SAHA) for refractory cutaneous T-cell lymphoma (CTCL). Blood 109:31–39

Escoubet-Lozach L, Lin IL, Jensen-Pergakes K et al (2009) Pomalidomide and lenalidomide induce p21 WAF-1 expression in both lymphoma and multiple myeloma through a LSD1-mediated epigenetic mechanism. Cancer Res 69(18):7347–7356, Sep 15

Gimsing P, Hansen M, Knudsen LM et al (2008) A phase I clinical trial of the histone deacetylase inhibitor belinostat in patients with advanced hematological neoplasia. Eur J Haematol 81(3):170–176, September

Guo HQ, Huang GL, Guo CC, Pu XX, Lin TY (2010) Diagnostic and prognostic value of circulating miR-221 for extranodal natural killer/T-cell lymphoma. Dis Markers 29(5):251–258

Hagelkruys A, Sawicka A, Rennmayr M, Seiser C (2011) The biology of HDAC in cancer: the nuclear and epigenetic components. Handb Exp Pharmacol 206:13–37

Hang Y, Nayak S, Jankowitz R et al (2011) Epigenetics in breast cancer: what's new? Breast Cancer Res 13:225

Leucci E, Onnis A, Cocco M, De Falco G, Imperatore F, Giuseppina A, Costanzo V, Cerino G, Mannucci S, Cantisani R, Nyagol J, Mwanda W, Iriso R, Owang M, Schurfeld K, Bellan C, Lazzi S, Leoncini L (2010) B-cell differentiation in EBV-positive burkitt lymphoma is impaired at posttranscriptional level by miRNA-altered expression. Int J Cancer 126(6):1316–1326

Li H, Zhao H, Wang D, Yang R (2011) microRNA regulation in megakaryocytopoiesis. Br J Haematol 155(3):298–307

Liu YL, Chiang YH, Liu GY, Hung HC (2011) Functional role of dimerization of human peptidylarginine deiminase 4 (PAD4). PLoS One 6(6):e21314

Lwin T, Lin J, Choi YS, Zhang X, Moscinski LC, Wright KL, Sotomayor EM, Dalton WS, Tao J (2010) Follicular dendritic cell-dependent drug resistance of non-Hodgkin lymphoma involves cell adhesion-mediated Bim down-regulation through induction of microRNA-181a. Blood 116(24):5228–5236

Mehnert JM, Kelly WK (2007) Histone deacetylase inhibitors: biology and mechanism of action. Cancer J 13(1):23–29

Merkel O, Asslaber D, Pinon JD, Egle A, Greil R (2010) Interdependent regulation of p53 and miR-34a in chronic lymphocytic leukemia. Cell Cycle 9:2764–2768

Moussay E, Palissot V, Vallar L et al (2010) Determination of genes and microRNAs involved in the resistance to fludarabine in vivo in chronic lymphocytic leukemia. Mol Cancer 9:115

Narducci MG, Arcelli D, Picchio MC, Lazzeri C, Pagani E, Sampogna F, Scala E, Fadda P, Cristofoletti C, Facchiano A, Frontani M, Monopoli A, Ferracin M, Negrini M, Lombardo GA, Caprini E, Russo G (2011) MicroRNA profiling reveals that miR-21, miR486 and miR-214 are upregulated and involved in cell survival in Sézary syndrome. Cell Death Dis 2:e151

Ohyashiki K, Umezu T, Yoshizawa S, Ito Y, Ohyashiki M, Kawashima H, Tanaka M, Kuroda M, Ohyashiki JH (2011) Clinical impact of down-regulated plasma miR-92a levels in non-Hodgkin's lymphoma. PLoS One 6(2):e16408

Olsen EA, Kim YH, Kuzel TM (2007) Phase IIb multicenter trial of vorinostat in patients with persistent, progressive, or treatment refractory cutaneous T-cell lymphoma. J Clin Oncol 25:3109–3115

Onnis A, De Falco G, Antonicelli G, Onorati M, Bellan C, Sherman O, Sayed S, Leoncini L (2010) Alteration of microRNAs regulated by c-Myc in Burkitt lymphoma. PLoS One 5(9):e12960

Paik JH, Jang JY, Jeon YK, Kim WY, Kim TM, Heo DS, Kim CW (2011) MicroRNA-146a downregulates NFκB activity via targeting TRAF6 and functions as a tumor suppressor having strong prognostic implications in NK/T cell lymphoma. Clin Cancer Res. 17(14):4761–4771

Park SW, Chung NG, Eom HS et al (2011) Mutational analysis of EZH2 codon 641 in non-Hodgkin lymphomas and leukemias. Leuk Res 35(1):e6–e7

Pfau R, Tzatsos A, Kampranis SC et al (2008) Members of a family of JmjC domain-containing oncoproteins immortalize embryonic fibroblasts via a JmjC domain-dependent process. Proc Natl Acad Sci U S A 105(6):1907–1912, Feb 12

Piekarz RL, Frye R et al (2009) Phase II multi-institutional trial of the histone deacetylase inhibitor romidepsin as monotherapy for patients with cutaneous T-cell lymphoma. J Clin Oncol 27:5410–5416

Rai D, Kim SW, McKeller MR, Dahia PL, Aguiar RC (2010) Targeting of SMAD5 links microRNA-155 to the TGF-beta pathway and lymphomagenesis. Proc Natl Acad Sci USA 107:3111–3116

Ralfkiaer U, Hagedorn PH, Bangsgaard N, Løvendorf MB, Ahler CB, Svensson L, Kopp KL, Vennegaard MT, Lauenborg B, Zibert JR, Krejsgaard T, Bonefeld CM, Søkilde R, Gjerdrum LM, Labuda T, Mathiesen AM, Grønbæk K, Wasik MA, Sokolowska-Wojdylo M, Queille-Roussel C, Gniadecki R, Ralfkiaer E, Geisler C, Litman T, Woetmann A, Glue C, Røpke MA, Skov L, Odum N (2011) Diagnostic microRNA profiling in cutaneous T-cell lymphoma (CTCL). Blood 118(22):5891–5900

Sasaki D, Imaizumi Y, Hasegawa H et al (2011) Overexpression of enhancer of zeste homolog 2 with trimethylation of lysine 27 on histone H3 in adult T-cell leukemia/lymphoma as a target for epigenetic therapy. Haematologica 96(5):712–719

Shakovich R, Melnick A (2011) Epigenetics and B-cell lymphoma. Curr Opin Hematol 18:293–299

Sharma S, Kelly TK, Jones PA (2010) Epigenetics in cancer. Carcinogenesis 31(1):27–36, Jan Epub 2009 Sep 13

Sneeringer CJ, Scott MP, Kuntz KW et al (2010) Coordinated activities of wild-type plus mutant EZH2 drive tumor-associated hypertrimethylation of lysine 27 on histone H3 (H3K27) in human B-cell lymphomas. Proc Natl Acad Sci U S A 107(49):20980–20985

Sureda A, Engert A, Browett PJ et al (2010) Interim results for the phase II study of panobinostat (LBH589) in patients (Pts) with relapsed/refractory Hodgkin's lymphoma (HL) after autologous hematopoietic stem cell transplant (AHSCT). J Clin Oncol 28(suppl):Abstract 8007):15s

Tsang DP, Cheng AS (2011) Epigenetic regulation of signaling pathways in cancer: role of the histone methyltransferase EZH2. J Gastroenterol Hepatol 26(1):19–27

van der Fits L, van Kester MS, Qin Y, Out-Luiting JJ, Smit F, Zoutman WH, Willemze R, Tensen CP, Vermeer MH (2011) MicroRNA-21 expression in CD4+ T cells is regulated by STAT3 and is pathologically involved in Sézary syndrome. J Invest Dermatol 131(3):762–768

van Kester MS, Ballabio E, Benner MF, Chen XH, Saunders NJ, van der Fits L, van Doorn R, Vermeer MH, Willemze R, Tensen CP, Lawrie CH (2011) miRNA expression profiling of mycosis fungoides. Mol Oncol 5(3):273–280

Velichutina I, Shaknovich R, Geng H et al (2010) EZH2-Mediated epigenetic silencing in germinal center B cells contributes to proliferation and lymphomagenesis. Blood 116(24):5247–5255

Wang L, Brown JL, Cao R, Zhang Y, Kassis JA, Jones RS (2004) Hierarchical recruitment of polycomb group silencing complexes. Mol Cell 14(5):637–646

Wang L, Pal S, Sif S (2008) Protein arginine methyltransferase 5 suppresses the transcription of the RB family of tumor suppressors in leukemia and lymphoma cells. Mol Cell Biol 28(20):6262–6277, Oct

Watanabe T, Kato H, Kobayashi Y et al (2010) Potential efficacy of the oral histone deacetylase inhibitor vorinostat in a phase I trial in follicular and mantle cell lymphoma. Cancer Sci 101:196–200

Yap DB, Chu J, Berg T et al (2011) Somatic mutations at EZH2 Y641 act dominantly through a mechanism of selectively altered PRC2 catalytic activity, to increase H3K27 trimethylation. Blood 117(8):2451–2459

Zain J, O'Connor OA (2010) Targeting histone deacetyalses in the treatment of B- and T-cell malignancies. Invest New Drugs 28(Suppl 1):S58–S78, Dec
Zain JM, Foss F, Kelly WK et al (2009) Final results of a phase I study of oral belinostat (PXD101) in patients with lymphoma. J Clin Oncol (ASCO annual meeting 2009) 27:15s
Zhang X, Chen X, Lin J, Lwin T, Wright G, Moscinski LC, Dalton WS, Seto E, Wright K, Sotomayor E, Tao J (2011) Myc represses miR-15a/miR-16-1 expression through recruitment of HDAC3 in mantle cell and other non-Hodgkin B-cell lymphomas. Oncogene. doi:10.1038/onc.2011.470 [Epub ahead of print]
Zhao JJ, Lin J, Lwin T et al (2010) microRNA expression profile and identification of miR-29 as a prognostic marker and pathogenetic factor by targeting CDK6 in mantle cell lymphoma. Blood 115:2630–2639

Index

J. Tao and E. Sotomayor (eds.), *Hematologic Cancers: From Molecular Pathobiology to Targeted Therapeutics*, Cancer Growth and Progression 14,
DOI 10.1007/978-94-007-5028-9, © Springer Science+Business Media Dordrecht 2012

M

MIX
Papier aus verantwortungsvollen Quellen
Paper from responsible sources
FSC® C105338

If you have any concerns about our products,
you can contact us on
ProductSafety@springernature.com

In case Publisher is established outside the EU,
the EU authorized representative is:
Springer Nature Customer Service Center GmbH
Europaplatz 3, 69115 Heidelberg, Germany

Printed by Libri Plureos GmbH
in Hamburg, Germany